THE GRAMINEAE

LONDON
Cambridge University Press
FETTER LANE

NEW YORK · TORONTO
BOMBAY · CALCUTTA · MADRAS
Macmillan

TOKYO
Maruzen Company Ltd

A Sod of Turf, with Meadow-grass and Cock's-foot-grass

"Das grosse Rasenstück"

ALBRECHT DÜRER, 1503

THE GRAMINEAE

A STUDY OF CEREAL, BAMBOO, AND GRASS

BY

AGNES ARBER, M.A., D.Sc.

SOMETIME FELLOW OF NEWNHAM COLLEGE
CAMBRIDGE

WITH A FRONTISPIECE
AND TWO HUNDRED AND TWELVE
TEXT-FIGURES

CAMBRIDGE
AT THE UNIVERSITY PRESS
1934

PRINTED IN GREAT BRITAIN

TO

THE MEMORY OF

DUKINFIELD HENRY SCOTT

1854–1934

PREFACE

IT is now more than thirty years since, at my suggestion, Ethel Sargant sowed the grains of wheat and maize which formed the starting-point of work upon the grass seedling which we did together. Before her death in 1918, I had turned to a more general examination of the monocotyledons; but in course of time I began to feel that it might be possible to get a clearer view of the questions arising out of this study, if I combined it with more intensive work upon some one of the great monocotyledonous orders. With this in view, I returned to the Gramineae, and for the last ten years they have seldom been far from my thoughts. My concern with them was at first purely morphological, but, as I came to visualise the nexus of problems presented by the life of bamboo and grass, as well as by the history of the cereals—so inextricably interwoven with the history of man—these wider fields began to fascinate me almost as much as the more limited and technical facies of my own special studies.

The grasses are a vast group, and a correspondingly vast mass of information about them has been accumulated by botanists. If this information were brought together and correlated, it would form a library rather than a volume. In the present book I have had no such encyclopaedic aim, but I have deliberately limited myself to those aspects of the subject which happen to make the greatest appeal to me personally. It is difficult to map out these aspects in a few words. I hope that some picture of them will emerge in the following pages, so I will only now say that I have sought, primarily, to detect the pattern and rhythm underlying that complex of plant types called the Gramineae; but, though I have tried never to let this central clue slip through my fingers, I have not hesitated to turn aside, from time to time, down any passage of the labyrinth which has taken my fancy. About my omissions it is easier to be explicit than about my aims. The reader will notice, for instance, that the references to ecology are incidental only; this is not due to any lack of appreciation of the work that has been done in this field, but to the fact that the ground has already been covered by an important recent volume—*The World's Grasses*, by Professor J. W. Bews—

which is written throughout "from the standpoint of ecological differentiation". It is hoped that the table on pp. 410–11 of the present book, which gives the taxonomic grouping of all the genera named in the text, will facilitate its use in conjunction with *The World's Grasses*, and with the standard systematic accounts of the family.

If the argument in the following chapters maintained an entirely logical order, it would begin with a rigidly scientific account of the Gramineae, while their relation to man would be ignored, or consigned to appendices. I found it easy to construct such a logical skeleton, but difficult to endow it with vitality. When I desisted from the attempt, and allowed the book, as it were, to write itself, I discovered that its spontaneous course pursued an historical rather than a logical path. Man, in his primitive state, thought of plants merely in connection with himself and his needs, and it was only gradually that this egocentric interest matured into the objectivity of pure botany. This book follows a corresponding sequence; it begins with the study of the grasses in relation to man, and the more strictly botanical aspect is treated as developing out of the humanistic.

For the last seven years I have done my research upon the grasses at home; I have to thank both Girton College, and my own college—Newnham—for the loan of apparatus, equipment and books which has made this possible. I am indebted to the Dixon Fund Committee of the University of London for grants towards the running expenses of my little laboratory.

Except for reproductions of the work of artists of the seventeenth century and earlier, the illustrations in this book are all from my own drawings, with the addition of a few by Ethel Sargant. The Oxford University Press, and the Editors of *The Annals of Botany*, have kindly allowed me to use seventy-nine blocks from a series of ten articles of mine upon the grasses, published in that journal, while one figure is taken from my previous book, *Monocotyledons*. The one hundred remaining illustrations, initialled [A.A.] in the legends, have been drawn specially for this volume. Albrecht Dürer's water-colour of a sod of turf ("Das grosse Rasenstück"), which forms the frontispiece, is used by kind permission of the Staatliche Kunstsammlung, Albertina, Vienna.

In the decades during which I have been concerned with grasses,

I have received so much assistance from botanists in different countries, that to record my gratitude in any complete fashion would be altogether impracticable. Some part of my indebtedness I have already acknowledged in my individual papers, and I will only now attempt to express my thanks to a few of those who have aided me during the actual preparation of this volume. I must name, in the first place, Mrs Agnes Chase, Associate Agrostologist in the Smithsonian Institution, Washington, who has allowed me to draw, time after time, upon her special knowledge; and Mr C. E. Hubbard, Temporary Botanist in the Herbarium, the Royal Botanic Gardens, Kew, who, by kind permission of the Director, Sir Arthur Hill, K.C.M.G., F.R.S., has repeatedly answered my enquiries and solved my problems. I am, moreover, particularly grateful to Mr F. L. Engledow, Drapers Professor of Agriculture in the University of Cambridge, for his help during my study of the cereals. Though there are many more to whom I should like to express my indebtedness in detail, I must content myself with a bare enumeration of those whom I have to thank for information and suggestions, guidance in questions of taxonomy, gifts of material, and help in connection with the literature: the Director, the Assistant-Director, and the Keeper of the Herbarium, the Royal Botanic Gardens, Kew; the Director and the Superintendent, the Cambridge Botanic Garden; the Directors of 's Lands Plantentuin, Buitenzorg, Java; of the National Botanic Garden, Kirstenbosch, South Africa; of the Botanic Gardens, Singapore, Straits Settlements, and Sydney, New South Wales; the Chief Conservator of Forests, Burma; Mr R. N. Aldrich-Blake; Mr J. Ardagh; Professor Ellsworth Bethel; Professor E. Blatter, S.J., of Bombay; Professor P. Bugnon of Dijon; Mr I. H. Burkill; Mlle Aimée Camus of Paris; Professor L. Corbière of Cherbourg; Professor G. N. Collins of Washington; Dr J. Burtt Davy; Mr and Mrs Foggitt; Dr F. W. Foxworthy of the Forest Research Institute, Federated Malay States; the late Mr J. S. Gamble, F.R.S.; Miss S. Garabedian of the South African Museum, Cape Town; Professor Ruggles Gates, F.R.S.; Mr B. L. Gupta of Dehra Dun; Mr Cecil Hanbury, M.P., of La Mortola, Ventimiglia; Professor A. S. Hitchcock of Washington; Miss Gulielma Lister; Professor F. W. Oliver, F.R.S.; Professor L. R. Parodi of Buenos Aires;

Mr R. T. Pearl; Professor J. Percival; Mr P. W. Richards; Mr H. N. Ridley, F.R.S.; Professor B. Sahni of Lucknow; Professor E. J. Salisbury, F.R.S.; Professor A. C. Seward, F.R.S.; Dr T. A. Sprague; the late Dr O. Stapf, F.R.S.; Mr W. T. Stearn; Miss Sidney M. Stent of South Africa (Dep. of Agric.); Miss E. L. Stephens of the University of Cape Town; Mr J. R. Swallen of Washington; Dr Göte Turesson of Lund; Mr W. B. Turrill; Mr T. G. Tutin; Professor N. I. Vavilov of the U.S.S.R. Institute of Plant Industry; Professor P. Weatherwax of Indiana; and Mr W. C. Worsdell. To the Editors of *The New Phytologist* I am indebted for permission to incorporate two papers of mine which have appeared in their journal. My brother, D. S. Robertson, Regius Professor of Greek in the University of Cambridge, has given me his aid in various questions of scholarship, while, in matters of expression, I owe much to my daughter Muriel's criticism. I also wish to thank the staff of the Cambridge University Press for the patient and helpful way in which they have brought their expert knowledge to bear upon every problem which has arisen during the production of this book.

I received my training in research from Ethel Sargant, who dated her own initiation from the period during which she worked at the Jodrell Laboratory under Dr D. H. Scott, F.R.S. His death at the beginning of this year has brought a deep sense of loss to botanists all over the world. As one of those who owe more to him than they can well express, I am venturing to dedicate the present study to his memory, with the wish that it were worthier of a connection—even so slight as this—with his name.

As a last word, I wish to stress the fact that this is only one of numberless possible books, for which, to different minds, the grasses might supply material. I have of the Grass, as Coleridge had of the Rose-tree, "a distinct Thought", but I am continually conscious of his unanswered question—"what countless properties and goings-on of that plant are there, not included in my *Thought* of it?"

AGNES ARBER

CAMBRIDGE
June 15, 1934

CONTENTS

LIST OF ILLUSTRATIONS

CHAPTER I

CEREALS OF THE OLD WORLD

THE culture of cereals by man dates back to time immemorial, so that our ideas about its origin can be but guesswork. The discovery that the seeds of certain wild grasses were specially good for food, may be pictured as the first step, and this would lead gradually to their regular collection. How, and when, man first

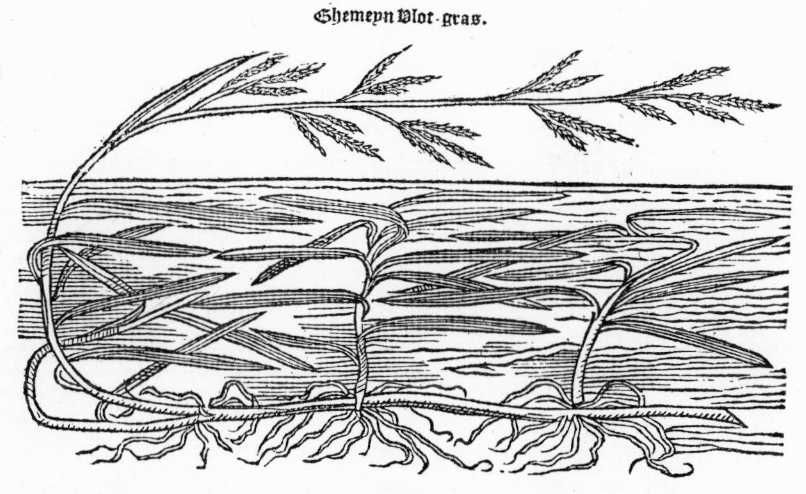

Fig. 1. Manna-grass, *Glyceria fluitans* R.Br. [Reduced from the *Kruydtboeck* of Mathias Lobel (de l'Obel), 1581.]

began to cultivate grasses, instead of relying on those that grew spontaneously, we do not know. It is clear, however, that the use of wild grasses did not come to a sudden end when systematic agriculture began; traces of it, indeed, have lingered on, side by side with cereal cultivation, into modern times. One of our British grasses, *Glyceria fluitans* R.Br. (Fig. 1), is called Manna-grass because a slight shake detaches its edible seeds in showers. I know of no record of their having been used in this country, but Benjamin Stillingfleet—whose blue worsted stockings are enshrined in our language, though his botany is seldom remembered—noted in 1759 that the seeds of this grass were said to be collected in Poland and carried thence into Germany, where they were much used "at the tables of the great on account of their nourishing quality and agreeable taste". A nineteenth-century writer,[1] also, describes the seeds as collected in Russia by the peasant, who "takes an old felt hat, and, wading in the water, skims the hat among the patches of glyceria when the grain is ripe". *Avena fatua* L., the Wild-oat, is another uncultivated grass which may serve for food; in the latter part of the eighteenth century it was still harvested in Sweden.[2] Similar traces of the use of wild grasses can be found outside Europe. In Africa the traveller Barth[3] noticed, in the mid-nineteenth century, that the Tagáma used a species of *Pennisetum*; the seeds were collected and pounded by the slaves. He also recorded that in Bagírmi, where Rice was not cultivated, it was obtained in great quantities, after the rains, from wild plants in the forest swamps. As an Asiatic example we may recall that in the Multan division of the Punjab, the poorer classes, especially the nomads, were accustomed, about seventy years ago, to use the seeds of wild plants for food, including those of various grasses, e.g. *Cenchrus* and *Pennisetum*. The fallen seed was swept up with a whisk into straw baskets resembling our dustpans.[4]

Corresponding examples can be found in the New World. The Cocopa Indians, for instance, eat the grain of *Uniola Palmeri* Vasey.[5] Until recent times, also, the wild Indian-rice, *Zizania aquatica* L.—a waterside plant with a slender grain 2 cm. long—was harvested in America. Before the seed was ripe, the Indian women went about in canoes and tied together the heads of as many plants as could be gathered in the arms. The grass was then left to ripen, when the women returned, and, holding the tied heads over their canoes, beat out the grain;[6] but this use of *Zizania* has something in common with actual cultivation, for it is said that the Indians are

[1] Archer, T. C. (1856).
[2] Koernicke, F. (1885).
[3] Barth, H. (1857–8).
[4] Edgeworth, M. P. (1862); see also Coldstream, W. (1889).
[5] Hitchcock, A. S. (1920).
[6] Hitchcock, A. S. and Chase, A. (1931).

careful also to scatter some of the seed to ensure further crops.[1] Another hint of a transition stage towards settled cultivation may be found in semi-nomadic agriculture. Tribes with wandering propensities can achieve a certain amount of cereal culture, if the grain required for sowing is not too cumbrous to accompany them on their treks, and if the plants are not exacting about treatment. *Panicum miliaceum* L., the Common-millet, fulfils these conditions, and it is grown, for instance, all over Afghanistan by nomads as well as by the stationary population.[2]

The epoch in which our ancestors knew of no cereal food except the seeds of wild grasses, and the conjectural succeeding phase of semi-nomadic husbandry, must both belong to a very distant past, for the period of settled agriculture, in which we now live, can be traced back to remote antiquity, and even the earliest records which we possess belong to a time when the cultivation of grain was already far advanced.[3] In the oldest Neolithic period, no part of Europe possessed one cereal only; where grain occurs at all, Wheat, Barley, and sometimes Millet are found. These Neolithic finds show, however, that Wheat, in frequency of occurrence and in the number of kinds in cultivation, far surpassed the other cereals. Unfortunately, in considering prehistoric Wheats, we have to bear in mind that, owing to the deterioration that takes place in the grain in course of time, there must always be an element of doubt in the discrimination of the different types. The dating, also, of the earlier records can have little absolute value, but we may recall that Wheat occurs in the oldest Lake-dwellings, whose age is estimated at 3000–4000 B.C., and it is also known from Egyptian tombs of a comparable era. A few years ago, in a Sumerian house at Kish, a fine red and black jar was found, containing a quantity of Wheat; this was reported to be the only cereal ever recovered from the early remains of Mesopotamian civilisation. It could be dated at about 3500 B.C. from the associated pottery and pictographic tablets.[4] The

[1] Hackel, E. (1887).

[2] Vavilov, N. I. and Bukinich, D. D. (1929).

[3] On the history of the cereals, see Candolle, A. de (1884); Schweinfurth, G. (1884); Koernicke, F. (1885), (1908); Hoops, J. (1905); Schulz, A. (1913).

[4] Langdon, S. (1927).

evidence, from the various early finds, thus allows us to say that Wheat cultivation was well established in the fourth millennium before Christ, and that it must have a history, still unknown, extending back to a much earlier period. The fact that its origin has been attributed to divinities, such as Isis and Ceres, attests its antiquity.

In England, Wheat seems to have been the chief bread corn up to and during the Roman occupation, while Rye was introduced later by the Anglo-Saxons.[1] Exact information about the relative status of different cereals at different periods is not easy to obtain, but from indirect indications it appears that, in the Middle Ages and later, Wheat was regarded as a luxury of the upper classes. A twelfth-century Latin poem, of which a translation is given in *The Bread of our Forefathers*,[2] sheds an amusing light on snobism eight hundred years ago, as evidenced by the connection between wheaten bread and social standing. A few lines of it may be cited here:

For a guest it is but meet
To offer bread that's made of wheat;
Lest there be no bread to take
This the clever plan they make.
With common purse they buy, 'tis said,
A single loaf of wheaten bread.
They put it under lock and key,
And if a guest they chance to see,
They bid the servant go for it—
But no one dares to cut a bit.

Even in the fourteenth century we find hints of a similar distinction, for the monks of Norwich Priory are known to have had wheaten loaves, while, on the other hand, there seems to have been more Rye than Wheat in the almoner's supply for the poor.[2]

At the present day the distribution of Wheat in the world is so wide that no month passes without a crop of this cereal being harvested in some region or other.[3] Rarely, e.g. in Colombia, it is grown at the Equator. It is cultivated in every country in Europe and Asia (except Siam) and also in Australia and New Zealand,

[1] Percival, J. (1934). [2] Ashley, W. (1928). [3] Percival, J. (1921).

Africa and North and South America.[1] This wide range is reflected in multifarious races.[2] Vavilov—who has 29,700 samples under cultivation at the U.S.S.R. Institute of Plant Breeding[3]—believes that the number of hereditary forms in existence must be reckoned in millions. All the known types fall naturally into three main groups.

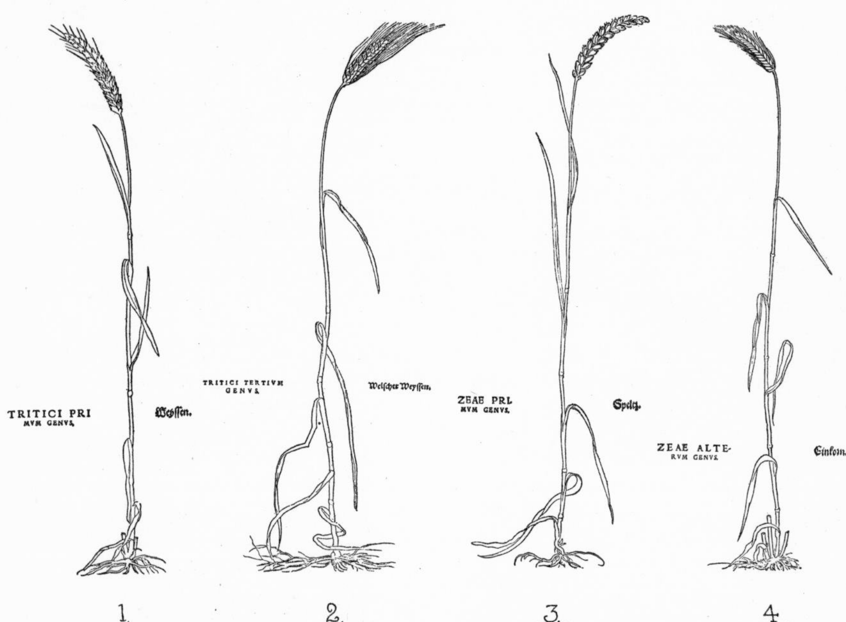

Fig. 2. *1*, "Tritici primum genus" (*T. vulgare* Host); *2*, "Tritici tertium genus" (*T. turgidum* L.); *3*, "Zeae primum genus" (*T. Spelta* L.); *4*, "Zeae alterum genus" (*T. monococcum* L.). [Reduced from Leonhard Fuchs, *De Historia Stirpium*, 1542. On the identifications, see Sprague, T. A. and Nelmes, E. (1931).]

The first or Small-spelt group contains only the various forms of *Triticum monococcum* L., also called St-Peter's-corn, or One-grained-wheat. It seems to have been one of the chief Wheats grown in Mid-Europe in the Neolithic period. Leonhard Fuchs in 1542 gave the earliest illustration of it, as "Zeae alterum genus" (Fig. 2, *4*). St-Peter's-corn produces bread of a dark brown colour, and the

[1] Percival, J. (1921).

[2] For the classification and history of the Wheats, see Percival, J. (1921) on which the following epitome is based; also Hoops, J. (1905); Koernicke, F. (1908) and Stapf, O. (1909).

[3] Information from Professor Vavilov by letter dated Dec. 28, 1933.

yield is low. It has its value, nevertheless, for it suffers little from the depredations of birds, and it is resistant to frost and disease; moreover it will grow on a poor rocky soil where other Wheats cannot survive.

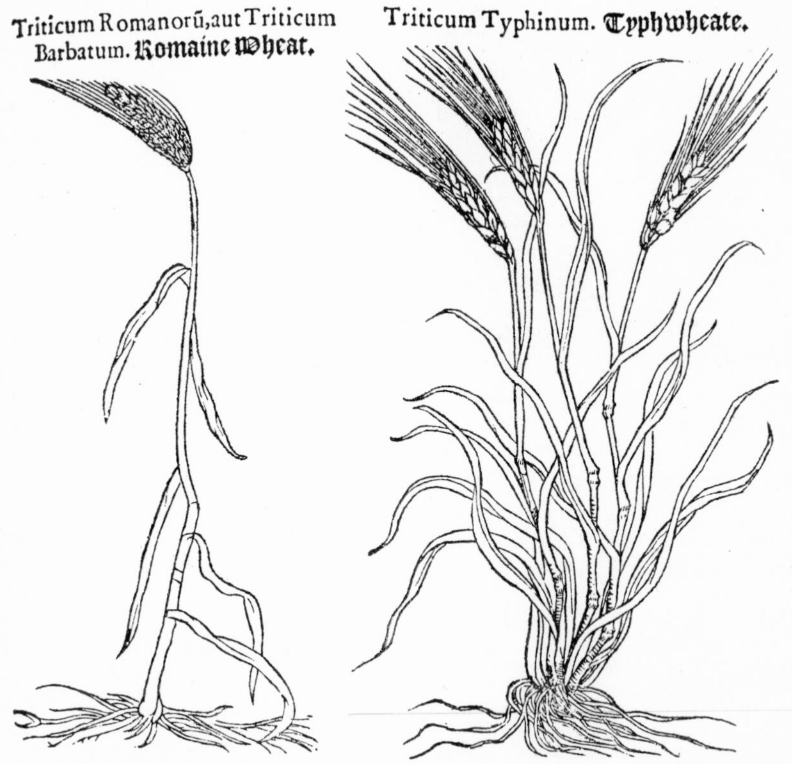

Fig. 3. "Romaine Wheat" (*Triticum turgidum* L., Rivet-wheat). "Typhwheate" (*T. durum* Desf., Macaroni-wheat). [Reduced from the *Nievve Herball* of Rembert Dodoens (translated by Henry Lyte), 1578. On the identifications, see Percival, J. (1921).]

The second or Emmer group includes a number of Wheats, that which gives its name to the series being *T. dicoccum* Schübl., one of the most anciently cultivated of the cereals. Ears of corn, carved in ivory and in wood, from an Egyptian tomb, prior to 5000 B.C., have been described as representing Emmer. Another important member of the same group is *T. durum* Desf., Macaroni-wheat, which was first mentioned by Rembert Dodoens in the sixteenth century. His illustration of it, which was used under the name of "Typhwheate" in Lyte's *Nievve Herball* of 1578, is shown in Fig. 3. Besides other less familiar Wheats, the Emmer series also includes *T. turgidum* L., Rivet- or Cone-wheat ("Romaine Wheat" in Fig. 3). The name 'English-wheat', which is sometimes used for this species, is unfortunate, for it did not arise here, nor is it specially characteristic of this country. There is some doubt as to whether Rivet-wheat was in existence in prehistoric times. The first published figure of it is to be found in Fuchs' herbal of 1542, under the name of "Tritici tertium genus" (Fig. 2, *2*, p. 5).

The third or Bread-wheat group is by far the most important from the economic standpoint. *Triticum vulgare* Host, the Common-bread-wheat, is the most widely grown of all the Wheat races, and it is also one of the most ancient of cereals, being known abundantly from Neolithic sites. It is extraordinarily rich in varieties; more than a thousand distinct forms are known. Another Wheat belonging to the same group as *T. vulgare* is *T. compactum* Host, the Club- or Hedgehog-wheat, which also was common in the Neolithic period. In addition there is another race, *T. Spelta* L., Spelt-wheat (Fig. 2, *3*, p. 5); unlike the Bread-wheat and Club-wheat, it is a relatively modern type.[1] It has the disadvantage of possessing brittle ears, so that the ordinary process of threshing detaches segments of the ear with the grain, and the grain itself has to be separated later by means of special mill-stones.

The classification of the Wheats into the three main groups, which we have enumerated, is consistent with the results of cytological study; for the number of chromosomes in the nuclei of ordinary vegetative cells, such as those of the root-tips, proves to be 14 (diploid) in *Triticum monococcum*, 28 (tetraploid) in *T. dicoccum*, *T. durum* and *T. turgidum*, and 42 (hexaploid) in *T. Spelta*, *T. compactum* and *T. vulgare*.[2] The Emmer and Bread-wheat series are thus polyploid, like the majority of cultivated plants.[3]

[1] The view put forward tentatively by Stapf, O. (1909), that the Spelt-wheats are derived from a wild grass, *Aegilops cylindrica* Host, is not accepted by Schulz, A. (1913), or Percival, J. (1921).

[2] Sakamura, T. (1918).

[3] Darlington, C. D. (1932²); see also Hurst, C. C. (1933), and for a review of Wheat species in the light of cytological work, Watkins, A. E. (1930).

The history of the vast complex of Wheat types is highly obscure, and there has been much difference of opinion, especially as to the origin of the third group—the Bread-wheats. One view is that these Wheats have not been derived from any wild prototype, but that they originated through hybridisation between Wheats of the Emmer series and grasses belonging to the genus *Aegilops*, which is related to *Triticum*.[1] Another view is that the ancestor of the Bread-wheats awaits discovery somewhere in the region of South-East Afghanistan and North-West India. It has been shown that no country in the world exceeds Afghanistan in the number and variety of Wheats belonging to *T. vulgare* and *T. compactum*; a hundred and ten distinct forms have been recorded here. It thus seems not unreasonable to regard this region as the world centre for these races.[2]

The first group—the various forms of *T. monococcum*—are less problematic. They appear to have originated from a wild grass, *Crithodium* (*Triticum*) *aegilopoides* Link, which ranges from the Balkans to the Mesopotamian region and the western border of Persia. There is little doubt that Mesopotamia has been one of the principal homelands of the cereals; indeed, through its extreme fertility, it seems marked out by nature for this rôle. Lower Mesopotamia consists entirely of rich alluvial soil, laid down by the rivers, especially the Tigris and Euphrates. A fresh layer of fertilising deposit is spread over the land every year, when these rivers overflow. Though alluvial soil is generally rich in certain plant foods, it often has its productivity reduced by lack of mineral elements. The Euphrates and Tigris, however, in their upper regions drain mountains whose bases are composed of disintegrating limestone; the alluvium, in presence of this supply of lime, gives a soil of incomparable richness.[3] The importance of Mesopotamia as a centre of distribution of the cereals has been recognised from early days. The Babylonian historian, Berossus, who flourished in the time of Alexander the Great, speaks of wild Wheat and wild Barley growing in the region between the Tigris and Euphrates.[4] Today, not only

1 On hybrids between *Aegilops* and Wheats, see Percival, J. (1926[2]), (1930).

2 Vavilov, N. I. and Bukinich, D. D. (1929).

3 Scott, J. A. (1920).

4 Müller, C. (1878), vol. II, p. 496; φύειν δὲ αὐτὴν πυροὺς ἀγρίους καὶ κριθάς.

is the wild form of St-Peter's-corn, *Triticum monococcum*, to be found—as we have seen—in Palestine, which is no great distance from Mesopotamia, but, in company with it, upon Mount Hermon, a second grass occurs, which has been regarded as the parent[1] of the Emmer series (*T. dicoccum*) and has been named *T. dicoccoides* Koern. Moreover the wild Barley, *Hordeum spontaneum* C. Koch, has been found near Lake Tiberius in association with *Triticum dicoccoides*. The occurrence in Palestine of two possible progenitors of Wheat, as well as a wild form of Barley, is in striking accord with the ancient mythology of the region.[2] Diodorus Siculus[3] tells us that "a City of *Arabia Fœlix*, named *Nysa*, which is neer adjoyning to Egypt", was the home of the god Osiris, "a great studier and lover of husbandry". There is a tradition that at Nysa a votive column was reared to Isis, the wife of Osiris, celebrating her discovery of corn, and it has been supposed that the site of this town may have been at the foot of Mount Hermon.[4] If this were so, the wild Wheat and wild Barley, which are still to be found there, carry the suggestion of an immemorial lineage. They fire the imagination to a vision of the days when the gods walked with men, and when Isis, among these very hills, revealed to her worshippers the kindly fruits of the earth.

The use of the grain of Wheat, and also of the 'barleycorn' (the Troy 'grain'),[5] as original units of weight, is indicative of the antiquity of the culture of both cereals; it is not possible to say whether one is older than the other.

In the structure of the ear, Barley (*Hordeum sativum* Jessen) differs considerably from Wheat. The spikelets[6] are arranged in triads, in two rows, one on either side of the axis. The principal races[7] of Barley are distinguished by the relative degree of development of the spikelets in the triad (Fig. 4, p. 10).

In *H. sat. hexastichon* L. ('Polystichum', Fig. 4) all three spikelets

[1] Vavilov, N. I. (1928) does not accept the view that *T. dicoccoides* is the ancestor of the other members of the group; see also Watkins, A. E. (1933).

[2] See the discussion of this coincidence in Stapf, O. (1909).

[3] Diodorus Siculus (H. C.) (1653).

[4] References in Hartmann, F. (1923).

[5] Ridgeway, W. (1892).

[6] The spikelets of the Gramineae are discussed in Chapter VIII.

[7] For an account of the different forms of Barley, see Hunter, H. (1926).

of each sub-opposite triad are fertile[1] and have long awns; such Barleys are hence known as 'six-rowed'. There are also races which are popularly called 'four-rowed' (*H. vulgare* L., *H. tetrastichon* Koern.),

Somer-Gerste van veel rijen/ghelijck de Winter-Gherste/in Griecks ghenaemt Polystichum.

Cleyn Gerste oft Somer-Gerst / heet in Griecks Distichō, dat is/ Gerste met twee rijen. In Hoochduytsch eñ Nederduytsch/Cleyne Gerste. In Walsch Nederlandt/Pamelle. In Engelsch/Comon Barly.

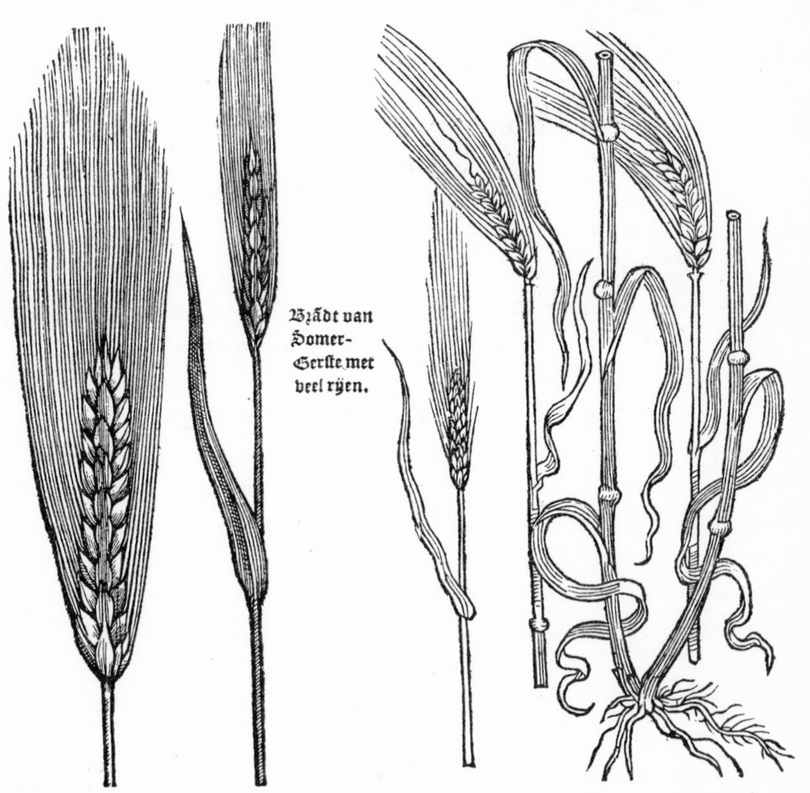

Fig. 4. Barley (*Hordeum sativum* Jessen) with more than two rows and with two rows only. [Reduced from the *Kruydtboeck* of Mathias Lobel (de l'Obel), 1581.]

but they are merely laxer varieties of the six-rowed type, in which the laterals of the sub-opposite triads intermingle to form, in appearance, one row. The 'bere' or 'big' of Scotland is a 'four-rowed' Barley.[2]

[1] See Fig. 77, D, p. 172, for the spikelet-triad of six-rowed Barley.

[2] In crossing experiments, Barleys have been produced in which the median spikelet is sterile and the two laterals fertile; it would be more reasonable to restrict the description 'four-rowed' to such forms as these; see Drabble, E. (1906).

In another important race, *H. sat. distichon* L., the median spikelet in each triad is fertile, but the laterals are staminate (infertile) and unawned. The grain is thus carried in two rows only, on opposite faces of the axis. The Barleys most commonly cultivated in the British Isles belong to the two-rowed type; *H. sat. zeocriton* L., Fan-, Peacock-, or Battledore-barley, is a specially dense, broad-eared variant of this class.

H. sat. deficiens Steud. represents a third type, in which the spikelet-reduction is carried further than in *H. sat. distichon*. The lateral spikelets of the triad produce no stamens and are merely vestigial.[1] It is cultivated in Abyssinia.

The wild Barley, *H. spontaneum* C. Koch, to which we have already referred, occurs in various localities from the Caucasus to South Persia and also in Arabia. It is generally accepted as the ancestor of all the two-rowed Barleys used in agriculture, with which it agrees in essentials. How and when the six-rowed type originated is unknown. It is undoubtedly of great antiquity; representations of it are to be found in Egyptian monuments attributed to the third millennium before Christ.[2] It also appears on Greek coins of the late Archaic period. Fig. 5 is from a coin of Metapontum in South Italy, dating probably from about 520 B.C.; it represents a form of the six-rowed type which can be matched today among living Barleys. The six-rowed type, indeed, appears to have predominated over the two-rowed in ancient agriculture. This fact makes it improbable that the *hexastichon* form originated from *distichon*; it seems more likely that six-rowed Barley had an independent origin from another wild form. It has been suggested that the ancestor of the *hexastichon* type may be *H. ithaburense* Boiss., var. *ischnatherum* Cosson.[3] This form, which occurs in the Mesopotamian region, is said to show a structure transitional between that of the two- and six-rowed Barleys. It is sometimes found in

Fig. 5. A six-rowed Barley represented on a coin of Metapontum (South Italy). [Adapted from Head, B. V. (1881), Pl. VII, fig. 10.]

[1] See Fig. 173, A6, p. 326, for a Barley in which the lateral spikelets are vestigial.
[2] Hoops, J. (1905). [3] Koernicke, F. (1908); Schulz, A. (1913).

company with *H. spontaneum*, and it is possible that it is a derivative of this species.

The Oat[1] is not, like Wheat and Barley, of immemorial antiquity; it appears to have been first cultivated in Mid-Europe in the Bronze Age.[2] There is much difference of opinion about the classification and origin of the different forms. The Oat commonly grown in this country (*Avena sativa* L., Figs. 6, p. 14 and 64, p. 147) is generally supposed to have been derived from *A. fatua* L., the Wild-oat, a worldwide weed, and recently some cytological evidence for this view has come to light.[3] Another polymorphic species, *A. byzantina* C. Koch, is held to have originated from *A. sterilis* L.,[4] which is found in countries bordering on the Mediterranean. A third species, *A. strigosa* Schreb., Horsehair-oat, is grown as a grain crop in the Shetlands and in Wales, under conditions unfavourable for other Oats; it seems to be a derivative of *A. barbata* Brot., which is native to Arabia, Asia Minor and the Mediterranean region.

Rye, *Secale cereale* L., is perhaps the least ancient of our cereals, though it may have been in cultivation in the seventh or sixth century B.C.[5] It is the only European grain-plant which has little tendency to produce constant varieties, a fact which is probably due to the self-sterility of most of its races.[6] It runs wild more readily than any of the other cereals, and is able to maintain itself for some time, if circumstances happen to be favourable. This creates a difficulty in deciding whether forms that appear to be wild are, indeed, genuinely native, or are merely feral descendants of cultivated plants.[7] The original source of Rye is uncertain. It may be derived from *Secale montanum* Guss.,[8] a species extending from Spain to Central Asia, or it may be a culture form from Turkestan, derived from *S. anatolicum* Boiss.[5] Afghanistan has been considered as an important centre for the diffusion of Rye. It is suggested that this cereal has been carried north from that country in company with the Bread-wheats, which may have origi-

[1] For a general account, see Hunter, H. (1924).
[2] Hoops, J. (1905).
[3] Percival, J. (1930).
[4] Hunter, H. and Leake, H. Martin (1933).
[5] Schulz, A. (1911).
[6] Rimpau, W. (1877); Koernicke, F. (1885).
[7] Hitchcock, A. S. and Chase, A. (1931).
[8] Hackel, E. (1887).

nated there. In this northward progress, Rye, being hardier than Wheat, gradually supplanted its companion;[1] but in course of time, with the improvement of methods of agriculture, the process has been reversed, and Wheat has more and more ousted Rye,[2] even in northerly countries. By the end of the eighteenth century, Wheat had long been the staple bread corn of the Mediterranean lands, but in Germany, the Low Countries and Denmark, Rye was the food of the great mass of the people; even in France, which was already becoming the land of fine wheaten bread, Rye, at this date, still predominated in most districts. In England, too, Rye had a longer reign than is generally recognised. As we have shown, there is evidence that, in the Middle Ages, it was more widely used than Wheat. The position does not seem to have been reversed until capitalist farming made it possible to raise the level of fertility of our soil.

A harvest of which, today, we have no experience in England, is the mixture of Rye and Wheat generally called 'maslin'. In earlier times, this corresponded, as a winter crop, to the combination of Barley and Oats sown in spring, and called 'drage' or 'dredge'. In Green's herbal of 1820, the sowing of Wheat and Rye together is noted as a common practice in the northern counties, but condemned as "bad husbandry", as their ripening synchronises imperfectly. It would be interesting to know whether this practice ever resulted in natural hybridisation, for Wheat-Rye hybrids[3] are known, and have been much studied, particularly in Russia, in recent years. The mixed crop of Wheat and Rye went under a variety of names. The Old French 'miscelin' gave the best-known term—'maslin' or 'maslen', but there were also certain English names of a Teutonic type—'mengcorn', 'muncorn', 'mancorn', 'monkcorn'. 'Mengcorn' in German signifies 'mingled grain'; the first syllable reappears in our word 'mongrel'. 'Monkcorn' is thus innocent of

[1] Vavilov, N. I. and Bukinich, D. D. (1929).

[2] The account of Rye and Maslin in these two paragraphs is taken from Ashley, W. (1928).

[3] Wilson, A. S. (1875); for modern references, see Hunter, H. and Leake, H. Martin (1933).

Fig. 6. *Avena sativa* L. var. β L., White-oat. [From the *De Stirpium* of Hieronymus Tragus (Jerome Bock), 1552.]

Fig. 7. "Siligo" (*Secale cereale* L., Rye). [From the *Herbarum vivae eicones* of Otto Brunfels, 1530. For the identification, see Sprague, T. A. (1928).]

monastic meaning, and 'mancorn' carries no reference to man; they are both merely rationalised corruptions of 'Mengcorn'.

A combined crop of Rye and Barley was used in the nineteenth century in the Upper Engadine,[1] but not in order to produce a mixed corn. In the spring, Winter-rye was sown, with Summer-barley above it. In the first season, the Barley overtopped the Rye and produced its ears. After the Barley was harvested, the Rye grew up and was cut in the autumn with the Barley stubble as green fodder. In the following spring, the Rye made a second start, and ripened in that year.

The mixture of crops attracts our thoughts to a curious bypath in the study of the cereals—the history of their fabulous, but widely credited, conversions into one another. The most notorious of these superstitions was the belief that Wheat could degenerate into Rye.[2] This belief arose out of a philological muddle. 'Siligo' in classical antiquity meant 'fine wheat'. It is so used in Philemon Holland's translation of Pliny's *Naturall Historie*, where we read that the "best manchet bread for to serve the table, is made of the winter white wheat Siligo and the most excellent workes of pastrie likewise are wrought thereof".[3] But this meaning was not always remembered, and siligo came to be treated as the ordinary, everyday, Latin word for Rye—perhaps because of a certain resemblance between its sound and that of 'secale' or 'seigle'. Fig. 7, which is from the herbal of Otto Brunfels of 1530, shows a woodcut of Rye labelled 'Siligo'. Columella, in his *De Re Rustica* (Lib. 2. 9) happens to state that "all Wheat on moist soil is changed after the third sowing into Siligo".[4] He undoubtedly means by this that it is changed into an improved form of Wheat, but subsequent writers misunderstood his use of 'Siligo', and credited him with the absurd opinion that Wheat changes into Rye. This belief had curious repercussions. In the thirteenth century St Thomas Aquinas,[5] in his decision respecting

[1] Christ, H. (1879).

[2] This account of Siligo and Rye is taken from Ashley, W. (1928).

[3] Pliny (Holland, P.) (1601), Book 18, Chapter IX.

[4] Columella (1541), "Nam omne triticum solo uliginoso post tertiam sationem convertitur in siliginem".

[5] Aquinas, St Thomas (1854), *Summa Theologica*, Part III, Qu. LXXIV, Art. III, p. 339.

the bread that might be used for the Host, ruled that the material ought to be *Triticum* (Wheat), but one of the concessions which he made was that Rye might be used, because Wheat degenerates into Rye, and therefore Rye is of the same species as Wheat. Similar degenerations were held to occur in other cereals. We read in Pliny that "The first and principall defect observed in bread-corne, and Wheat especially, is when it doth degenerate and turne into Otes: and not onely it, but Barley also doth the like".[1] Moreover the weed grasses were sometimes regarded as degraded descendants of the crops with which they were so closely associated. Darnel, of poisonous repute (p. 50), is a widespread cornfield weed, which in old days was believed to arise through the transformation of Rye and other cereals. Mattioli, in the description accompanying the woodcut reproduced in Fig. 8, speaks of it as arising from the seeds of Wheat and Barley corrupted by moisture. The belief in such metamorphoses died hard; indeed, it would be more correct to say that it still drags on a lingering existence. In the nineteenth century the idea that the cornfield weed called 'Cheat', *Bromus secalinus* L., was a degenerate form of Rye, still prevailed in some parts of England,[2] and I have been told of an instance in Cambridgeshire, as late as 1920, in which the belief that Oats might give rise to 'Cheat' was firmly held.[3] Until recent times, certain farmers in the United States regarded both good clean Wheat seed and Timothy-grass (*Phleum pratense* L.) as capable of giving rise to 'Cheat' under certain circumstances;[4] and, even to the present day, the belief that Barley can become transmuted into Oats may still be found surviving in England.[3] To the botanists of the nineteenth century such superstitions seemed of dream-like absurdity; but, strangely enough, recent work has shown that something not altogether unlike 'metamorphosis' is actually possible, so that these traditional beliefs are perhaps a degree less ludicrous than they appeared to our former ignorance. The occurrence of weeds called 'fatuoid' Oats (resembling Wild-oats) is a not unusual cause of trouble to growers of high-

[1] Pliny (Holland, P.) (1601), Book 18, Chapter XVII.
[2] Johnson, C. (n.d. [1857–61]).
[3] I am indebted to Professor Engledow for telling me of these examples.
[4] Lamson-Scribner, F. (1898); Garman, H. (1900).

grade seed Oats. For a long time the source of these forms remained a mystery, but it has now been shown[1] that they actually arise from

Lolium.

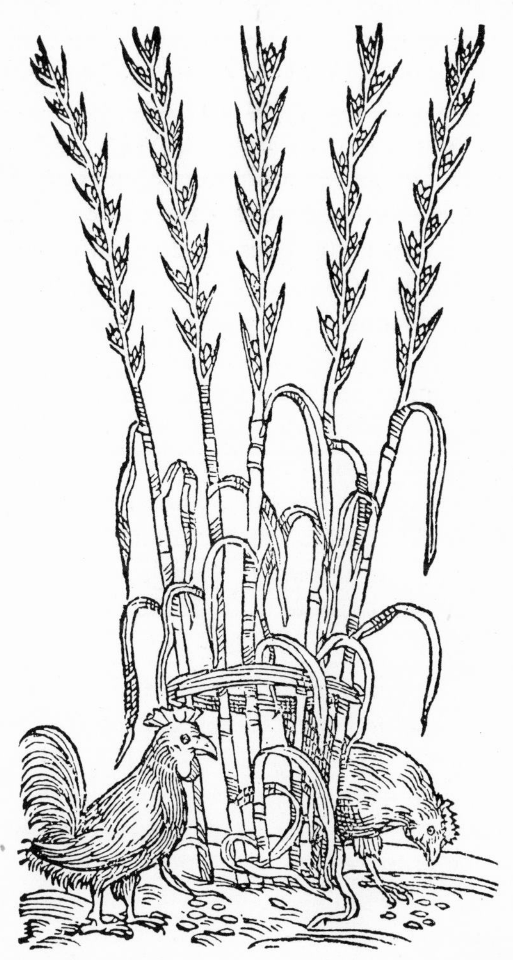

Fig. 8. Darnel (*Lolium temulentum* L.). [From the *Commentarii* of Pierandrea Mattioli, 1560.]

normal Oats, neither by crossing, nor by gene mutation, but by an irregularity in meiosis—the critical nuclear division preceding spore

[1] Huskins, C. L. (1927).

formation.[1] The result of this abnormal behaviour is to duplicate one particular chromosome, bearing the factors responsible for the fatuoid character, while a chromosome bearing factors for the normal type is correspondingly absent. This remarkable conclusion may offer a clue to the interpretation of other points in the process of down-grade evolution.[2]

Although, in Europe, Wheat, Barley, Rye and Oats are the cereals *par excellence*, in other parts of the world, different series of grain-plants take the predominant place. Among these, the Millets rank high. One of the ambiguities that confuse the study of the cereals is the unfortunate fact that the name 'Millet' is used for quite a number of distinct grasses. I shall only refer here to four of the most important—Common-millet, Italian-millet, Indian- or Giant-millet, and Pearl-millet.

The Common-millet, *Panicum miliaceum* L. (Fig. 9), which is the cereal generally meant when the name Millet is used in English, has been cultivated since prehistoric times in the south of Europe, and in Egypt and Asia. In Lyte's *Nievve Herball* of 1578 we are told that "at the highest of the stemmes come foorth the bushie eares, very muche severed and parted like the plume or feather of the Cane or Polereede [*Phragmites communis* Trin.], almost like a brushe or besome to sweepe withall". The wild prototype of Common-millet is unknown,[3] but it seems possible that it originated in the Egypto-Arabian region.[4] In early records and finds, it is often indistinguishable from the so-called Italian-millet, *Setaria italica* (L.) Beauv. (*Panicum italicum* L.), Figs. 10, p. 20, and 11, p. 21. The specific name of *S. italica* is misleading; it is not a native of Italy and is seldom cultivated there. It has been very largely grown for food in Northern China, where Rice is too dear for the poorer classes.[5] The original wild form of the Italian-millet is believed to be *S. viridis* Beauv.[3]

Durrha,[6] and Indian-millet, Giant-millet, or Kaffir-corn, are cereals

[1] For a slightly different view of the fatuoid Oat, see Jones, E. T. (1930), and, for an extension of Huskins's work to the speltoid Wheats, see Huskins, C. L. (1928).

[2] On 'down-grade' evolution, see Arber, A. (1925), p. 218.

[3] Koernicke, F. (1885).

[4] Candolle, A. de (1884).

[5] Bretschneider, E. (1870).

[6] Various spellings are used for this name.

Fig. 9. "Milium" (*Panicum miliaceum* L., Common-millet). [Reduced from the *De Historia Stirpium* of Leonhard Fuchs, 1542. On the identification, see Sprague, T. A. and Nelmes, E. (1931).]

Fig. 10. "Panicum" (*Setaria italica* (L.) Beauv., Italian-millet). [Reduced from the *De Historia Stirpium* of Leonhard Fuchs, 1542. On the identification, see Sprague, T. A. and Nelmes, E. (1931).]

which are much used by the African natives; they are varieties of cultivated Sorghum—the *Holcus Sorghum* of Linnaeus, now called *Andropogon Sorghum* (L.) Brot.[1] (Fig. 12). Sorghum is not confined

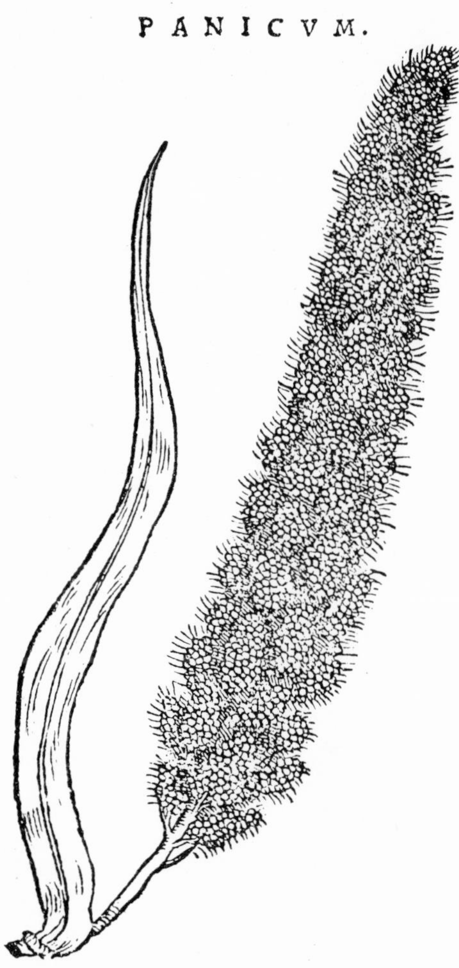

Fig. 11. "Panicum" (*Setaria italica* (L.) Beauv., Italian-millet). [From the *Commentarii* of Pierandrea Mattioli, 1560.]

to Africa, but has also been used in the East since a period long preceding the Christian era. It is generally agreed that all the cultivated Sorghums are derived from annual forms related to *Andropogon*

[1] On the cytology of the Sorghums, see Huskins, C. L. and Smith, S. G. (1932).

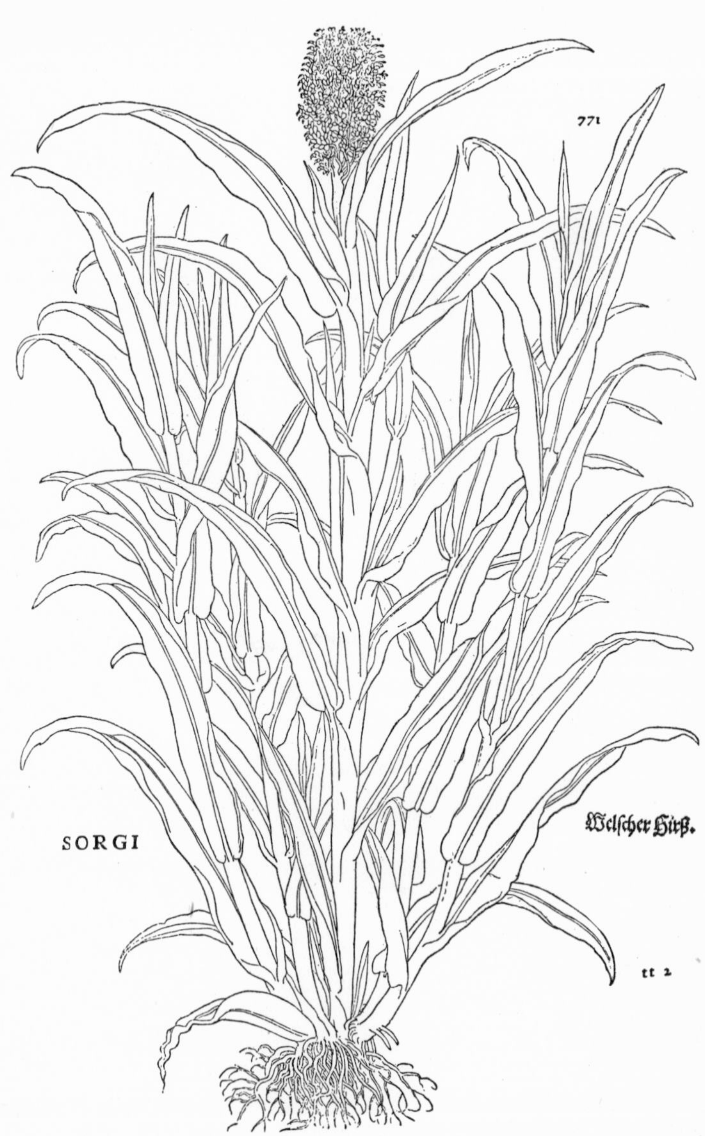

Fig. 12. 'Sorghi" (*Andropogon Sorghum* (L.) Brot. var. *vulgaris*; *Sorghum vulgare* Pers.). [Reduced from the *De Historia Stirpium* of Leonhard Fuchs, 1542.]

halepensis (L.) Brot., Johnson-grass, a wild species found abundantly in the tropical and sub-tropical parts of the Old World, whose seeds have sometimes been used in India in times of famine. Possibly the cultivated Sorghums have originated independently from the wild forms in India and in Africa.[1] The Sorghums are well adapted to primitive agriculture; they need less care and attention than any other important cereals.[2] Not only do the seeds provide human food, but the plant, which has been known to grow to a height of 20 ft. or more, furnishes, in addition, abundant fodder for animals. Its tall grace is celebrated in the Palestinian folk-verse:[3]

> I am the shining dura
> Like the tall spears,
> My height is that of a lance,
> My beauty is above all beauties.

The boastful saying, "I am so strong that I can sprout through the sole of a fellah's shoe", is also attributed to this plant.[3] Among the sculptures from Nineveh, which are now in the British Museum, there are reliefs portraying Sorghum. One of these shows a sow and her young in a field of a tall cereal, which apparently belongs to a type of Giant-millet still grown in Mesopotamia (Fig. 13, p. 24). These Nineveh bas-reliefs seem to be the earliest delineations of this cereal known; the claim that it was grown and depicted in ancient Egypt is probably based on a confusion with Flax.[4]

The Sorghums are not used as cereals and fodder plants only. A form called Sweet-sorghum, or Chinese-sugar-cane, has a sugary pith, for which it has been cultivated in America; the farmers in the Middle West used to make molasses from it.[5] The branching heads of another race, var. *technicus*, Broom-corn, have long been used in industry; one may sometimes find Millet grains still attached to a new broom.

Though the Giant-millet is so valuable to man, there is another side to its character; for young Sorghum pasture, under some con-

[1] Ball, C. R. (1910). [2] Montgomery, E. G. (1920).
[3] Crowfoot, G. M. and Baldensperger, L. (1932).
[4] Piédallu, A. (1923).
[5] Hitchcock, A. S. and Chase, A. (1931).

ditions, becomes virulently poisonous. The plant contains a glucoside, dhurrine, which on crushing is hydrolysed to prussic (hydrocyanic) acid. The production of the poison seems to be favoured by hot, dry weather, which induces stunted growth.[1] Millet is not unique in this respect, for it has been shown that prussic acid is widely, though erratically, distributed among grasses.[2]

Fig. 13. Reduced outline from a limestone slab with carved reliefs, showing a sow and her young in a field of Giant-millet (*Andropogon Sorghum* (L.) Brot.), from the palace of Sennacherib, Nineveh, 705–681 B.C. No. 56, Ḳuyunjiḳ Gallery, Department of Egyptian and Assyrian Antiquities, British Museum.

Pearl-millet, Duchn, Dochan, or Negerhirse, is another cereal of great importance, cultivated over the whole of the African continent and also in Arabia, Afghanistan, parts of India, and the West Indies.[3] The history of the names, which have been applied to it,

[1] Montgomery, E. G. (1920).

[2] For references, see Alsberg, C. L. and Black, O. F. (1915) and Bews, J. W. (1929), pp. 361 and 370.

[3] Leeke, P. (1907).

forms an ironic commentary on the botanist's claim that his use of a Latin terminology results in precision of meaning;[1] for this plant, early in the nineteenth century, possessed almost as many Latin names as there were floras. It has been called *Pennisetum typhoideum*, *P. alopecuroides*, *P. americanum*, *Penicillaria spicata*, and *Panicum spicatum*; of these, *Pennisetum typhoideum* Pers. has been most popular in recent years. This welter of names is difficult to defend, but perhaps it can claim more excuse than appears at first glance. Systematists may have been influenced subconsciously by an exceptional complexity in the constitution of the species; for it has been suggested that Pearl-millet may have arisen polyphyletically from a number of wild forms native to Tropical Africa.[2] If this theory is sound, the plant would be an interesting subject for a thorough genetic analysis.

[1] This point is discussed by Chase, A. (1921[1]), where the synonymy will be found.
[2] Leeke, P. (1907).

CHAPTER II

CEREALS OF THE EAST AND OF THE NEW WORLD: GENERAL CONCLUSIONS

BEFORE considering certain general points arising out of the study of cereals, we have to discuss two other important crop plants—Rice and Maize—as well as a few minor members of the Gramineae, whose seeds supply food for man. Rice, *Oryza sativa* L. (Fig. 14), is probably the staple food of more people in the world than any other cereal. It is essentially a swamp plant, and the 'deep-water Rices' will succeed in 5 or 6 ft. of water. Certain varieties called 'mountain Rices', may, however, be cultivated with no more water than other cereals.[1] The distribution of *Oryza*, like that of other aquatic plants, is very wide. At the present day it is apparently native to India, Australia and Africa, and it thus becomes very difficult to decide what was its country of origin. The earliest record we have of it is in old Chinese writings, and it has been cultivated continuously in that country since the remote past. Indeed there are fields in China where Rice is believed to have been grown for four thousand years uninterruptedly[2]—a state of things rendered possible by the high pitch to which the use of manures has been brought in that country. The traditional Chinese ceremonies[3] associated with the sowing of the five kinds of 'corn' at the vernal equinox, have a history stretching back into antiquity. The first spring sowing is attributed to the Emperor Shên-nung, the Father of Agriculture and Medicine, who reigned about 2700 B.C. We cannot tell whether all the plants sown in the nineteenth century at the spring ceremonies—Rice, Wheat, Sorghum, Italian-millet, and the Soya-bean—were actually those cultivated in the third millennium before Christ, for Chinese commentators disagree about the identifications. There is, however, no reason to doubt that Rice was sown

[1] Hunter, H. (1931).
[2] Copeland, E. B. (1924).
[3] Bretschneider, E. (1870).

from the first, and that it was considered the chief of these gifts to men, for the Emperor himself handled it, whereas the other grains were left to the princes and officials.

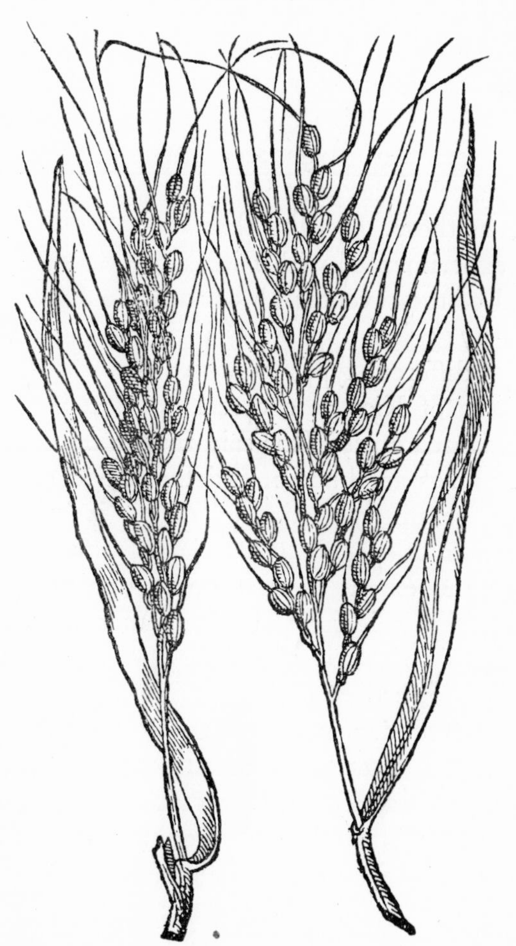

Fig. 14. Rice (*Oryza sativa* L.). [From the *Commentarii* of Pierandrea Mattioli, 1560.]

South-Eastern Asia has been claimed as the primaeval home of Rice.[1] On the other hand, the possibility that Tropical Africa was its fatherland is perhaps not excluded; but, if its source was in

[1] Copeland, E. B. (1924).

Africa, it must have been carried to Asia very early indeed, and re-distributed thence to the north of Africa and to Europe.[1] It is said that medieval Europe acquired it from the Saracens.[2] Even the details of its comparatively modern introduction into North America are somewhat obscure. According to tradition,[3] Carolina received Rice from two different sources—a bag of the unhusked grain, known as 'paddy', given as a present by a treasurer of the East India Company to a Carolina trader, and a second supply, obtained from a Dutch vessel from Madagascar. Modern research,[4] however, does not confirm the idea that its introduction was due to happy accidents such as these. It seems that, from the first, the promoters of the colony experimented with Rice, and that it was in cultivation before the end of the seventeenth century; in Carolina, the period from 1695 to 1715 was apparently a time of great activity in the trial of Rices of different types from various parts of the world.

The great age of Rice as a cereal is attested by the immense number of cultivated races. It is said that in the Philippines alone, 3500 varietal names are known, and that 2000 of these represent strains that are genuinely distinct; "*Oryza sativa* L." is clearly a complex aggregate of specific and varietal forms.

Considering the extent of mankind's dependence upon Rice, it is regrettable that this cereal is not a more complete foodstuff. It is lacking in nitrogenous matter and also in fat; the so-called 'glutinous' Rices owe their special quality to a sticky carbohydrate (amylodextrin).[5] The deficiencies of the grain from the food standpoint become virtues, however, under tropical conditions, since they improve its keeping capacity.[6] Rice culture has been carried on in Southern Europe, but the fact that it requires standing water, which is liable to render the neighbourhood malarious, has been an obstacle.[1,2]

Maize or Indian-corn (*Zea Mays* L.) occupies an isolated position

[1] Koernicke, F. (1885). [2] Copeland, E. B. (1924).
[3] See, for instance, Green, T. (n.d. [1816–20]).
[4] For a full discussion with many references, see Gray, L. C. (1933).
[5] Parnell, F. R. (1921).
[6] Finch, V. C. and Baker, O. E. (1917).

among the cereals.[1] It undoubtedly originated in the New World, but it became diffused with great rapidity through Europe after the discovery of America.[2] It thus has no ancient history in the Old World; the 'Zea' of certain herbalists of the sixteenth and seventeenth centuries (Fig. 2, *3* and *4*, p. *5*) was not Maize, but Wheat. The first European illustration of Indian-corn is that of Leonhard Fuchs, who published a woodcut of it in 1542 as 'Turkisch Korn' (Fig. 15, p. 30). He describes it as having been "recently introduced into Germany from Turkey, Asia and Greece. Now fairly common and cultivated in many gardens".[3] The names,[4] by which Maize has been called, show that it is a hopeless task to search in the nomenclature of cultivated plants for light on their geographical origin. In Lorraine and the Vosges, this cereal has been known as Roman-corn; in Provence, Barbary-corn; in the Pyrenees, Spanish-corn; in Turkey, Egyptian-corn; in Egypt, Syrian-dourra; in Persia, Wheat-from-Mecca; and in Abyssinia, Millet-from-the-sea. These names prove nothing, except that in all these regions, Maize was an alien whose source was unknown. But the name 'Turcicum frumentum', Turkish-corn, first recorded by Ruellius in 1536, was more widespread in Europe than any of those just enumerated, and, though misleading, it is not so meaningless as the rest. Ruellius says that Maize is so named because it came "in the days of our grandfathers" (avorum nostrorum aetate) from Greece or Asia.[5] Turkish-corn as a name for Maize, corresponds with 'Turkey' as a name for the farmyard bird. Both plant and bird were so designated because the New World, of recent discovery, was confused with the Indies of Asia, and Turkey lay on the trade route to the *East* Indies.

Setting aside certain problematic references in the 'Wineland' sagas of Iceland,[6] the European knowledge of Maize may be said to date from November 5, 1492. On that day its existence was reported to Columbus by the men whom he had landed to explore

[1] For a further study of this cereal, see Chapter XVI.

[2] Candolle, A. de (1884).

[3] Sprague, T. A. and Nelmes, E. (1931).

[4] Candolle, A. de (1884); Bretschneider, E. (1870); Harshberger, J. W. (1893).

[5] Ruellius, J. (1536).

[6] Sturtevant, E. L. (1885); Gathorne-Hardy, G. M. (1921) considers that "the identification of the wild corn" of these sagas "will always be an insoluble problem".

Fig. 15. "Turcicum frumentum" (*Zea Mays* L.). [Reduced from the *De Historia Stirpium* of Leonhard Fuchs, 1542.]

the interior of Cuba; they brought back word of a grain called "*Maiz*, which was well tasted, either boiled whole, or made into flower".[1] When the Spaniards arrived in Peru, they found magazines stored with Maize. The fields surrounding the temple of the Sun God at Titicaca were sacred, and the Maize from these fields was distributed among the public granaries, so that it should convey a blessing to the rest of the harvest.[2] Varieties with white, yellow, blue, purple, red and black grains were grown by the Mexicans.[3] All the main types of *Zea*—the Pop-corns, Flint-corns, Dent-corns, Soft-corns, Sweet-corns and Starchy-sweet-corns—were known to the American Indians, and were taken over from them by the European settlers.[4] The existence of these numerous and well-defined races of unknown origin, can be understood only if we suppose that, like many of the cereals of the Old World, Maize had a long career in prehistory. The importance of this grain in ancient America is reflected in the Maya tradition that, after man had been created out of the earth, it was by means of Maize that he was transmuted into a being of flesh and blood.[5]

There is reason to think that Maize may have originated in Mexico, near the ancient seats of the Maya tribes;[6] but speculations as to its ancestry are rendered peculiarly difficult by the fact that none of its known forms can maintain themselves except under the guardianship of man.[7] All other cereals may lead a feral existence, at least temporarily, but this never happens with Maize; it cannot propagate itself spontaneously, for the grain does not escape from the husks without help. It follows that if a wild ancestor of Maize is still extant, it must be something very different from its descendant. It is clearly useless to search for it in the Old World, where no members of the Gramineae, bearing any close affinity with Maize, are to be found. In Mexico, however, there is a grass, Teosinte (*Euchlaena mexicana* Schrad.), which recalls Maize in many respects, and from which—possibly by a process of hybridisation with some

[1] Churchill, A. and J. (1732).
[2] Prescott, W. H. (1847).
[3] Clavigero, F. S. (1787).
[4] Sturtevant, E. L. (1894).
[5] Payne, E. J. (1892).
[6] Harshberger, J. W. (1893).
[7] Candolle, A. de (1884); see also Collins, G. N. (1923).

species unknown—Maize may have arisen.[1] Whatever its origin, it is at least clear that Maize happened to be peculiarly well adapted to the conditions of primitive agriculture in America. The Indians had no farm animals and no ploughs, so they were necessarily confined to crops of which the individual plants were large enough for hand culture. Maize, Cassava, and the Potato, all fall into this category.[1] The Pilgrim Fathers learned this form of individual culture from the Indians, who taught them to plant the grain sparsely in 'hills' analogous to those used in hop gardens, and to bury two fish in each 'hill' to fertilise the soil.[2] Agriculture of this personal type renders possible the survival of strains which large-scale methods might eliminate. An American writer[1] has recorded that among "the Navajo Indians there is a very distinct variety [of Maize] that shows through breeding experiments that it is uncontaminated by other sorts". He adds that "this variety was in the custody of one Indian, Lone Cedar Tree, who received the variety from his father. A peculiar colour pattern, which has recently figured in genetic literature, is derived entirely from this variety and owes its preservation in a pure state to this one Indian".

The early appearance of Maize in China is a puzzle.[3] According to one view, it was introduced into India from the New World by the Portuguese, and then travelled, viâ Sikkim, Bhutan and Tibet, to Western China. This opinion has been summarised as follows: "Counting a generation as, on an average, thirty years, we might well say that, during the first generation after the discovery of America, maize became known and planted in Europe; at the end of this period it must have reached India; and during the second generation it spread all over China, so that, after about seventy or eighty years, its wanderings to the farthest East were completed".[4]

The Portuguese introduced Maize into the Congo in 1560. Their

[1] Collins, G. N. (1912); for a further discussion of the origin of Maize, see Chapter XVI. [2] Hitchcock, A. S. and Chase, A. (1931).

[3] Bonafous, M. (1836); Bretschneider, E. (1870); Collins, G. N. (1909).

[4] Laufer, B. (1907); it is to be wished that scholars of Chinese nationality would take up this problem.

efficiency in bringing to the African natives food-plants from east and west, forms some small counterpoise to set against the hardships they imposed upon those who came under their heel.[1] The Portuguese word 'Milho', which, though derived from the Latin *milium* (Millet), is applied to any cereal used for human food, has been corrupted in Afrikander-Dutch to 'Mielie', and has become the usual South African word for Maize.[2] The spelling 'Mealie' is a further corruption—a rationalisation of the term for British ears, by analogy with the Teutonic word 'meal', which is connected, not with '*milium*', but with the English 'mill', and its equivalents in Latin and Greek—*mola* and μύλη.

The most familiar types of Maize have a starchy grain, but in Sugar-maize the greater part of the starch is replaced by an amorphous substance, which shines in fracture like gum arabic.[3] There are also strains of Maize in which the endosperm has a hard, 'flinty' texture, and it is these flinty Maizes which are used for 'pop-corn'. It seems that, when they are heated, the water contained in the very hygroscopic starch-grains attempts to expand as steam, but the desiccated flinty remains of the cell protoplasm resist the expansion. When the point is reached at which this resistance is overcome, the starch-grains all burst simultaneously, and, in this way, the spongy, air-filled texture of the 'popped' Corn is produced.[4] In India, Rice is popped like Maize, by soaking the grain and throwing it on to plates of iron heated over the fire.[5] The modern manufacture of 'puffed' cereals is said to depend upon the same principle as the 'popping' of Corn, but an air-tight metal drum exercises the controlling influence, which is due in Maize to the natural pressure of the hardened relics of the cell protoplasm.[4]

Three other cereals of minor importance call for a brief mention—Tef, Korakan and Job's-tears. Tef is *Eragrostis abessinica* L., a delicate-looking plant with seeds so small that it is somewhat surprising to find that they can be an effective source of food. It is reckoned that one hundred and forty-eight of these tiny grains are needed to counter-

[1] Johnston, H. H. (1913).
[2] Davy, J. Burtt (1914).
[3] Koernicke, F. (1885).
[4] Weatherwax, P. (1922[1]).
[5] Archer, T. C. (1856).

poise one grain of Wheat.[1] The original stock of Tef is *Eragrostis pilosa* Beauv.[2] This wild species is one of those which, according to the traveller Barth,[3] were used in his time by the natives of Bórnu. He saw a woman collecting the seeds of this and other grasses "by swinging a sort of basket through the rich meadow ground".

Korakan, *Eleusine coracana* Gaertn., is cultivated in Africa, India, the Malay Archipelago, China and Japan. It is believed to have originated from *E. indica* (L.) Gaertn.,[2] a grass which is widely distributed in tropical and sub-tropical regions of Asia, Africa and America, as well as in Spain and the Pampas of South America. Korakan is said to be capable of yielding a crop when sown on mere stones and shingle, but, to set against this advantage, it has the drawback of a bitter taste. The grain has the reputation of being immune from insect attacks; perhaps it is the bitter principle which protects it.

A member of the Gramineae which Europeans seldom think of as a cereal, but which is of some consequence in the East, is Job's-tears, *Coix lacryma-Jobi* L.[4] In India this species is fairly widespread, and its cultivation seems to have had an ancient origin; it is believed to have been one of the cereals grown at a very early period on the hill slopes of the Himalayas. We have only the most fragmentary knowledge of its history, but it must have been widely diffused in the seventeenth century, for Rumphius[5] states that in his day it was planted in Java and Celebes on the margins of Rice fields. Its introduction into China seems to have come about through what is now French Indo-China. In the first century of the Christian era, a Chinese general conquered Tongking, where this cereal was in use, and where it was still in recent years known as 'the grass of life and health'. It is said that he became so fond of it that he carried back several cartloads of the seeds to his own country.

In the forms of *Coix lacryma-Jobi* L. cultivated for food, the 'shell' is soft and the kernel sweet.[6] A large number of differing races are grown, in some of which the grain is adapted for parching or boiling,

[1] Braun, A. (1841). [2] Koernicke, F. (1885). [3] Barth, H. (1857–8).

[4] This account of *Coix* is taken almost entirely from Watt, G. (1904).

[5] Rumphius, G. E. (1750); though not published until the middle of the eighteenth century, the MS. dated back to the seventeenth century.

[6] Hooker, J. D. (1854).

Fig. 16. "Lachryma Iobi" (*Coix lacryma-Jobi* L.). [Reduced from the *Hortus Eystettensis* of Basil Besler, 1613.]

while in others it can be milled, ground and baked into bread. It seems probable that these races are as diverse in chemistry and morphology as are the different strains of Wheat or Rice. The leaves are used in parts of India as fodder; elephants are said to be particularly partial to them.

Coix is unlike the other cereals in one respect—that its reputation depends on its appeal to man's desire for decoration even more than to his desire for food; its 'seeds'[1] seem made to string into ornaments. In Philemon Holland's version of Pliny,[2] which is practically contemporary with the engraving reproduced in Fig. 16, p. 35, there is an account of Job's-tears; in turn of expression, both description and picture are redolent of the early seventeenth century. We read that "of all hearbes that be, there is none more wonderfull...; some call it in Greek Lithospermon....It bringeth forth close joining to the leaves, certain little beards one by one, and in the top of them little stones white and round in manner of pearls, as big as cich pease, but as hard as very stones....And verely of all the plants that ever I saw, I never wondered at any more: So sightly it groweth, as if some artificiall goldsmith had set in an alternative course and order, these pretie beads like orient pearles among the leaves: and so rare a thing it is and difficult to bee conceived, that a very hard stone should grow out of an hearbe".

Many of the names by which the plant has been known refer to these "pretie beads"—the polished enclosures of the seeds—which in the largest forms may approach an inch in length. The Arab travellers, who learned to know the plant in the East, called it 'David's-tears' and, afterwards, 'Job's-tears'. Gerard[3] tells us that "in English it is called *Iobs* Teares or *Iobs* Drops for that every graine resembleth the Drop or Teare that falleth from the eie". In Assam it is known by a name meaning 'Crow's-jewel'. Various ornaments are made from the 'tears', such as the crown-like headdresses of the women of certain Burmese tribes, in which *Coix* seeds are used in combination with silver, beetles' wings, and squirrels'

[1] The nature of these 'seeds' will be considered later (p. 196).

[2] Pliny (Holland, P.) (1601), Book 27, Chapter XI.

[3] Gerard, J. (1597).

tails. The seeds are well adapted for making rosaries, and have been used for this purpose in the West Indies, under the name of 'St-Mary's-tears'; while in Italy they have been called 'Lachryma Christi'. It is found that under cultivation the shells soon lose their hard, pearly quality and rich gloss, and become relatively soft. This fact is fully recognised in Burma, and as soon as deterioration sets in, a fresh stock is obtained from the jungle.

Now that we have passed the principal cereals individually in brief review, we may next discuss certain points which emerge when they are considered broadly as a single group. One of these points is that the cereals, though belonging to a number of strikingly different types, yet show certain definite convergences. It is found, for instance, that the grasses grown for their seeds are nearly all one-year plants. It is true that, in Africa, Sorghum sometimes has its existence continued vegetatively from the stock for some years,[1] and that, in South Russia, Rye is cultivated as a perennial,[2] while, in Senegal, the natives use a wild Rice which has perennating rhizomes;[3] but these rare exceptions do not disprove the rule that cereals are typically annual. Among the many advantages of this character, from the standpoint of the cultivator, is the fact that the one-year habit ensures a more or less simultaneous harvest, in place of the gradual and lingering seeding of perennials.[4] Despite their tendency to an annual character, the cereals, however, include 'winter' varieties, which provide a transition to perennials; these forms, when sown in spring, produce a great number of tillers in the first season, but do not flower until the next year. Spring or summer varieties, on the other hand, lack the capacity of the winter varieties for resisting winter damage, and their life-history is more rapid; when sown in spring, they flower and fruit in the same year. It should be noticed that the difference between winter and summer forms, as regards tillering, is rather one of degree than of kind. At a certain stage, in both forms, a lull takes place in the obvious growth, while root and shoot formation are in progress underground. In summer forms, the pause is short, but in winter forms, it is prolonged.[1] The

[1] Koernicke, F. (1885). [2] Schulz, A. (1911).
[3] Ammann, P. (1910). [4] Leeke, P. (1907).

marked differences in seasonal behaviour between spring and winter cereals[1] are found to exist side by side with certain structural differences. In Wheat, for instance, the spring forms have a broader leaf of a lighter green, and the plant has a more upright habit in the early stages of growth. It would be of interest to know whether these visible features are the result of the physiological factors which render the plants liable to winter damage. The problem of these seasonal races has been attacked from the chemical side, and it has been found, by study of the juice expressed from plants of various Wheat races, that the more winter-hardy the race, the richer is the juice in water-holding colloids. It seems probable that this is a 'cause-and-effect' relation, and that this chemical distinction indicates the field which further research will have to explore, if we are to arrive at a full understanding of the divergences between winter and spring races.

It is commonly said that the wild species nearly allied to cultivated Wheat, Barley and Rye—*Triticum dicoccoides* Koern., *Hordeum spontaneum* C. Koch, and *Secale montanum* Guss.—are all winter plants, the last-named being even a perennial. But a closer investigation of these wild species reveals the existence of spring varieties among them. So it is not exactly a case of man changing winter and perennial varieties into spring varieties, but of his having selected spring races out of a mixture of spring and winter plants.[2] In England Spring-wheat appears to have come into cultivation quite late, and Winter-wheat is even now the prevalent form.[3] On the other hand, Barley is mostly sown in spring and harvested in the same year.

Besides the annual character, another common feature of the cultivated cereals is a certain toughness of the spindle, and a tendency towards the retention of the grains. This character is demonstrated when a bunch of fully ripe ears of Wheat or Barley is placed with the stalks in a jar of water. Under these conditions, the seeds will often germinate *in situ*; the spindle shows no inclination to break up, nor do the grains scatter. The toughness of the spindle and the retention

[1] This statement of the difference between the winter and spring types is taken from Engledow, F. L. (1927).

[2] Vavilov, N. I. and Kouznetsov, E. S. (1922). [3] Percival, J. (1934).

of the seed have the great advantage that the crop can be harvested without loss, and the grain subsequently threshed out. But this prolonged connection of the seed with the parent would, in nature, be a serious drawback to dissemination, and we find, indeed, that the wild species most closely related to the cereals have a spindle which breaks into joints at maturity. This contrast between the 'shattering' wild species, and the tough-spindled cultivated races, is common to a great many cereals; it occurs, for instance, in Rye, Barley, Rice, Sorghum and most of the Wheats. The tough spindle seems to have arisen through loss of the dominant character of fragility, for it has been shown that, when the Wild-emmer is crossed with a grass of the genus *Aegilops* (related to *Triticum*) the shattering rachis of the Wild-emmer proves to be dominant to the tough rachis of *Aegilops*.[1]

From the botanical point of view, it would be of the greatest interest if we could know the exact mode of origin of the countless races of cereals which have arisen under the hand of man; but only an infinitesimal proportion of the historical facts are now available. In classical times there are records of seed selection of a somewhat crude type being practised in order to improve the crops. In Holland's translation of Pliny,[2] we read that "The corne that settleth to the bottome of the mowgh in a barne towards the floore, is ever to be reserved for seed. And that must needs be best, because it is weightiest". This kind of continuous *mass selection* was used until the nineteenth century, and belief in it was reinforced by the Darwinian theory, which laid stress on the supposed possibility of accumulating and stereotyping variations in any wished-for direction, by breeding in each generation from those individuals which possessed the desired qualities in the highest degree. On the other hand, even before the end of the eighteenth century, practical men had arrived instinctively at a procedure of a different type, which did not receive scientific recognition until more than a hundred years later.[3]

[1] Percival, J. (1926[2]).

[2] Pliny (Holland, P.) (1601), Book 18, Chapter XXIV. Mowgh (mow) is an old word for a heap of corn in a barn.

[3] Johannsen, W. (1903) is the basis of the scientific work on 'pure lines', which has meant so much to the practical breeder.

This new method consisted essentially in the selection of *an individual plant*, showing desirable characters, and the cultivation of a pure lineage from this plant. By this method the breeder did not himself 'improve' the cereal, but he isolated and protected a strain offered by Nature, which happened to be an improvement from the farmer's point of view. An early and memorable example of pure line selection dates from 1788, when a fine individual Oat plant was found growing in a Potato patch in Cumberland. Its seeds were kept and the progeny cultivated, and the race became famous under the name of 'Potato-oat'. A second valuable variety was discovered nearly a generation later in the parish of Rhynie in Aberdeenshire —a place which has a special claim on the interest of botanists, owing to the fossil plants of Devonian age, which have been found there. In the summer of 1824 or 1825, a cow-herd, Alexander Thomson, observed an Oat plant of unusually luxuriant growth on a new-made bank of earth on the Rhynie farm where he was employed. His master preserved the seed and carried on the race. Like the Potato-oat, it proved very successful; the farm hand, who made the discovery, is immortalised in its name—'Sandy'.[1]

The 'Potato' and 'Sandy' were, as we have seen, the products of casual finds. Later on, races of Oats were isolated by Patrick Shirreff,[2] a Haddington farmer, who began in 1862 to make a systematic examination of Oat fields, with a view to finding individual plants to breed from. Shirreff did not work merely by rule of thumb; on the contrary, he analysed his own procedure, and he arrived at perfectly clear ideas about the improvement of the cereals. "Many people believe", he wrote, "that some plants can be altered by skilful treatment, but my experience has tended to show that there is no way of permanently improving a species but by new varieties." Shirreff not only selected and propagated hopeful variants which he found in nature, but he also tried *hybridisation* with a view to initiating such variants. His Oat crosses for various reasons were unsuccessful, but he seems to have been the first in this country to produce hybrid Wheats which were an improvement upon

[1] Lawson, P. and Son [C.] (1852).
[2] Shirreff, P. (1873).

existing races. Work on these lines continues actively to the present day in connection with all the principal cereals.

Since the cereal races often owe their preservation to man, and since their annual character puts them well within his control, the geographical distribution of the cultivated Gramineae is closely interwoven with the history of the human race. An account of cereal geography would be outside the scope of this book,[1] and we can only touch upon certain theoretical conclusions, which have been deduced from an intensive study of these plants in the various parts of their range.

There is now some agreement as to the position of the geographical centres for the various cereals, and a remarkable fact is revealed by a comparative study of the species and varieties met with in these putative foci of development. It is that these centres show an accumulation of forms characterised by *dominant genes*, whereas, as we pass from the principal genetical bases towards the periphery, the cultivated plant types show a loss of dominant characters. We owe this discovery to the Russian biologist, Vavilov, and the following account is derived from his work, and expressed, in part, in his own words.[2] As an example we may take Abyssinia, which appears to be the centre of origin of the vast 'Emmer' group of Wheat species, whose cell nuclei contain 28 chromosomes. In the Abyssinian Wheats of this group, we find not only the varietal characters met with in European and Asiatic forms, but also many races unknown in Europe, whose features are clearly dominant, e.g. purple-grained and pubescent Wheats. Again, the whole genus *Secale*, including cultivated Rye, may have originated in Asia Minor and Transcaucasia, and among the numerous Rye forms which occur there, many dominant characters are found, such as red, brown, or black ears, and pubescence. On the other hand, in Europe, where we are remote from the ancient agricultural centres, the majority of the cereals have light-coloured

[1] The distribution can best be understood from graphic presentments, such as those found in Finch, V. C. and Baker, O. E. (1917) and Messer, M. (1932).

[2] Vavilov, N. I. (1927); on the centres of origin of cultivated plants, see also Vavilov, N. I. (1928), (1931¹), (1931²), (1931³), and Vavilov, N. I. and Bukinich, D. D. (1929).

ears, which have lost their pigmentation and other dominant features. For details Vavilov's papers must be consulted, but these instances will show the nature of the evidence on which he relies. The interest of his generalisation is increased by his extension of it to include not only the domesticated animals but also man. On comparing the map of the main centres of origin of the principal crop plants with that of the distribution of the human races prior to the migrations of modern times, it will be found that there is, broadly speaking, a coincidence between the regions where the coloured races are concentrated, and the primary centres of agriculture. Now the pigmentation of the darker races, like that of certain strains of cereals and of domesticated animals, seems to be determined by a series of dominant genes; and, just as the cereals become lighter in colour as we pass outwards from their primary centres of distribution, so the human type loses its pigmentation, owing to the falling out of dominant genes. The European is, as it were, to some extent emancipated from the tyranny of dominance.

If these conclusions of Vavilov's are soundly based, and stand the test of further critical work, they must prove of the utmost significance in connection not only with the study of domesticated plants and animals, but also with the history of man.

CHAPTER III

PASTURE, SUGAR, AND SCENT

IN the preceding pages, attention has been concentrated on those Gramineae whose *seeds* are used by man for food. In this chapter we must turn to those from whose *vegetative shoots* he takes toll, indirectly or directly. The significance of grass in the life of man long ago received symbolic recognition. Pliny tells us that in Rome no Coronets were "better esteemed...to give testimonie of honour and reward for some notable service performed for the Commonweale, than those which were made simply of greene grasse". He also refers to an ancient tradition that the greatest sign "of yeelding to the mercie of the enemie, was this, If the vanquished did take up grasse, and tender it unto the conqueror: for this served as a confession and protestation, That they rendered up all their interrest which they might challenge in the earth (the mother that bred and fed them)".[1] Grass means as much to man today as it did in antiquity, but town life tends to dull his consciousness of the part it plays in his existence. This, however, becomes sufficiently clear when we turn to statistics and learn that in 1932 the number of arable acres in England and Wales was less than 10 millions, while the acreage of pasture and rough grazing exceeded 21 millions.[2]

Considering the immemorial antiquity of cereal culture, it might have been expected that the culture of chosen species of grass would also date back to the remote past. We do not, however, find evidence of it. It is true that Columella[3] mentions hay seed, which was probably sown to some extent from early days, but, until a comparatively recent date, the farmers of Europe seem, in the main, to have been contented to rely upon their ancient natural grass lands. Possibly the earliest grass definitely cultivated in England was Rye-grass, *Lolium perenne* L., which was sown for agricultural purposes in

[1] Pliny (Holland, P.) (1601), Book 22, Chapter IV.

[2] I am indebted to Professor F. L. Engledow for this information.

[3] Columella (1541), Lib. 2. 18.

Gramen uulgo cognitum.
Gemein Graß.

Fig. 17. "Gemein Grass" (probably *Poa annua* L.). [From the *De Stirpium* of Hieronymus Tragus (Jerome Bock), 1552.]

the seventeenth century; the first mention of its sowing, which I have been able to find, is in Dr Plot's *Natural History of Oxford-shire* (1677), where it is called "*Ray*[1] or *bennet-grass*" or "*Gramen Loliaceum*". The colonisation of the New World gave an impetus to grass culture, for the pioneers found that, when they cleared the forests, the native species failed to provide the grass lands they needed. As Captain John Smith wrote in 1612, "*Virginia* doth afford many excellent vegitables and liuing Creatures, yet grasse there is little or none but what groweth in lowe Marishes: for all the Countrey is overgrowne with trees". It thus became necessary to import grass seed from Europe.

One of the most valuable of these importations was *Phleum pratense* L., which was introduced from England into Carolina,[2] or Maryland,[3] in 1720 by a certain Timothy Hansen, and afterwards into Virginia. It is said that, at a later date, when this grass had proved its utility, seeds were sent back to England;[4] under the name of Timothy-grass, it still keeps green the memory of its American sponsor. This to-and-fro migration across the Atlantic sounds rather objectless, but recent cytological work[5] has thrown light upon it. It appears that the large 'hay' type of Timothy-grass is hexaploid (42 chromosomes), while the small 'wild' British form is diploid (14). America, then, in returning this grass to England, was not giving back merely what she had received, but was sending a new (possibly hybrid) race, which had originated on her own soil, and which had the enhanced value which so often attaches to polyploid forms.

Timothy is not the only grass of European provenance which plays a great part in the agriculture of the United States. The famous Kentucky-blue-grass (*Poa pratensis* L.), for instance, much as it flourishes on the Kentucky limestone, and widely as it is naturalised elsewhere in the States, is native only to the Old World.[6] In the "Sod of Turf", which Albrecht Dürer drew in Germany in 1503—

[1] The modern name 'Rye-grass' is probably a rationalisation of 'Ray-grass'—'Ray' being a corruption of the French name 'Ivraie', given to another member of the genus (*L. temulentum* L., Darnel), in allusion to the intoxication produced by its poisonous grain.

[2] Jones, L. R. (1902).

[3] Lyon, T. L. and Hitchcock, A. S. (1904).

[4] Stillingfleet, B. (1811).

[5] Gregor, J. W. and Sansome, F. W. (1930).

[6] Hitchcock, A. S. (1920).

reproduced as the frontispiece of the present book—this grass in flower forms a conspicuous feature. It has taken very kindly to its American home; one author records that he measured a plant growing under favourable conditions, which was 40 in. to the top of the six-inch panicle.[1] The analogy of Timothy-grass makes one wonder whether here, also, a grass of European origin has given rise, in the States, to a polyploid form.

Among the many native forage grasses of the plains of North America, one of the best is *Buchloë dactyloides* Engelm. (*Bulbilis dactyloides* (Nutt.) Raf.). Formerly vast hordes of buffalo subsisted on it through the winter; it owes its high food value to an unusual richness in starch.[2] It was employed by the early settlers in the construction of their sod houses.[3] Its success as a pasture grass and also as a sod-former is no doubt partly due to its good root development; it has been found to possess many roots from 4 to 6 ft. long, and lengths of over 7 ft. have been recorded.[4] This grass also furnishes lawns in the Southern States of America, while the same purpose in the Northern regions is served by Kentucky-blue-grass.[3] In Britain we do not think of members of the genus *Calamagrostis* as pasture grasses, but they have a certain value in South America. Species of *Deyeuxia* Beauv. (a sub-genus of *Calamagrostis*), of which some sixty are concentrated in the Andes, reach an elevation of about 5000 m. in the Cordilleras, thus approaching the extreme limit of altitude for phanerogamic vegetation; they provide irreplaceable pasturage for the Vicunas and Guanacos.[5]

The production of artificial pasture has been so late a development among civilised races that one hardly expects to find it among less advanced peoples. It is thus somewhat surprising to learn that Kikuyu-grass (*Pennisetum clandestinum* Chiov.), a good grazing grass, now well distributed in Kenya, owes its wide range to a migratory movement among the natives some two or three generations ago. The members of each family, when they left their home, carried with them a bundle of the rhizomes of this grass with which to stock

[1] Garman, H. (1900).
[2] Smith, J. G. (1888).
[3] Hitchcock, A. S. (1920).
[4] Weaver, J. E. (1920).
[5] Weddell, H. A. (1875).

new pastures.[1] Kikuyu-grass is interesting also in another way. It shows very vigorous vegetative development of runners and stolons, but its reproductive organs tend to be reduced and stunted.[2] This concentration on vegetative life enhances its value as a pasture grass.

One of the reasons why grasses form more efficient forage plants than any Dicotyledon, lies in their special mode of growth. Elongation takes place at the leaf-base, so that grazing does not destroy the active region. Critical experiments on the growth of the sheath and limb of grass leaves require delicate technical methods, but anyone who wishes to do so, can easily satisfy himself of the truth of the broad statement that the extreme base of grass leaves is a growing region. In ordinary grass shoots, the bases of the younger leaves are so closely enwrapped in the sheaths of the older leaves that it is difficult to expose them. The best plan, therefore, is to sow a few Oats in damp earth, and to dig up the seedlings carefully when the first green leaf is about 3 in. high. The outer colourless scale can then readily be removed, and the enclosed green leaf marked from base to apex by means of fine but distinct horizontal lines in Indian ink, 2 mm. or $\frac{1}{16}$ in. apart. If the seedling be replaced, and examined again after three days, the basal distribution of the growth will be at once revealed by the changes in the spacing of the lines.

From the point of view of grazing, an even more important feature than the basal growth of the leaf, is the basal *tillering* of the grass plant. When the main shoot is cropped, lateral shoots from the buds in the axils of the lower leaf-bases take their place, and this process may be repeated time after time. This repetitive growth is due to the abbreviation of the basal internodes, and the consequent close packing of successive leaf-bases. A whole series of axillary buds are thus grouped near the level of the soil surface, where adventitious roots are readily formed to supply the needs of the lateral branches (tillers), into which these axillary buds develop, under the stimulus of mutilation of the shoot. Both basal leaf-growth and tillering, as essential economic features of the grass plant, were clearly recognised by Edward Lisle before 1722, though his terminology differs from that of a modern botanist. "The reason", he says, "why many

[1] Watt, W. L. (1925). [2] Stapf, O. (1921).

plants are to be killed by often cropping, and yet the natural pasture-grass no wise suffers by it, I conceive, is, because the leaf of the natural grass is a continued spire, and, when it is bit, lengthens itself out again by growth...; and in case it could be bit below the leafy spire into the ground sheath, yet in the tuft of the same root, are a multitude of issues [buds] monthly and weekly breaking out."[1]

A feature closely associated with tillering, which adds greatly to the value of the grasses from the point of view of man and his animals, is the tolerant way in which many of them accept trampling underfoot. Long ago Benjamin Stillingfleet[2] wrote that he had observed on Malvern Hill that a "walk that was made there for the convenience of the water drinkers, in less than a year was covered in many places" with *Poa annua* L., Suffolk-grass (Fig. 17, p. 44), "tho' i[2] could not find one single plant of it besides in any part of the hill. This was owing no doubt to the frequent treading, which above all things makes this grass florish". This species is indeed an example of that extreme vigour of growth which commends the grasses to man. Elsewhere Stillingfleet notes that he has "counted forty-three flowering stems besides a great number of radical leaves from one root of this kind without particularly searching for a vigorous plant, and this plant was not above three weeks growth".

It is sad that many of our meadow, pasture, and weed grasses[3] have no folk-names sufficiently widespread and distinctive for practical use. Most of the book names were deliberately invented in the eighteenth century by Stillingfleet. He explains that, as "there has reigned hitherto the greatest confusion in the English names of these most valuable plants,...i shall...give new generical names with trivial ones to distinguish the species of all our English grasses". Examples of his genera are "Hair-grass" for *Aira*, "Dog-tail grass"

[1] Lisle, E. (1757); I quote from the second edition of this book on farming, since I have been unable to see a copy of the first (1756). Lisle died in 1722 and his work was posthumous.

[2] Stillingfleet, B. (1759). An idiosyncrasy of this writer was to adopt a small 'i' for the first person singular, in order to express his sense of personal modesty.

[3] For a general account of agricultural grasses, as well as cereals, see Percival, J. (1926[1])

Of Grasse. Chap.xliiij.

✠ The Kindes.

A Man shal finde many sortes of grasse, one lyke another in stemme, and leaues, but not in the knoppes or eares: for one hath an eare like Barley, the other lyke Millet, another like Panick, another lyke Juray, and such vnprofitable weedes that growe amongst corne. Some haue rough prickley eares, and some are soft and gentle, others are rough & mossie lyke fine downe or cotton, so that there are many sortes and kindes of grasse: whereof we will make no larger discourse, but of suche kindes onely, as haue bene vsed of the Auncient Physitions, and are particularly named Agrostis and Gramen.

✠ The Description.

Gramen. Couche grasse.

THE grasse whereof we shall nowe speake, hath long rough leaues almost lyke the Cane, or Pole reede, but a great deale lesser, yet muche greater & broder then the leaues of that grasse which groweth cõmonly in medowes. The helme or stemmes are small, a foote or two long, with fiue or sixe ioyntes, at the vppermost of ẏ stalkes there grow soft & gentle eares, almost like ẏ bushy eares of ẏ Cane or Pole reede, but smaller and slenderer. The roote is long and white, full of ioyntes, creeping hither & thither, & platted or wrapped one with another, & putting forth new springs in sundry places, & by the meanes hereof it doth multiplie and increase exceedinly in leaues and stalkes.

✠ The Place.

This grasse groweth not in medowes & lowe places, lyke the other, but in the corne feldes, & the borders therof, & is a noughty & hurtful weede to corne, the which the husbandmen would not willingly haue in their lande, or feeldes: & therfore they take much payne to weede, and plucke vp the same.

✠ The Names.

This grasse is called in Greeke ἄγρωστις. Agrostis, bycause it groweth in the corne

Fig. 18. "Couche Grasse" (identified tentatively by Britten, J. and Holland, R. (1886) as *Poa pratensis* L., but it seems best not to attempt to name it). [From the *Nievve Herball* of Rembert Dodoens (translated by Henry Lyte), 1578.]

for *Cynosurus*, and "Brome-grass" for *Bromus*; his names survived owing to their adoption in Hudson's *Flora Anglica*.[1]

The grasses, considered as pests, do not fall within the scope of the present book; but Fig. 18, p. 49, which is a reproduction of part of a page of Lyte's *Nievve Herball* of 1578, reflects the age-long lament of the farmer over the "noughty and hurtful weede", which adds so grievously to his labours. Fig. 19 shows the particular feature of Couch-grass which condemns it in the eyes of the gardener—its gift for producing vigorous rhizomes, whose penetrating apices are protected by thick fibrous scales. Fig. 20, p. 52, is the woodcut of Darnel (*Lolium temulentum* L.) from Lyte's *Nievve Herball*; he describes it as "a vitious grayne that combereth or anoyeth corne". In this grass the mycelium of a fungus is frequently to be found between the seed-coat and the aleurone layer.[2] In such grains a poisonous alkaloid is present, while those that are free from fungus are non-poisonous.[3] Since individual seeds on infected plants may show no fungus, while there are also races which are wholly fungus-free, the apparently capricious onset of Darnel poisoning[4] finds its explanation. The mycelium has been detected in good preservation in specimens of Darnel from an Egyptian tomb, which is believed to date from 2000 B.C.; the study of this material shows that the fungus has retained its mode of life unchanged for 4000 years.[5]

The pasture grasses, important as they are, serve man only indirectly; but there is one grass, the Sugar-cane, whose vegetative organs he lays under direct contribution on a colossal scale. This plant, *Saccharum officinarum* L. (Fig. 21, p. 53), offers a marked contrast to the cereals in the type of development that is encouraged by cultivators. The cereals are annuals, selected for fertility, and encouraged

[1] Hudson, W. (1762).
[2] On the structure of grass seed, see Chapter XI.
[3] Hannig, E. (1907).
[4] Cf. Wilson, A. S. (1873[1]), (1873[2]); poultry are seen eating the grain in the woodcut from Mattioli reproduced in Fig. 8, p. 17.
[5] Lindau, G. (1904); on the fungus associated with *Lolium*, see also Guérin, P. (1898[1]); Hanausek, T. F. (1898); Nestler, A. (1898), (1904); Freeman, E. M. (1904); Buchet, S. (1912); Fuchs, J. (1912); McLennan, E. I. (1920), (1926); Rayner, M. C. (1927).

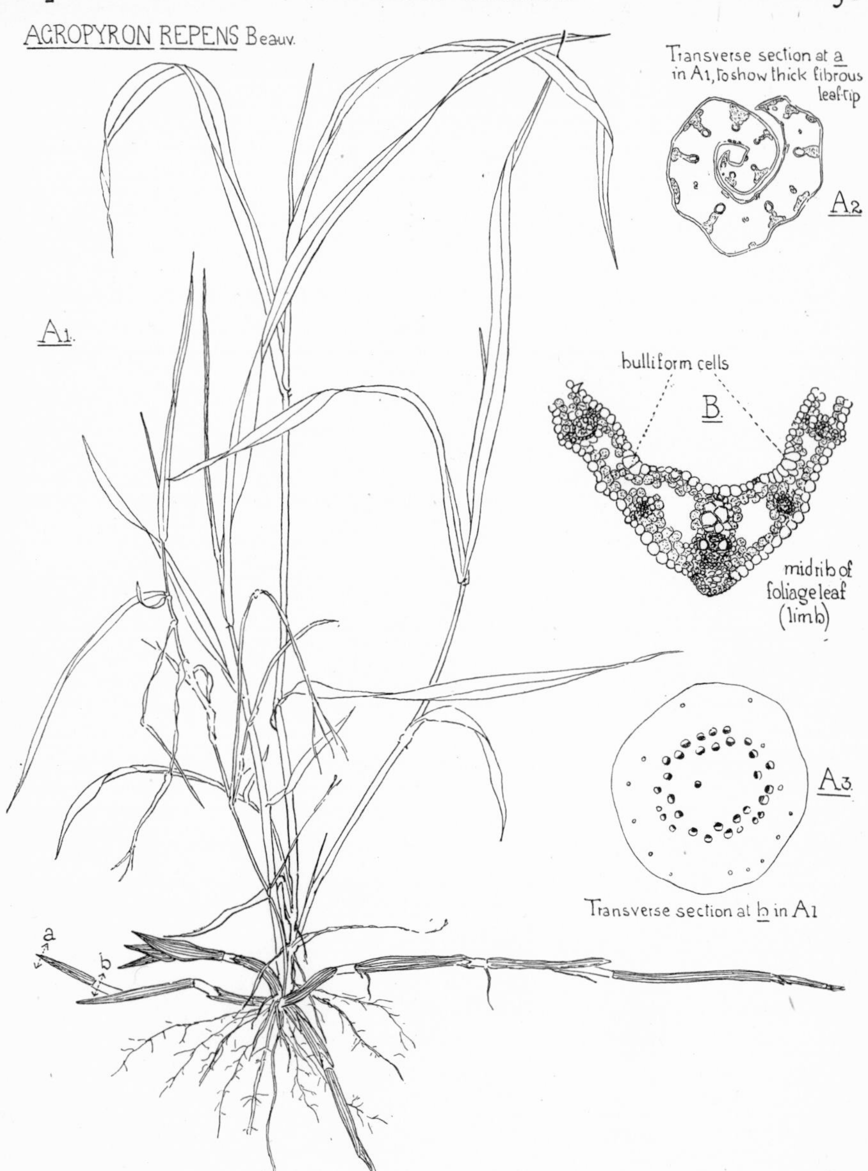

Fig. 19. *Agropyron repens* Beauv. A 1, plant in vegetative stage (× ½), May 9. A 2, transverse section near apex of sheath of terminal bud of stolon at *a* in A 1 (× 23). A 3, transverse section of axis of stolon at *b* to show arrangement of bundles (× 14). B, transverse section of midrib of a leaf to show the three bundles in the midrib, with a group of bulliform cells on either side. Except in the midrib, the bulliform groups alternate regularly with the bundles (× 47). [A.A.]

Fig. 20. "Ivray or Darnell" (*Lolium temulentum* L.). [From the *Nievve Herball* of Rembert Dodoens (translated by Henry Lyte), 1578.]

to produce an excessive supply of endosperm. The Sugar-cane, on the other hand, is the only important perennial grass grown for human food,[1] and it is selected for sterility, since the store of sugar in the stem would be exhausted by the act of flowering.[2] Man tyrannises over both cereal and cane, inducing them to pervert their life-histories to serve his purposes.

The Sugar-cane is now grown in all tropical and sub-tropical regions, provided that the rainfall is adequate.[2] Its original home is a

Fig. 21. "Arundo Indica Saccharifera" (*Saccharum officinarum* L., Sugar-cane). [From the ninth and last edition of an anonymous herbal published at Lyons in 1766 (Pritzel, 10768). The block appears to be an adaptation of one of those issued by Christophe Plantin of Antwerp two hundred years earlier.]

matter of speculation, but it is held that it may have had two centres of origin—one in Oceania, and possibly New Guinea, from some species unknown, and the other in India, from *Saccharum spontaneum* L., which may itself be a species-complex.[3] There are indications that the cultivation of the Sugar-cane began very early in India. In the *Institutes of Manu*[4] (dating back to some time in the first 500 years of the Christian Era, if not earlier), we are told that

[1] Hitchcock, A. S. (1929). [2] Hitchcock, A. S. and Chase, A. (1931).

[3] See Hunter, H. and Leake, H. Martin (1933) for references and an account of the cytological relations of the various forms.

[4] Burnell, A. C. and Hopkins, E. W. (1884), Lecture VIII, 341, Lecture XII, 64; Watt, G. (1889–96).

the stealer of sugar will be punished in the hereafter by becoming a bat; nevertheless, if a 'twice-born' man, while on a journey, finds his provisions exhausted, it is lawful for him to take two Sugar-canes from the field of another. From India the Cane spread east and west. Sugar is first mentioned in Chinese writings in the second century B.C., while there is a record, dating from A.D. 286, of sugar being sent as tribute into China from a part of India beyond the Ganges.[1] As evidence of its migration towards the west, we learn that it was cultivated on the shores of the Persian Gulf in the ninth century, and a writer in 1108 notices that the "Crusaders found sweet-honeyed reeds in great quantity in the meadows about Tripoli, which reeds were called *sucra*".[2] It is supposed that the Saracens originated sugar cultivation in the islands of Sicily, Rhodes and Cyprus. The Moors brought it to Africa and Spain; in Granada this cultivation has been continuous ever since they introduced it. Early in the fifteenth century, the Portuguese carried the Sugar-cane to Madeira, and, somewhat later, the Spaniards brought it to the Canary Islands.[3] Henry Lyte's *Nievve Herball* of 1578 speaks of the "Sugar Reede or Sugar Cane" used to "make Sugar, in the Ilandes of Canare, and elswhere" (Fig. 105, § 8, p. 218). From Madeira it was conveyed to the West Indies and Brazil. It has sometimes been supposed that the Sugar-cane was already growing in various parts of America before the arrival of Europeans, and that the Mexicans and Peruvians understood the manufacture of cane sugar. It seems more probable, however, that the sugar in question was made from the stems of Maize. It was not until the seventeenth century that sugar manufacture was begun by the Dutch and English in the West Indies.

Hitherto we have spoken only of the grasses of which man takes toll for food. There are some, however, from which he gets satisfaction for a more sophisticated craving—the desire for perfumes. Our British grasses are not, as a rule, strongly fragrant, but there are certain exceptions. The roots of *Melica* are said to exhale a Cowslip-like fragrance when bruised.[4] Those of *Anthoxanthum odoratum* L., on

[1] Bretschneider, E. (1870).
[2] Quoted in Watt, G. (1889–96).
[3] Green, T. (n.d. [1816–20]).
[4] Royer, C. (1883).

the other hand, are described as having an unpleasant odour; amends are made, however, by the delicious smell both of the fresh flowers and of the dried leaves, which has earned for it the name of Sweet-vernal-grass. It is very common in meadows, and the coumarin, which it contains, is partly responsible for the familiar fragrance of new-mown hay. Northern-holy-grass, *Hierochloë borealis* Roem. et Schult.,[1] is a related grass with a similar but stronger vanilla scent. In Britain it grows only in Caithness, where it was discovered by Robert Dick, the baker-geologist of Thurso.[2] In America the Indians make scented baskets of it;[3] it is said to induce sleep if bunches of it are hung over the bed. It was formerly used to strew church pavements in Germany and Scandinavia.

Though in *Anthoxanthum* and *Hierochloë* we have examples of scented grasses, we have none in this country which are of economic importance. In other parts of the world, however, there are certain Gramineae which produce scented oils on a considerable scale; they belong mainly to the genus *Andropogon* and its allies in India and Ceylon.[4] A number of oils of the Gramineae, known under the names of Palmarosa oil, Lemon-grass oil, Ginger-grass oil, Vetiver oil, Citronella oil and Camel-grass oil, are recognised in commerce; but the exact provenance of the products sold under these names is often obscure. The leaf anatomy has been investigated in *Andropogon laniger* Desf.[5] The oil is enclosed in special cells of which there may be fifty or more in a single transverse section of the leaf. These oil cells, which are 40–50μ in diameter, and 80–200μ long, form vertical files associated into strands.

The chief of the oil grasses, Camel-hay, the 'schoenanthus' of the ancients, is *Cymbopogon Schoenanthus* Spreng. (*Andropogon Schoenanthus* L.). It has been found in Egyptian tombs of a millennium or

[1] This grass is also called *Torresia odorata* (L.) Hitchc., *Savastana odorata* (L.) Scribn., and *Hierochloë odorata* Wahl.

[2] Smiles, S. (1878).

[3] Hitchcock, A. S. (1920).

[4] Stapf, O. (1906); this important paper, from which the following account of oil-grasses is taken, should be consulted for a fuller treatment, and for details of taxonomy.

[5] Hoehnel, F. von (1884); this author describes *Andropogon laniger* Desf. but under the incorrect name of *A. Schoenanthus* L.; see Stapf, O. (1906).

more before Christ, and even after this length of time it retains its odour to some degree.[1] It was sought after by the Orientals, Greeks and Romans, and an oil expressed from it was largely used. It figured, also, in the herbals of the renaissance; but it has now fallen into neglect. In the words of Otto Stapf—"This is then all that is left of the once much-prized drug; a few dusty bundles of hay in oriental bazars, a few ounces of oil, and the ancient name under cover of which other grasses have found their way into the pharmacopeias and the chemical industry of our day".

Another well-known oil-grass is Vetiver or Khas Khas, *Vetiveria zizanioides* Stapf (*Andropogon muricatus* Retz.), whose aromatic roots are woven into screens, mats and fans, which retain their delicious scent for years, giving it off especially when sprinkled with water. Vetiver is native to India, Ceylon and Burma, and it has been introduced also into the West Indies, Brazil and Louisiana.[2]

Citronella oil, made from *Cymbopogon Nardus* Rendle (*Andropogon Nardus* L.), is much used, particularly in the tropics, because its scent keeps insects at bay. *C. Nardus* is known only in the cultivated state, and it seems to be partially sterile. Sterility is encouraged by the growers of this and other scented grasses, since it is the vegetative organs which are required, and they are liable to be exhausted by flowering. In this respect these oil-grasses recall the Sugar-cane and Kikuyu-grass.

Cymbopogon citratus Stapf (*Andropogon citratus* DC.) is another grass which is known only in cultivation; it gives Lemon-grass oil. A certain Dr Samuel Browne, surgeon at Fort St George, Madras, in an account of plants collected in 1696, described *Cymbopogon citratus* as "a most delicate sort of *Fragrant Grass*, which being rubb'd smells like *Baume* and *Lime* or *Limon-peel* together....Certainly so excellent a Plant of such *Fragrant* and *Aromatick* taste must have many *Vertues*....While I was writing this, in came a Person, who says, that about 30 years ago, *viz.* about 1666. one *Antonio Palia* brought 3 Pots of this *Grass* from *Batavia* to *Paliacut*, one of which he sent to a *Garden* here at Madrass".[3]

[1] Schweinfurth, G. (1883) and (1884).
[2] Lamson-Scribner, F. (1898).
[3] Petiver, J. (1702).

A pleasant aromatic infusion can be made from the leaves of *Cymbopogon citratus*. It was recorded a hundred years ago that Dr Maton, Physician Extraordinary to Queen Charlotte, was repeatedly "treated with a dish of Lemon-grass tea by Her Majesty, who used to be very fond of it, and was supplied with the plant from the Royal Gardens at Kew".[1]

In the eighteenth century[2] a letter was communicated to the Royal Society from an Englishman at Lucknow, who wrote: "travelling with the Nabob Visier, upon one of his hunting excursions towards the northern mountains, I was surprised one day, after crossing the river Rapty,...to perceive the air perfumed with an aromatic smell; and upon asking the cause, I was told it proceeded from the roots of the grass [*Cymbopogon Jwarancusa* Schult.] that were bruised or trodden out of the ground by the feet of the elephants and horses of the Nabob's retinue". The writer goes on to recall that on Alexander the Great's expedition into India, the air was perfumed by Spikenard, which was trampled under foot by the army. The Spikenard of the ancients is generally held to have been *Nardostachys* (Valerianaceae), but one would like to think that the 'Spikenard' that scented the air for Alexander, was perhaps the very same grass that gave forth its fragrance beneath the tread of the "Nabob Visier's" elephants, more than two thousand years later. It is sad that this picturesque theory does not seem to have gained acceptance.

[1] Wallich, N. (1832). [2] Blane, G. (1790).

CHAPTER IV

BAMBOO: VEGETATIVE PHASE

KTÊSIAS, the Knidian, who wrote a treatise on India about four hundred years before Christ, speaks of reeds growing there, of "a height to equal the mast of a merchant ship of the heaviest burden".[1] This is believed to be the earliest European reference to the tribe of the Gramineae called Bambuseae, or bamboos. It seems to have been long before any exact knowledge of these "reeds" penetrated to the West, for there is little notice of them by European writers, between Pliny and the sixteenth-century herbalists, who make some slight allusion to them; Jerome Bock in 1552, for instance, mentions reeds which "in India...in arboream magnitudinem excrescunt". The origin of the name bamboo is obscure.[2] It may possibly be a trade corruption of the Malay word 'Samámbu', used for the Malacca-cane. Though the bamboo meant nothing to Western civilisation in early days, its extreme importance to the peoples of tropical countries is reflected in the position which it occupies in their folk-lore. It is said, for instance, that the Kings of Boeton[3]—a small island near Celebes—claimed that they sprang originally from a giant bamboo. The story[4] ran that in old days, when the people of Boeton had no king, a man entered the forest to fell bamboos for his own use. He was just attacking a fine stem, when a voice cried, "Man! do not destroy my foot, but insert your axe a little higher; I am in bondage here". The man, desirous of seeing who was thus confined, split the bamboo lengthways, when out stepped a perfect man, whom all the people acclaimed for their king. Another bamboo obligingly yielded a woman to be his queen.

Even in temperate countries, where they are rare, bamboos are of

[1] Ktêsias (McCrindle, J. W.) (1882).

[2] For a discussion of the origin of the word, see Yule, H. and Burnell, A. C. (1903).

[3] Also spelt Buton.

[4] Recorded in Rumphius, G. E. (1750) as told in 1654 by a Dutch Admiral, who knew it as a tradition.

considerable value to man. This is true of the only member of the Bambuseae found wild in North America, *Arundinaria tecta* (Walt.) Muhl.,[1] which may grow to 25–30 ft., forming dense thickets, called canebrakes, in the alluvial river bottoms in the Southern States. Its stems are used for light scaffolds, fishing-rods, pipe-stems, baskets, and mats. Stock are fond of the leaves, and the young shoots are also sometimes cooked as a pot-herb.[2] This use recalls the very wide employment of bamboo sprouts as a vegetable in China, Korea and Japan.[3] They are described as looking like giant stalks of asparagus, some of them being 3 or even 5 in. in diameter, and a foot in length, at the stage at which they are ready for cutting. They are sent in large quantities to provinces where they do not grow, or in which they are out of season.

It would hardly be possible to exaggerate the importance of bamboos to the life of man in the tropics. The Javanese, for instance, makes his house and his furniture from bamboo, and on his journeys he cooks his rice, or his young bamboo shoots, in bamboo stem segments, over a bamboo fire,[4] while the capacity of a bamboo joint has been taken as a measure of quantity.[5] A list of the uses to which the bamboo is put in the East Indies would be well-nigh endless[6, 7]. Aqueducts, oars, masts, baskets, fish-hooks, spear-shafts, bows and arrows, knives, ladders, rafts, pails, and churns, all come from the same source. The joints of bamboo stems root as readily as Willows, and are thus invaluable in hedge-making. Split bamboos, tied with silk, form roll curtains. The bamboo also supplies fibre for ropes and cordage; tiles for roofs; axles and springs for carts; scarecrows, consisting of a long bamboo stem with a bamboo wind-wheel at the end; beehives; fans; walking-sticks; umbrella-frames; bird-cages; tiger-cages; paper; and chop-sticks. Moreover the bamboo ministers not only to man's material wants, but also to his artistic pleasures. In the Malay peninsula, "The bamboos in a village clump or far away in the jungles are perforated here and there in such a way as to keep whistling in all tones at once as

[1] On this species, see Brown, C. A. (1929). [2] Hitchcock, A. S. (1920).
[3] King, F. H. (1927). [4] Jagor, F. (1866).
[5] Ridgeway, W. (1892). [6] Watt, G. (1889–96). [7] Kurz, S. (1876).

the wind blows through the culms.[1] The sound produced in this way has been described as at times soft and liquid like the notes of a flute, and again deep and full like that of the organ".[2]

Certain uses of the bamboo depend upon a chemical peculiarity characteristic of the Gramineae—the presence of an unusual amount of silica in the tissues. This silica sometimes makes the wood so hard that it can be used as a whetstone. Not only is the substance of the wood impregnated with silica to such a degree that the ash of burnt bamboo stems shows distinctly the silica skeleton of the structure,[3] but a silica residue[4] may even be found in the hollow internodes. This residue is apparently left over from the watery fluid which collects in the cavities between the nodes, but there is still a good deal of obscurity about its origin and history, and about the extent to which it is found in different bamboos. Various references in early writers might equally well relate to these silica concretions, or to cane sugar—a fact which has been a stumbling-block in the study of the history of economic plants. Bamboo silica has long been known in Indian *materia medica* under the names of 'bamboo-manna', '-milk', '-sugar', '-salt', or '-camphor', or the 'ornament of the bamboo'. By Avicenna and other Arab physicians it was reputed to be a valuable medicine; the Arabic name, Tabasheer or Tabashir, is commonly used for it. From *Bambusa arundinacea* Willd. it is obtained in large pieces, like fragments of shells, but softer. It may be opaque or semi-transparent, firm or powdery, and in colour it may approach white, or it may be azure-blue, yellow, brown or black. Some varieties have been compared to the semi-opals, others to jasper, and others to chalk. Tabasheer consists chiefly of silica (70–90 per cent.) with a small quantity of lime, potash, organic matter and water.

The great variety of uses which can be made of the bamboos, in comparison with other Gramineae, is chiefly due to the arboreal

[1] The word 'culm', which is used for the aerial stems of the Gramineae, is from the Latin 'culmus', which means a stalk or stem, especially of grain; see p. 70, note 1.

[2] Watt, G. (1889–96).

[3] Kurz, S. (1876).

[4] On bamboo silica, see Russell, P. (1790); Brewster, D. (1828); Turner, E. (1828); Fingerhuth, K. A. (1839); Brandis, D. (1887); Watt, G. (1889–96).

habit of this tribe. We will leave the significance of this habit to be discussed in Chapter v, and we will consider here only the growth phenomena involved. It must not be assumed, however, that all Bambuseae are trees. Some, indeed, are quite small. *Arundinaria densifolia* Mun. has a stem hardly 3 ft. (0·9 m.) high at most, with a diameter of $\frac{1}{3}$ in. (0·8 cm.),[1] while certain pigmy bamboos of Japan and the Kuriles may be only 4–6 in. (10–15 cm.) in height, and hardly as thick as a crowquill.[2] An extreme case of departure from the arboreal habit is that of a climber, *Arthrostylidium sarmentosum* Pilg., in which the stems are not perennial but herbaceous—dying down every year. Its largest culms are only 3 mm. in diameter, and not more woody than those of some species of *Panicum* and *Andropogon*.[3] But these herb-like forms are the exceptions, and at the other end of the scale we have bamboos in which heights of 120–30 ft. (36·6–39·6 m.) have been measured.[4] It has been recorded, for instance, that when a clump of *Bambusa arundinacea* Willd., with 112 shoots, was felled in Central Travancore, one culm was found to be 121 ft. 6 in. (over 37 m.) in height, and six others reached 118 ft. (nearly 36 m.).[5] A height of 120 ft. (36·6 m.) has also been measured for *Dendrocalamus giganteus* Mun.[6] The climbing bamboos, *Dinochloa MacClellandii* Kurz and *D. andamanica* Kurz, may mount trees to a height of 100 ft. (30·5 m.), which implies a great length of axis.[7] The longest internodes produced are said to be those of *Teinostachyum Helferi* Mun., which may reach 52 in. (132 cm.).[8] Even the tallest bamboo culms always remain relatively slender, the greatest diameters named in trustworthy records being from 10 to nearly $11\frac{1}{2}$ in. (25·4–29·1 cm.);[9] in the relation of height to diameter, some of these culms appear, indeed, to approach the theoretical limit.[10]

As might be expected, from the height which bamboos may attain, their growth is often remarkably rapid. The quickest piece

[1] Gamble, J. S. (1896). [2] Kurz, S. (1876). [3] Chase, A. (1914).
[4] See records cited in Schroeter, C. (1885). [5] Pillai, M. V. (1905).
[6] Gamble, J. S. (1896); Macmillan, H. F. (1908).
[7] Kurz, S. (1876). [8] Gamble, J. S. (1896).
[9] Cited by Schroeter C. (1885); Yule, H. and Burnell, A. C. (1903).
[10] See discussion in Thompson, D'Arcy W. (1917), pp. 19, 20.

of individual growth observed is, apparently, the increase of a culm of *Bambusa arundinacea* Willd., cultivated in a glasshouse at Kew, which grew 91 cm. in twenty-four hours, i.e. 0·63 mm. per minute.[1] This rate of elongation is only surpassed by another 'record' held by the Gramineae, that for the elongation of the filaments of certain grass stamens, which has been described as reaching 1·6 mm. per minute.[2] Next to the Kew record for *Bambusa arundinacea* Willd., comes a Japanese account of *Phyllostachys mitis* Riv., which has been observed to grow as much as 88 cm. in twenty-four hours,[3] while Robert Fortune noted that, in the same period, a healthy plant of the bamboo 'Mow-chok', cultivated in China, generally grew from 2 ft. to 2 ft. 6 in. (61–76 cm.).[4] In twenty-four hours, also, *Bambusa Balcooa* Roxb. in Bengal has been reported to grow 29·8 in. (about 76 cm.);[5] *B. Tulda* Roxb., in Bengal, 73·3 cm.;[1] *Dendrocalamus* sp., 57 cm.;[6] *D. giganteus* Mun., 13 in. (33 cm.) at Dehra Dun.[7] In other records, the striking feature is the length of time over which a high average rate of growth has been kept up. According to measurements made long ago in the Calcutta Botanic Garden by a Hindoo head gardener, Mooty-Dollah,[8] *Bambusa gigantea* Wall. (*Dendrocalamus giganteus* Mun.) grew 25 ft. 9 in. (7 m. 85 cm.) in thirty-one days; that is to say it continued to make, for a period of a month, an average increase of nearly 10 in. (25·4 cm.) every twenty-four hours. *Bambusa baccifera* Roxb., again, grew 9 ft. 9 in. (3 m.) in twenty-eight days, or an average of over 4 in. (10 cm.) every day during four weeks. Kurz observed that *Gigantochloa robusta* Kurz grew 5 m. 90 cm. in a month, which would give about 19 cm. a day during this period.[9]

We have some knowledge of the periodicity of growth in *Dendrocalamus giganteus* at Dehra Dun.[7] The young culms were found to appear early in August, and growth in height was completed by the end of November. The growth-rate was at first very low; it gradually

[1] Schroeter, C. (1885).
[2] Rimpau, W. (1882).
[3] Shibata, K. (1900).
[4] Fortune, R. (1857).
[5] Kanjilal, U. (1891).
[6] Kraus, G. (1895).
[7] Osmaston, B. B. (1918).
[8] Martius, C. F. P. de (1848).
[9] Kurz, S. (1876).

rose for four to six weeks, until the bamboo reached about 12 ft. in height, when a maximum was attained which was uniform for several weeks. After this, a gradual diminution in the rate of growth set in, leading finally to cessation. The period when the culms began to develop was towards the middle of the rainy season, and they did not complete their height growth until a couple of months or so after the rains normally cease. The maximum rate of growth was attained when the atmosphere was saturated, and this was the condition at night both during and shortly after the rains. At dawn, there was a rise in temperature, and, unless it was actually raining, the relative humidity fell. The result was that evaporation occurred from the surface of the growing part of the culm, and the reduction in turgescence lowered the rate of growth. Normally the growth that took place in the night was nearly double that in the day, but it was observed that growth would take place as rapidly by day as by night, provided there was a sufficient degree of humidity. One culm, whose growth was measured, took three and a half months to reach its full height of 71 ft.

An observer[1] who measured a species of *Dendrocalamus* at Buitenzorg, recorded that, after the early stages, he found the growth of each axis to be surprisingly irregular. These irregularities seemed to bear no relation to external conditions, and, as axes of the same plant behaved simultaneously in quite different ways, it appeared that the cause of this capricious behaviour must be sought within the individual shoots themselves.

The culms of bamboos seem generally to be of a single season's growth, but this is not universal. It has been recorded[2] that in 1926, a year in which the rains ended early, the culms of *Bambusa burmanica* Gamb., at Lachiwala, did not complete their growth, but went on developing in the next season, 1927.

Young, growing bamboo shoots sometimes exude water on a considerable scale.[3] It was noticed in the Hamma garden in Algiers, that when shoots were just about to emerge from the ground, the soil became damp, while, in the two or three days after emergence,

[1] Kraus, G. (1895). [2] Trevor, C. G. (1927).
[3] Rivière, A. and C. (1878).

the bud was found to be wet, especially in the early morning. At first this moisture was mistaken for condensed mist or dew, but it was found that it appeared even when the buds were protected, so it became evident that it was a natural exudation. This wetting effect was seen especially in *Bambusa macroculmis* Riv., *B. vulgaris* Schrad., *Dendrocalamus Hookeri* Mun. and *Phyllostachys mitis* Riv. The secretion of water is not confined to the shoots in their bud state.

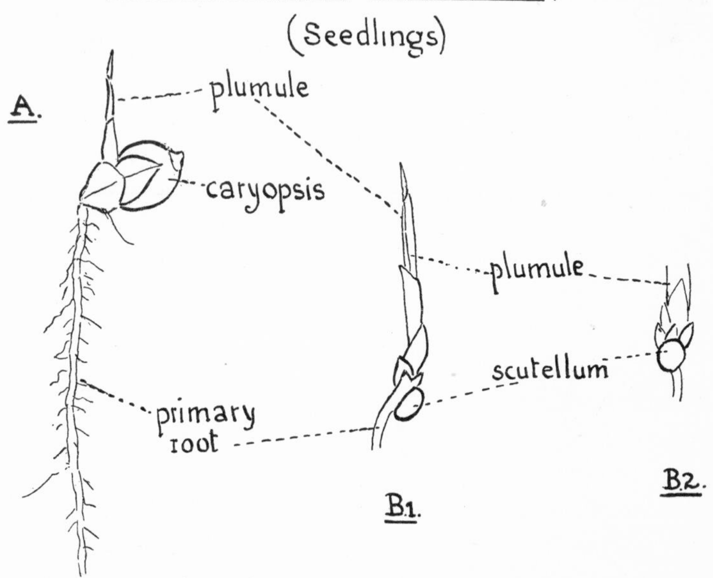

Fig. 22. *Dendrocalamus sikkimensis* Gamb. A, seedling (nat. size), root incomplete. B 1, side view, and B 2, front view, of another seedling, in which the plumule, cotyledon and radicle have been dissected out (slightly enlarged). [A.A.]

On certain August evenings, a regular rain was observed to fall from the foliage of clumps of bamboo in the Hamma garden.

Owing to the infrequency of seed production in many bamboos, our information about the growth and development of members of the tribe generally has reference to vegetatively produced shoots rather than to seedlings. Such knowledge as we have about sexual reproduction indicates that bamboo fruits as a rule drop readily from the plant, and germinate usually within the first week.[1] Fig. 22

[1] Kurz, S. (1876).

shows a *Dendrocalamus* seedling, in which the vigorous first root has burst through the glume surrounding the caryopsis.[1] A more thorough knowledge of early seedling stages in the Bambuseae is much to be desired. Considering the rapidity of normal germination in the bamboos, it is perhaps not surprising to find that the rather rare condition of true vivipary occurs in two genera—*Melocalamus* and *Melocanna*. A writer[2] in *The Indian Forester* found *Melocalamus compactiflorus* B. et H. (*Pseudostachyum compactiflorum* Kurz), at the fruiting stage, at a height of 6000 ft., on the hills between Burma and China. Some large caryopses could still be seen attached to the parent plant, but with the seed already germinated, so that a tuft of roots and a shoot protruded. Others again had fallen and the seed had taken root. *Melocanna bambusoides* Trin. shows a similarly anomalous germination *in situ*. An eye-witness describes that from the fruit hanging on the tree "springs a young bamboo leaf and also a bunch of roots; when the young shoot is some 6 in. long, the whole thing drops off the tree, and apparently plants itself in the ground by the roots".[3]

On the post-germination history of normal bamboo seedlings we possess a few observations, chiefly by Brandis,[4] but more modern work is needed. Brandis pointed out that in his day the remarkable process of the development of the seedlings into a clump had not been sufficiently studied, and unfortunately this still holds true. *Bambusa arundinacea* Willd. is the species about whose seedlings we have the fullest information. It is said[5] that in the first stages of their existence the young plants are very delicate, and, except under the influence of plenty of moisture, they are unable to resist the scorching effect of the sun's uninterrupted rays; on the other hand, excess of water about their roots causes them to die off rapidly. Moreover they are incapable of competing with the minor grasses, by which they are easily and speedily choked and destroyed. Brandis's account of the stages in their development is that in March 1882 he

[1] 'Caryopsis' is the name given to the somewhat peculiar fruit of the Gramineae, which will be discussed later; see p. 162.

[2] "B., F. G. R." (1902).

[3] Cited in Stapf, O. (1904[2]).

[4] Brandis, D. (1899).

[5] "D., J. C." (1883).

found large patches of young seedlings, from seed which had been produced in 1881, and had germinated during the rains of that year. The youngest plant consisted of one shoot, about 6 in. long, bearing two or three leaves at the tip, and, below these, a sheath with a small imperfect blade. Near the ground, the shoot bore a short, membranous-pointed sheath, at the base of which were two rootlets, about 3 in. long. At a later stage, several conical side shoots made their appearance, just below the surface of the ground; they were bent, first downwards, then upwards, and were covered with numerous membranous, white sheaths. These side shoots, which would ramify later, were the beginnings of the rhizome. They were destined to turn upwards at the tip, thus forming leaf-bearing stems, each rooting from the bend. Besides these underground side shoots, with short internodes, others arose which had moderately long internodes, and rooted at the nodes, sending up leaf-bearing stems from these points also. In this manner it came about that seedlings, not quite a year old, had an underground rhizome of complicated build, pushing numerous rootlets into the soil, and bearing a number of shoots, of which the first to be formed were short-lived. The other bamboos which Brandis examined showed a general similarity to *Bambusa arundinacea*. In a plant of *B. Tulda* Roxb., the rhizome at the end of the second year consisted of six short branches, 2 in. in diameter. Four of these rhizome branches terminated in leaf-bearing shoots.

Bamboo rhizomes sometimes dive deep into the soil and they may show great penetrating power. Under cultivation in Algiers they have been found at a depth of 80 cm., and they have even been known to pierce macadamised roads.[1]

It is always difficult, psychologically, to pay enough attention to those organs of the plant which are hidden underground. The result is that our mental picture of a bamboo clump seldom extends to an adequate image of its immensely complex subterranean system. It is indeed rarely possible to get more than a glimpse of this system, but fortunately a good specimen showing the rhizome plexus of a clump of *Bambusa vulgaris* Schrad., from Calcutta, is preserved in one of the

[1] Rivière, A. and C. (1878).

Museums of the Royal Gardens at Kew. Fig. 23 is a sketch of part of this plexus on a much reduced scale; some idea of the actual size may be gained from the fact that the segment marked X was about 14 cm. in length. The specimen is hard to understand at first glance, because the rhizome complex is seen *from below*—an unaccustomed point of view.

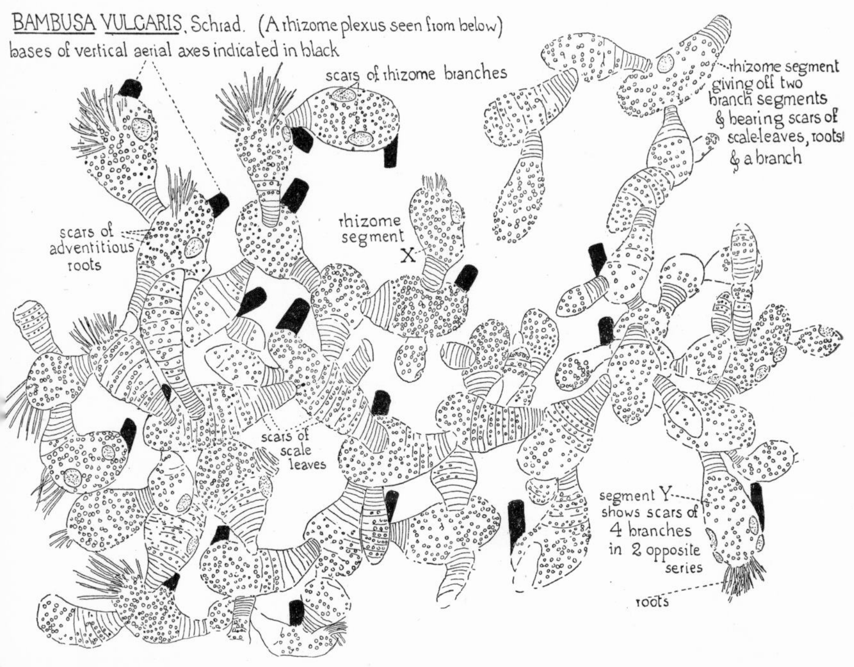

Fig. 23. Rhizome plexus of a clump of *Bambusa vulgaris* Schrad., seen from below. Specimen No. 205, from Calcutta, Museum 2, Royal Gardens, Kew (much reduced). [A.A.]

Each rhizome segment is pear-shaped—narrower at the base than n the distal region. The narrow neck at the base of each segment is :haracteristic of various other bamboos. In *Phyllostachys mitis* Riv., .he 'neck' has been examined in detail; it has been shown that ;everal of the lowest internodes unite at an early stage to a solid, voody structure (2–4 cm. long and only 1·0–1·5 cm. in diameter), in vhich all internal distinction between internodes and nodes is lost.[1] The apices of many of the rhizome segments in Fig. 23 turn

[1] Shibata, K. (1900).

upwards to form the erect aerial axes. These erect axes cannot always be seen in this view from below, but where they are visible, I have indicated them in black in the sketch to distinguish them from the underground parts. The rhizome segments bear two or more branches (rhizome segments of the next order) in two series. In Fig. 23 the scars marking the detachment of the rhizome segments have been dotted. The segment Y shows scars of four branches. Each rhizome segment is marked by horizontal scale-leaf scars, which are conspicuous in the basal, but less so in the distal

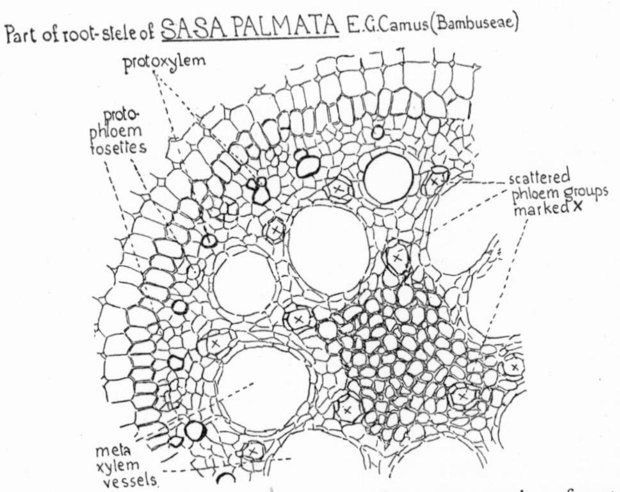

Fig. 24. *Sasa palmata* E. G. Camus. Part of transverse section of root stele (× 193 *circa*) to show supernumerary phloem groups. [A.A.]

region. The adventitious roots, on the other hand, are chiefly developed from the distal regions of the rhizome segments. They have mostly been cut off in the specimen illustrated, and only their scars remain. In the lower part of each segment, the roots are distinctly seen to originate in rows between the scars, but in the distal, more swollen part, the number of roots is so great that all obvious order is lost.

The roots of many bamboos are peculiar in possessing supernumerary phloem groups.[1] Fig. 24 shows part of the central cylinder of a root of *Sasa palmata* E. G. Camus. The protophloem groups,

[1] Ross, H. (1883).

which alternate with the protoxylems close to the surface of the stele, take the form of irregular rosettes. In addition, there are scattered phloem groups, each consisting of a relatively large sieve tube surrounded by companion cells; these groups occur between the metaxylem vessels and also in the pith. At the attachment of each lateral root to the central cylinder, there is much anastomosis of the phloem strands.[1]

A number of examples of mycorrhizal roots have been described in the bamboos and other Gramineae.[2]

In comparison with the complexities that sometimes occur in the rhizome system of the Bambuseae, that sketched in Fig. 23, p. 67, is a simple example. In certain old bamboo clumps, the newer rhizomes are found to have doubled on themselves, since they have been prevented from diving into the earth by the thick layer of pre-existing rhizomes. They become piled upon one another, until they may reach a metre's height above the surface of the ground. In such examples, the rhizomes, and the débris of other members, form a dense compact block, which cannot be separated into its component parts.[3]

The rhizomes of some bamboos have a curiously undulating course. Those of *Phyllostachys mitis* Riv., for instance, may emerge from the soil and rise up in a slight curve and then plunge into the soil again, re-emerging in a short distance, so that "lorsque l'on passe à travers une plantation de cette espèce, faut-il prendre garde que les pieds ne s'engagent sous ces sortes d'arceaux qui provoqueraient une chute".[4] This 'croquet-hoop' growth, which has also been observed in bamboos growing in Japan,[5] can be paralleled outside the Gramineae. In July 1916, on Braunton Burrows, North Devon, I found a patch of the Sea-purslane, *Arenaria peploides* L., which showed the same peculiarity. The plants had slender elongated stolons, which took a course below the surface for most of their

[1] Shibata, K. (1900).

[2] Kirchner, O. von, Loew, E. and Schroeter, C. (1908, etc.); Rayner, M. C. (1927) for references.

[3] Rivière, A. and C. (1878).

[4] Rivière, A. and C. (1878); cf. also Freeman-Mitford, A. B. (Lord Redesdale) (1896).

[5] Takenouchi, Y. (1931).

length, but many of them were arched up into the air in a region a little behind the apex. The effect was striking, as the arched part was green, and the remainder of the stolon was whitish. The apex itself was never found emerging, and from this, and from the disturbance of the sand surface about the arch, it was clear that the arching was a secondary effect, due to intercalary growth.

The causes which determine the conversion of horizontal rhizome branches into erect haulms,[1] deserve fuller investigation than they have hitherto received. When rhizomes develop under normal conditions, presumably it is internal forces which determine the moment at which the tips shall turn up vertically and take on the characters of aerial axes; but it has also been observed that meeting an obstacle in the soil may induce a permanent upward curvature, and the conversion of the rhizome tip into a true haulm. The same change can be induced experimentally.[2] If the rhizome apex of *Phyllostachys reticulata* C. Koch is made to enter a tube with a right-angled bend, it continues its upward course when it emerges into freedom, and thus becomes a haulm. It would be of great interest if the Japanese botanist, to whom we owe this experiment, would carry out a more complete study of the relative parts played by internal and environmental conditions in determining the continuance of rhizome growth, or the conversion of the horizontal rhizome into an erect haulm.

The bamboo haulm is generally more or less cylindrical, but the internodes may be marked by a groove, divided into two by a median ridge. This channelling is initiated by the pressure of the bikeeled prophyll of the axillary bud, which in the earliest stages is as long as the internode. The imprint is retained and extended, even after the bud has been left behind by the elongation of the internode.[3] A greater divergence from the cylindrical form is shown by *Bambusa angulata* Mun.,[4] the Square-bamboo. The haulm of this species resembles a rod trimmed with a knife into a four-square shape. It is

[1] The word *haulm* for a stem or stalk of grass is Teutonic, and probably cognate with the Greek κάλαμος, a reed, and the Latin *culmus* (p. 60, note 1). *Haulm* and *culm* are thus related words.

[2] Takenouchi, Y. (1931).

[3] Rivière, A. and C. (1878).

[4] Also called *B. quadrangularis* Fenzi and *Phyllostachys quadrangularis* Rendle.

apparently a Chinese plant, and its origin is ascribed traditionally to "Ko Hung, the most famous of alchemists (fourth century A.D.)", who "thrust his chopsticks (slender bamboo rods pared square) into the ground,...which by thaumaturgical art, he caused to take root and to appear as a new variety of bamboo—square".[1] Staves cut from the stems of this species are used by Buddhist monks and village elders, and hence it is known in Chinese by a name which may be translated as Bamboo-supporting-the-old.[2]

Ktêsias, to whose early notice of this tribe we have already referred (p. 58), was aware that some bamboos are hollow and pithless, while others are solid. He distinguished the solid and hollow forms as 'male' and 'female' respectively.[3] The name Male-bamboo is now confined to those rare, almost solid stems of *Dendrocalamus strictus* Nees, which were formerly used for spear-shafts in the East Indies.[4]

The bamboos share with the other Gramineae the character of having a localised region of curvature at each node. In many grasses it is the swollen base of the leaf-sheath which forms the whole of this pulvinus, but in the bamboos the part of the cushion formed from the sheath withers early, and it is the stem pulvinus, almost exclusively, which is concerned with bending.[5]

The leaves of the Bambuseae, like those of other grasses, alternate with each other, in two opposite rows. They consist, in general, of a sheathing base and an elongated limb, but sometimes a petiole is intercalated between the two. The foliage leaves with the largest blades are said to be those of *Planotia nobilis* Mun., which are 4·5 m. long and 30 cm. wide, although the haulm itself is only 3 m. high.[6] At the other extreme from the broad-leaved bamboos, we may set *Arthrostylidium capillitifolium* Griseb., whose leaf-blades, as the specific name implies, are almost hair-like.

The mesophyll cells of the bamboo leaf may show curious invaginations of the walls. In *Bambusa Simoni* Carr., the indented cells have been described as tabular and relatively flat, with their roofs

1 Dyer, W. T. Thiselton (1885).
2 Porterfield, W. M. (1925).
3 Ktêsias (McCrindle, J. W.) (1882).
4 Smythies, A. (1881).
5 Lehmann, E. (1906).
6 Schroeter, C. (1885).

and bases at right angles to the long axis of the leaf. The wall towards the lower face of the leaf shows two or three folds.[1] In Fig. 25, these invaginations are seen in *Sasa disticha* E. G. Camus; A 1 shows them in a transverse section of the leaf, while in A 2 the plane is tangential, with the result that each cell looks like a cluster of separate elements. Fig. 25, B, shows epidermal features which are repeated also in Gramineae outside the bamboos—'ripple-walls', papillae, and contrasted 'long' and 'short' elements.[2]

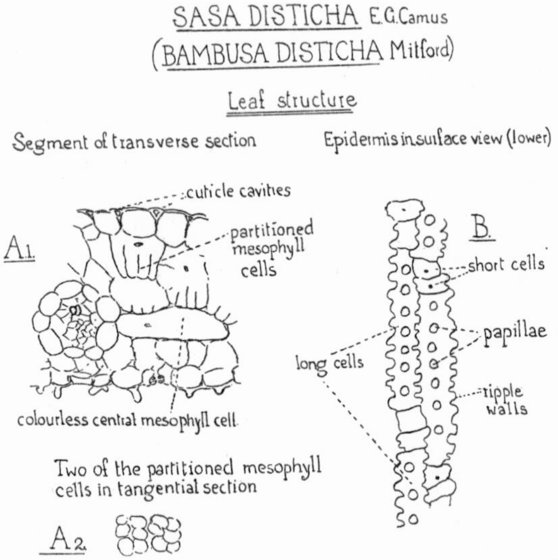

SASA DISTICHA E.G.Camus
(BAMBUSA DISTICHA Mitford)

Leaf structure

Segment of transverse section

Epidermis in surface view (lower)

cuticle cavities

partitioned mesophyll cells

A1.

colourless central mesophyll cell

Two of the partitioned mesophyll cells in tangential section

A2.

B.

short cells

papillae

long cells

ripple walls

Fig. 25. *Sasa disticha* E. G. Camus (*Bambusa disticha* Mitford). A 1, small part of transverse section of a leaf limb (× 318). A 2, two of the partitioned mesophyll cells as seen in a section parallel to the leaf surface (× 318). B, cells of lower epidermis of leaf in surface view (× 318). [A.A.]

The early leaves borne on bamboo haulms are of a leaf-sheath nature, with the limb reduced in varying degrees. This reduced blade sometimes has a curiously crisped and twisted form—a peculiarity which is apparently a mechanical result of the conditions prevailing in the apical bud. The nodes of the young shoot are so densely crowded in the neighbourhood of the apex that the development of a whole series of leaves is practically simultaneous; packed

[1] Karelstschicoff, S. (1868); Haberlandt, G. (1882); on invaginated walls in other Gramineae, see p. 304.

[2] On the epidermis of the Gramineae, see pp. 298–304.

as their blades are within the sheaths of the outer scales, they have no room to develop properly, and so they become wrinkled and puckered.[1]

In most families, scale-leaves are relatively small objects, but in the bamboos they may attain a large size. Those of *Bambusa macroculmis* Riv. may reach 44 cm. in length, and 52 cm. in width.[2] As it is often difficult or impossible to obtain the flowers, or even the mature foliage of bamboos, the characters of the scale-leaves—since they are readily accessible—have been much studied and are often used in distinguishing species.[3] The leaf-sheaths of various bamboos bear dark, sepia-brown hairs, sharp-pointed and easily shed. They are highly irritating to the human skin,[4] and anyone who has to pass, for instance, through clumps of *Gigantochloa Scortechinii* Gamb., may suffer acute discomfort.

Bamboos which are native to river banks and the shade of tropical forests are apt to be evergreen, while those whose foliage leaves are normally deciduous may become evergreen when transferred to moister climates.[5] The scale-leaves of bamboo haulms may either disarticulate, or they may persist and finally disintegrate in place, like those of the subterranean rhizomes.

The nature of the branch system in bamboos is indicated, in a very condensed form, in Fig. 26, p. 74, which is a diagram drawn from transverse sections of one of the vegetative buds which sometimes take the place of spikelets in the inflorescences. It will be seen that each shoot begins with a bikeeled prophyll. Axes of three orders are included in the diagram. Since the leaves are distichous throughout, and lateral in each axillary shoot, the plane of symmetry of the scales borne by Shoot II is perpendicular to that of the scales on Shoot I; the plane of symmetry of the scales of shoots III *a*, III *b*, and III *c* is perpendicular to that of the scales of Shoot II; and so on, in alternating sequence. The successive buds illustrate a character of some interest—the right-over-left or left-over-right nature of the overlap of the leaf-margins. Of the three axillant leaves, whose overlap can

[1] Takenouchi, Y. (1931).
[2] Rivière, A. and C. (1878).
[3] For details, see Gamble, J. S. (1896) and Camus, E. G. (1913).
[4] Ridley, H. N. (1924).
[5] Kurz, S. (1876).

be traced, one is identical in this respect with the prophyll of its own axillary bud, while, in the other two, the relation is reversed. In four cases the overlap of the prophyll and of the succeeding leaf can be determined; in three the overlap is the same, and in the fourth

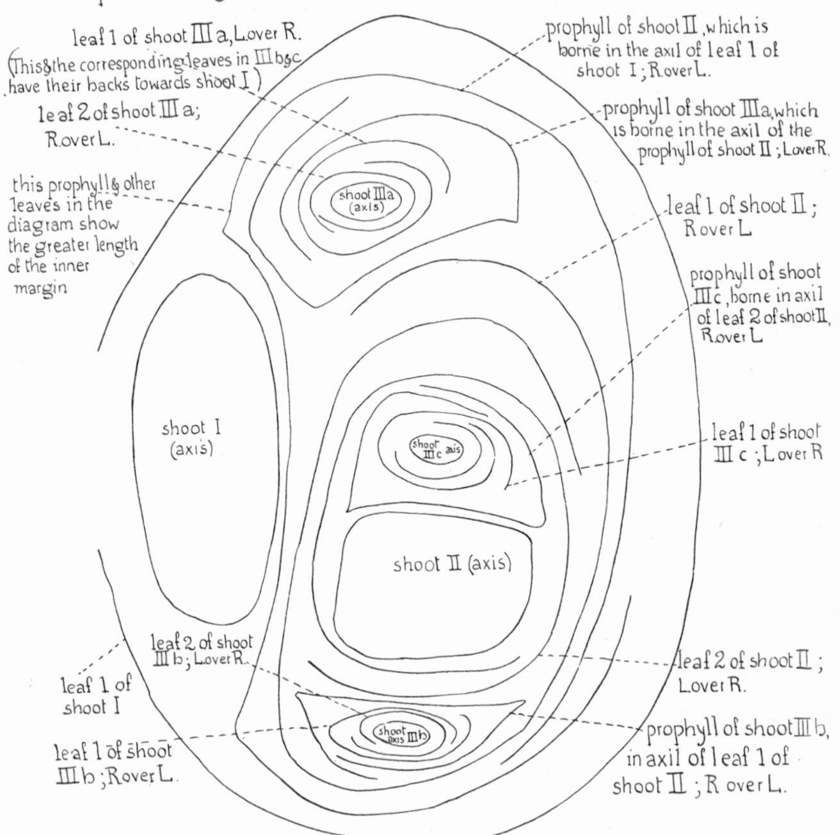

Fig. 26. *Bambusa bambos* Back. Diagram of a vegetative bud from an inflorescence of which one section is shown in Fig. 106, A, p. 221. [A.A.]

it is reversed. In three shoots, the second leaf after the prophyll can be seen, and in each it is reversed as compared with the preceding leaf. An alternating overlap-relation of the leaves (after the prophyll) seems, indeed, to be the rule in Gramineae shoots, but it is evident from the present example that there is less regularity in the overlap-relation of the axillant leaf to the prophyll, and of the prophyll to the succeeding leaves. Another point brought out in these diagrams is the greater breadth of that leaf-margin which falls inside.[1]

The general appearance of the bamboo shoot depends upon variations in the number of lateral branches at those nodes at which branching occurs.[2] The simplest possibility is the development, at each node, of a single axis which does not branch precociously (e.g. *Sasa* and *Yadakeya*). The next stage is that of *Phyllostachys*, in which the main lateral axis produces a branch of the next order from the lowest node, so early that it seems to arise simultaneously with its parent shoot; the result is that there is a pair of apparently twinned branches at each node. In the third stage (represented by *Semiarundinaria*, *Sinobambusa*, etc.) the two lowest nodes of the lateral branch produce precocious secondary branches, so that there seem to be three laterals at each node. A further complexity is produced in *Shibataea Kumasana* Mak., in which the secondary branches each give rise precociously to one branch of the third order, so that there is a set of five branches at each node. Still richer branching occurs in *Dendrocalamus* and *Bambusa*, in which the two secondary axes each bear two tertiary branches, so that quite a cluster of branches is produced at once. This basal telescoping and precocious branching of axes of successive orders is not confined to the bamboos. We shall see later that it is one of the factors responsible for the characteristic form of the inflorescence in many grasses.

The result of the mode of development described is that the branch system of many bamboos is, at maturity, very difficult to analyse. In *Arundinaria Simoni* Riv., for instance, there may be, at each node, as many as twenty-five terminal branchlets,[3] which seem to arise from a verticil of branches; but when the axillant scale-leaf is removed, it is seen that, however wide the tuft of branches may be, its actual origin is from the scale-leaf face of the node, while the opposite side of the

[1] On these subjects, see Schoute, J. C. (1910); Bremekamp, C. E. B. (1915); Seybold, A. (1925).

[2] Takenouchi, Y. (1931); this well-illustrated article contains a great deal of information about the development of Japanese bamboos.

[3] Bureau, E. (1903).

node is naked. It is only the breadth of its base, due to precocious branching, which gives the axillary tuft a whorl-like appearance. If this mode of branching is carried to an extreme, it produces a singular effect. For instance, the climbing bamboo, *Arthrostylidium angustifolium* Nash, has linear leaf-blades crowded on the dense pseudo-whorls of short branches at the far-apart nodes of the slender culms; the appearance is that of "great pompons strung along the stems at distant intervals".[1] In dense forests, without undergrowth, these liane Bambuseae may become of practical importance to travellers. It has been recorded,[2] for instance, that, in the forest of Quindiù on the western slopes of the Cordillera, *Aulonemia Queko* Goudt. is almost the only resource for any animal which has to be taken through. It serves as fodder, and, to get it, the haulms, which hang from the trees like cords, are dragged down with a strong pull, so that they bring their leaves with them. An account by Agnes Chase[1] of the climbing bamboos of the island of Porto Rico—species of *Arthrostylidium* and *Chusquea*—may be quoted for the picture which it offers of these strange Gramineae. "The three species of *Arthrostylidium* have much the same habit, climbing high, repeatedly branching, and in their greatest development swinging down in great curtains from the trees overhanging trails or streamlets. They love the glints of sunlight along the trails or water courses and are very rarely found in deep shade". In *A. angustifolium* Nash, which has already been mentioned for its pompons of foliage, the "culms hang straight 20 or 30 feet from trees 40 or 50 feet high, or festoon themselves over lower growth". In the third species, *A. multispicatum* Pilg., we meet with another peculiarity—the development of prickles. The "slender naked growing ends of the culms and branches are beset with very short, sharp-pointed retrorse prickles. These ends, 5 to 12 feet or more long, swing in the breeze like whip lashes until they strike a place to take hold. Only after attaching themselves to some support do the leaves and branches develop from the clusters of short, sharp, radiating, scale-covered branch buds". The grappling branches form "an inextricably tangled mass that draws blood at every foot of one's progress through it". Fig. 27

[1] Chase, A. (1914). [2] Goudot, J. (1846).

illustrates the structure of the downwardly directed prickles. They consist of single epidermal cells with immensely thick walls, jacketed at the base by other epidermal cells, whose development is less exaggerated. They eventually fall off, leaving the old culms smooth.

Though climbing is one of the established modes of life among bamboos, habitual epiphytism is almost unknown. A clump of *Bambusa arundinacea* Willd. has, indeed, been observed[1] growing upon a tree of *Cassia fistula* L., the lowest point of attachment of the bamboo being more than 4 ft. from the ground; but such records seem to be rare.

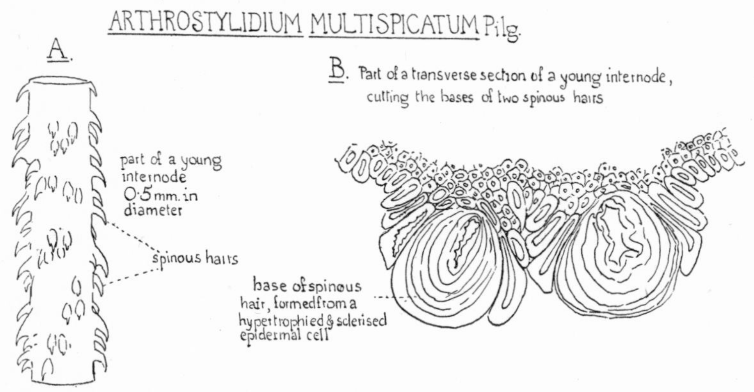

Fig. 27. *Arthrostylidium multispicatum* Pilg. Herbarium material collected by Agnes Chase, east of Adjuntas, Porto Rico, November 9, 1913 (No. 6470). A, part of an internode towards the apex of a shoot with spinous hairs (× 14). B, part of the edge of a transverse section of an internode such as A (× 193 *circa*). [A.A.]

The prickles of *Arthrostylidium multispicatum* Pilg. are, as Fig. 27 shows, merely thickened hairs. In others members of the tribe, solid thorns occur, of a totally different type from these minute but vicious prickles. The bamboo thorns belong to two classes—shoot thorns and root thorns. The shoot thorns consist of recurved spinescent buds, which are either dormant or permanently arrested.[2] Their shoot nature may be revealed by the presence of vestiges of leaves. The formidable spines of *Bambusa arundinacea* Willd. (including *B. spinosa* Roxb.) and of *B. teba* Miq. (*Arundarbor spinosa* Rumph.) belong to this category, as well as those of *Guadua angustifolia* Rup. illustrated in Fig. 28, A, p. 78. On the other hand, many of the clump-

[1] Menon, K. G. (1918[1]). [2] Kurz, S. (1876).

growing bamboos produce, on the lower half of the culm, a whorl of rootlets at each node; these rootlets harden into spinescent bodies.[1] Certain bamboos, again, give off, from the lower nodes of old stems, downward branches which are almost solid, and produce numerous adventitious roots from all their nodes; these partly penetrate the earth, and partly become thorns.[2,3] Nodes of a species of *Arundinaria*, each bearing a whorl of root thorns, are sketched in Fig. 28, B. This metamorphosis of roots into thorns can be paralleled in a few other Monocotyledons, such as *Moraea ramosa* Ker-Gawl;[4] *Dioscorea prehensilis* B. et H.[5] and other Yams;[6] and certain palms.

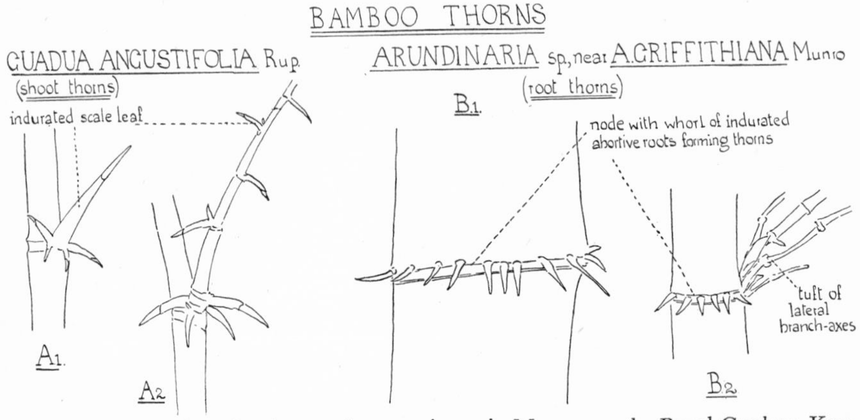

Fig. 28. Thorns of the Bambuseae, from specimens in Museum 2, the Royal Gardens, Kew. A 1 and A 2, thorn-bearing nodes of *Guadua angustifolia* Rup. B 1 and B 2, nodes of *Arundinaria* sp., near *A. Griffithiana* Mun., bearing root thorns. (All × ½.) [A.A.]

Although a great many facts about the vegetative anatomy of the bamboos have been recorded, especially by Japanese botanists, there is still need for a full comparative survey, dealing more particularly with the difficult problems bound up with the *relations* of the vascular strands. We know that, in general, the course of the bundles in the haulm follows the palm type; the large median leaf-trace may run down through five to six internodes before fusing with other leaf-bundles.[7] It may pass through an amphivasal phase in the node

1 Watt, G. (1889–96).
2 Kurz, S. (1876).
3 Schroeter, C. (1885).
4 Scott, D. H. (1897); Arber, A. (1925).
5 Scott, D. H. (1897).
6 Burkill, I. H. (1923).
7 Shibata, K. (1900).

of exsertion, though it is collateral above and below (Fig. 29). The traces in the inner part of the ground tissue are apt to show considerable torsion, so that the plane of symmetry of each strand, instead of being radial to the stem, may form an angle with the radial plane; indeed, this twisting may be carried so far that the phloem may be turned inwards.[1]

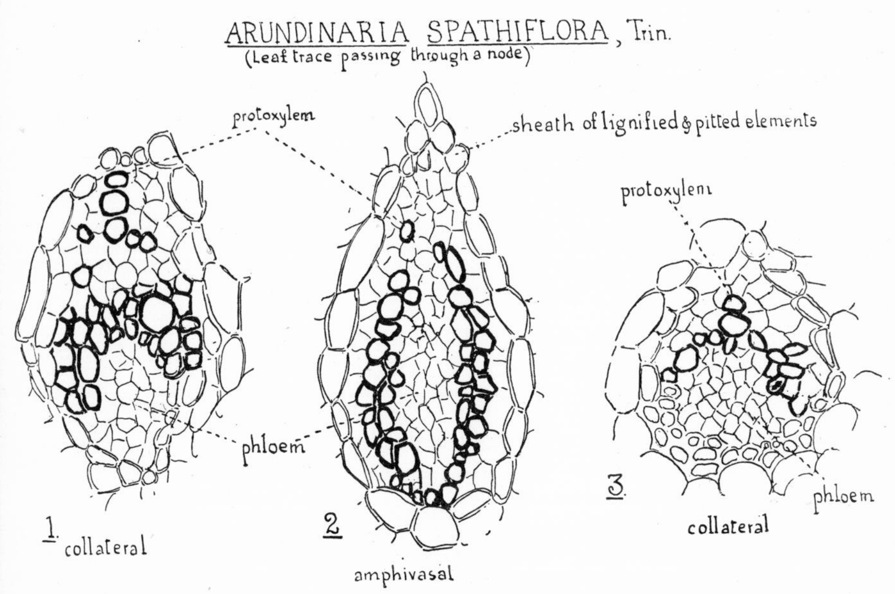

Fig. 29. *Arundinaria spathiflora* Trin. Transverse sections (× 424) from a series passing upwards from below through a node, showing the phases passed through by the median bundle of the leaf exserted at this node. Amphivasal structure in *2*; collateral above and below. In the sheath of lignified pitted elements, which is specially conspicuous in *2*, many of the elements are quite short. [Arber, A. (1930[1]).]

At the rhizome nodes, junction takes place between leaf-traces and leaf-traces, and also between leaf-traces and bundles of axillary buds.[2] In a number of bamboos, the phloem of the bud-bundle at its attachment to the leaf-trace is spindle-shaped in section and much swollen. In this region it does not show differentiation into sieve-tubes and companion-cells, but consists of homogeneous cambiform tissue. This fact raises the whole question of the function of the various types of phloem element; for here we have a junction region

[1] Takenouchi, Y. (1931). [2] Shibata, K. (1900).

in a conducting channel, in which sieve-tubes and companion-cells seem to be efficiently replaced by mere cambiform elements. These 'junction' phloem swellings are highly developed in the rhizome nodes of *Phyllostachys*, where their diameter is as great as that of a large vascular bundle. These bud-bundles, applying themselves to the rhizome strands, may be compared with haustorial sucking organs.

The individual elements of the vascular bundles of bamboos may be on a large scale. The sieve-tubes of the rhizome of *Phyllostachys mitis* Riv. are often 0·15 mm. in diameter, while the lateral vessels have a maximum width of 0·2–0·3 mm.[1] A peculiar feature of the vessels has been described in a bamboo which is probably *P. nidularia* Mun. The annular tracheae in the young leaf-sheath bundles show transverse bars, grouped in planes at right angles to the long axis of the vessel. These bars seem to be the thickened remains of the end walls of the elements which formed the vessels.[2]

In the mature haulm, it is the fibrous tissue, rather than the xylem or phloem, which is the most conspicuous feature.[3]

In the intercalary meristem of the haulm ground tissue of a number of bamboos, it is found that the daughter cells do not all behave alike.[4] Some of them elongate immediately and divide no more, while others undergo further division. The result is a differentiation of the ground tissue into 'long' and 'short' cells, recalling those of the leaf epidermis (Fig. 25, B, p. 72). The fact that we have no clear notion of the inner meaning of even so simple a differentiation as this, brings home to us the superficial nature of our laborious and detailed knowledge of plant structure.

[1] Shibata, K. (1900). [2] Porterfield, W. M. (1923).
[3] Cf. Schwendener, S. (1874), Pl. VII, fig. 1.
[4] Takenouchi, Y. (1931).

CHAPTER V

BAMBOO: TREE HABIT

AMONG the Gramineae, the distinction between the towering, woody bamboos, and the herbaceous grasses and cereals, is so striking that an effort of the mind is needed before we can think of them together, as members of a single and seemingly natural family. In Chapter IV we treated the tree habit of the bamboo merely as an observed fact; in this chapter we shall make an attempt to understand something of its significance. It cannot, however, be interpreted at all, if it is considered as an isolated phenomenon; in order to get any glimmering comprehension of it, we shall be obliged now to turn to the problem of the dendroid and herbaceous type in the flowering plants as a whole.[1] From the earliest times at which a desire to pigeon-hole the facts of the world led man to attempt some kind of classification of the plants which he saw around him, the difference between the tree and the herb has seemed to him an important one. Theophrastus (born *circa* 370 B.C.), the earliest scientific botanist of whose opinions we have a record, divided plants into the tree, the shrub, the undershrub and the herb.[2] His classification thus makes a fundamental distinction between the arboreal and herbaceous habits, though his shrubs and undershrubs to a certain extent form a transition between the two. Actually to define a tree is a very difficult matter, as one may see by looking at the attempted definitions in the text-books; what we need is to penetrate beyond these definitions, and to arrive at some answer to the question—what is the true inwardness of the tree habit? We must bear in mind that the difference between a tree and a herb is not

[1] This chapter is based, though with much modification, upon Arber, A. (1928[3]), which contains a fuller treatment of some aspects of the subject, with additional references. I am indebted to Dr Irène Manton for drawing my attention to the fallacy in the argument from geographical distribution (p. 73 of the 1928 paper), which I have omitted in the present version. For a further study of the literature of the subject, see Bancroft, H. (1930).

[2] Theophrastus (Hort, A.) (1916).

merely a question of size. A plant, such as one of the huge Gunneras, with its vast leaves, soft rhizome, and lack of secondary thickening, is indeed a giant among herbs, but it is obviously not a tree. One might, at first glance, be disposed to wonder why more herbs do not develop into plants approximating to trees in size. I think the answer is that, both in physical and biological matters, *scale* is in itself significant.[1] We have to remember that, with increase of linear dimensions, volume increases more rapidly than sectional area, so that increase of size does not confer a proportionate increase of strength; and hence a magnified copy of a herb would be unlikely to make an efficient type of plant. S. T. Coleridge, with uncanny acumen, described himself in one of his letters as having "A sense of weakness, a haunting sense that I was an herbaceous plant, as large as a large tree, with a trunk of the same girth, and branches as large and shadowing, but with pith within the trunk, not heart of wood".[2] He thus arrived intuitively at the botanical truth that the most essential feature of the tree is not so much its size as the *woodiness* or fibrousness that means strength. In our mental picture of a tree, this feature should take precedence of height, and the production of a single trunk, which are the points usually emphasised in definitions.

In the present chapter, I shall adopt for simplicity the generally accepted assumption that the Angiosperms (flowering plants) are monophyletic—that is to say, that they are descended from a single ancestral stock—though I am aware that we do not know whether this is true. Now the Angiosperms (reckoning by species) include approximately equal numbers of trees and herbs,[3] and if they are all descended from one primaeval stock, we are faced with the problem of whether this remote ancestor was arboreal or herbaceous. In the present century many botanists have offered answers to this question, but very few have made a definite pronouncement in favour of the view that the ancestral Angiosperm was a herb. On the other hand we have a large body of opinion endorsing the general idea that the

[1] Thompson, D'Arcy W. (1917).
[2] Coleridge, S. T. (E. L. Griggs) (1932).
[3] Sinnott, E. W. and Bailey, I. W. (1914).

primaeval members of this group were trees. A brief statement of this view in one of Sinnott's papers[1] may perhaps be taken as representative. He writes that "a considerable body of evidence derived from the study of the history, structure and distribution of the Angiosperms,...indicates that the most ancient members of the group were woody, and that herbaceous vegetation has made its appearance in comparatively recent geological time". Most of the arguments which have been adduced in favour of the arboreal ancestry of the flowering plants, have been stated by Sinnott and Bailey.[2] It is impossible here to do more than select one or two of the most salient points for discussion. The fossil evidence is naturally the first line to which we turn in studying such a problem. We find that these authors deduce, from the record of the rocks, that the woody Angiosperms preceded the herbaceous; but they make it clear that they regard the evidence as anything but conclusive. Seward, though with caution, expresses the same view in a recent book.[3] Sinnott, in another paper,[1] goes so far as to say that "There is evidence that the herbaceous type, save perhaps as a negligible portion of the flora, did not arise until the early Tertiary". It seems to me that there is a serious fallacy here. This writer assumes that we know what the earliest Angiosperms were like—but this is, in fact, a point on which we are entirely ignorant. Although the group is first found in the late Jurassic or early Cretaceous, there is no reason to suppose that it took its origin at that time. Everything seems, on the contrary, to indicate that the flowering plants originated aeons before their first appearance in the fossil record as we now have it; so that, by the time the late Jurassic was reached, the class was already an old one. There is indeed nothing primitive or synthetic about the earliest known fossil Angiosperms; the group, when its existence is first revealed to us in the rocks, is full-fledged and has long left all traces of its babyhood behind. It has even been suggested—I think with reason—that the Angiosperms must have had unknown representatives even in the Palaeozoic. If these representatives were herbaceous, and thus less likely to be preserved in the rocks, our

[1] Sinnott, E. W. (1916). [2] Sinnott, E. W. and Bailey, I. W. (1914).
[3] Seward, A. C. (1933).

ignorance of them ceases to be surprising. In this lack of records of their earliest phases, there is a parallelism between the flowering plants and the higher animals. The origin of Mammals, for instance, seems to recede further and further into antiquity the more the palaeontology of Vertebrates is studied. Smith Woodward[1] supposes that the Mammals, which in the Eocene replaced the Reptiles of the Cretaceous, "had actually originated long ages before, and had remained practically dormant in some region which we have not yet discovered, waiting to burst forth in due time". Now if we are to regard the Angiosperms, like the Mammals, as already a mature group when they first come to our knowledge, it is clear that the earliest fossils, with which we are acquainted at present, cannot decide for us whether the original progenitors of the class were trees or herbs.

Another argument derived from the fossil record, though concerned with the Cryptogams (flowerless plants), is suggestive in connection with the general problem of the relation of herb and tree. Sinnott and Bailey state that in the Lycopodiaceous and Equisetal series, "the herbaceous forms have, without much question, been derived from ancestral woody types". Other writers take the same view, and it has become one of the picturesque commonplaces of the text-books that the little Club-mosses and Horsetails of the present day are the reduced and humble descendants of the aristocratic, arboreal Lepidodendra and Calamarieae of the Palaeozoic. This may be so, but the validity of such a pedigree is open to doubt, and there is at least a possibility that our Lycopodiums, Selaginellas and Equisetums are descended from herbaceous forms which may have existed in the Palaeozoic, even if they have seldom been preserved. In this connection it is worth while to scrutinise those modern 'herbs' whose ancestry, we have reason to think, was arboreal; here the work of Burtt Davy[2] is of special importance. This author shows that under the peculiar environmental conditions of the High-veld of the Transvaal, certain genera which elsewhere consist only of trees or shrubs, are represented by sub-shrubs or even by 'herbs', and he regards these plants as forming stages in the evolution of an

[1] Woodward, A. Smith (1909). [2] Davy, J. Burtt (1922).

herbaceous from an arborescent type; but the examples which he adduces are on the whole suffrutescent, or else, if externally herb-like, they have woody underground parts, or in other respects show at least a ghost of arboreal characters. One of them he himself describes as resembling "the leafy, floriferous end of a tree branch...cut off and stuck into the ground". The 'herbaceous' members of the Bambuseae fall into much the same category as Burtt Davy's plants. Some, though not all, of them retain a tree-like habit, despite their reduced dimensions. *Chusquea andina* Phil., which in the Andes of Chili reaches the limit of the eternal snows, and *Bambusa Fortunei* van Houtte, in Japan, are said to be dendroid on a miniature scale, rather than herbaceous, though neither of them exceeds 2 ft. in height.[1] These examples seem to me to indicate that a type resembling a herb may undoubtedly arise through reduction from an arboreal form, but that the resemblance does not amount to identity. Though the reduced tree may come within the conventional definitions of a herb, it is still something a little different from the herb in the strictest sense. Dollo's *Law of Irreversibility*[2] throws light upon the 'shrublets' of the High-veld, and the 'herbaceous' Bambuseae. According to this law, no species in the course of its evolutionary development ever really retraces its steps; it may return to something superficially resembling a stage which it has passed through, but the later is never an exact reincarnation of the earlier stage. On this law we should suppose that, if the primaeval Angiospermic stock were herbaceous, a tree lineage arising from this stock might eventually be reduced again to something resembling a herb, but it would not achieve the herbaceous habit with any exactness—and this is precisely what, in fact, we find.

In considering the theory of the arboreal ancestry of the Angiosperms, we must now turn to the causes which have been invoked to account for the presumed transition from the arboreal to the herbaceous type. Sinnott and Bailey hold that the herbaceous form of vegetation was developed in the north temperate zone, mainly as a response to progressive refrigeration. It is indeed true that trees

[1] Schroeter, C. (1885). [2] Arber, A. (1918–19) and (1919[1]).

are, broadly speaking, characteristic of warm climates and that herbs seem better able than woody plants to withstand a climate with extremes of cold; but because herbs are proportionately more abundant in cold, and trees in hot climates, it does not follow that it was a cold environment that originally produced the herb. It almost seems as if there were something inherent in human nature that inclines us to magnify the effect of environment; and botanists are peculiarly under the influence of this tendency. I believe it is because the plant is quiescent and lets one manipulate the conditions of its life; the zoologist, on the other hand, has as a rule a much stronger sense of the importance of innate trends, because they are less masked, in the creatures with which he deals, than in the botanist's more passive subjects. It seems to me probable, on general grounds, that the change of habit from the herb to the tree (or *vice versa*) is regulated by some fundamental tendency in the plant itself, rather than by external forces; but it is a matter hardly within the range of proof, and Sinnott and Bailey's hypothesis, and my opposition to it, are both equally speculative.

We must now turn to the other side of the shield. Something has already been said on the incompletely convincing nature of the evidence for the arboreal ancestry of the herb, but it remains to be seen how far the probabilities favour the opposite view. To begin with, the first question that naturally enters one's mind is whether any individual instances can be cited in which there is a strong presumption that an herbaceous stock has given rise to tree forms. This question may, I think, safely be answered in the affirmative. Among the Cryptogams, the Tree-ferns afford a striking example. Bower[1] is of opinion that the Cyatheaceae and the Dicksonieae represent two distinct phyletic lines, in both of which dendroid plants have sprung from an ancestor whose habit was creeping. Turning to the Monocotyledons, we find that the Liliiflorae as a whole are characteristically herbaceous, but in the individual families, Liliaceae, Amaryllidaceae, Iridaceae and Dioscoreaceae, certain tree forms occur, such as *Dracaena*, the Dragon-tree.[2] There appears to be a general agreement among botanists that these arboreal forms, with their

[1] Bower, F. O. (1918). [2] Arber, A. (1925), pp. 41–3, 226.

unusual type of secondary thickening, have been derived from an herbaceous stock. Among the Dicotyledons, also, where woody forms occur in families which are prevailingly herbaceous, there seems at least to be a strong probability that these forms have had an herbaceous ancestry. It must be admitted, however, that such theories as these, though suggestive, are open to the same criticisms as the putative derivation of 'herbs' from trees, which we have already considered.

In discussions on the tree and herb habits in Angiosperms, a mode of statement is often used, which seems to me to be the root of much morphological evil: I mean that in comparing various groups of flowering plants, botanists are apt to speak of some as *more ancient* than others. Every one of the various lineages must have lived through all the time that has elapsed since the original stock came into being, and it is hence a misuse of language to talk of one lineage being 'older' than another. This usage is sometimes defended on the ground that 'old' or 'ancient' in this sense means 'having primitive characteristics', but I do not think that the word 'old' should be treated as a synonym for 'old-fashioned'. Now there is a time-relation, to which attention has been drawn in Sinnott's paper on *Comparative Rapidity of Evolution in Various Plant Types*, which is of the first importance when considered in connection with the equality of evolutionary age of all existing stocks of flowering plants. This point is the *shortness of generations* in herbs as compared with trees. Sinnott estimates that in herbaceous species we may reckon fifty to a hundred generations to a century, and in most trees, only four or five. It is possible that he exaggerates this difference in general, but in some Gramineae it is undoubtedly even more striking than his figures suggest. In the next chapter we shall consider certain arboreal bamboos, which seed, for the first and only time, at the age of thirty years or more. There is an enormous difference between such life-histories and those of small herbaceous grasses, which run through their whole course, from germination to seeding, in a few weeks. The dictum, that "the single step in evolution is not a year but a generation",[1] must be accepted, if we

[1] Bidder, G. P. (1927).

believe that evolutionary development depends on the changes in the germ cells which come about in connection with sexual reproduction. It is thus obvious, *a priori*, that, in a given period of years, the herb will have more numerous chances than the tree of undergoing the changes in its genic outfit which condition variations.[1] This theoretical conclusion is confirmed by the statistical facts concerning the number of species in genera and families of trees and herbs; for it is clear that the existence of a relatively high number of species in a genus or family means that it has been the scene of correspondingly active evolutionary development. Sinnott shows that in the Dicotyledons the average number of species in the woody genera is 12·5, while in the herbaceous genera it is 15. When we reckon by families, the difference is more striking, for the woody families average 310, and the herbaceous, 510 species to a family. The existence of this *evolutionary lag*—if we may so name it—among trees as compared with herbs, is well demonstrated by Sinnott, and yet he hardly seems to appreciate its full significance. In a previous article (with Bailey) he deals at length with the distribution of trees among the families, and presents a case that seems to have convinced many botanists that herbs are derivative, but which is open to another explanation, in the light of the relative rapidity with which herbs can suffer evolutionary change. We may take as an example the Leguminosae (the family of the Pea, the Wistaria, etc.), on which Sinnott and Bailey lay special stress. They point out that, in the Leguminosae, two of the sub-families, Mimoseae and Caesalpineae, with their regular or nearly regular flowers, are without question more primitive in type than the Papilionatae, with their curious butterfly-shaped blossoms. Now of the 121 genera in the first two sub-families, 113 are wholly woody, and the other eight contain both woody and herbaceous forms; there are no entirely herbaceous genera. In the Papilionatae, on the other hand, 29 per cent. of the genera are herbaceous. It is evident, therefore, that the more primitive members of the Leguminosae are almost all woody, but

[1] This statement is, of course, only broadly true; the idea needs working out in detail, with special reference to those herbs which take a number of years to reach sexual maturity, and to other intermediate types.

that there is a much higher percentage of herbs in the part of the family with more specialised flowers. Sinnott and Bailey interpret these facts as showing that the family was originally arboreal, and that a subsequent reduction from the tree to the herbaceous type in certain genera has been accompanied by the development of specialised flowers. But, bearing in mind the evolutionary lag in the tree form, it is surely more logical to suppose that the Leguminosae were originally herbaceous, and that those lineages which adopted the tree habit, put a break, as it were, upon their own evolution, and

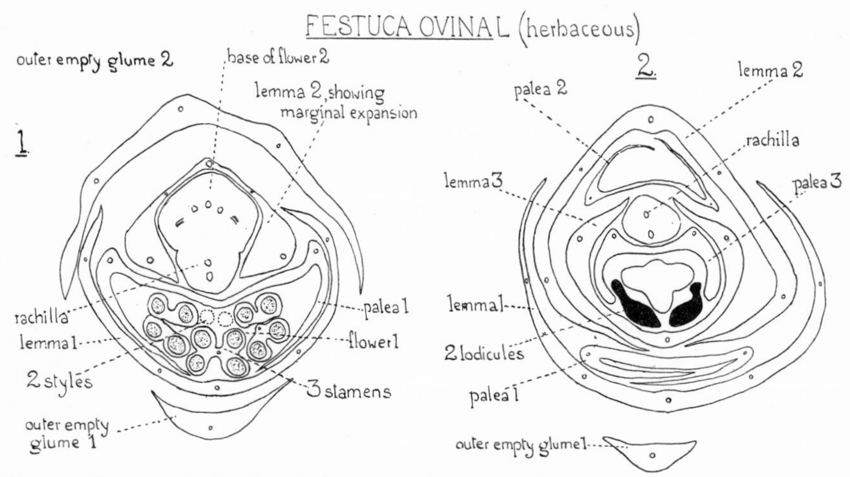

Fig. 30. *Festuca ovina* L., Sheep's-fescue. Sections (× 47) from a transverse series from below upwards through a spikelet to show reduced construction of flower characteristic of herbaceous grasses. Styles dotted in *1*, *flower 1*, as they do not reach to this level. Second outer empty glume not seen in *2*. Lodicules indicated in black. [A.A.]

thus dropped behind the lineages which, by remaining herbaceous, retained the power of mounting the ladder of floral specialisation at a relatively rapid rate. The Leguminosae are by no means the only family in which the same process can be followed. The example which chiefly concerns us here, but to which Sinnott and Bailey do not refer, is the Gramineae. The herbaceous grasses have, as a rule, extremely reduced flowers (Fig. 30), but the flowers of the dendroid Bambuseae approximate much more closely to the usual Monocotyledonous type, as we shall show in Chapter VII. They often have a complete apparatus of six stamens in two whorls, instead of the three in one whorl which characterise the other grasses. Moreover, in addition to the two lodicules of the herbaceous forms, there may be a third in the bamboos, so that one whorl of the perianth is complete (Fig. 31). Instances from other families might be cited, and indeed the array of evidence is altogether so striking that Sinnott and Bailey are probably right when they say (with certain reservations) that in practically every case in which there are herbs and woody plants in the same group, *the woody plants show a more primitive floral structure than the herbs.*

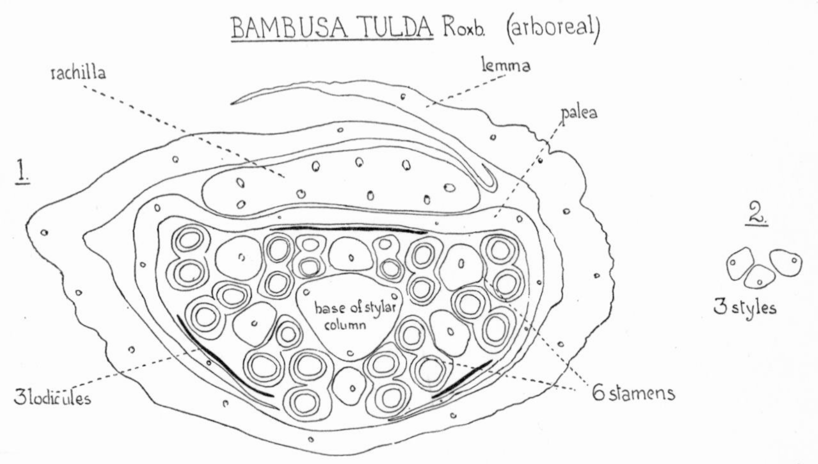

Fig. 31. *Bambusa Tulda* Roxb. from the Buitenzorg Garden. Transverse sections (× 47) from a series through a very young flower; anthers at pollen-mother-cell stage. In *1*, the gynaeceum is cut above the ovule; in *2*, the gynaeceum alone is drawn at a slightly higher level to show the three stigmas. Lodicules indicated in black. The sections show the typical construction of the bamboo flower. [A.A.]

For the importance of evolutionary lag in trees, as an inhibiting factor in phyletic development, we can find a zoological parallel. Andrews,[1] in his work on the Proboscidea, observed that, as evolution proceeds, the length of time taken by the individual to attain sexual maturity is liable to increase, and a "necessary consequence of the longer individual life will be that in a given period fewer generations will succeed one another, and the rate of evolution of the stock will be lowered in the same proportion". A remoter analogy may be found in the history of mechanical invention, for we may perhaps

[1] Andrews, C. W. (1903).

compare the difference in the rate of evolution of trees and herbs with a corresponding difference between the development of airships and aeroplanes. The relatively large number of 'generations' in the latter may well be one of the reasons why the evolution of the aeroplane has been so much more rapid than that of the airship.

Just as an animal analogy can be found for the evolutionary lag in trees, so help can also be had from zoology towards understanding the significance of the two tangible features which most markedly characterise the tree habit—*size* and *woodiness*. Let us first consider size. Zoologists have had more opportunity than botanists of following the history of definite lineages in the fossil records, and they have come to some very interesting conclusions about the general trends exhibited. One of these is the *Law of the increase of size in phyletic branches*. On this subject Smith Woodward[1] writes, "towards the end of their career through geological time, totally different races of animals repeatedly exhibit certain peculiar features, which can be described as infallible marks of old age. The growth to a relatively large size is one of these marks, as we observe in the giant Pterodactyls of the Cretaceous Period, the colossal Dinosaurs of the Upper Jurassic and Cretaceous, and the large Mammals of the Pleistocene and the present day". We might, then, expect by analogy that gigantism would prove to be a symptom of racial old age in plants as well as animals, and for this there certainly seems to be some evidence. It has been suggested, for instance, that gigantism in the Calamites of the post-Westphalian period was a premonition of extinction.[2] It has also been imagined that the large size of some Cycads, and of their sperms, may point in the same direction. In applying this conception to plants, may we not carry it into a wider field and suppose—as a working hypothesis—that the tree form in general is a development analogous to the colossal end-products of those animal and plant lines which achieve an excessive size in the period preceding their extinction.[3]

[1] Woodward, A. Smith (1909); on gigantism, see also Larger, R. (1917).

[2] Seward, A. C. (1933)

[3] There is evidence that large size is in itself a physiological drawback, and that the tree pays the penalty for its large-scale development in reduction of "incremental efficiency"; see Aldrich-Blake, R. N. (1927).

We now come to the second distinguishing feature of trees—*woodiness*—a feature even more involved in perplexity than the character of size. I think, however, that the germ of an answer to the question, "What is the significance of woodiness?" may be found in Church's memoir, *Thalassiophyta and the Subaerial Transmigration*. This author holds that the transference of plant life from the sea to a subaerial environment, with its intensive insolation, and its relatively inadequate supply of salts, leads to an accumulation of photosynthetic products; the plant under these changed conditions stores sugars, starches and celluloses, because it manufactures them in undue quantities and does not know how to dispose of them. On Church's view, "The problem of how to get rid of the excess and redress the balance is seen in...the timber tree". It is a familiar notion that the excessive cork production occurring, for instance, in *Quercus suber* L., may be regarded as a form of disease; but the idea that trees in general, with their large size and their masses of woody tissue, are organisms laden with waste products—that they are, one might almost say, examples of a chronically pathological state—is startling and, perhaps, repellent. I think, nevertheless, that it ought not to be dismissed without serious consideration. In judging of evolutionary phases, it requires an active effort to free ourselves from the hypnotic effect which mere size produces upon us. We are inclined, subconsciously, to despise the herbaceous plant, because it fails to reach to man's six feet, and to look upon the tree as the fullest expression of plant life, because it is so much taller than we are. This anthropocentric point of view emerges amusingly in Asa Gray's *Structural Botany*, where we read that one of the essential characters of a tree is that it attains "four or five times the human stature". One's admiration for the beauty and splendour of trees is so intense a feeling, that it seems an act of disloyalty to suggest that their grandest characteristics may be, in one sense, pathological; but we do not admire pearls the less because we have learned that they are the tragedy rather than the 'treasure' of the oyster. If then we can do so much violence to our instinctive feeling of respect as to face a consideration of the subject, we shall find in the animal kingdom certain analogical evidence bearing on this conception of the tree. The accumulation of non-living material in the body is recognised by palaeontologists as a frequent mark of old age in animal races. Smith Woodward[1] points to the tendency among all animals with skeletons to produce a superfluity of dead matter, which accumulates in the form of spines or bosses, when the race passes its prime and begins to be on the down-grade. He mentions, among other examples, the spiny Graptolites of the end of the Silurian. Moreover, W. D. Lang, in a remarkable paper on the evolution of a group of Cretaceous Polyzoa,[2] has instanced corresponding accumulations of calcium carbonate, which characterise the ends of many lineages and lead to their extinction. "It seems", he writes, "that, when once the habit of secreting Calcium Carbonate is established, it becomes increasingly constitutional, and the Polyzoa have discovered no means of checking this tendency, which finally overwhelms and obliterates the lineage."

If we are willing to regard the tree habit as an expression of racial senile degeneration, we may perhaps go a step further, and interpret it, not as an isolated pathological phenomenon, but as the final expression of a certain fundamental tendency in plant life. If the power of photosynthesis, with which the first green plants were endowed, carried with it, as the defect of its quality, the risk of the accumulation of excess polysaccharides, may not the earliest outward sign of this tendency perhaps have been the deposit of a surface layer of excreted cellulose? Such a cell-wall would automatically form a barrier against the accumulation of fluid in the intercellular spaces. It would thus prevent the existence of that controlled, internal medium bathing the tissues, which, in animal evolution, has made possible not only extracellular digestion[3] and excretion, but also the development of the blood[4]—processes without which the higher animals, as we now know them, could never have come into being. When the primaeval plant protoplast suffered self-imprisonment within a cellulose enclosure, it paid for the privilege of photosynthesis by the loss of these possibilities. It renounced, moreover, much of its sensitiveness and plasticity, and laid upon itself a weight, "Heavy as frost, and deep almost as life". Movement, responsiveness, and the free

[1] Woodward, A. Smith (1909). [2] Lang, W. D. (1920).
[3] Yonge, C. M. (1932). [4] Pantin, C. F. A. (1932).

expression of vitality, were left to the animal cell, which was less encumbered with the products of its own metabolism. The development of the cell-wall seems, indeed, to have been the most important individual factor which has played a part in controlling and inhibiting the evolution of the plant as compared with that of the animal. May we not then visualise the tree habit as the ultimate expression of the liability to the accumulation of inert organic matter—a tendency which can be kept within bounds in the youthful phases of a race, but which is apt to pass out of control when senescence is reached?

CHAPTER VI

BAMBOO: REPRODUCTIVE PHASE

THE flowering of bamboos, like their vegetative growth, is liable to be on a grand scale. Large quantities of pollen may be formed, which sometimes induce a kind of hay fever.[1] The scale of the seed production can be gauged from the statement that, from one clump of *Dendrocalamus strictus* Nees, the crown of which covered an area of about 40 square yards, grain was collected to the amount of 160 seers (330 lb.), besides a quantity naturally shed, which resulted in a dense mass of seedlings around the clump.[2] Another record describes the wild tribes of the Assa forest gathering the seed of the same species in March 1901; the outer culms of each clump were cut, one by one, at a height of about 4 ft., and each was laid on the ground which had previously been cleared and swept. The culm was beaten with stout sticks until all its caryopses had fallen; they were then carefully winnowed by children. One adult could collect 2–3 seers (4–6 lb.) of seed in a day.[3]

The periodicity of flowering in the bamboos seems to vary within wider limits than in any other homogeneous group of plants.[4] In South America, annually flowering species are common; they belong to *Arundinaria*, *Bambusa*, *Guadua*, and other genera.[5] In India, on the other hand, only a limited number of species flower every year, e.g. *Arundinaria Wightiana* Nees, *Bambusa lineata* Mun. and *Ochlandra stridula* Thwaites.[5,6] Many of the Asiatic bamboos have a more prolonged life-cycle, and show themselves, in Rivière's phrase, "assez avares de leurs fleurs".[7] The time interval, from one flowering to the next, shows a wide range between different species; a few examples may be cited here. The flowering phase is reached in three

[1] Nisbet, J. (1895).
[2] "D., J. C." (1883).
[3] Ryan, G. (1901).
[4] For a general account, see Gamble, J. S. (1904), and for details, see under the individual species in Gamble, J. S. (1896) and Blatter, E. and Parker, R. N. (1929).
[5] Brandis, D. (1899).
[6] Schroeter, C. (1885).
[7] Rivière, A. and C. (1878).

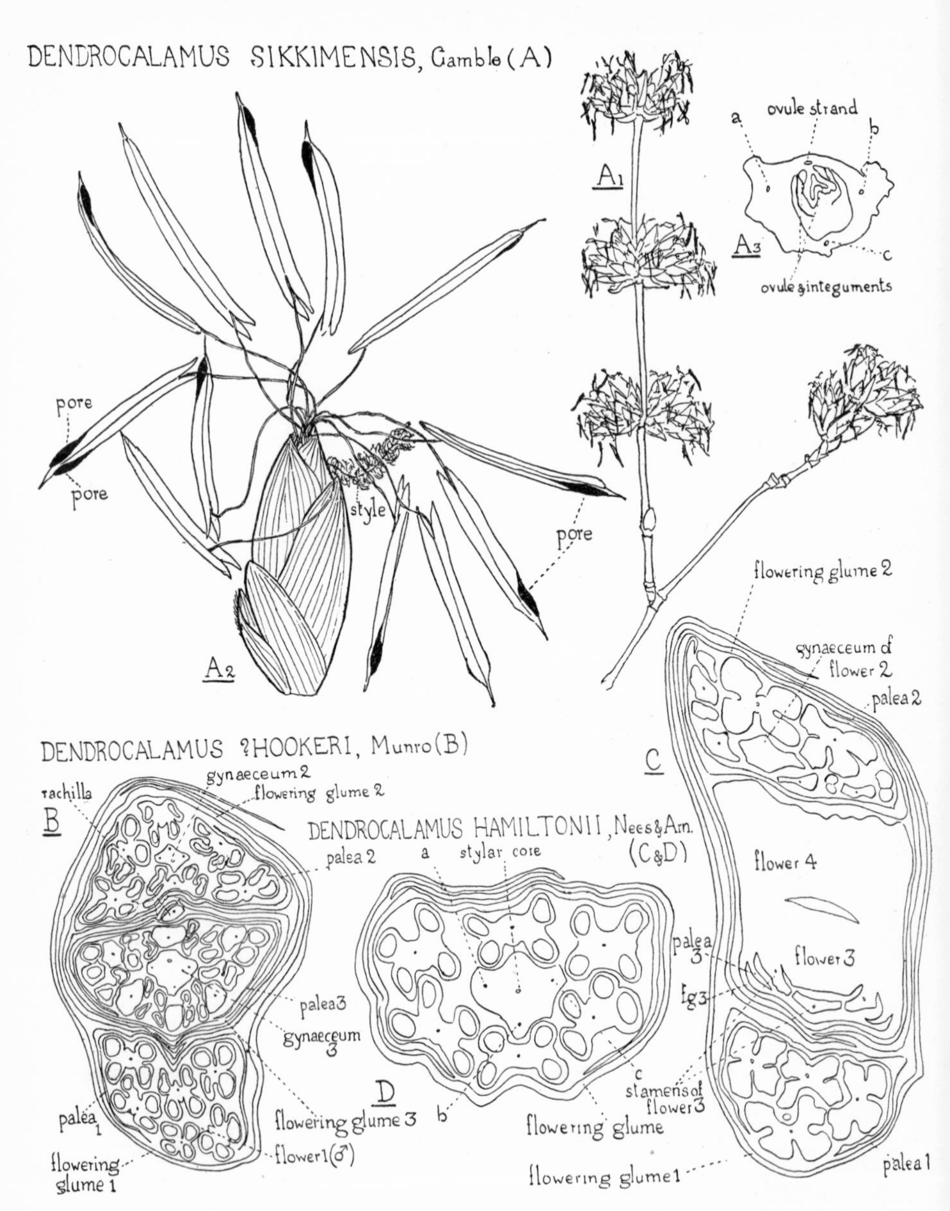

Fig. 32. The spikelet structure of *Dendrocalamus*. [For full legend, see Arber, A. (1926).]

years in *Schizostachyum elegantissimum* Kurz;[1] seven years in the Elephant-reed, *Ochlandra travancorica* Benth. et Hook.;[2] fifteen years in certain bamboos of Southern India;[3] sixteen or seventeen years in the Ringal-bamboo, *Thamnocalamus spathiflorus* Mun.;[4] twenty to twenty-five years in *Arundinaria Falconeri* Benth. et Hook.;[5] periods up to some thirty years in *Dendrocalamus strictus* Nees[6] and *D. Hamiltonii* Nees et Arn.;[7] thirty-one years in *Arundinaria racemosa* Mun.;[8] and thirty-two years in *Bambusa Tulda* Roxb.[9] Twenty-five to thirty-five years was regarded by Kurz[1] as the general age at which the commoner kinds of Malayan and Indian bamboo, belonging to *Bambusa* and *Dendrocalamus*, come into flower. In Africa, also, it is said by the Ashanti negroes that bamboos flower once in thirty years, and then die.[10] The tendency to a thirty-year period is met with again in South America, where it even finds a reflection in the animal world. It has been recorded[11] that plagues of rats occur in Brazil at intervals of about thirty years, and that these plagues synchronise with the dying of a bamboo, 'Taquara',[12] which abounds in the Brazilian forests. The popular explanation is that every cane of bamboo contains a grub—the germ of a rat—and that when the bamboo dies, the rat is, so to speak, hatched out. The more rational, though less picturesque, explanation relates the plague of rats to periodicity in the food supply offered by the bamboo. Though the Taquara generally comes to maturity in thirty years, the process of flowering is not actually confined to one season, but may spread over five years. Each cane bears about a peck of edible seed resembling rice. The total quantity produced is enormous, and large areas are often covered to a depth of 5 or 6 in. with the fallen grain. After seeding, the cane dies, breaks at the base, and falls to the ground.

[1] Kurz, S. (1876).
[2] "B., T. F." (1887), (1893).
[3] Buchanan, F. (afterwards Hamilton) (1807).
[4] "S." (1882).
[5] Gamble, J. S. (1900); see also Rogers, C. G. (1901).
[6] Smythies, A. (1901).
[7] Cavendish, F. H. (1905); see also p. 102.
[8] "H., H. H." (1892).
[9] Brandis, D. (1906[1]).
[10] Schroeter, C. (1885).
[11] Derby, O. A. (1879).
[12] Mrs Agnes Chase tells me that 'Taquara' is the Brazilian vernacular name for any large, strong bamboo, fit for use in constructing mud-huts, carts and fences.

The process of decay is hastened by the boring of larvae, which appear to be particularly abundant at seeding time. It is under the influence of this five-year period of exceptional food supply, due to the bamboo harvest, that the inordinate increase of the rats takes place.

When the intervals between the flowerings of the bamboo are thirty years or more, it becomes very difficult to arrive at precise dates. One writer tells us[1] that an Indian, whom he questioned about the flowering age of the Great-bamboo (*Bambusa arundinacea* Willd.), replied that between its successive seedings a child would grow to be a man, and his son, again, would reach manhood. Another native, speaking of an aged friend, declared him to be old beyond computation—a hundred years or more—but if the Sahib was not content with that, surely it was enough to say that his friend had twice in his lifetime eaten the seed of the Great-bamboo. The old man was able to date the first eating of the seed as being at the time that a certain Rajah lost his kingdom (1818), while the second was between 1865 and 1870, thus giving a period of about fifty years. Elsewhere, however, a period of thirty[2] or thirty-two years[3] has been claimed for *B. arundinacea*, so it seems as if separate study were required for each district.[4] *B. polymorpha* Mun., which extends over vast tracts in Burma, flowered gregariously in the eighteen-fifties.[5] In May 1933, the Chief Conservator of Forests, Burma, was so kind as to answer my enquiries about this bamboo; he wrote that sporadic flowering, and gregarious flowering over small areas, occur to some extent year by year, but that there has been no general flowering since 1852–8. The time-cycle of *Bambusa polymorpha* must thus exceed eighty years.

When the flowering periods of a bamboo are far apart, the onset of this phase may cause the utmost consternation. An example occurred in 1904 when, in a certain district in China, the local bamboo flowered. This was such an unprecedented event that it

[1] Nicholls, J. (1895).
[2] Footnote by "B., D." in "D., J. C." (1883); data collected by Gamble, J. S. (1896).
[3] Kurz, S. (1876) citing Beddome.
[4] Brandis, D. (1899).
[5] "Wathôn" (1903); Bradley, J. W. (1914); Seifriz, W. (1923).

aroused the same sort of alarm as a comet, and was held to portend failure of the crops and other disasters.[1]

There is no doubt that when a bamboo approaches the flowering phase, its whole constitution is profoundly modified. It is said, for instance, that in *Bambusa polymorpha* Mun., no new shoots are put forth by the clump in the year before flowering.[2] A more remarkable fact is that, in some bamboos, both old and young shoots of the same clump,[3] as well as offsets from the parent clump,[4] all break into flower simultaneously. When cut clumps of *B. arundinacea* Willd. (*B. spinosa* Roxb.) were buried lengthways along a garden boundary in order to make a fence, they gave rise to vigorous shoots in the second year, which burst into flower contemporaneously with the uncut parent clumps.[5] Moreover plants transported to different localities—even to different continents—have been observed to bloom at the same time as those still growing in their native habitats. Plants of *Chusquea abietifolia* Griseb., sent to Europe from the West Indies and cultivated in the Gardens at Kew, flowered at the same time as undisturbed wild plants in Jamaica.[6] *Arundinaria japonica* S. et Z. bloomed in the Bois de Boulogne, and in various gardens in Europe, in 1867 and 1868, contemporaneously with the flowering of the same species in the Jardin d'Essai du Hamma, Algiers. In Algiers all the stems—old and young—and the buds as they came out of the earth, flowered at once.[3] *Arundinaria falcata* Nees, again, flowered in 1875–6 in Brittany, Normandy, Angers, Nantes and Algiers, while in the Luxembourg Gardens even pot plants took part in this blooming. The Chinese bamboo, *Arundinaria Simoni* Riv., flowered simultaneously in 1903 in Algiers, France, and the Isle of Wight. Flowering was even observed in a plant only 70 cm. high, and this little plant bloomed persistently, through summer and winter, for four years.[7]

It is remarkable that when a clump of bamboo is in the repro-

[1] Tingle, A. (1904).
[2] Bradley, J. W. (1914).
[3] Rivière, A. and C. (1878).
[4] Brandis, D. (1899).
[5] "B., A. H." (1882).
[6] Information by letter from Sir Daniel Morris, March 17, 1924.
[7] Bureau, E. (1903).

ductive phase, ill-treatment is powerless to stop the flowering. For example, a single clump of *Bambusa arundinacea* Willd. was observed to flower and seed in the Cochin State Forests in March and April 1918, among others which remained vegetative. This clump was pulled down bodily by wild elephants, and the panicle eaten up. After the advent of May showers, small leafless shoots came up from the rhizome of this clump and all these shoots flowered and seeded by the end of June.[1] In Burma, dwarfed plants of *Cephalostachyum pergracile* Mun., which had been burnt down continuously by jungle fires and were only 6 in. to 1 ft. high, were observed to flower, together with their uninjured companions, which were 30–40 ft. in height.[2]

The gregarious flowering of bamboos becomes of practical consequence, because it generally involves the death of the parent plants. It is possible that the exhaustion of flowering does not always completely destroy the rhizomes, but the extent to which they survive is uncertain.[3] A vivid picture of the flowering of a bamboo forest in Brazil, and of its resulting annihilation, was given long ago by Auguste Saint-Hilaire.[4] "Il faut à ces herbes immenses", he wrote, "plusieurs années pour qu'elles puissent...parvenir à l'époque de leur floraison. Mais, quand elles ont porté des fruits, elles se dessèchent et meurent comme la Graminée la plus humble de nos climats si froids,...La première fois que j'entrai dans une forêt entièrement formée de l'espèce de Graminée appelée vulgairement *Toboca* [? *Guadua paniculata* Mun.], j'éprouvai un véritable ravissement en voyant ces tiges d'un aspect presque aérien, qui, hautes de quarante à cinquante pieds, se courbaient en arcades élégantes, se croisaient en tous sens, entremêlaient leurs immenses panicules et laissaient entrevoir l'azur foncé du ciel à travers un feuillage étalé comme un tapis à jour; alors la plante était en fleur; je repassai quelques mois plus tard, la forêt avait disparu: dans l'intervalle les fruits avaient succédé aux fleurs; ils avaient mis un terme à la végétation de la plante; ses tiges s'étaient desséchées, elles s'étaient brisées, et il n'en restait plus que les débris gisant sur le sol."

1 Menon, K. G. (1918[2]).
2 Kurz, S. (1876).
3 Rivière, A. and C. (1878).
4 Saint-Hilaire, A. de (1847).

Beside this picture from Brazil, we may set one of Brandis's descriptions of a similar occurrence in the East. He tells us that in April 1856 he found *Bambusa polymorpha* Mun. in full leaf forming a dense underwood in a certain district in Burma. In the hot season of 1859 it flowered, and in March 1861 the hills a little to the north were quite impassable, because the dead, dry stems of the bamboo had fallen, forming a tangled mass which had blocked up every pathway. The jungle fires of that year swept away the dead stems, and after the rains the ground was everywhere covered by millions of seedling bamboos, which soon grew up into slender plants, 2 or 3 ft. high, forming dense, waving green masses on the ground under the trees.[1] Another writer describes how, in January 1880, he came upon a considerable stretch of bamboo forest in the Chittagong Hill Tracts (Eastern Bengal), clothing the entire slope of a hill. Here the ground was so closely covered with the dead rotting stems of *Bambusa Brandisii* Mun. as to render walking difficult, while a full crop of seedling shoots, about 10 ft. high, had already taken the place of the old trees.[2]

The bamboo seedlings do not, however, always have it all their own way. It has been observed in Burma[3] that, after the simultaneous flowering and death of the parent stocks, the seedlings of other light-loving herbs and trees—formerly kept in abeyance by the dense shade of the bamboos—seize the opportunity to spring up in competition with the seedlings of the bamboos themselves. It then becomes a matter of uncertainty whether there will again be a pure bamboo jungle, or whether the other competitors will get the upper hand. How the bamboo seedlings may be assisted in their struggle for existence, is shown by Wallich, who reported that the celebrated bamboo grove surrounding the city of Rampur blossomed universally in 1824; after this, among its millions of stems not a single one was to be seen which was not dead and prostrate. "I observed", he adds, "with peculiar pleasure that the Nawab had adopted a very effectual and judicious plan of defending the tender age of the myriads of seedling bamboos which were seen growing on

[1] Brandis, D. (1899).
[2] "C., E. G." (1881).
[3] Kurz, S. (1877).

the site as thickly as you can conceive it possible, by not allowing a single one of the old and withered stems to be cut or in any way disturbed. I was told by some old inhabitants that...similar renewals have succeeded each other for ages past."[1]

Though the death of the bamboo after flowering may be a most serious matter from the practical standpoint, it is said that, at least in Burma, it is accepted fatalistically. *Bambusa polymorpha* Mun. has charming purple bracteoles and anthers, and the Burman has been described as gazing at the majestic flowering culms, and saying, "Te hla de" ("It is very beautiful"), with an aesthetic appreciation undimmed by the loss involved.[2] Such is the immemorial contrast between East and West: the saying, "If thou hast two loaves, sell one and buy some flowers of the Narcissus", could never have originated in this country. By the forester, indeed, whether Eastern or Western, the economic aspect cannot be ignored, and there has been much discussion about possible plans for preventing the gregarious flowering of bamboos. Divergent opinions are expressed as to its cause. Some think that it is brought about by external conditions, while others hold that it invariably supervenes at a certain phase of the plant's maturity. A review of the literature seems to show that there is truth in both views.[3] Undoubtedly certain bamboos have a definite periodicity, from which there is no known means of deflecting them. Others, however, adhere less rigidly to a predictable plan. *Dendrocalamus Hamiltonii* Nees et Arn. flowers sporadically almost every year,[4] but may also flower gregariously at thirty years.[5] *D. strictus* Nees, again, is another of those bamboos whose periodicity is not irrevocably fixed. It is liable to some sporadic flowering every year,[6] and it has even been recorded that cultivated seedlings have begun to flower at thirteen to fourteen months after sowing.[7] It seems, moreover, that, in a given area, there may be a gradual spreading of fertility in a certain geographical direction. For instance, in the Siwalik forests of Dehra Dun and Saharanpur, be-

[1] Cited by "B., A. H." (1882).
[2] Bradley, J. W. (1914).
[3] In addition to the review in the present chapter, see Arber, A. (1925), pp. 206–7.
[4] Rogers, C. G. (1900).
[5] Cavendish, F. H. (1905).
[6] Ellis, E. V. (1907).
[7] Brandis, D. (1899).

tween 1883 and 1886, the onset of flowering spread gradually, from the south-east, westward.[1] On the other hand, this bamboo has been described elsewhere as coming into flower in a typically gregarious fashion, practically all over an area of 1200 square miles—not only mature clumps, but slender seedlings of six or seven years' growth, blooming simultaneously. This flowering seems to have occurred at a period of about thirty years from the last.[2] But the onset of the reproductive phase may be much more erratic. In a report from another district it is said that, though this bamboo bore flowers fairly extensively in a particular year, yet one could travel through large areas without finding a single flowered stem. Here the flowering seemed to be affected by the situation, the bamboo being more disposed to bloom in exposed and hot places than in those that were more sheltered and cooler.[3] A somewhat different relation between habitat and flowering in *Dendrocalamus strictus* has been recorded by another writer.[4] He observed that in Orissa this bamboo, when growing in coarse-grained dry soils, generally flowered only sporadically in isolated clumps. On the other hand, in moister soils, provided they were not too wet to allow it to thrive tolerably well, flowering might sometimes take place over hundreds of acres. He found, moreover, that in, or immediately after, abnormally dry years, gregarious flowering might occur on all soils. Examples have also been recorded in which other bamboos have been stimulated to flowering by lack of water. It was found, for instance, that the digging of a drainage trench induced anthesis in *Arundinaria tecta* (Walt.) Muhl., the 'Cane' of the Southern and Eastern United States.[5] It has indeed been maintained by many writers that drought is at least one of the principal incitements to the blooming of bamboos.[6] It has sometimes been taken for granted that this is a providential arrangement by which bamboo grain is furnished to replace other cereal food in times of extremity. It was recorded, for instance, that in February 1812, "a failure occurred in the rice crops in the Province of Orissa.

[1] Broun, A. F. (1886).
[2] Smythies, A. (1901).
[3] Bruce, C. W. A. (1904).
[4] Nicholson, J. W. (1922).
[5] Brown, C. A. (1929).
[6] E.g., Kurz, S. (1876); "D., J. C." (1883); Rebsch, B. A. (1910).

Much distress was the consequence, a general famine was apprehended, and would, no doubt, have taken place, but for the merciful interposition of Providence in causing a general flowering of all the bamboos of the thorny kind [*B. arundinacea* Willd.], both old and young, throughout the country. The grain [not unlike wheat] obtained from these bamboos was most plentiful and gave sustenance to thousands....So great was the natural anxiety that was evinced to obtain the grain, that hundreds of people were on the watch day and night, and cloths were spread under every clump to secure the seeds as they fell from the branches".[1]

The assumption that the flowering of the bamboo is causally connected with drought, has not remained unchallenged. An American writer[2] has brought forward a number of facts which cannot easily be reconciled with it. He points out that, though the wholesale flowering of bamboos in India has sometimes synchronised with a water famine, this may well be a mere coincidence, not surprising in a country where droughts are so frequent. He considers that the evidence for a relation between drought and flowering requires more critical scrutiny than it has usually received. For instance, the monsoon rains in India failed in 1899, and as a result there was a disastrous drought in the early months of 1900. At about this time, large bamboo forests in Northern India burst into flower, but examination of the dates reveals that the onset of flowering, though sometimes following the drought, sometimes *preceded* it; thus the failure of the rains could not be held responsible for the flowering. The same writer himself observed a gregarious blooming of *Chusquea abietifolia* Griseb. in Jamaica in 1918. Here the flowering occurred after two unusually wet years; moreover, in a moist cool ravine, he found plants that had seeded, although those on a neighbouring dry spur showed no sign of flowering. He concludes that the most that can be said for the drought theory, is that, when bamboos are nearing their time of reproduction, an unusually dry season may have the effect of accelerating the formation of flower-buds.

In addition to climatic factors, mutilation seems to have a marked

[1] Blechynden, C., cited by "B., A. H." (1882).
[2] Seifriz, W. (1923).

influence upon the flowering of bamboos. Examples have been cited to show that injury, so far from inhibiting flowering, may even be claimed to have a stimulating effect. One observer describes[1] how, in clearing the ground to build a small bamboo hut on a hillock at Palaung (Burma), all the bamboos (*Dendrocalamus strictus* Nees) were cut in December quite, or almost, flush with the ground. In the first week of March, the writer stayed in the hut, arriving after dusk. Next morning, while his elephants were being loaded, he noticed on the ground, close to the hut, "several small heaps, as it were, of bamboo flowers about 7 inches diameter and 3 to 4 inches high. On examination they proved to be almost solid rounded masses of Myinwa flower [*D. strictus* Nees] in full bloom, apparently growing out of the solid earth, without leaves or stem; here and there...were further little buttons of flowers pushing through the ground, resembling more than anything which I can think of, the way in which a young mushroom pushes its way up. On further examination these small compact masses of flower were found to be growing from the bases of the culms just below the surface of the ground". This account of *D. strictus* Nees recalls another description of *D. giganteus* Mun., devastated by wind and axe. "The whole mass of stumps is literally smothered with the inflorescence issuing in masses from almost under the ground surface. Mixed with the surface dead roots of the stumps are masses of sessile spikelets forming a cushion around the node."[2] In *Bambusa Tulda* Roxb., a similar response to maltreatment has been observed. A clump in the Calcutta Botanic Garden, which had been partly blown down and half uprooted, produced flowers, while neighbouring clumps, which had not so suffered, were flowerless.[3] Again, in the American bamboo, *Arundinaria tecta* (Walt.) Muhl., when a brake was cleared in November, flowers appeared in March and April on the lower nodes of the cane stubble.[4]

This brief summary of our existing information about the flowering of bamboos—information which we owe in great part to contributors to *The Indian Forester*—shows that the problem is a

[1] Branthwaite, F. J. (1902).
[2] Macmillan, H. F. (1908).
[3] Gamble, J. S. (1890).
[4] Brown, C. A. (1929).

most intricate one; additional detailed field records, extending over prolonged periods, as well as experimental work, are needed before we can hope for anything approaching a final solution. Each species, and to a great extent each area, requires separate consideration. Meanwhile, certain provisional and tentative conclusions may be drawn. It is clear that many species only reach the phase of sexual reproduction after a prolonged 'juvenile' period, which may find expression in a definite term of years. The onset, however, of flowering—the outward symptom of the attainment of maturity—may, in certain species, be speeded up or delayed by something in the environment. Moreover, certain species, whose general flowering occurs at long intervals, may also show some sporadic flowering every year, which we can only attribute to internal or external conditions peculiar to the individuals involved. The interpretation of these obvious conclusions has been carried a step further by Brandis.[1] His discussion has the great merit of not treating the flowering of bamboos as an isolated phenomenon, but of linking it up with features in the life-history of plants belonging to other groups. He points out that various European forest trees (e.g. Beech, *Fagus sylvatica* L., and Spruce, *Picea excelsa* Link) do not flower and seed regularly every year, but at intervals which vary in each species with soil, elevation, climate, etc. In this connection he cites the work of Robert Hartig.[2] It was formerly believed that the reserve materials in the sap wood were in great measure used up every year in the new formation of leaves, branches and annual increment of the stem tissues, and that a fresh accumulation of reserves took place at the end of summer and in autumn. Hartig's work, however, did not confirm this view. He found that, in the Beech tree, only a minimal part of the reserve material in the more recent annual rings was withdrawn temporarily in connection with spring growth, and then collected again in autumn. On the other hand, when there was a year of abundant seeding, the starch largely disappeared from the sap wood. In all the Beeches which he examined after seeding, from one-third to one-half of the starch had vanished from the sap wood, while in many trees the depletion had gone still further, only traces of starch being left. The

[1] Brandis, D. (1899). [2] Hartig, R. (1888).

loss of nitrogenous material in the wood of the trees which had seeded was even more remarkable. He concluded that the periodicity in the recurrence of seeding years was related to the gradual accumulation of reserves in the wood; and that these reserves were, to a great extent, consumed at each periodic seeding. Brandis held that it was probable that a similar process took place in all trees which flower and seed at intervals of several years, and he suggested that, before periodically-flowering bamboos were in a position to form flower-buds in place of leaf-buds, a certain concentration of reserve materials must have been attained in the rhizome. Provided this precondition were fulfilled, the exact date of flowering might be determined by other agencies. It is to be hoped that some day Brandis's theory will receive a thorough testing with the aid of biochemical methods. If it could be developed on modern lines, it might give the answer to some, at least, of the riddles connected with the reproductive phase in the bamboo.

CHAPTER VII

BAMBOO: SPIKELET AND FRUIT

IN considering the arboreal habit, we noticed that the flowers of the Bambuseae were developed on a fuller plan than those of the other grasses, and approached more nearly to the complete Monocotyledonous type. This contrast will be realised on comparing Fig. 33, which represents the flower of a bamboo, with Fig. 34, p. 111, which shows the relatively reduced flower of Rye. It is thus obvious that the best way to understand the flower of the Gramineae is to start with the bamboo, but, unfortunately, the historical process has followed the reverse direction. Because analytical botany began in temperate regions, there has been a tendency to treat the exiguous European grass as the type form for the Gramineae. This has led to an exaggerated idea of the degree of peculiarity of the flower in this family, which has found expression in a special and elaborate terminology. Much of this terminology, and also of the controversy which has raged about the interpretation of the 'palea' and 'lodicules', might have been spared, if the study of the Gramineae had proceeded from the tropics to the temperate regions, and not *vice versa*. So, in order to follow the logical course, we will now examine the flowers of the bamboos,[1] leaving those of the 'grasses', in the limited sense, to be dealt with in later chapters.

In the bamboos which flower annually, the inflorescence terminates the leaf-bearing culm; but in those which are gregarious and periodic flowerers, the leaves fall, and the whole culm becomes one huge reproductive shoot.[2] A bamboo in this phase, leafless and entirely given over to flowering, may tower to 40 or 50 ft.—a truly gargantuan inflorescence.

In the inflorescences of the Gramineae, the leaves axillant to the branches are as a rule absent, though they may, in rare cases, be

[1] For a more detailed treatment of the bamboo flower, see Arber, A. (1926) and (1929[3]).

[2] Brandis, D. (1906[2]).

present but vestigial (e.g. *Luziola Spruceana* Benth., l_2 in Fig. 166, B 4, p. 317). On this point, however, as in the flower structure, a more completely developed type can be illustrated from the Bambuseae than from the other tribes. The South American bamboo, *Glaziophyton mirabile* Franch., may be unique among Gramineae in

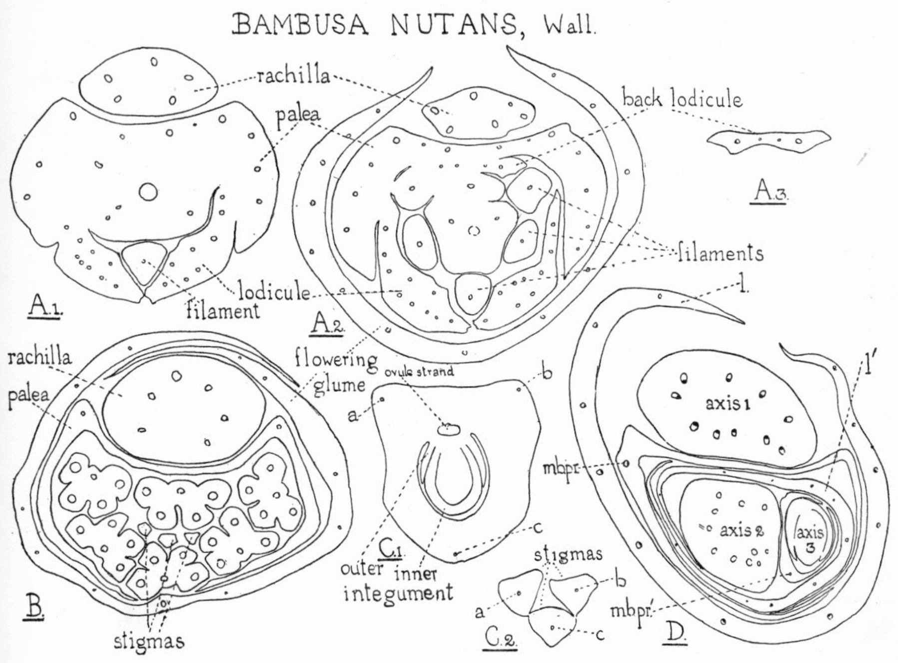

Fig. 33. *Bambusa nutans* Wall. A 1–A 3, transverse sections from series passing upwards through a flower (× 77). A 3, back lodicule after detachment. B, transverse section of another flower at a higher level (× 77). C 1 and C 2, transverse sections (× 77) through a very young gynaeceum. D, transverse section (× 47) of a small part of an inflorescence to show vegetative budding; *axis* 1, which is fertile at a higher level, bears at a lower level axis 2 and axis 3, which are sterile; *m.b.pr.*, median bundle of prophyll borne on *axis* 2, which arises in the axil of leaf *l* on *axis* 1; *m.b.pr'.*, median bundle of prophyll borne on *axis* 3, which arises in the axil of leaf *l'* on *axis* 2. [Arber, A. (1926).]

that each branch of the inflorescence arises in the axil of a bract, and itself bears a bikeeled prophyll.[1]

The partial inflorescences of the bamboos, like their vegetative branches, are often aggregated into complex tufts at the nodes of the reproductive shoots. Fig. 32, A 1, p. 96, shows a twig of an

[1] Franchet, A. (1889).

inflorescence of *Dendrocalamus sikkimensis* Gamb. about a month after full anthesis. The most conspicuous objects are the anthers, hanging at the ends of the slender filaments. The partial inflorescences—the units of which the flower clusters are built up—are called *spikelets*; one of them is shown in Fig. 32, A 2. The number of spikelets in a cluster may be very high; in *Bambusa macroculmis* Riv., which flowered in Algiers in 1872, the tufts at each node of the inflorescence were found to be composed of as many as 120 spikelets.[1] The special term 'rachilla'[2] has been coined for the axis of the individual spikelet in the Gramineae. The rachilla is often markedly flattened, and its anatomy tends to symmetry of a dorsiventral rather than radial character. This is shown in Fig. 35 for *Gigantochloa*, and in Fig. 31, p. 90, for *Bambusa Tulda* Roxb.; but the best example I have seen is *Arthrostylidium longiflorum* Mun., Fig. 169, B, p. 320. The rachilla is sometimes 'jointed' below each flower in a way that suggests a one-sided cupule. I have figured this structure under the name of 'rachilla flap' in Fig. 115, p. 241, which represents *Cephalostachyum virgatum* Kurz.

At its base the rachilla bears a variable number of sterile bracts or glumes. In *Phyllostachys aurea* Riv. and in *P. flexuosa* Riv., the outermost glume terminates in a little foliaceous limb;[3] I cannot recall any other case in the Gramineae in which the homology of glume and foliage leaf finds so clear an expression. The *outer empty glumes* (*sterile glumes*, *Hüllspelzen*) are succeeded by bracts, each of which has a flower in its axil. There is no reason why any name other than *bract* should ever have been used for them, but the idea that the grass flower is unique, and requires a special vocabulary for its description, has led to these fertile bracts being distinguished by a series of names, such as *gluma florifera*, *palea inferior*, *flowering glume* and *lemma*, of which the two latter are the more generally familiar in England and America, and will be used in this book. In France the term used is *glumelle inférieure* (or *glumelle extérieure*, *imparinervée*, or *carénée*) and in Germany, *Deckspelze*.

[1] Rivière, A. and C. (1878).
[2] Suggested in the form "rachillus" by Mortier, B. Du (1868).
[3] Daveau, J. (1922) and Camus, E. G. (1913).

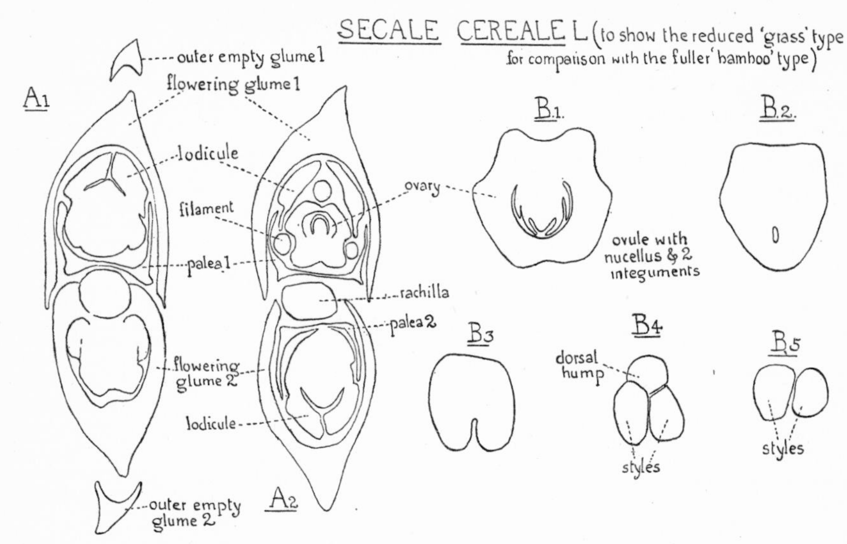

Fig. 34. *Secale cereale* L. A 1 and A 2, sections from a transverse series through a spikelet from below upwards (× 23); the lemmas are slightly reconstructed; the outer empty glumes in A 1 are added from another spikelet. B 1–B 5, sections of a gynaeceum from a series from below upwards (× 47). The inequality of the styles in B 5 is due to obliquity in the sections. [A.A.]

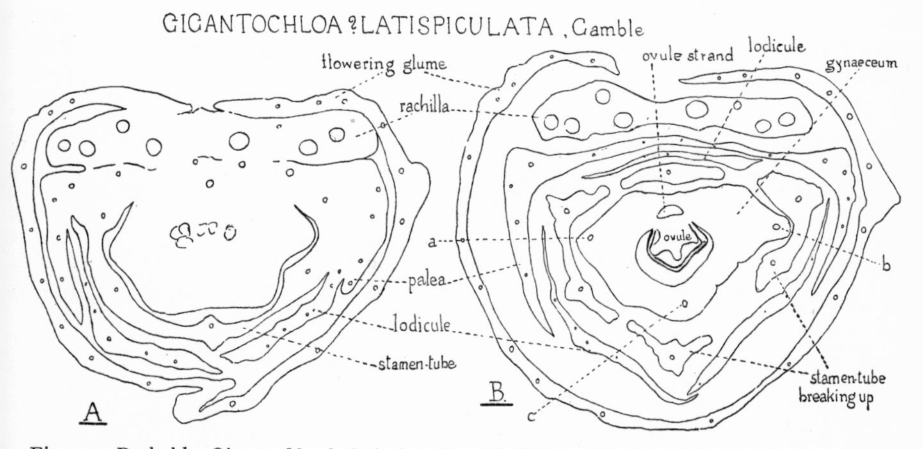

Fig. 35. Probably *Gigantochloa latispiculata* Gamble (Jelebu District, N.S., Federated Malay States). A and B, transverse sections at the base and higher up through one spikelet (× 43). [Arber, A. (1926).]

The base of the lemma is sometimes carried down the rachilla as a cushion or flap, which may be fimbriated, e.g. *Bambusa arundinacea* Willd., Fig. 106, E 1–E 4, p. 221. The lemma, as we have said, is the *bract borne on the spikelet axis*; the prophyll or bracteole, the *first leaf borne on the flower axis* (the axis lateral to the spikelet axis), faces the lemma. For this bracteole, as for the bract, a series of terms has been invented, of which *palea* is the most usual and will be adopted here; *upper palea* and *palet* are also employed by some writers. In

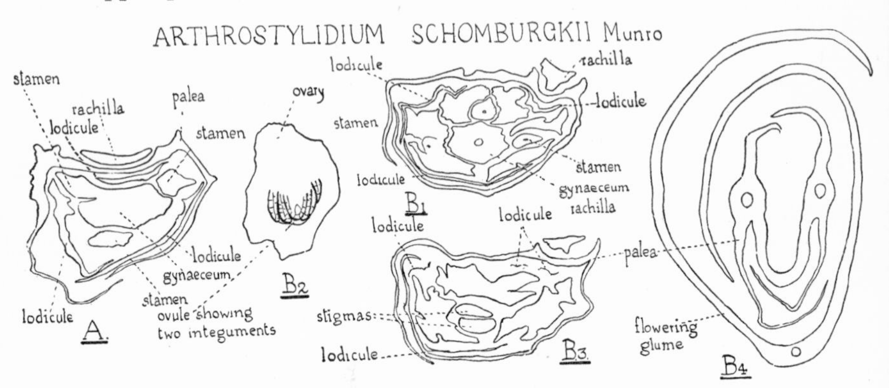

Fig. 36. *Arthrostylidium Schomburgkii* Munro (*Arundinaria Schomburgkii* Bennett), Cambridge Botany School Herbarium, collected by R. H. Schomburgk, British Guiana. A, transverse section low in flower, showing the third stamen just originating (× 47). B 1–B 4, transverse sections of one flower; B 1, cut through the level of the anthers, and of the gynaeceum above the ovule (× 47). B 2, ovary at a level slightly lower than B 1, showing the ovule with two integuments (× 77); B 3, cut at a higher level than B 1, to show the two stigmas (× 47); B 4, the flowering glume and palea at a level above the flower (× 47). [Arber, A. (1926).]

France it is called *glumelle supérieure* (or *glumelle bicarénée*, or *parinervée*) and in Germany, *Vorspelze*. The palea, like other Monocotyledonous bracteoles, is bikeeled. An example with two very marked keels is shown in Fig. 36 (*Arthrostylidium Schomburgkii* Mun.). The keels may continue above into a pair of awn-like apices (Fig. 37, D and E). The general relation of flowering glumes and paleas in a spikelet of *Dendrocalamus flagellifer* Mun. can be seen in Fig. 38, p. 114, in which the lemmas are dotted and the paleas indicated in solid black. This diagram also illustrates the suggestion that the bikeeled form of palea is due to space conditions in the bud. The flowers are so closely packed that the palea is, as it were, moulded between the flowering glumes, on one side, and the stamens on the other. The keels are formed in the only regions in which there is space for any protrusion of tissue, that is to say, in the angle between the corresponding lemma and the margins of that of the succeeding flower. A similar close packing of the flowers in other species of *Dendrocalamus* is shown in Fig. 32, p. 96. In certain bamboos, in which the

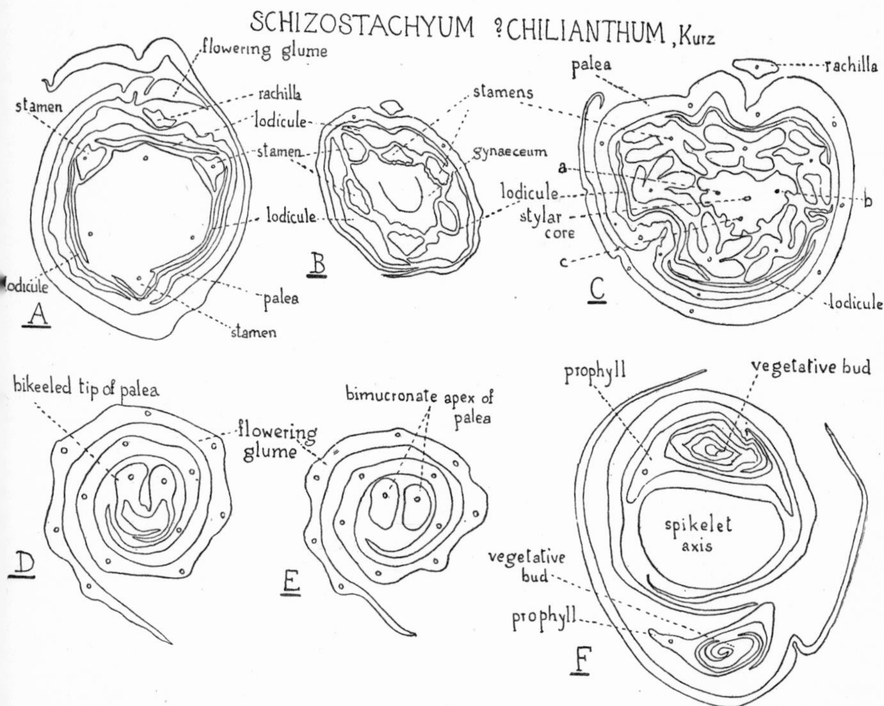

Fig. 37. Probably *Schizostachyum chilianthum* Kurz (Jelebu District, N.S., Federated Malay States). All transverse sections, × 47. A, base of a flower; vascular tissue shown only in stamens. B, near base of a flower (same spikelet as E) to show three lodicules and bases of stamens; flowering glume omitted. C, a flower at level of anthers; *a*, *b*, *c*, carpel midribs; flowering glume omitted. D, near tip of spikelet passing through bikeeled palea. E, bimucronate termination of the palea whose lower region is shown in B. F, two vegetative buds occurring at the base of a fertile spikelet. [Arber, A. (1926).]

rachilla is relatively more important, it is this on which the palea is moulded, so that the keels are formed in the angles between the flowering glume and the rachilla: e.g. *Phyllostachys aurea* Riv., Fig. 169, A 2 and A 3, p. 320; and *Arthrostylidium multispicatum* Pilg., Fig. 169, C 1 and C 2. Spikelets in which there is a terminal flower are interesting in this connection. For instance in *Oxyten-*

anthera albociliata Mun. (Fig. 40, B, C, p. 116) the lower of the two flowers drawn has a bikeeled palea with the keels far apart. This wide separation of the keels is due to the fact that the flower does not possess the usual slender rachilla, but is cut off from a relatively solid mass, the remainder of which will at once differentiate into a second

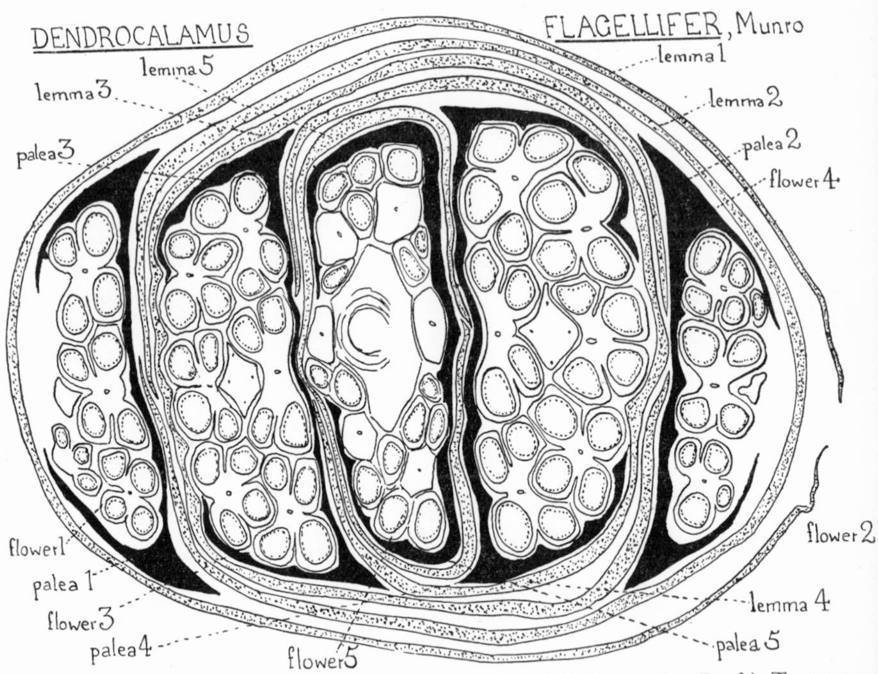

Fig. 38. *Dendrocalamus flagellifer* Munro (collected by F. W. Foxworthy, Perak). Transverse section (× 62) from a series through a young spikelet, showing five flowers; two outermost glumes somewhat broken and reconstructed from other sections lower in the spikelet. Several of the stamens in the lowest flowers (1 and 2) do not reach to the level of the section. Flower 5, the highest, is cut through the ovule, and the other flowers through the styles. Lemmas dotted and paleas indicated in solid black. [Arber, A. (1929[3]).]

lemma with its axillary flower. There is no trace of axis above the detachment of the flowering glume of the second flower, and its palea is curved and keelless in outline. The absence of a keel in this terminal flower, which is free from the pressure either of an axis or of another flower, indirectly confirms the idea that the bikeeled form of the more normal type of palea is due to pressure.

The theory that the form of the palea is determined by the sur-

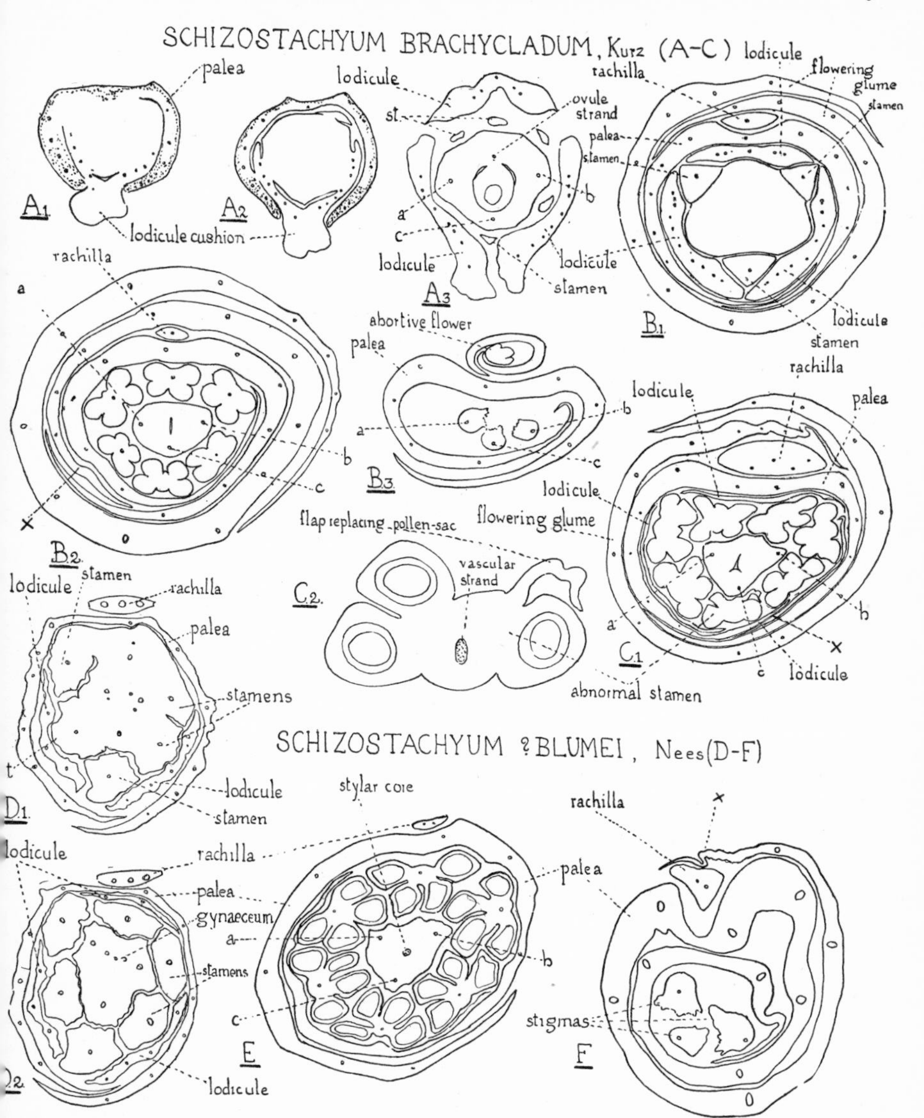

Fig. 39. *Schizostachyum.* A–C, *S. brachycladum* Kurz (in cultivation at Singapore). A 1–A 3, sections from a transverse series through a flower whose stamens were exserted (× 23); *a, b, c,* midribs of carpels. B 1–B 3, sections from a transverse series through a younger flower than A (× 47). C 1 and C 2, transverse sections of a flower from the same spikelet as B (× 47) passing through the gynaeceum just above the ovule. The androeceum is abnormal. C 2, anterior stamen in C 1 (× 193). D–F, *S.? Blumei* Nees. D 1 and D 2, transverse sections from a series through a flower (× 47). E, transverse section of another flower (× 77); F, transverse section of another flower passing through the stigmas (× 77). [Arber, A. (1926).]

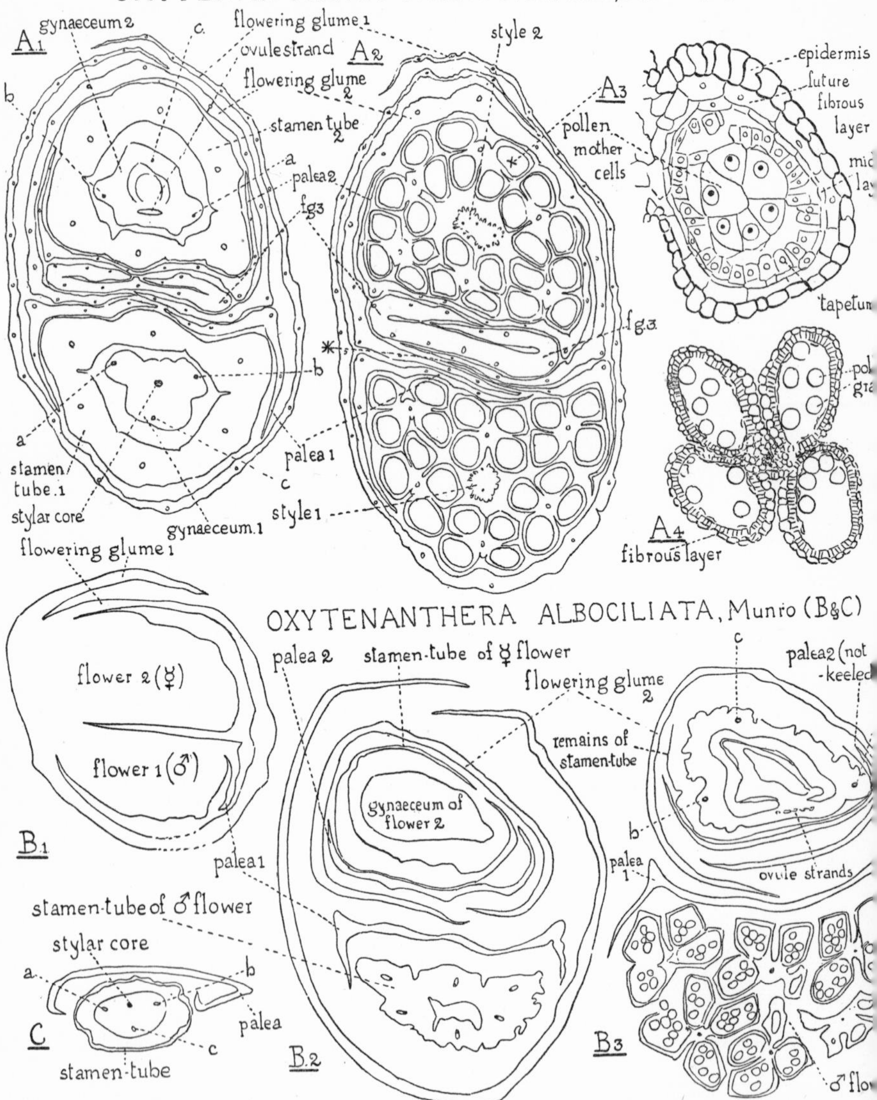

Fig. 40. *Oxytenanthera*. A. *O. nigrociliata* Mun. (Perak, Malay Peninsula). A 1 and A 2, transverse sections (× 47) through a spikelet cut at two levels, showing two functional flowers (1 and 2) and a third flower reduced to the lemma, *f.g.* 3; *a*, *b*, *c*, carpel midribs. A 3, one lobe of the anther marked with a cross in A 2 (× 318). A 4, transverse section of an anther from an older flower (× 77). B and C, *O. albociliata* Mun. (Burma). B 1–B 3, transverse sections (× 47) through a spikelet showing a male flower below and hermaphrodite above. C, transverse section of another hermaphrodite flower (× 47). [Arber, A. (1926).]

rounding members, may seem at first sight unconvincing, because the pressure[1] due to them must cease to operate at a very early stage in the development of the spikelet. Various instances are, however, known, in which a pressure temporarily exerted may influence form permanently; there is much truth in the old proverb, "As the twig's bent, so the tree's inclined". For instance, the cavity hollowed by an axillary bud in the adjacent face of the parent axis in a bamboo shoot, may persist and extend long after the pressure which initiated it has become a thing of the past.[2]

In the completest type of bamboo flower, the palea (bracteole) is immediately succeeded by the outermost members of the flower itself—the so-called *lodicules* (*glumellules*, *paléoles*, *Schüppchen*)—which are, however, absent in certain genera, e.g. *Dendrocalamus*, Fig. 38, p. 114, and *Oxytenanthera*, Fig. 40, p. 116. If the bamboo flower is to be described on the typical Monocotyledonous scheme, the lodicules may be held to be an inner whorl of perianth members, the outer whorl being absent. One member of the whorl is median and posterior, while two are lateral and anterior. The three lodicules are shown for *Bambusa Tulda* Roxb. in Fig. 31, p. 90, and *B. nutans* Wall. in Fig. 33, A 2, A 3, p. 109. The idea that the lodicules are of perianth nature is confirmed by the fact that members intermediate between stamens and lodicules may sometimes be found; the external appearance and sectional structure of these intermediate organs (stamen-lodicules) are sketched in Fig. 41, p. 118.

In *Bambusa arundinacea* Willd. the front lodicules have, in section, a 'fish-tail' shape (Fig. 106, F, p. 221), which recurs in certain grasses outside the bamboos (e.g. Fig. 66, B, p. 150); this shape seems to be due to their extension in the direction in which the opening of the palea gave them freedom. The posterior lodicule is in general the least developed; in *Chusquea* (Fig. 42, p. 119) it may be so small at the base as to look, in section, like a stamen filament. The majority of the bamboos conform to the usual Monocotyledonous scheme in having

[1] In Fig. 30, *1*, p. 89, an example is illustrated, from a grass, in which a lemma approaches a bikeeled palea in its form; here, also, space conditions seem to be responsible.

[2] For a further consideration of the part played by pressure in the flowers of the bamboos and grasses, see p. 176, et seq.

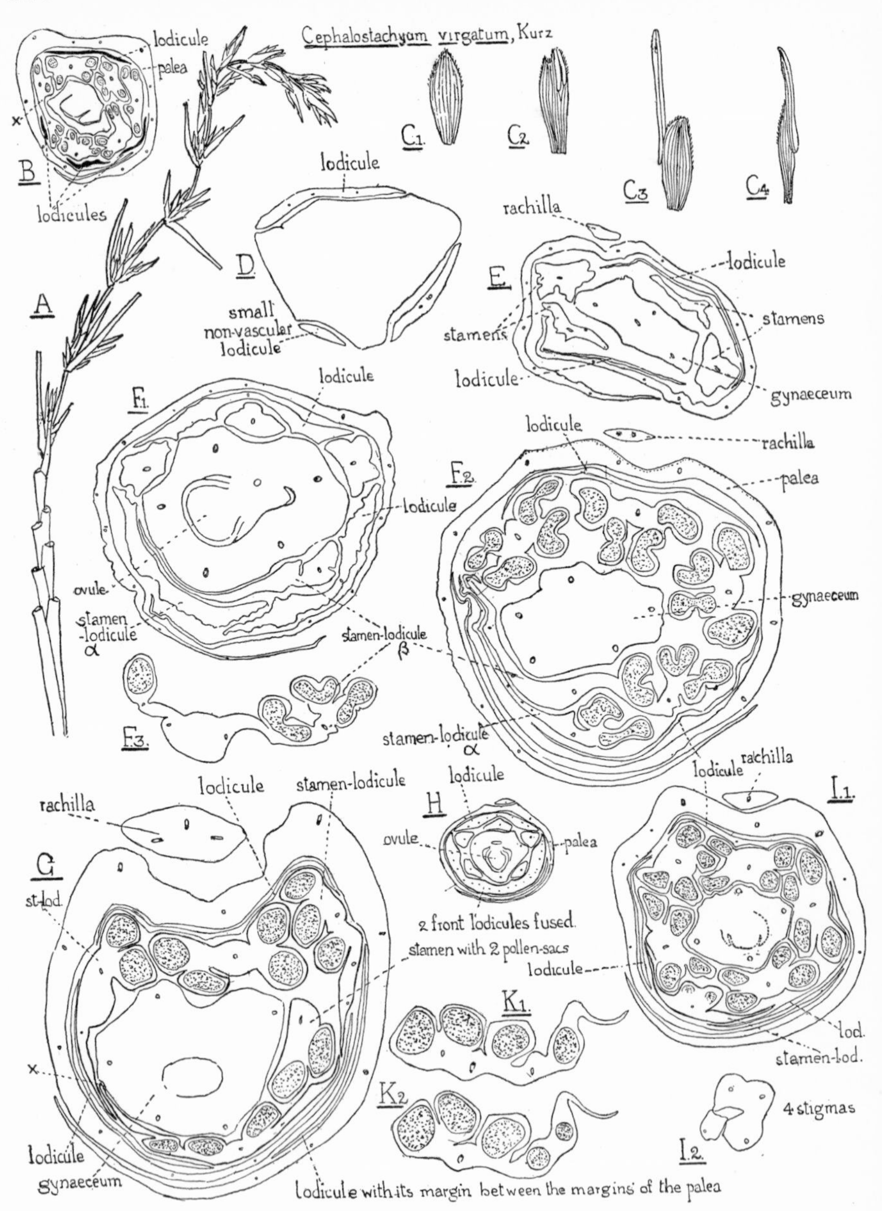

Fig. 41. *Cephalostachyum virgatum* Kurz. A–C and F–K, Calcutta Bot. Gard. A, part of a flowering shoot (× ½). B, transverse section of a flower (× 47) to show four lodicules (black) and a stamen (×) with a two-bundled connective. C 1–C 4, normal lodicule and stamen-lodicules (enlarged). D and E, Bhamo, Burma; transverse sections of flowers (× 47). F–K, sections of flowers with stamen-lodicules. [For fuller legend, see Arber, A. (1927[1]).]

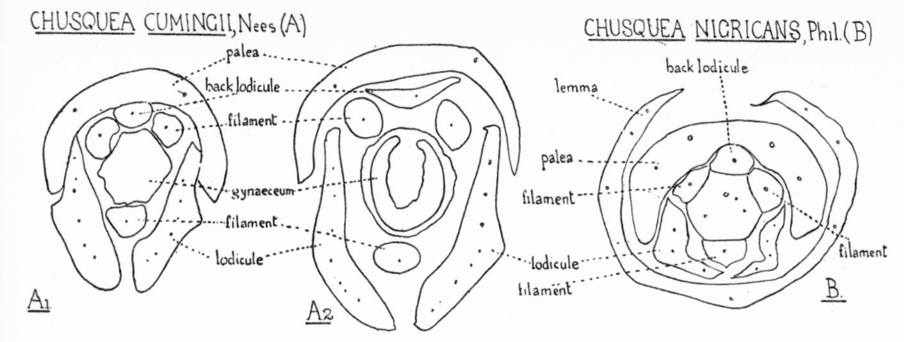

Fig. 42. *Chusquea.* Diagrams of sections from transverse series from below upwards through spikelets (× 47). A 1 and A 2, *C. Cumingii* Nees, from near Valparaiso, Chili; A 2, from more than one section. B, *Chusquea nigricans* Phil., from Valdivia. [Arber, A. (1929[3]).]

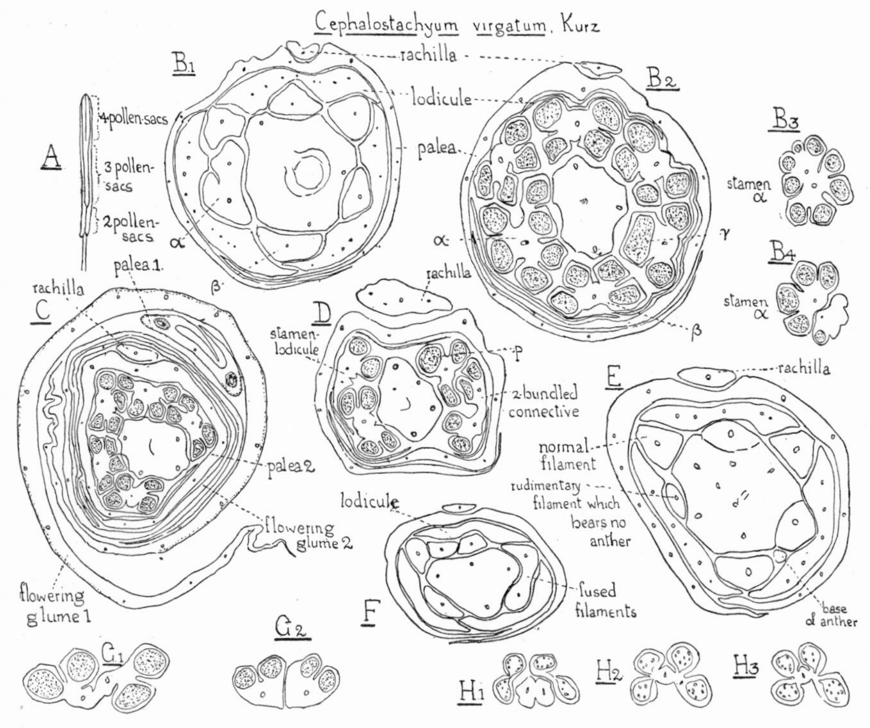

Fig. 43. *Cephalostachyum virgatum* Kurz, Royal Botanic Garden, Calcutta. A, stamen from a dissected flower showing pollen-sacs varying in number at different levels. B 1–B 4, series of transverse sections through a flower showing stamen-abnormalities (× 47). C, transverse section of a spikelet (× 47). D, transverse section (× 47) of a flower with one stamen-lodicule and one stamen with a two-bundled connective. E, transverse section (× 77) at level of filaments of a flower in which one stamen is small and develops no anther. F, transverse section of a flower at the level of the filaments to show irregularly placed stamens, two of which are united by their filaments (× 47). G 1 and G 2, transverse sections lower (G 1) and higher (G 2) through a back outer stamen with a two-bundled connective whose anther bifurcates (× 47). H 1–H 3, transverse sections (× 23) of the anther with two-bundled connective shown in Fig. 202, B, p. 396, which is at a level between H 1 and H 2. [Arber, A. (1927[1]).]

six stamens. Some, however, recall the more familiar grass type in having three stamens only (e.g. *Chusquea*, Fig. 42, p. 119; *Phyllostachys* and other genera illustrated in Fig. 169, p. 320). It is interesting that these three stamens belong to the *outer* whorl; for, as an abnormality, we sometimes find six-stamened bamboos showing a tendency to lose the members of the *inner* whorl—thus suggesting a transition to the three-stamened type. One of these reduction stages is shown in Fig. 43, E, p. 119, in which a lateral stamen of the inner whorl is represented by an abortive filament only. A more striking example is seen in Fig. 44, B 1 and B 2; here the three outer stamens are normal, but the inner whorl consists of two stamens, each with two pollen-sacs only, and a third, in which there is no anther. A peculiarity, which occurs in some species with six stamens, is the union of their filaments into a tube of variable length (e.g. *Gigantochloa*, Fig. 45, p. 122, and *Oxytenanthera*, Fig. 40, p. 116).

The gynaeceum,[1] though best interpreted as formed of three carpels, has an ovary with only a single cavity, and with one ovule attached to its back wall. Besides the strands supplying the ovule, there are usually three more bundles, two lateral and one anterior, marked *a*, *b*, *c*, in Fig. 45, p. 122. These three strands, which alternate with the inner stamens, represent the midribs of the three carpels; in the typical bamboo they pass into three styles (Fig. 31, p. 90). The marginal carpellary bundles are not, as a rule, differentiated except at the back, in the neighbourhood of the ovule, but in *Gigantochloa* the strands at the junction of the front carpel with the two others may also develop (*x* and *y* in Fig. 45, B 1, etc.). The vascular system then becomes typically that of a tricarpellary ovary, in which the carpels are united edge to edge.

A feature often noticeable in the stylar column of bamboos, is the presence of a central small-celled strand, which is continuous with the extreme apex of the ovary cavity. There is little doubt that it is the conducting tissue for the pollen-tubes, but, as I have never actually detected their presence in it, I have labelled it with the non-committal name of 'stylar core'. It may be seen in my sketches of

[1] I use the spelling "gynaeceum" instead of "gynoecium", as Kraus, G. (1908) has shown that it is the preferable form.

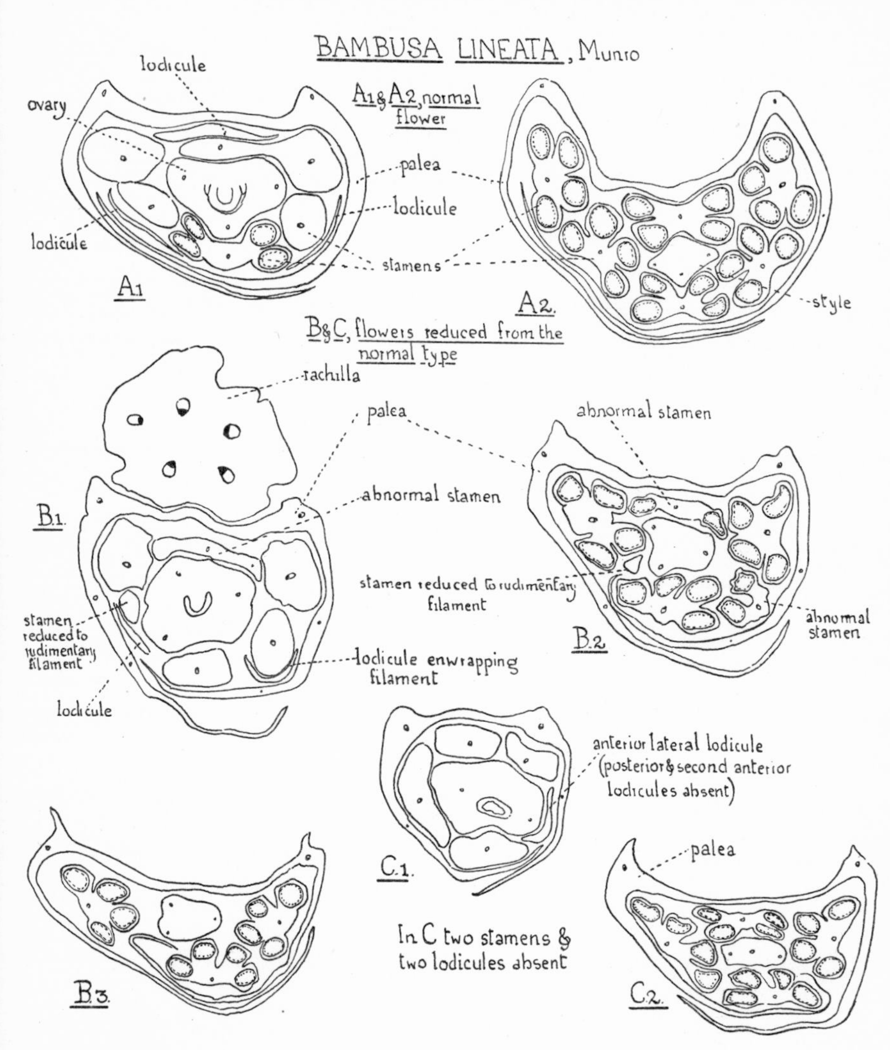

Fig. 44. *Bambusa lineata* Munro. All diagrams from transverse series passing upwards from below through spikelets (× 62). A 1 and A 2, normal form from Buitenzorg Garden; in A 1 the section passes so near the flower-base that two of the filaments are united. B and C, abnormal forms from the Calcutta Gardens; the drawings B 1–B 3 are each generalised from more than one section. [Arber, A. (1929[3]).]

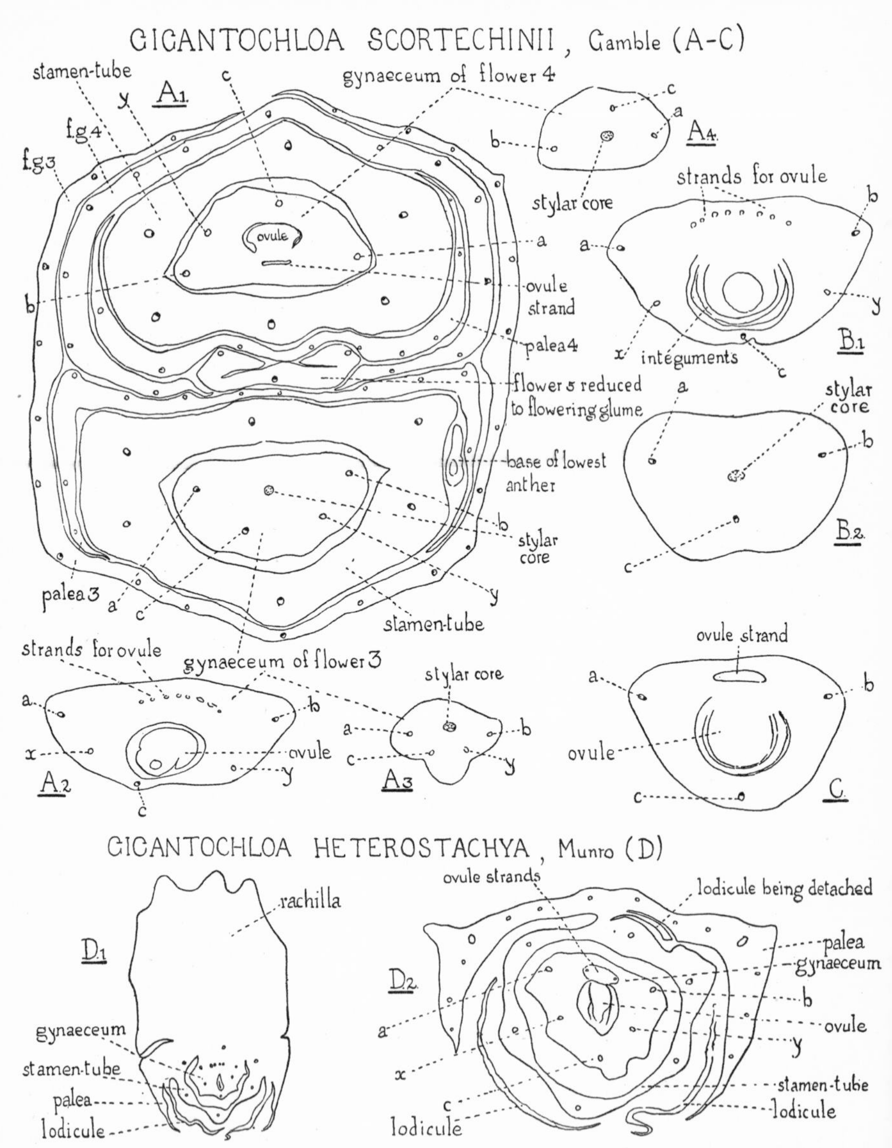

Fig. 45. *Gigantochloa*. Throughout the figure *a*, *b* and *c* are the carpel midribs, while *x*, *y*, and the ovule strand, are the fused marginal veins of the carpels. A–C, *G. Scortechinii* Gamble, transverse sections through spikelets (all × 47). A 1, third, fourth and fifth flowers of a spikelet; *f.g.*, lemma. B 1 and B 2, gynaeceum from another spikelet. D, *G. heterostachya* Mun. D 1, base of flower (× 23); D 2, higher in flower (× 47). [Arber, A. (1926).]

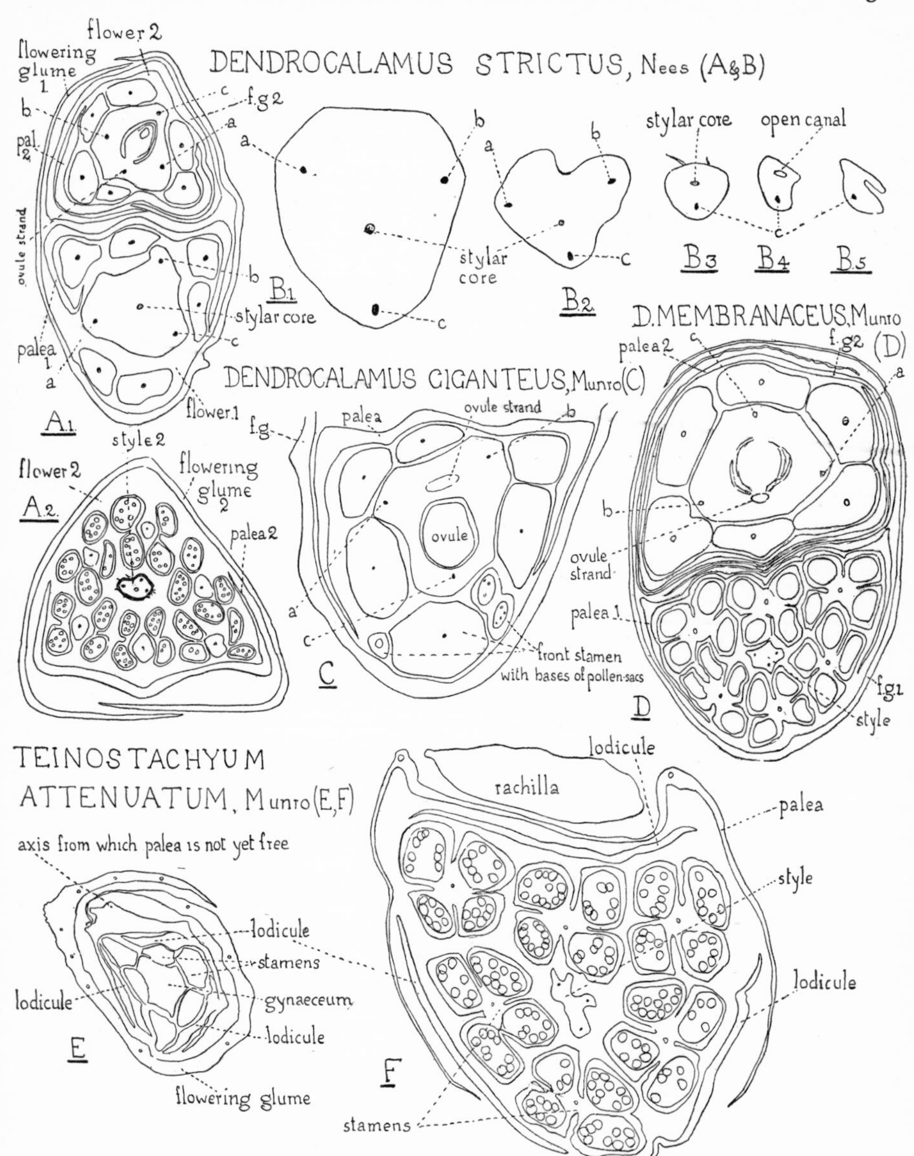

Fig. 46. A and B, *Dendrocalamus strictus* Nees. A 1 and A 2, transverse sections through a spikelet (from Dehra Dun) (× 23); bundles omitted, except in gynaeceum and stamens; *a*, *b*, *c*, carpel midribs. B 1–B 5, successive transverse sections of a style (× 77). C, *D. giganteus* Mun. Transverse section of a flower (× 47). D, *D. membranaceus* Mun. Transverse section o two flowers (×47). E and F, *Teinostachyum attenuatum* Mun. Transverse sections of two flowers; the distortion is due to imperfect recovery of the herbarium material used. [Arber, A. (1926).]

Dendrocalamus, Fig. 46, A 1, A 2, B 1–B 5, p. 123; *Gigantochloa*, Fig. 45, A 1, A 3, A 4, B 2, p. 122; *Oxytenanthera*, Fig. 40, A 1 and C, p. 116; *Schizostachyum*, Fig. 37, C, p. 113 and Fig. 39, E, p. 115.

In two of the subdivisions of bamboos—the Arundinariae and Eubambuseae—the fruit developed from the gynaeceum resembles a grain of Wheat or Barley;[1] the distinction between pericarp (fruit-wall) and seed-coat is lost, and the food material for the embryo is stored in the endosperm. But in the third subdivision, sometimes called Bacciferae, the ovary wall remains distinct from the seed-coat, and may become thickened, producing a comparatively large fruit, completely unlike that of the Gramineae in general. It is fortunate that we possess a description of the gynaeceum of a remarkable example of the tribe, *Melocanna bambusoides* Trin.,[2] a bamboo which normally flowers only once in a generation, so that its fruits are rarely obtainable. The cells of the pericarp have thickened membranes, and they are packed with food materials. The endosperm, on the other hand, is a mere relic, and the most conspicuous feature of the embryo is the huge scutellum (the sucking part of the seed-leaf) which fills up the cavity of the pericarp. The whole fruit may reach 5 in. in length and 3 in. in breadth. In one of the other Bacciferae, *Melocalamus compactiflorus* B. et H., the fruit is said to resemble an edible Chestnut, though it is less sweet.[3] *Ochlandra*[4] is another bamboo with a thick pericarp (Fig. 47, D 1–D 3). The one-flowered spikelets of this genus are unusual in the numerousness of their parts. The stamens range from six to the very high upper limit of 120, while there are a variable number of lodicules, and from three to six styles. The structure of several members of the genus is illustrated in Figs. 47 and 48. A notable feature is the multifasciculate character of the filament (Fig. 47, A 1, A 2, C 1, and Fig. 48, C 2). I know of no other genus among the grasses in which a filament with more than one bundle has been found. The gynaeceum is interesting in showing a tendency to loss on the posterior face, due presumably to pressure against the rachilla in early stages. The poor development

1 For a consideration of this type of fruit or 'seed', see p. 224.

2 Stapf, O. (1904[2]).

3 "B, F. G. R." (1902).

4 On *Ochlandra*, see Arber, A. (1929[3]).

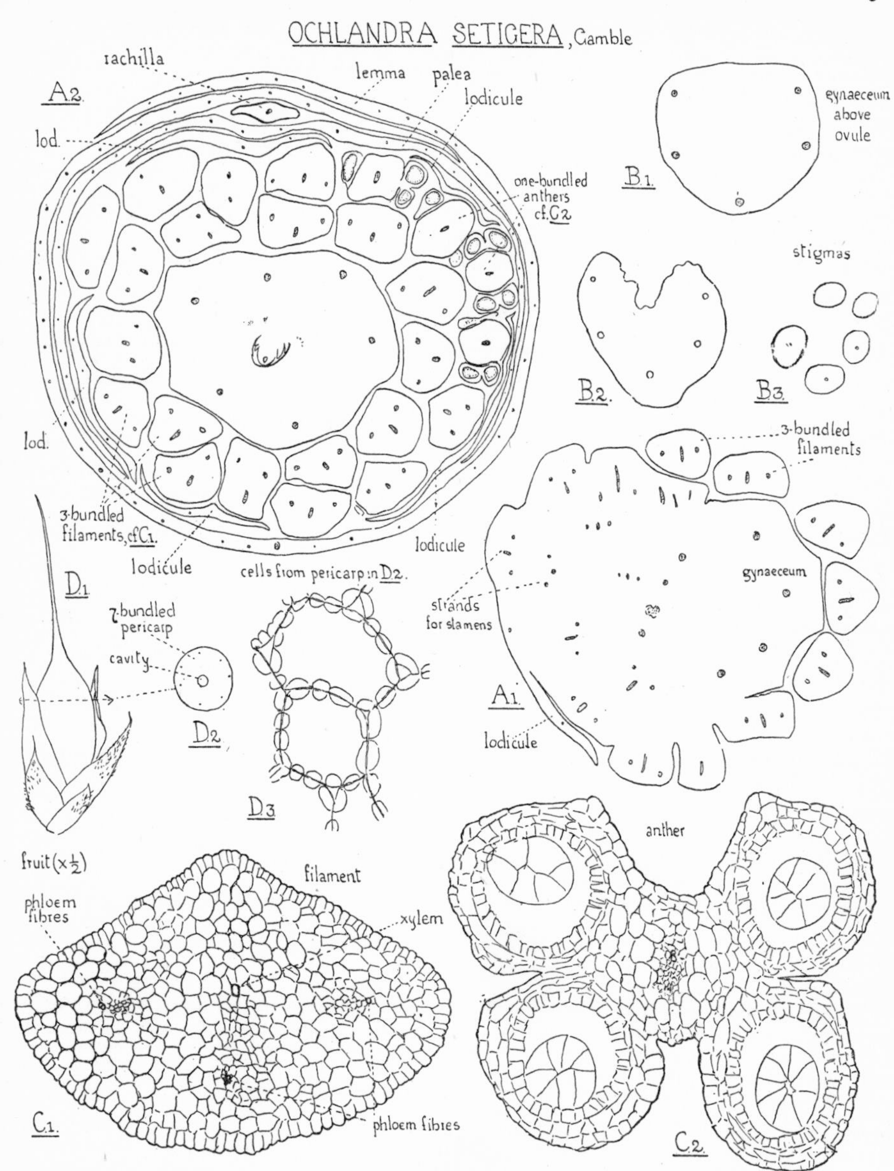

Fig. 47. *Ochlandra setigera* Gamble, Buitenzorg Garden. A 1 and A 2, slightly oblique transverse sections from a series from the base upwards through a young flower (× 47). A 1, base of ovary and associated structures only. B 1–B 3, passage from top of ovary, which was six-bundled below B 1, to stigmas (B 3), continuing the series in A, but from transverse sections of another flower, on a larger scale (× 77). C, transverse sections through filament (C 1) and anther (C 2) of a stamen (× 193). D 1, fruit enclosed in glumes (× ½). D 2, transverse section of fruit at level of arrow (× ½). D 3, ground-tissue cells from D 2 (× 193); the starch-grains which filled the cells are omitted. [Arber, A. (1929[3]).]

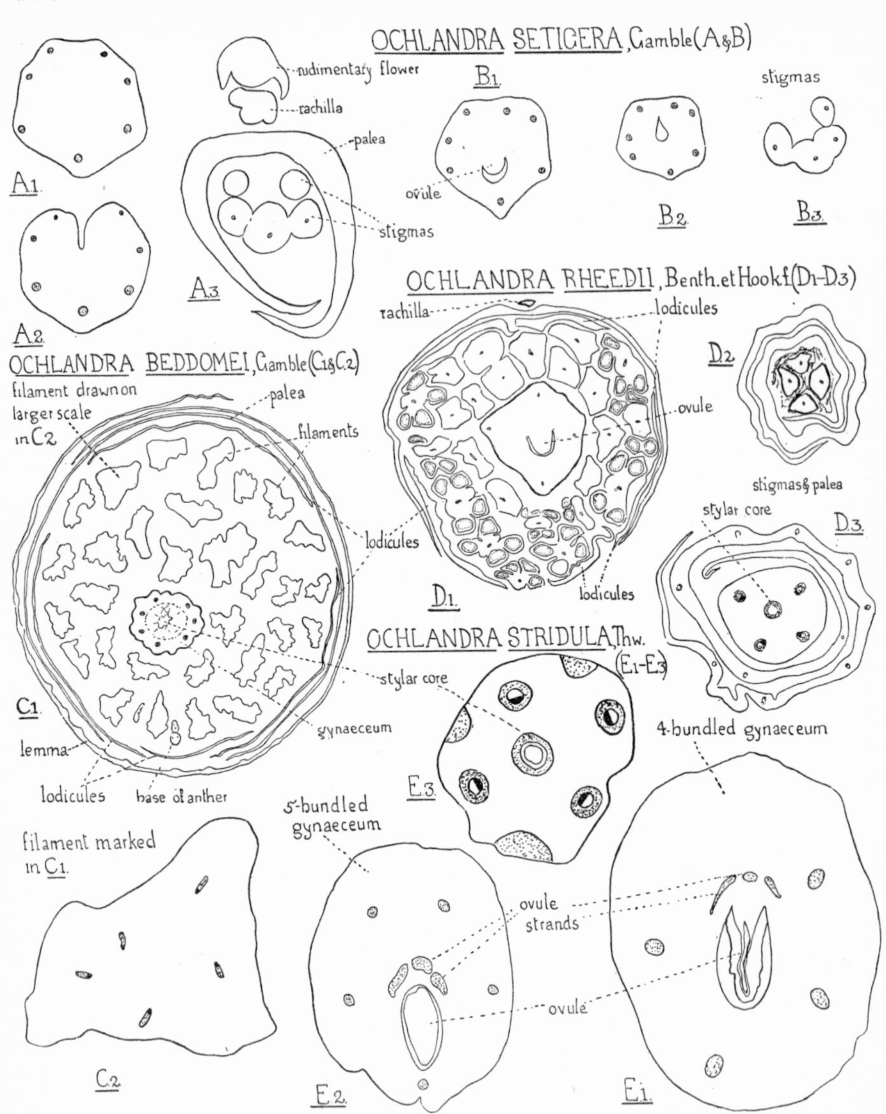

Fig. 48. *Ochlandra*. A and B, *O. setigera* Gamble. A 1–A 3, sections (× 77) from a transverse series from below upwards through the top of a seven-bundled gynaeceum to show origin of five stigmas. A 3 includes palea and rachilla, and shows a rudimentary flower above the normal flower. B 1–B 3, sections (× 77) from a transverse series from below upwards through a very young gynaeceum. C, *O. Beddomei* Gamble; C 1, transverse section (× 14) through a spikelet above level of top of rachilla. C 2, transverse section of the filament marked in C 1 (× 77). D, *O. Rheedii* Benth. et Hook. D 1 and D 2, transverse sections of a flower (× 47). D 3, upper part of gynaeceum of another flower (× 47). E, *O. stridula* Thw.; E 1 and E 2, transverse sections through two ovaries; E 1 (× 23); E 2 (× 47); E 3, transverse section (× 77) through upper part of a gynaeceum. [Arber, A. (1929[3]).]

of the stigmas and their bundles, at the back, in comparison with the anterior face, can be traced in Fig. 47, B 1–B 3, p. 125, and Fig. 48, A 1–A 3. *Pseudocoix*,[1] one of the remarkable bamboo genera described in recent years from Madagascar, is said to show features which recall *Ochlandra*.

We will now consider the peculiar Brazilian genus *Streptochaeta*,[2] though its position is uncertain, and it is possible that it is better placed near the tribe to which Rice belongs, than with the bamboos. This grass was regarded with much interest by botanists of the nineteenth century, since it was supposed to be an ancient and primitive type. I have been able to examine its two species, and their structure is illustrated in Figs. 49–51. The spikelet has an abbreviated axis, clothed with five sterile glumes, irregularly placed. These sterile glumes are followed by an awned flowering glume or lemma, and then by a palea, which is bifid almost to the base (Fig. 49, C 4, C 5, p. 128; Fig. 51, A 6 and A 7, p. 130). It is succeeded by three lodicules, which in *S. spicata* Schrad. show a convolute arrangement (Fig. 49, C 5); this, though unusual in the Gramineae, can be paralleled in the bamboo, *Guadua latifolia* Kunth. Except for the irregular placing of the sterile glumes, *Streptochaeta* shows no fundamental departure from other grasses with one-flowered spikelets, and I see no reason for treating it as a primitive form.

Though *Ochlandra* and *Streptochaeta* both diverge a good deal from the ordinary gramineous type, they are out-distanced in this respect by another plant which is so peculiar that one is tempted to doubt whether it is really a member of the family at all—*Anomochloa marantoidea* Brongn.[3] A gathering of this monotypic genus from Brazil was described by Brongniart[4] more than eighty years ago, but no one has ever succeeded in finding it since. Plants from the seeds of the original collection were grown for a time in the Jardin des Plantes, Paris, and I have had a dried spikelet from one of these specimens for examination. From this spikelet, I was able

[1] Camus, A. (1924); on other endemic Madagascan bamboos, see Camus, A. (1925) etc.

[2] On *Streptochaeta*, see Arber, A. (1929[1]).

[3] On *Anomochloa*, see Arber, A. (1929[1]).

[4] Brongniart, A. (1851).

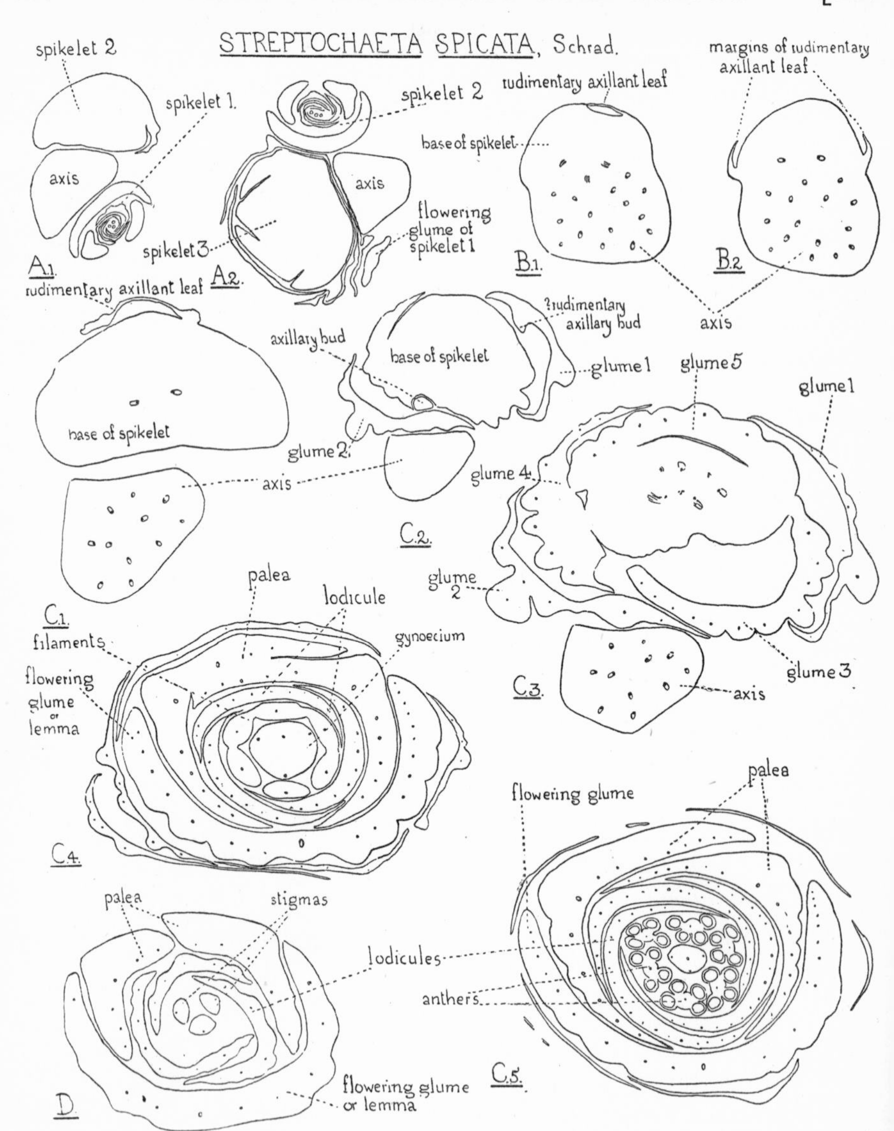

Fig. 49. *Streptochaeta spicata* Schrad. All figures from material collected January 1927 near Rio de Janeiro. A 1 and A 2, two sections (× 13·7) from a transverse series through part of an inflorescence to show succession of three flowers in spiral arrangement. B 1 and B 2, two sections (× 22) through the inflorescence axis just below detachment of a spikelet to show rudimentary axillant leaf. C 1–C 5, sections from a series through a spikelet, minor axis bundles omitted; C 1, C 3, C 4 (× 22), C 5 (× 22 *circa*), C 2 (× 13·7). D, transverse section through the apex of another spikelet to show stigmas (× 46). [Arber, A. (1929[1]).]

to cut more than a thousand serial sections; its structure is shown in Figs. 52, p. 131 and 53, p. 132. Two glumes, corresponding, ap-

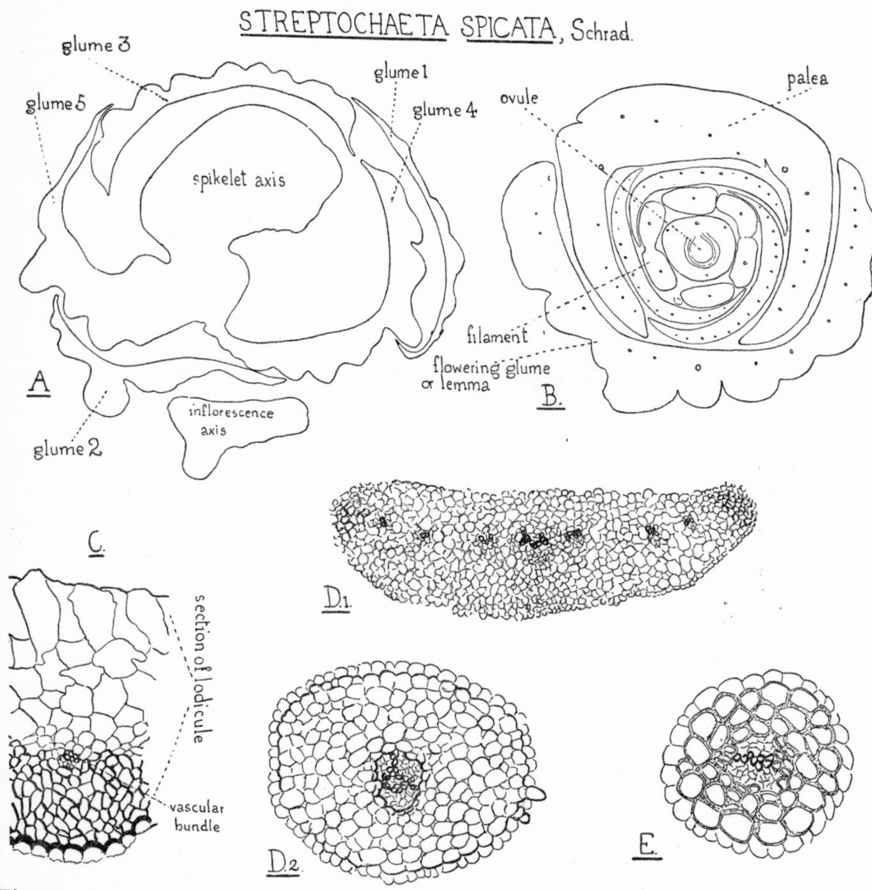

Fig. 50. *Streptochaeta spicata* Schrad. Except Fig. C, from material collected near Rio de Janeiro. A, transverse section (× 22) at the base of a spikelet with glumes numbered according to the order of detachment. B, transverse section of a spikelet (× 22) to show the base of the palea with the two lobes connected. C, small part of a transverse section of a lodicule (× 189), dried material, Southern Brazil. D 1 and D 2, transverse sections through lemma. D 1 (× 75) at the level where it is narrowing to the awn. D 2 (× 189) through awn. E, transverse section of another awn (× 311). [Arber, A. (1929[1]).]

parently, to the lemma and palea, occur immediately below the flower. They are succeeded by a hairy wreath, whose exact structure and interpretation remain uncertain; it has been called a perigon. Within this

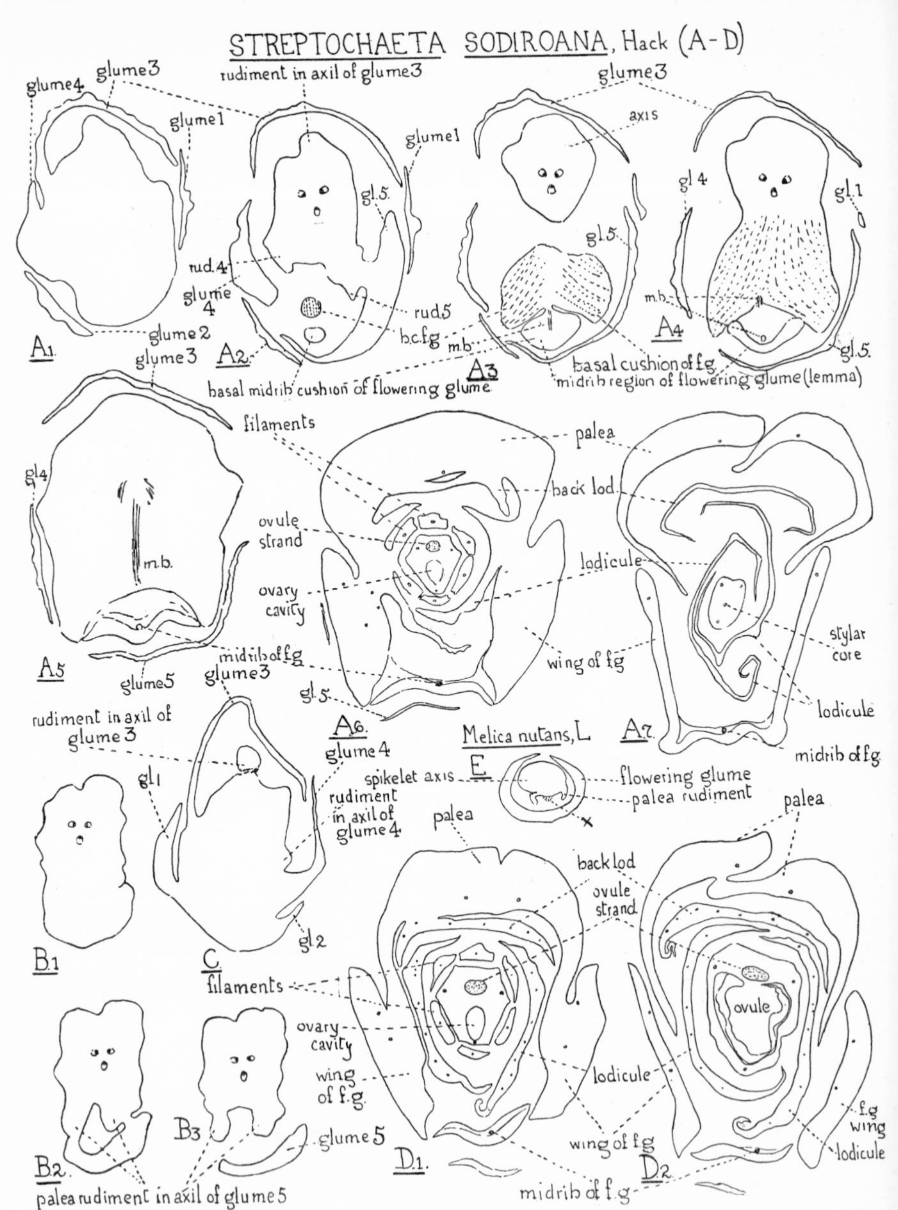

Fig. 51, A–D, *Streptochaeta Sodiroana* Hack. All drawings from transverse microtome sections (× 23) cut from dried specimens, Colombia: *gl.* 1–*gl.* 5, the sterile glumes preceding the lemma, *f.g.*, whose basal cushion is marked *b.c.f.g.* A 1–A 7, B 1–B 3, and C, sections from transverse series from below upwards through spikelets. D 1 and D 2, sections from another spikelet to supplement series A 1–A 7; they correspond to levels between A 6 and A 7. E, *Melica nutans* L. Transverse section (× 23) of the second abortive flower which succeeded two normal flowers; the flower itself is represented by a minute patch of degenerating tissue marked × (for comparison with B 3). [Arber, A. (1929[1]).]

are four stamens, surrounding the gynaeceum, which has a single style supplied by a posterior bundle. The stamens show a peculiarity, illustrated in Fig. 53, for which I know no exact parallel, either among the

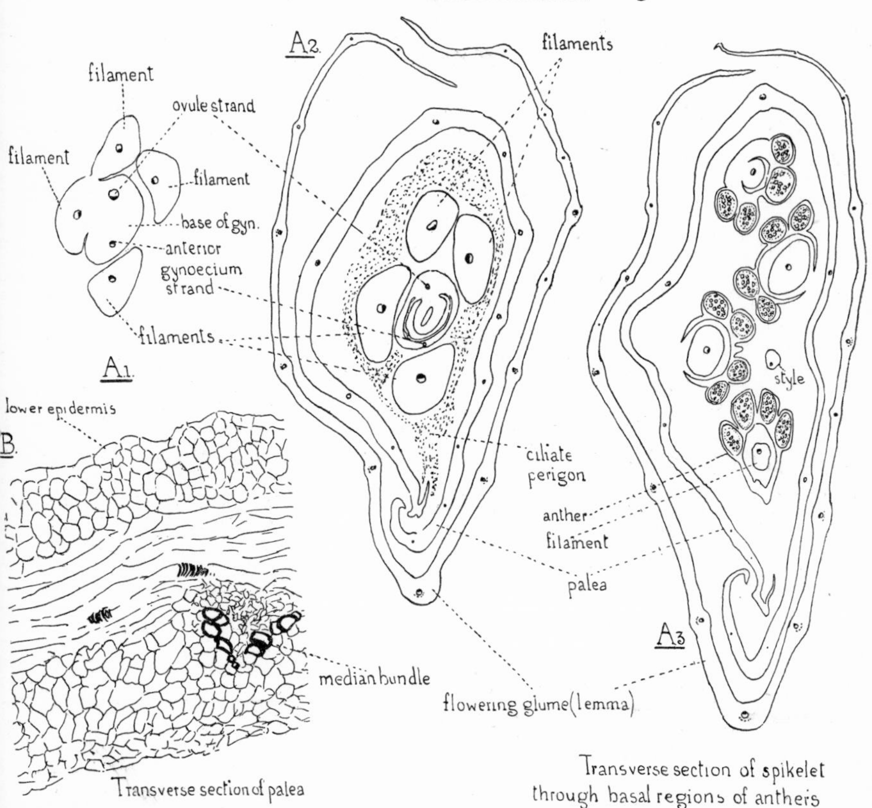

Fig. 52. *Anomochloa marantoidea* Brongn. The diagrams in Figs. 52 and 53 are all from sections of an herbarium spikelet from a plant cultivated in the Jardin des Plantes, Paris, 1860. A 1–A 3, transverse sections (× 22 *circa*) from a series from below upwards through the spikelet. A 1, the filaments and gynaeceum only, cut at their extreme base. A 2, higher up, to show the two glumes and the hairy wreath. A 3, through the style and the basal region of the anthers. B, transverse section of the midrib region of the palea (× 189) at the base of the flower below the level of A 1. [Arber, A. (1929[1]).]

grasses or elsewhere—namely that the anther, at a certain level, is ring-shaped in section and completely encloses the filament (Fig. 53, 3). It will be recognised even from this brief description, that

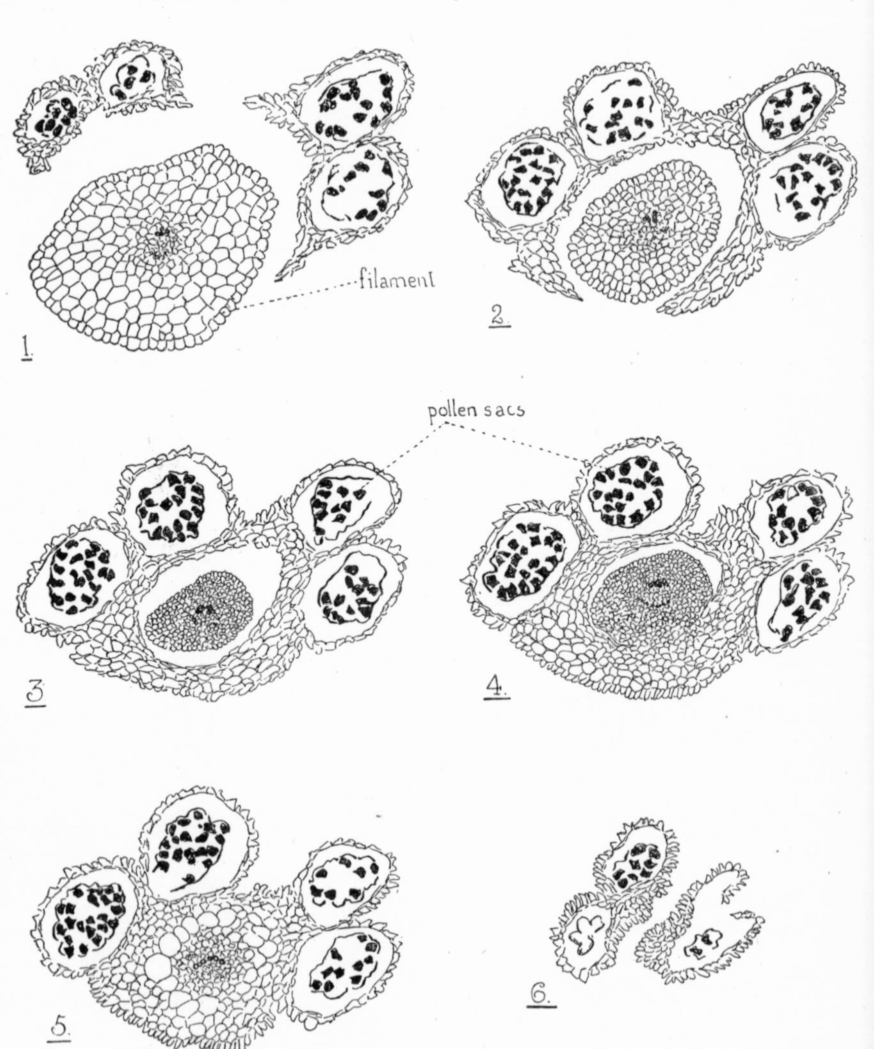

Fig. 53. *Anomochloa marantoidea* Brongn. Transverse sections from a series from below upwards through the back (northerly) anther in Fig. 52, A, p. 131 (× 77); the anther is not orientated as in Fig. 52, A, but has the dorsal surface downwards. [Arber, A. (1929[1]).]

Anomochloa, if a member of the Gramineae, is a most erratic one. Its rediscovery is much to be desired; a complete study of its structure and development would be of the deepest interest.

If, excluding *Anomochloa*, we review the construction of bamboo flowers in general, we find that they are all derivable (in theory) from such a type as *Bambusa nutans* Wall. (Fig. 33, p. 109), with its 3 perianth members, 3 + 3 stamens and 3 styles. In some bamboos (e.g. *Ochlandra*) there is an *increase* in the number of lodicules, stamens and styles. It is much commoner, however, to find a *reduction* in the number of parts. There may be no lodicules (e.g. *Dendrocalamus*) or the stamens may be reduced to a single whorl (e.g. *Chusquea*). Reduction may go so far as to result in unisexual flowers; part of a spikelet of *Oxytenanthera albociliata* Mun., including a male and an hermaphrodite flower, are shown in Fig. 40, p. 116. That unisexuality may be a reduction effect is indicated by an observation made long ago in the botanic garden at Buitenzorg; it was noticed that in one season *Gigantochloa maxima* Kurz (*G. verticillata* Mun.) bore many hermaphrodite flowers. In the next year, however, unisexual flowers preponderated on the same stocks—a change which seemed to be due to the exhaustion induced by anthesis.[1]

It is the spikelets showing varying degrees of reduction which link up the bamboos with the types of grass flower familiar in temperate countries. Before turning to these better-known forms, it may be worth while to ask if we can detect any cause for the reduction which seems to be so widely operative. We have noticed that each bamboo flower develops under pressure from the adjacent spikelets of the complex inflorescence, and from the enclosing glumes of its own spikelet; I think that this pressure must be at least one of the factors responsible for loss and incomplete development of parts. Attention has already been drawn to this factor in connection with the palea of the bamboos, as well as with the special case of the asymmetry of the gynaeceum in *Ochlandra*; we shall be in a better position to discuss it when we have considered the flowers of the other Gramineae.

[1] Kurz, S. (1876).

CHAPTER VIII

THE REPRODUCTIVE SHOOT IN GRASSES: STRUCTURE AND ANTHESIS

IN grasses the spikelets form the ultimate divisions of a type of branched raceme often vaguely described as a panicle.[1] It is very difficult to define a grass panicle, but the features which give it its characteristic appearance are, firstly, the absence of axillant leaves, and, secondly, the abbreviation of the basal internodes of the branches, associated with precocious development of lateral axes; this second factor results in a confused crowding of branches of different orders. The lowest of the secondary branches may arise at the very base of the primary branch, and may, again, have a tertiary branch at its base, so that three branches, arising apparently as a triplet, may yet be, in reality, of three different orders. Owing to this precocious branching, the secondary and later axes may form a pseudo-half-whorl at each node, these half-whorls alternating along the primary axis. Examples of secondary and tertiary axes arising at the same node, and thus giving 'twinning' of branches of different orders, may be seen in Fig. 54, A 1 and A 2 (*Briza media* L.). The hair-like lateral axes, at the ends of which the spikelets dance in the slightest breeze, have earned for this grass such dialect names as Earthquakes and Doddering Jockies.[2] In some Gramineae, for instance, the Fescues, the panicles are one-sided; for though the primary branches stand alternately to right and left, they do not lie in one plane, as one would expect of branches axillary to distichous leaves, but their planes meet at an angle which is less than 180°, so that they all stand out to one side of the axis. Moreover, the first of the secondary branches is developed on that side of the panicle towards which the primary branches converge, giving an effect of complex branching on one face, and bareness on the other. The

1 Godron, D. A. (1873) and Hackel, E. (1882).
2 Britten, J. and Holland, R. (1886).

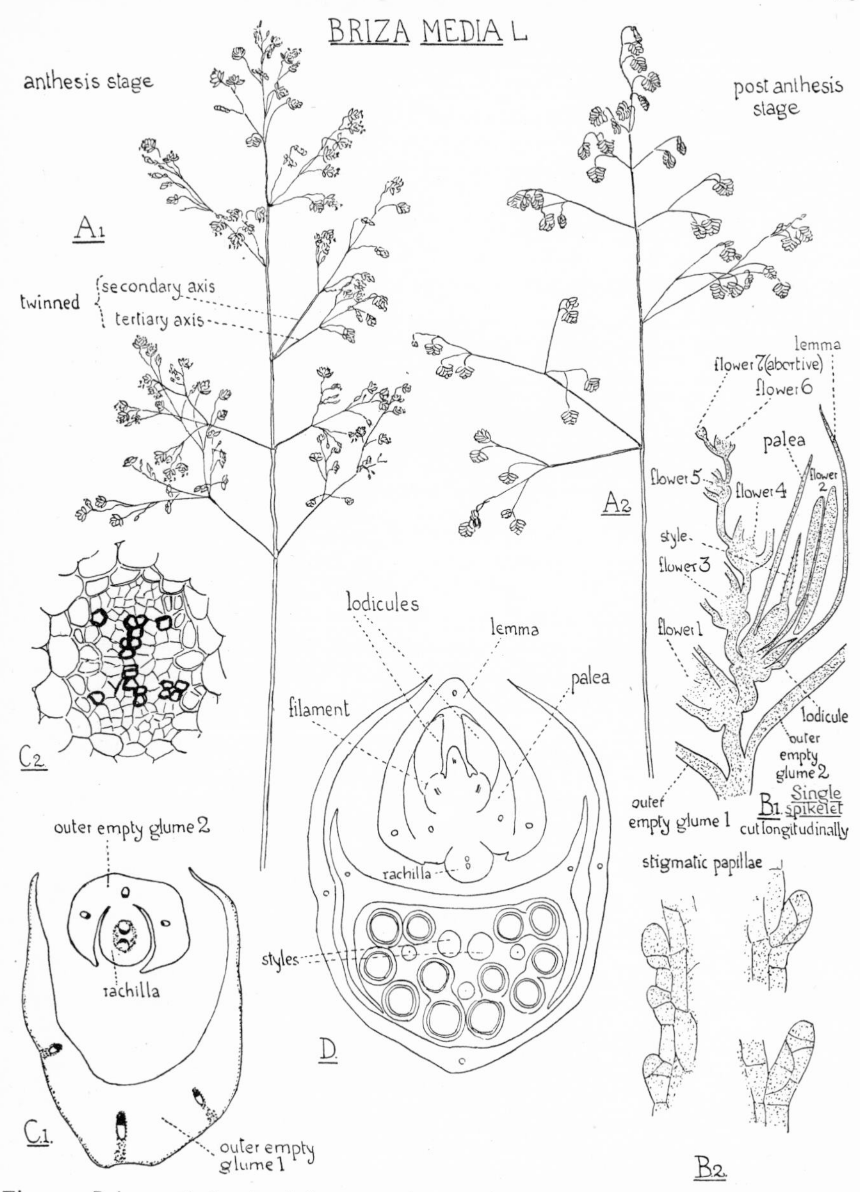

Fig. 54. *Briza media* L. A 1, inflorescence in anthesis; A 2, after anthesis (× ½). B 1, longitudinal section of a spikelet, details shown only in the lowest flower on the right-hand side (× 14). B 2, details of stigmatic papillae from a stylar longitudinal section such as that shown in B 1 (× 318). C 1, transverse section of the base of a spikelet (× 47); C 2, the two rachilla bundles at a level higher than C 1 (× 318). D, transverse section of a spikelet, showing two flowers, combined from two neighbouring sections (× 47). [A.A.]

panicle of the Cock's-foot-grass, *Dactylis glomerata* L. (Fig. 55), is an example of such an inflorescence, in which the symmetry is dorsiventral rather than radial. Dorsiventrality may also be shown by the inflorescence axis itself. In *Paspalum epilius* L. R. Par.[1] and *Phyllorachis sagittata* Trim.[2] the rachis of the inflorescence is flattened and bears its branches on the face of the thickened median rib.

If the nodes of the grass inflorescence are very close together, and the lateral axes are abbreviated, the whole thing may be called a spiciform panicle. Such panicles may be quite spike-like externally, and the telescoping of the axes makes them difficult to analyse. Fig. 96, B 1, p. 197, shows one of these apparent spikes in the Fox-tail-grass, *Alopecurus agrestis* L., while in Fig. 83, A 1, p. 179, a section is drawn through the base of a spikelet-fascicle of *A. pratensis* L. This fascicle consisted of a triplet of branches, *x*, *y*, and *z*; when they were followed upwards, it was found that *x* and *y* each gave rise to two spikelet axes, while *z* produced seven. The reduced panicles of *Molinia*, *Cynosurus*, *Phalaris* and *Lagurus* are drawn in Fig. 56, p. 138, and other illustrations of the variants on this type of inflorescence will be found scattered through this book.

The ultimate term in the simplification of the reproductive shoot of the Gramineae is a 'spike', in which spikelets take the place of the individual flowers of a spike *sensu stricto*. A species in which this type of construction is displayed diagrammatically is *Lolium perenne* L., Rye-grass (Fig. 73, A, p. 167). The grass inflorescence may even, in the last resort, be represented by a single spikelet (e.g. Fig. 102, A, p. 210, *Bromus hordeaceus* L.).

When we pass from the inflorescence as a whole to the units of which it is made up, we find that the structure, which is really quite simple, was confused out of recognition in the nineteenth century by a mass of controversial literature dealing with the interpretation of the spikelet. This literature is now of purely antiquarian interest, so I do not propose to summarise its content, nor even to catalogue the names bestowed by rival exponents upon the parts of the spikelet.[3] It is true that there are some genuine difficulties of interpretation,

[1] Parodi, L. R. (1926). [2] Trimen, H. (1879).
[3] A few of these were enumerated in Chapter VII (pp. 110, 112, 117).

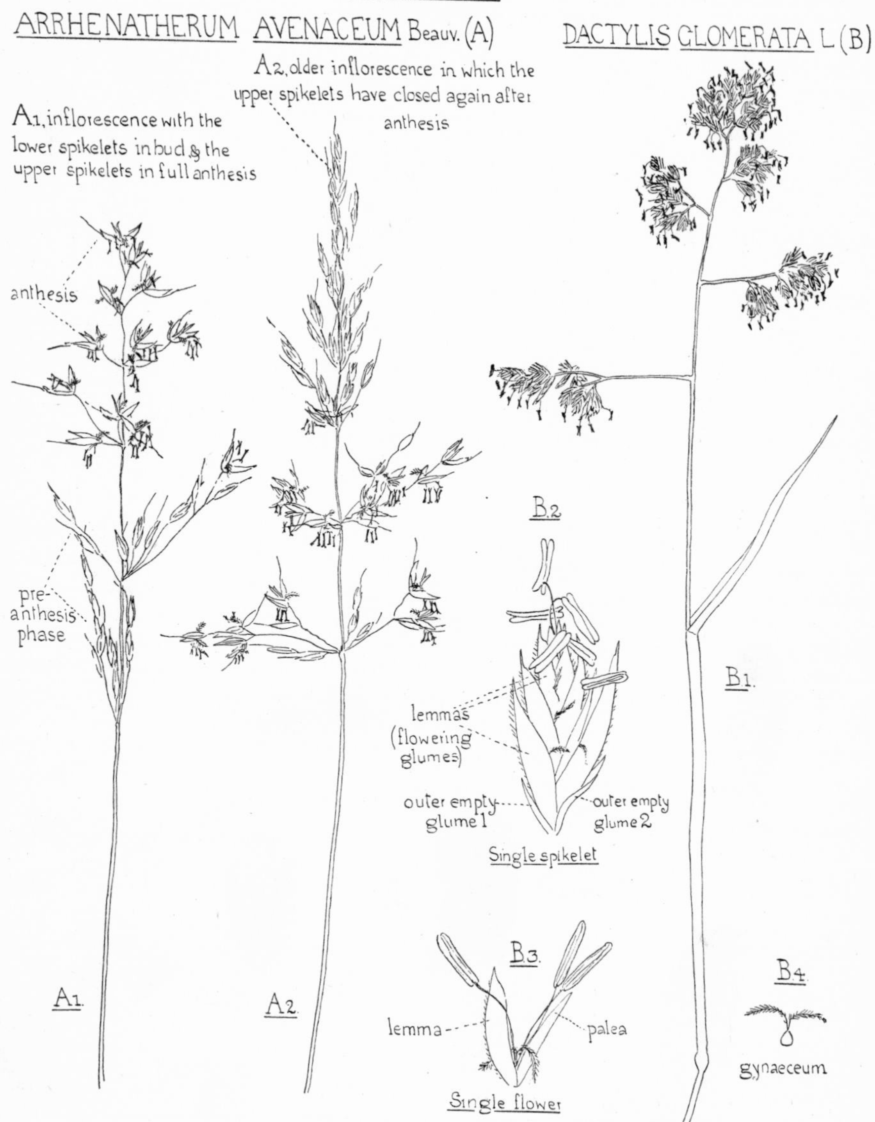

Fig. 55. Anthesis. A, *Arrhenatherum avenaceum* Beauv. A 1 and A 2, younger and older inflorescences (× ½). B, *Dactylis glomerata* L. B 1, inflorescence (× ½). B 2–B 4, details magnified. [A.A.]

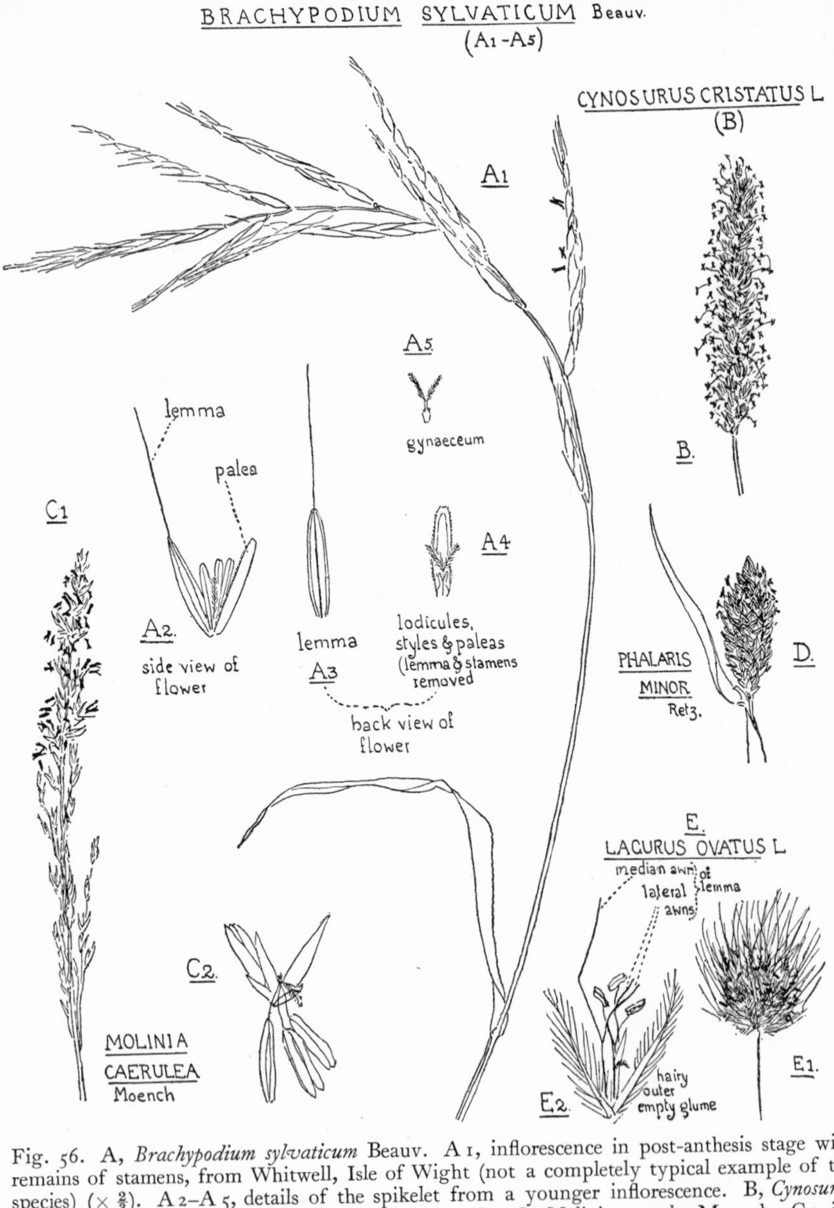

Fig. 56. A, *Brachypodium sylvaticum* Beauv. A 1, inflorescence in post-anthesis stage with remains of stamens, from Whitwell, Isle of Wight (not a completely typical example of the species) ($\times \frac{2}{3}$). A 2–A 5, details of the spikelet from a younger inflorescence. B, *Cynosurus cristatus* L., 'spike' in full anthesis, Wharfedale ($\times \frac{2}{3}$). C, *Molinia caerulea* Moench. C 1, inflorescence in anthesis, Wharfedale ($\times \frac{2}{3}$). C 2, spikelet enlarged. D, *Phalaris minor* Retz. Cambridge Botanic Garden, inflorescence ($\times \frac{2}{3}$). E, *Lagurus ovatus* L. Cambridge Botanic Garden. E 1, small inflorescence ($\times \frac{2}{3}$). E 2, single spikelet magnified to show hairy outer empty glumes, and lemma with median and lateral awns. [A.A.]

due to the small size of many grass flowers, and the abbreviation of the associated axes; but these difficulties can be overcome at the present day by the use of microtome technique. The study of serial sections of young inflorescences, in which the parts still retain their relations undisturbed, often throws light upon structures which could hardly be understood when a pocket-knife and pocket-lens were the only aids to analysis. Many of the illustrations in this book are based upon microtome series through reproductive shoots.

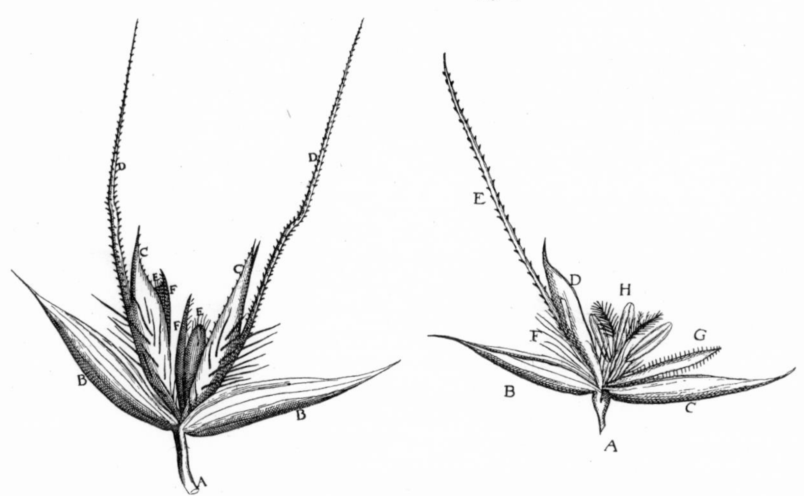

Fig. 57. Oat (*Avena sativa* L.). The spikelet and flower. [Reduced from the *Anatome Plantarum* of Marcello Malpighi, 1675.]

Fig. 57 is from the engraving of the Oat from Malpighi's great seventeenth-century work—*Anatome Plantarum*. It proves that he had a remarkably clear idea of the parts which he observed, though he happened to choose for illustration an Oat spikelet which was unusual in being one-flowered; the structure is better understood from spikelets in which more than one flower occurs. In some grasses, the spikelets are of considerable length, and include a whole series of flowers; examples of long spikelets are shown in Fig. 56, A 1, p. 138 (*Brachypodium sylvaticum* Beauv.) and in Fig. 70, p. 156 (*Glyceria aquatica* Wahl.). In the genus *Eragrostis* it is said that there may be fifty flowers in a spikelet.[1] Whether the individual spikelets are many-, few-, or one-flowered, their structure in general follows the same plan, which is illustrated schematically in Fig. 58. The rachilla, or spikelet axis, bears its members distichally. At the base

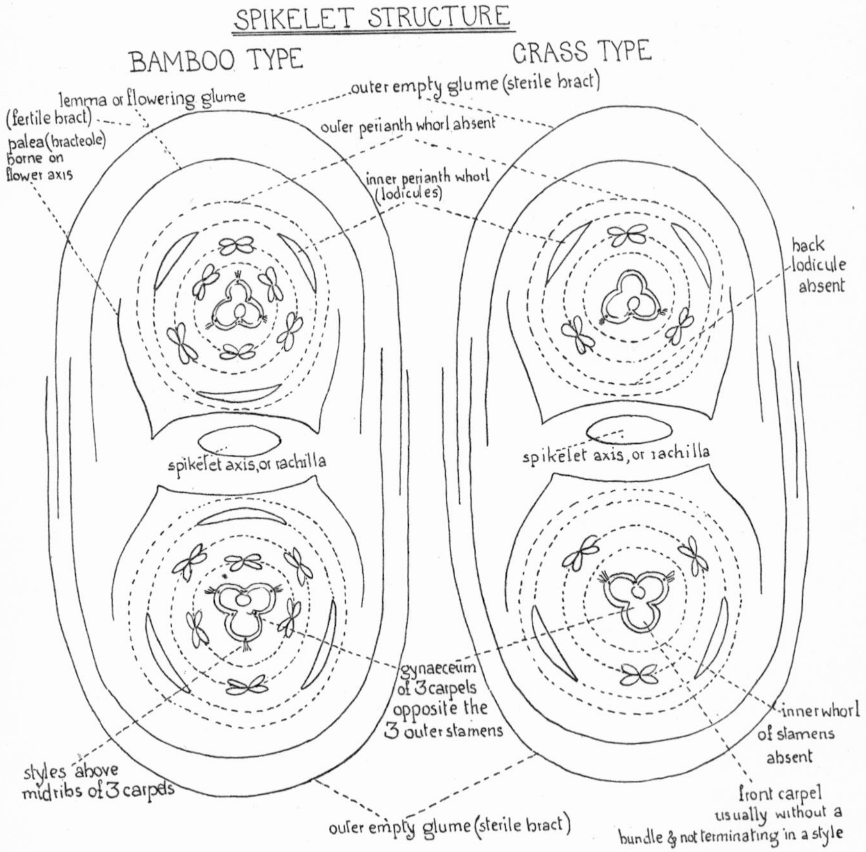

Fig. 58. Diagram showing the interpretation of spikelet structure in bamboo and grass adopted in the present book. [A.A.]

there are two bracts, with no flowers in their axils, called the *outer empty glumes* (B, B, in left-hand drawing, Fig. 57). These glumes are seen in longitudinal section in Fig. 54, B 1, p. 135, which shows the general construction of a spikelet of Quaking-grass. In most grasses the outer glumes are of simple sheathing form; sometimes they are

[1] Beal, W. J. (1900).

hairy, e.g. *Lagurus ovatus* L. (Fig. 56, E, p. 138). Rarely they are connate at the base, e.g. *Alopecurus agrestis* L., Foxtail-grass (Fig. 96, B2, p. 197). The less normal forms of outer glume will be best considered when we are dealing with some of the more aberrant grasses in the next chapter.

The rachilla above the outer empty glumes is often slender, and but lightly supplied with vascular tissue. The three-bundled, dorsiventral[1] rather than radial structure (shown in Fig. 59, D, p. 142, and in more detail in Fig. 81, B2, p. 177), is quite characteristic; indeed, the vascular skeleton of the rachilla may even be reduced to a single collateral strand, e.g. *Alopecurus pratensis* L. (Fig. 83, p. 179), *Poa annua* L. (Fig. 100, p. 202), and *Luziola Spruceana* Benth. (Fig. 168, C, etc., p. 319). Below each flower there is sometimes a slight cupule-like extension of the rachilla, with a free edge on the face opposite the flower, which in Fig. 59, p. 142, is labelled 'rachilla-flap'. It is also shown for the bamboo *Cephalostachyum virgatum* Kurz in Fig. 115, p. 241.

As in the bamboo, each flower arises in the axil of a bract—the flowering glume or lemma. This glume is often prolonged down the rachilla to form a basal cushion, such as that drawn in Figs. 59 and 60. In *Ichnanthus* the lemma may show remarkable wing-like appendages at the base, suggesting stipules (Figs. 61, p. 144, and 62, p. 145). Like the empty glumes, the lemmas may be connate. In *Lygeum Spartum* Loefl., in which the spikelet is two-flowered, the hairy bases of the lemmas are joined in this way, so that they enclose the flowers, whose paleas are similarly united at the base, but back to back (Fig. 63, p. 146).

A characteristic feature of many lemmas is the production of a spike-like 'awn', which is a continuation of the midrib above the main part of the glume. The Oat (Fig. 64, p. 147) is the example which comes first to one's mind in thinking of awned glumes. The awns of *Avena pratensis* L. are shown in Fig. 96, A, p. 197. In a Mediterranean Oat, *A. barbata* Brot., illustrated in Fig. 65, A, p. 148, the ripe spikelet, which includes two flowers with awned lemmas, falls at maturity, leaving the empty glumes behind. The awn of *Avena*

[1] On dorsiventrality in the rachilla of the bamboos, see p. 110.

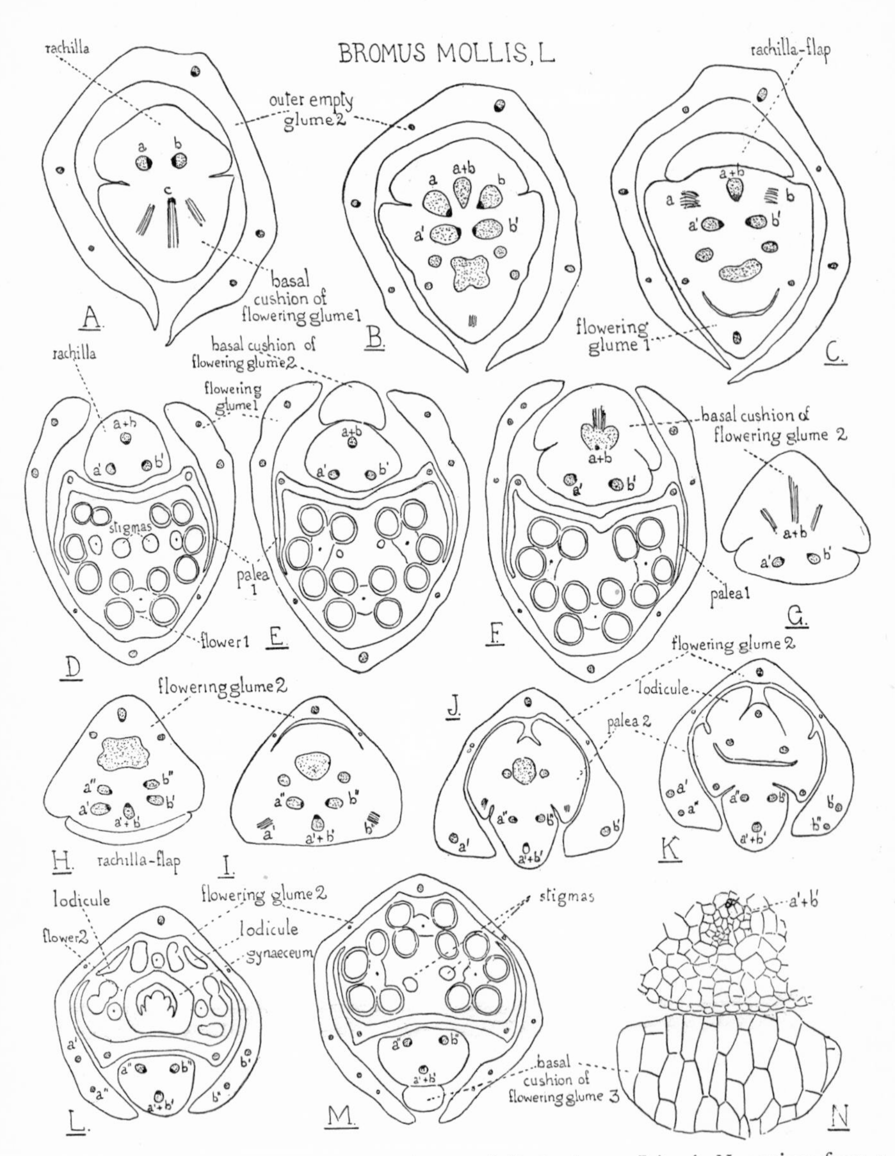

Fig. 59. *Bromus mollis* L. (more correctly named *B. hordeaceus* L.). A–N, sections from a series from below upwards through a spikelet (× 47, except N, which is × 193); vascular bundles treated somewhat diagrammatically. [Arber, A. (1927[2]).]

protrudes from the back of the glume before the apex is reached, so that the tip of the glume has no midrib. The awn is divided into two segments by a kink or knee. The lower part—the column—twists spirally[1] on drying, and even the upper part—the subule—which is generally straight, may give one or two spiral turns in the opposite sense, if the drying is extreme.[2] The two awns, as they twist on losing water, come in contact with one another with such force that the whole fruit-complex may have a jumping action, as well as more gradual movements; its antics are amusing to watch. The spikelets may travel, in this lively way, several paces from the parent

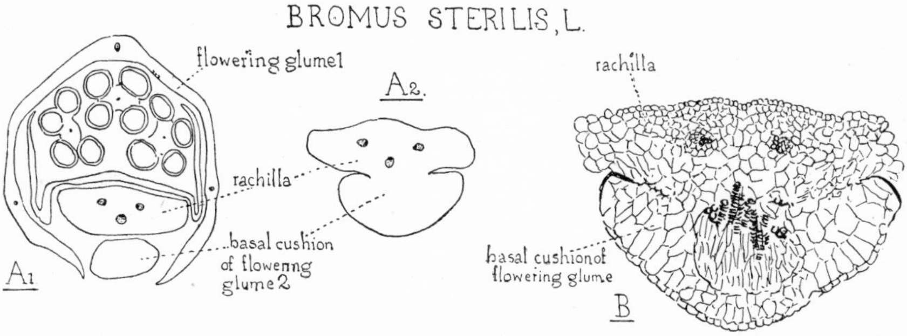

Fig. 60. *Bromus sterilis* L. A 1 and A 2, transverse sections (× 47) from a series from below upwards through a spikelet. A 1 shows *flowering glume* 1 (not the lowest of the spikelet) enclosing a flower cut at the level of the anthers. B, transverse section (× 77) of another flower from the same spikelet, to show the third bundle of the rachilla entering the basal cushion of a flowering glume and beginning to divide into three. [Arber, A. (1927[2]).]

plant.[3] *Avena sterilis* L. is sometimes called the Fly- or Animated-oat, from its resemblance to a winged insect; it is occasionally used as a bait for salmon-hooks,[4] since its movements, as the awns uncoil in the water, give an air of verisimilitude. Fish are not the only creatures that have been duped by "the Beard of the wild Oat". As Robert Hooke[5] wrote in the seventeenth century: "This so oddly constituted Vegetable substance, is first (that I have met with) taken notice of by *Baptista Porta*, in his *Natural Magick*, as a thing known to children and Juglers, and it has been call'd by some of those

[1] In *Anthoxanthum*, *Danthonia* and *Stipa*, the twist of the column is dextral; see Macloskie, G. (1895[2]).

[2] Royer, C. (1883).

[3] Hildebrand, F. (1872).

[4] Lawson, P. and Son [C.] (1836).

[5] Hooke, R. (1665).

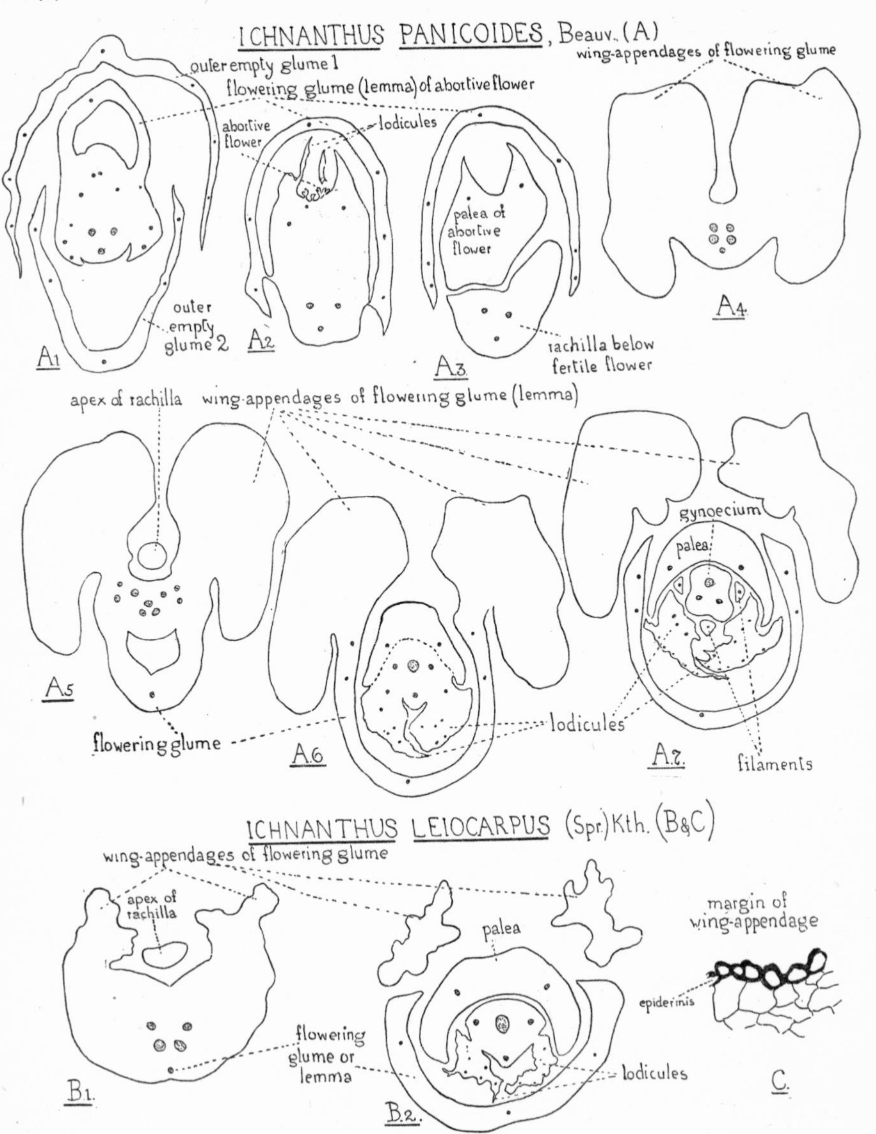

Fig. 61. A 1–A 7, *Ichnanthus panicoides* Beauv. British Guiana, floor of forest, vicinity of Penal Settlement, west side of Essequibo River, A. S. Hitchcock (17121), December 6, 1919. Transverse sections from a series (× 22) passing through a spikelet from below upwards, to show the wing-appendages of the flowering glume of the fertile flower, which is preceded by a sterile flower. B and C, *Ichnanthus leiocarpus* (Spr.) Kth., shady brushy slope, summit of Morro do Urca, Rio de Janeiro, Agnes Chase, February 9, 1925 (8402). B 1 and B 2, two sections from a series from below upwards (× 46) through a flower, to show the appendages of the flowering glume below and above their level of detachment. C, small part of the edge of a transverse section of a wing-appendage (× 189). [Arber, A. (1929[1])]

last named persons, the better to cover their cheat, the Legg of an *Arabian Spider*, or the Legg of an inchanted *Egyptian Fly*, and has been used by them to make a small Index, Cross, or the like, to move round upon the wetting of it with a drop of Water, and muttering certain words".

The awns of *Avena barbata* Brot. are represented in Fig. 65, p. 148, at their natural size. Those of some species of *Stipa*[1] are much longer;

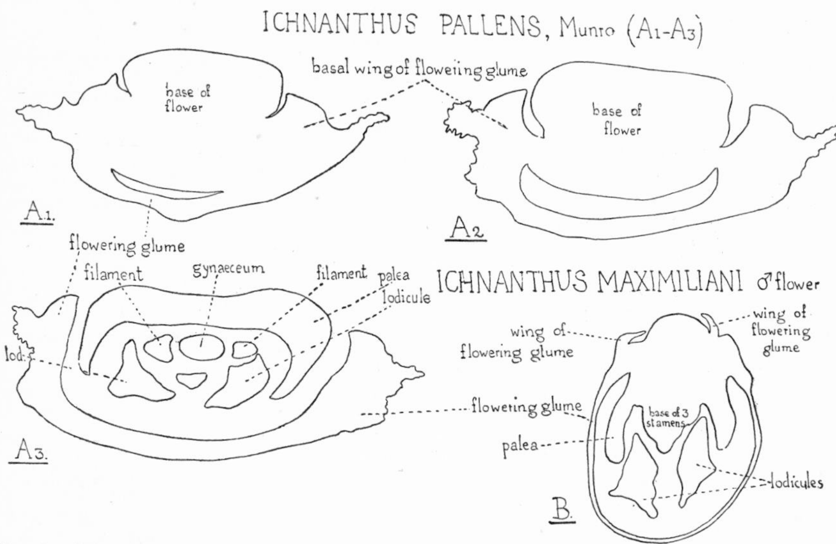

Fig. 62. *Ichnanthus*. A 1–A 3, *I. pallens* Mun. (Khasia); transverse sections from a series from below upwards through a flower and its lemma (× 77). B, *I. Ruprechtii* Doell, var. *glabratus* Doell (labelled *I. maximiliani*), Organ Mts., Brazil; transverse section (× 47) through the base of a male flower to show marginal wings of flowering glume (lemma), which is not yet detached except in the median region. [Arber, A. (1927[2]).]

in *S. pennata* L. the awn may reach nine inches, six of which are delicately feathered.[2] The length of awn in other species, and the sharp-pointed character of the spikelet-base, has led to such names as Needle-and-thread and Devil's-darning-needles. *Stipa setacea* R.Br., Spear-grass or Corkscrew-grass, is a serious pest on account of the penetrating power of its awned fruits. It has been said that "of all grasses and weeds [in Australia], spear-grass seeds are the most damaging to sheep and wool. Being straight, and with sharp-pointed

[1] On *Stipa* awns, see Murbach, L. (1900).

[2] Ridley, H. N. (1930).

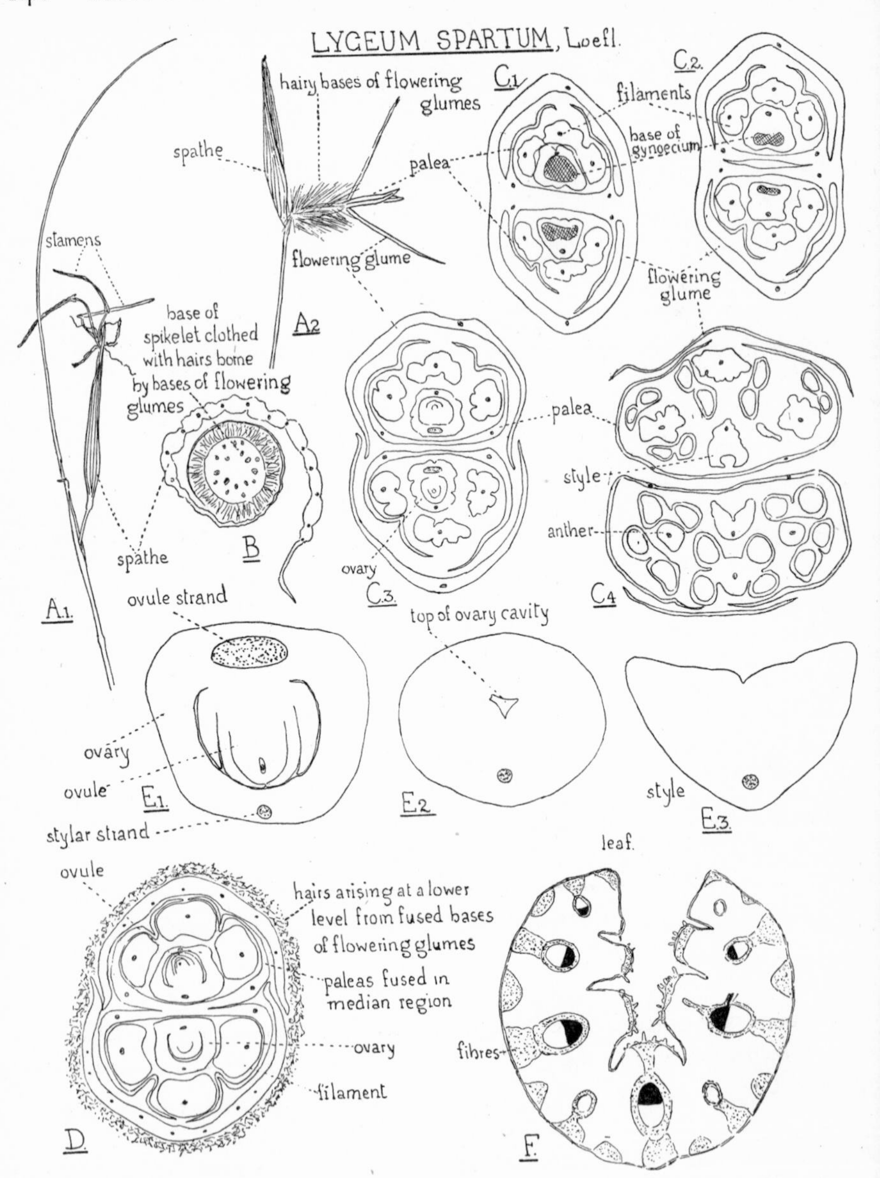

Fig. 63. *Lygeum Spartum* Loefl. (Cambridge Botanic Garden). A 1, inflorescence, June 2 (× ½); A 2, inflorescence, July 6 (× ½). B, transverse section through spathe surrounding the base of the spikelet (× 14). C 1–C 4, transverse sections from below upwards from a series (× 23) through a young spikelet; hairs on lemmas omitted. D, transverse section of a spikelet (× 23). The mat of hairs growing from the lemmas at a lower level has not been cleared away. E 1–E 3, transverse sections (× 77) from a series from below upwards through a young gynaeceum (× 77). F, transverse section of the limb of a leaf (× 47). [Arber, A. (1928[2]).]

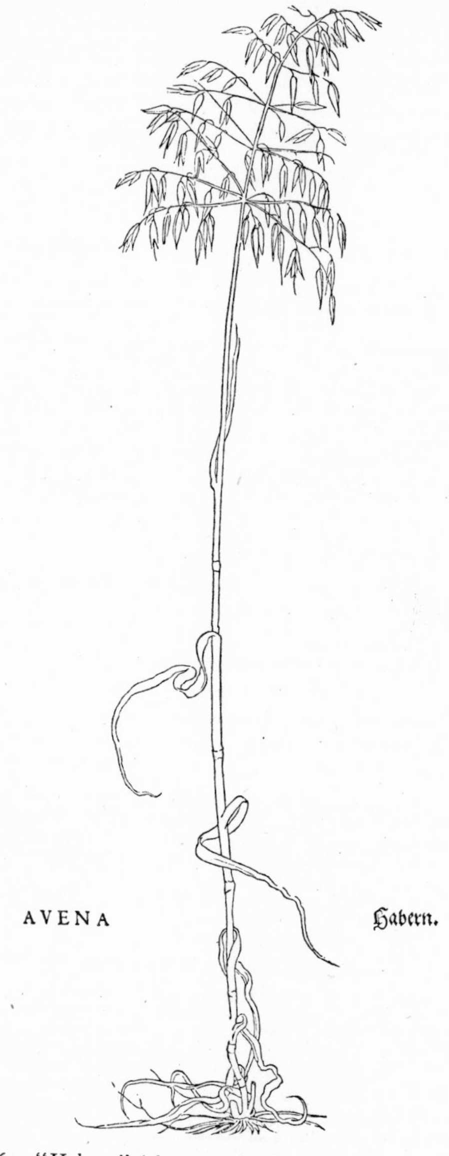

Fig. 64. "Habern" (*Avena sativa* L. var. β L., White-oat).
[Reduced from the *De Historia Stirpium* of Leonhard Fuchs, 1542.]

ends, when once they get attached to the wool, they lie parallel with the staples and fibres, and by the movement of the animal they work their way on to the skin. In extreme cases the fleece is composed of fully 75 per cent. of spear-grass seeds, so persistently do they hold on to the wool. When once they get a hold, they never fall out. In the same way, when these sharp-pointed seeds enter the skin,

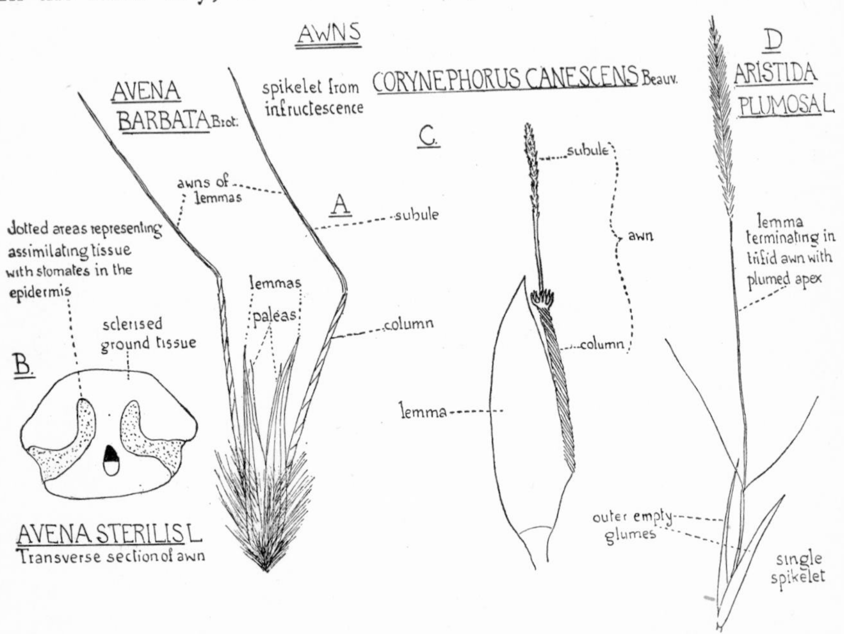

Fig. 65. Awns. A, *Avena barbata* Brot., spikelet from infructescence (nat. size). B, *Avena sterilis* L., transverse section of awn (× 47). C, *Corynephorus canescens* Beauv., awned lemma (magnified; total length of awn is 2–3 mm.). Material from Caister, July 1920, collected by T. J. Foggitt. D, *Aristida plumosa* L., single spikelet, hairs omitted (× 23). [A.A.]

they work through it, right into the sheep, until they come in contact with the vital organs, which results in certain death. I have seen them in the heart of a sheep, and even having a hold on the bones, from which they could not be pulled".[1]

The elongated awns of *Stipa* offer a contrast in scale with those of one of our rarer British grasses, *Corynephorus canescens* Beauv., in which the awns (Fig. 65, C), though of peculiarly complex shape, are only 2 or 3 mm. long. They each bear a corona of spines at the

[1] Cited in Maiden, J. H. (1898).

upper limit of the column, while the top of the subule is club-like and delicately toothed. In *Aristida plumosa* L., the awn, which is shown more enlarged in Fig. 65, D, is trifid, with two lateral branches, and a plumose apex to the median axis. The anatomical structure[1] of the awn of *Avena sterilis* L. is indicated in Fig. 65, B. It is sclerised in the main, but a curved plate of assimilating tissue occupies either flank.

The lemma, whether simple, or awned as in the species just described, is followed by the *palea*, or *bracteole* borne on the flower axis, which is generally a bikeeled member. The meaning of this unusual leaf form was considered in discussing the bamboos (p. 112, et seq.).

In analysing the grass flower, the 'typical' Monocotyledonous floral diagram may be used as a mental framework on which to arrange one's ideas, but it is not necessary to assume that it portrays any ancestral reality.[2] Taking this diagram as a basis, we may regard the lodicules as representing the inner perianth members—the outer whorl being unrepresented. The back member of the inner whorl is nearly always absent, but a few of the grasses resemble the typical bamboo in having a complete whorl of three members. *Stipa pennata* L. (Fig. 66, D, p. 150) is an example, but the back lodicule in this species seems on the way to extinction, as it is poorly developed and non-vascular.

Lodicules are usually small, greenish members, with a capacity for turgescence which we will consider later in connection with anthesis; this capacity is correlated with the absence of stomates.[3] The lodicules tend to be tumid near the base, but flatter and more leaf-like above. The scheme of their vascular supply is shown in Fig. 67, C, D, p. 151, for Nepaul-barley (*Hordeum trifurcatum* Jacq.). One strand enters each lodicule (D 1), and by repeated branching produces a large number of bundles scattered through the mesophyll (C). In other grasses there may be fewer bundles, restricted to

[1] On awn anatomy, see Duval-Jouve, J. (1871).

[2] Engler, A. (1892) points out that this diagram was called into existence by the imagination of the older systematists; but this, though true, does not make it any less valuable as a working tool.

[3] Zuderell, H. (1909).

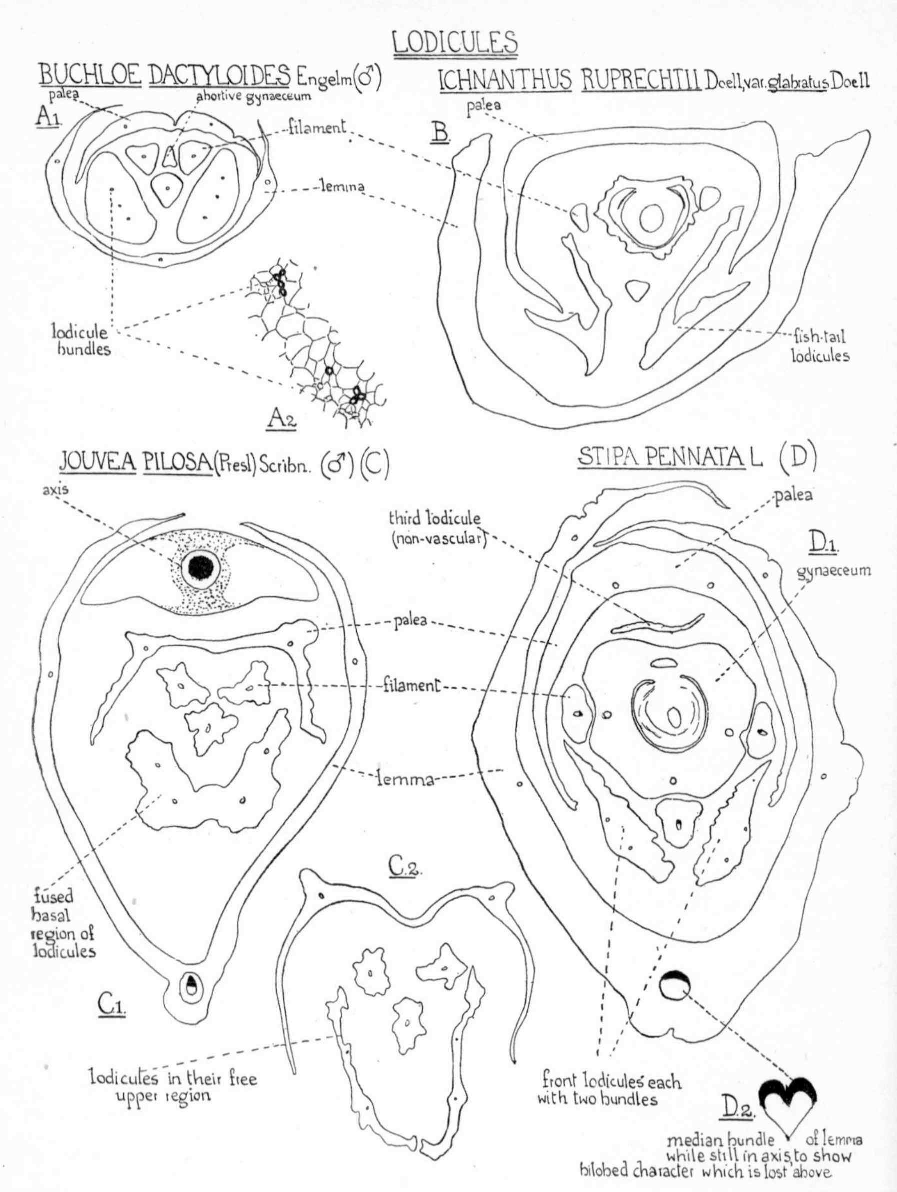

Fig. 66. Lodicules. A, *Buchloë dactyloides* Engelm. A 1, transverse section of a male flower passing through the filaments (× 77 *circa*); lemma and palea slightly broken and reconstructed. A 2, the three bundles from the left-hand lodicule in A 1 (× 318) to show differentiation of xylem and phloem. B, *Ichnanthus Ruprechtii* Doell, var. *glabratus* Doell, Organ Mountains, Brazil; transverse section of flower (× 77 *circa*). C, *Jouvea pilosa* (Presl) Scribn. C 1 and C 2, transverse sections from a series from below upwards through a male flower, collected by Professor A. S. Hitchcock from Corinto, Nicaragua (× 77 *circa*). C 1 shows lodicules in fused basal region, while in C 2 they have become free. D, *Stipa pennata* L. D 1, transverse section of a flower (× 77 *circa*) to show the third (posterior) lodicule which is non-vascular [A.A.]

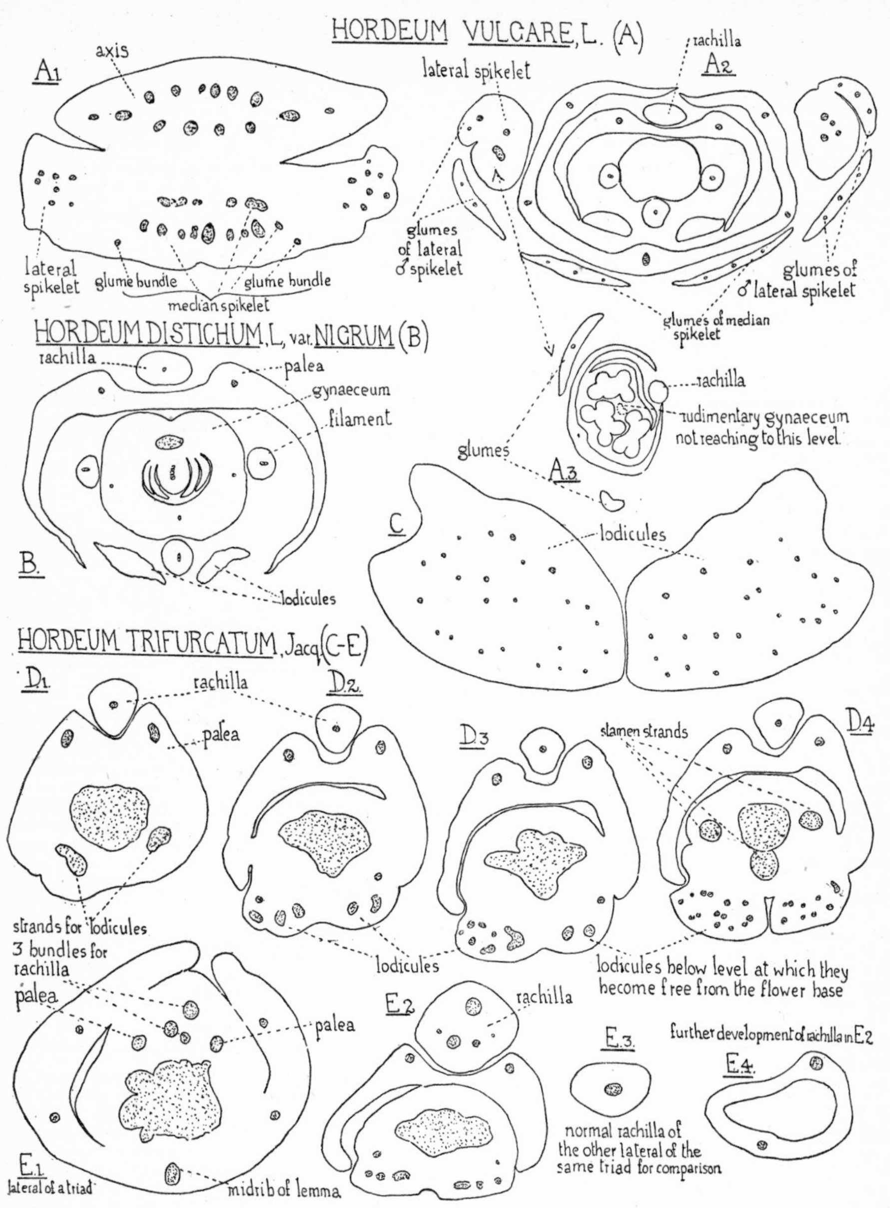

Fig. 67. *Hordeum*. A 1–A 3, transverse sections through a young inflorescence of a two-rowed Barley; the label, *H. vulgare* L., is incorrect. B, transverse section of a spikelet of *H. distichon* (*distichum*) L. var. *nigrum*, lemma omitted (× 48). C–E, *H. trifurcatum* Jacq.; C, transverse section of two lodicules from a flower (× 80). D 1–D 4, series of transverse sections from below upwards through the base of a flower from another spikelet to show vascular supply of lodicules (× 48). E 1, E 2, and E 4, series of transverse sections through a spikelet (× 48) to show an abnormal rachilla (with several bundles) which opens out above into a leaf-like structure (E 4). Lemma omitted in E 2. E 3, the normal one-bundled rachilla of the other lateral of the same triad (× 48). [Arber, A. (1929[2]).]

one plane. A botanist[1], who examined the lodicule anatomy of about fifty species of Gramineae, concluded that the strands consisted of small tracheal elements without phloem. I have, however, found what appears to be clear differentiation of both xylem and phloem in the lodicules of the male flowers of *Buchloë dactyloides* Engelm. (Fig. 66, A2, p. 150), and of *Zea Mays* L. (Fig. 185, C2, p. 361).

Occasionally the two lodicules of the grass flower may be united by their front margins. Those of the male flower of *Jouvea pilosa* (Presl) Scribn. are fused at the base, but free above (Fig. 66, C1 and C2), while the union in *Melica nutans* L. is complete (Fig. 193, C, p. 384). The spikelet of Rice (*Oryza sativa* L.) shows the unusual feature that the lodicules are joined to the margins of the palea at its base, so that they look deceptively as if they were related to it as stipules (Fig. 193, A, B1 and B2, p. 384).

The lodicules are succeeded by the stamens, which are usually limited to the three members of the outer whorl; but in certain grasses, e.g. Rice, both outer and inner whorls are developed. In *Festuca bromoides* Sm., on the other hand, the androeceum[2] is reduced to a single stamen—the front member of the outer whorl (Fig. 96, D, p. 197).

When we turn to the gynaeceum, we find that we receive no help from the ontogeny in deciding how many carpels are present. In its early stages the female organ is merely an open crescent on the anterior face of the apex of the flower axis; this is shown for *Bromus sterilis* L. in Fig. 68, D, and for *Festuca ovina* L. in F. The mature structure, however, harmonises perfectly with the descriptive convention that it consists of three carpels united edge to edge. In the ovary wall there may be three vascular strands, representing the midribs of the three carpels, in addition to the ovule supply at the back of the gynaeceum, formed from the marginal veins of the adjacent lateral carpels. Such a vascular skeleton is seen in *Stipa pennata* L., Fig. 68, A1. The vein of the front carpel, though present here, is absent in most grasses. Each of the median bundles of the

[1] Zuderell, H. (1909).

[2] I use this spelling instead of "androecium", because Kraus, G. (1908) has shown that it is the preferable form.

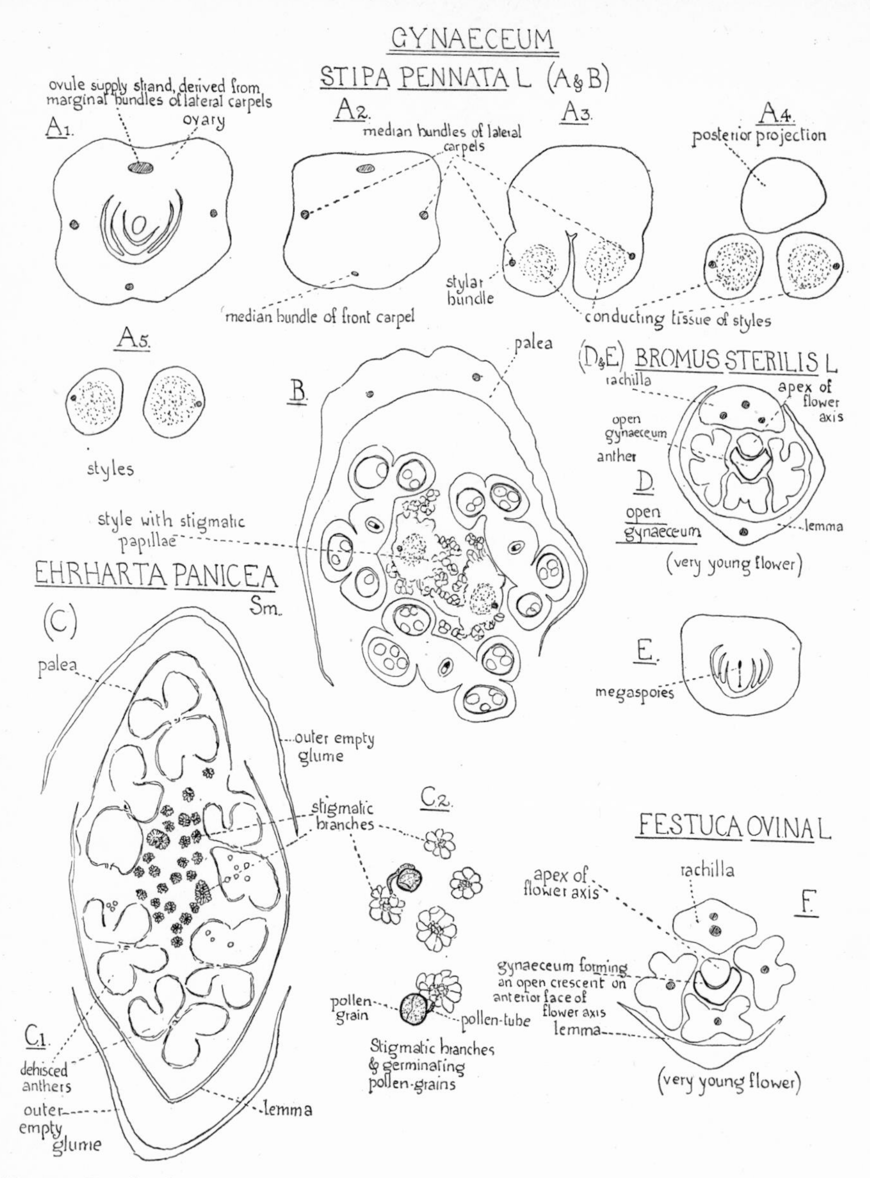

Fig. 68. Details of gynaeceum. A and B, *Stipa pennata* L. A 1–A 5, sections from a transverse series from below upwards through a gynaeceum (× 77 *circa*) from ovary to styles. B, transverse section of a flower through anthers and styles, lemma omitted (× 77 *circa*). C, *Ehrharta panicea* Sm. (*E. erecta* Lam.), collected near Port Elizabeth, South Africa. C 1, transverse section of a flower in which the anthers have dehisced (× 47). C 2, a few of the stigmatic branches and two germinating pollen grains from another section of the same flower (× 193 *circa*); see Fig. 87, p. 183, for general flower structure of this species. D and E, *Bromus sterilis* L. D, transverse section of a very young flower cut above the level of the top of the palea to show the open character of the gynaeceum at this stage (× 77 *circa*). E, gynaeceum from a transverse section of an older flower, showing nucellus and megaspores (× 77 *circa*). F, *Festuca ovina* L., transverse section of a very young flower; the rudiment of the palea does not reach to this level (× 77). [A.A.]

two lateral carpels passes into one of the styles, where it lies parallel to the small-celled core, which forms the conducting tissue for the pollen-tubes (Fig. 68, A 5). The styles produce papillae (cf. Fig. 54, B 2, p. 135), which grow out into slender branches, rosette-like in section. These are sketched in Fig. 68, B, C 1 and C 2. The rosettes are not always as complex as those shown for *Ehrharta panicea* Sm.; in many grasses the stigmatic branches consist of only four rows of cells.[1] In C 2, two germinating pollen-grains are seen; they have put out tubes which are applying themselves to the stigmatic branches. The specimen of this South African grass, from which these drawings were made, had existed for at least eight years as a dried plant in an herbarium, and yet such delicate structures as the stigmatic branches and the germinating pollen-grains were still in good preservation.[2] In Wheat, in which the anthers and pollen have been specially studied, the outer wall of the pollen-grain is smooth and has one pore, surrounded by a ring-shaped ridge.[1] A minute lid is pushed off this pore as the pollen-tube emerges.[3]

We have so far described the reproductive shoot from the structural standpoint only. It remains to consider anthesis as a dynamic process, and briefly to sketch the fruit which is the final result.

In many grasses a notable change takes place in the disposition of the inflorescence at flowering time. In the early stages the branches are erect and contracted, but during anthesis they spread apart, while, in some grasses, they contract again in the post-anthesis phases. The two first stages can be seen, for instance, in Fig. 69, which shows the successive poses of the inflorescence of Bermuda-quick, *Cynodon dactylon* Pers. These changes of attitude are due to the behaviour of a callosity at the base of the branch. In *Dactylis glomerata* L., the tissues both of the branch and of the main axis take part in the development of the pulvinus.[4] At the time of anthesis, this

1 Goliński, S. J. (1893).

2 Before sectioning, this herbarium material was treated for twelve days with dilute potash; on this (McLean's) method, see Arber, A. (1926).

3 Percival, J. (1921).

4 For diagrams, see Kirchner, O. von, Loew, E. and Schroeter, C. (1908, etc.), pp. 98 and 99.

cushion expands, so as to force the branch outwards. After anthesis it wilts, and the branch again rises. Inflorescence branches in all three phases are to be seen in *Arrhenatherum avenaceum* Beauv. (Fig. 55, A 1 and A 2, p. 137). Each spikelet in this species includes a lower male and an upper hermaphrodite flower. The palea and

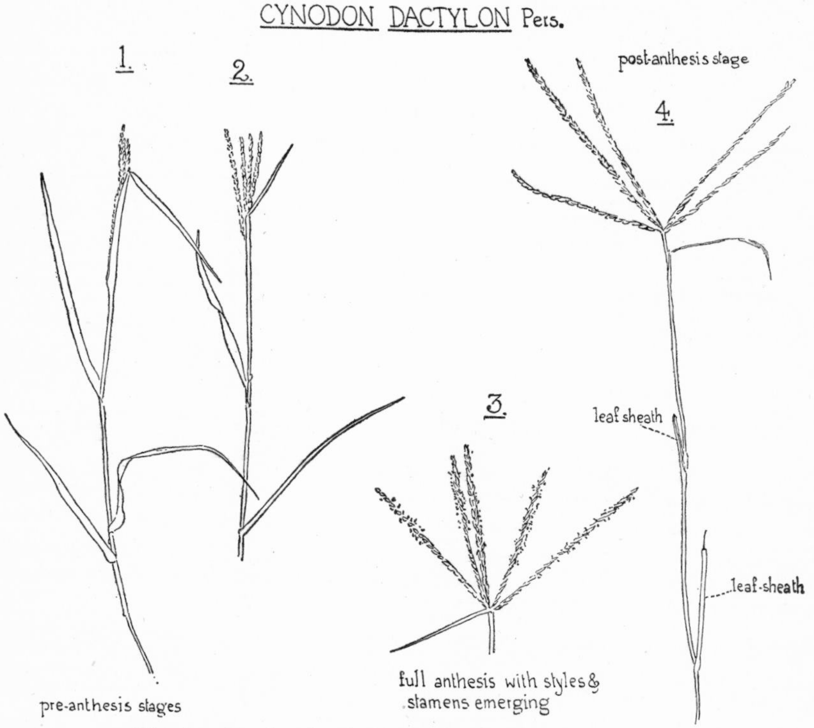

Fig. 69. *Cynodon dactylon* Pers., Carteret, Normandy. 1–4, inflorescences from bud to post-anthesis stages ($\times \frac{1}{2}$). [A.A.]

lemma separate by an angle of about 45°, and the paleas come into contact, back to back. Each anther first opens by a pore, and then along its entire length. The stigmas spread out and curve downwards, one on either side of the spikelet. They are easily reached by the pollen from the flowers above them in the panicle. The pollinated stigmas soon re-enter, and roll up inside the flower, which closes on them.[1]

In *Glyceria aquatica* Wahl. (Fig. 70), which is common in wet

[1] Godron, D. A. (1873).

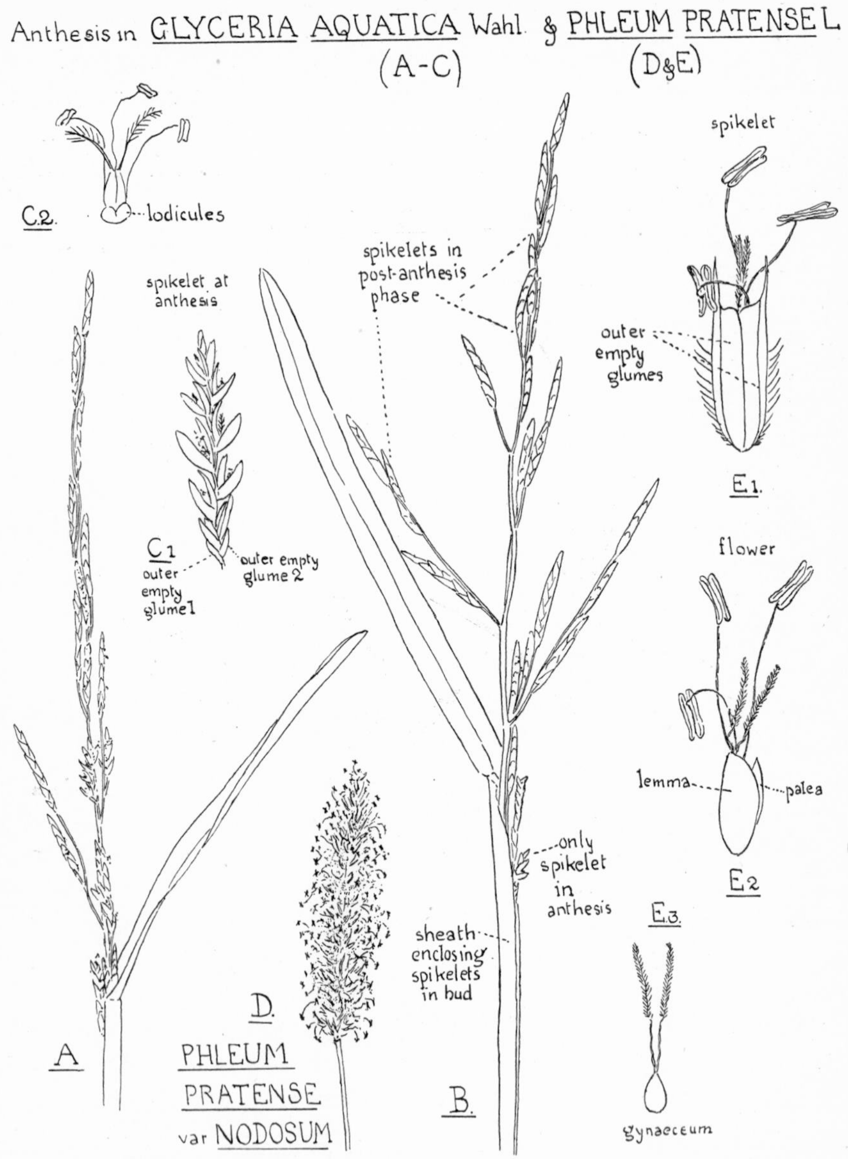

Fig. 70. Anthesis. A–C, *Glyceria aquatica* Wahl. A and B, inflorescences ($\times \frac{2}{3}$). C 1 and C 2, details, enlarged. C 1, spikelet at anthesis; C 2, gynaeceum, stamens and lodicules. D and E, *Phleum pratense* L. D, spike of var. *nodosum* L. in very full anthesis ($\times \frac{2}{3}$). E 1–E 3, details of spikelet in type form (magnified). E 1, spikelet, showing the outer empty glumes, which are green and hairy in the midrib region, but white and membranous in the marginal region. In E 2, the outer glumes have been removed. The lemma entirely conceals the palea, but in E 2 the latter has been pulled a little way out of the lemma to make it visible. E 3, gynaeceum. [A.A.]

places in England, the inflorescence grows out of the leaf-sheath gradually, and anthesis generally takes place in each spikelet just as it emerges. In the specimen drawn in Fig. 70, B, there was almost as much again of the inflorescence still hidden within the sheath. Inflorescences may be found in which the apical spikelets are already withered, while the basal spikelets are still in full anthesis. In grass panicles in general, this early opening of the uppermost spikelets is the rule, and in the individual branches the same order is followed. The privileged position of the top spikelets often seems to result in better nutrition. In Rice, Oats and Tef[1] the grains in the apical spikelets are the best, while, in Wheat, a difference between the grains in different regions of the ear has long been recognised. It was ordained in the reign of Henry VII that the "sterling" or pennyweight should "be of the weight of thirty-two grains of wheat *that grew in the midst of the ear* of wheat according to the old laws of this land".[2]

In most grasses the actual opening of the flower is due to the lodicules[3] becoming suddenly swollen at the base, so that the lemma is forced outwards. That the swelling is a turgescence effect, is shown by the fact that a drop of liquid exudes from a lodicule pricked with a fine needle while in the tumid phase. This turgescence is comparable with that of the basal pulvini of the panicle branches, which synchronises with flowering. When anthesis is over, the lodicules are found to be in a collapsed condition.

Protogyny—the maturation of the stigmas before the anthers—is rare in the Gramineae, but it occurs in certain species which have no lodicules. Examples are *Anthoxanthum odoratum* L. (Fig. 84, p. 181), *A. gracile* Bivon (Fig. 71), *Alopecurus pratensis* L. (Fig. 83, p. 179), *Nardus stricta* L. (Fig. 75, p. 169) and *Cornucopiae cucullatum* L. (Fig. 72, p. 165). It seems probable that, in these cases, protogyny and the absence of lodicules are causally connected, and that the earlier emergence of the styles is due to the difficulty which the anthers find in escaping out of the glumes, when the way is not opened for them by the lodicules. This idea is confirmed by the fact

[1] Koernicke, F. (1885). [2] Ridgeway, W. (1892); the italics are mine.
[3] Hackel, E. (1880); Rimpau, W. (1882); and Zuderell, H. (1909).

that in Pearl-millet (p. 24), the only cereal which is protogynous, the flower does not actually open, but the sex organs merely squeeze out, as it were, at the top.[1]

In addition to the turgescence of the inflorescence pulvini and of the lodicules, a third turgour effect is associated with anthesis, namely, the sudden extension of the stamen filaments, by which the anthers are carried out of the flower. This extension is amazingly

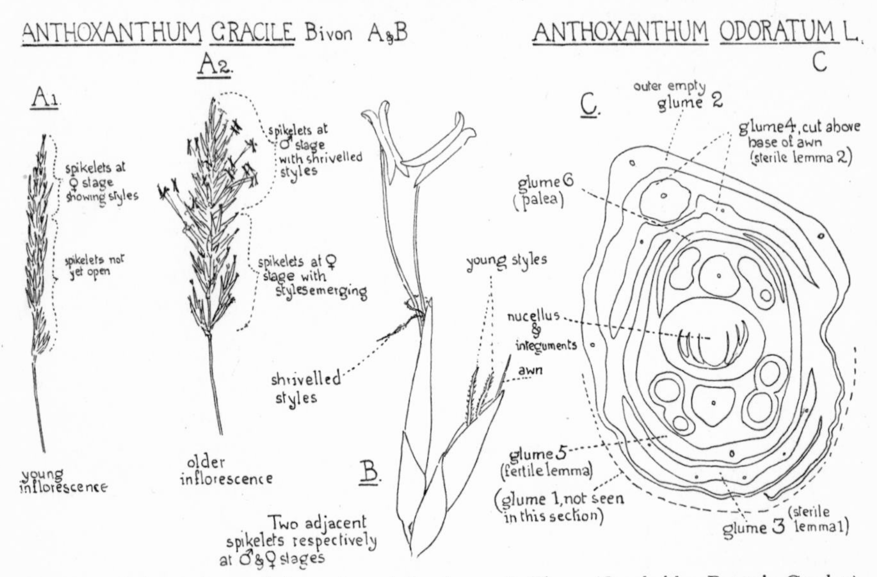

Fig. 71. *Anthoxanthum* spikelets. A and B, *A. gracile* Bivon (Cambridge Botanic Garden). A 1 and A 2, inflorescences at successive stages (× ½). B, two adjacent spikelets (enlarged), of which the lower is at the ♀ stage, and the upper at the ♂ stage. C, *A. odoratum* L., transverse section of a spikelet (× 77 *circa*) at a level above Fig. 85, A 7, p. 182. [A.A.]

rapid; a maximum rate of elongation of 1·6 mm. per minute has been recorded for Rye (*Secale cereale* L.).[2] That this lengthening is due to cell stretching and not to cell division, is proved by the observation that the change in length of the individual epidermal cells is proportional to the change in length of the entire filament.[3] It is the withdrawal of water from the anthers which not only brings about the extension of the filaments, but also induces the dehiscence of the pollen-sacs.

[1] Koernicke, F. (1885); Koernicke points out that his observations relate only to plants cultivated in Germany. [2] Rimpau, W. (1882). [3] Askenasy, E. (1879).

Anthesis is of such importance in the life-history, that it may be worth while to cite descriptions of a few individual instances. In Wheat, as soon as the glumes begin to separate owing to the pressure of the turgescent lodicules, the styles curve apart, and the feathery stigmas spread out. The filaments increase in length from 2–3 mm. to 7–10 mm., and in two to four minutes the anthers are thrust out of the gap between the glumes. The glumes generally open for a period of five to fifteen minutes, and then slowly close again.[1]

A Bohemian writer[2] has given a picturesque account of the process of anthesis in Rye (*Secale cereale* L.). At the beginning of June, he put a number of ears, which were ready to flower, in a curtained south window, and then suddenly at midday he withdrew the curtains and exposed the ears to the full rays of the sun. His description of the succeeding events is that "in the next half-minute, a peculiar stir occurs, a delicate crackling; the glumes begin to relax their connection: it is the signal for the ensuing enchanting phenomenon! Here and there, up between the glumes, the tips of the pretty violet anthers peep shyly out, and forthwith a general commotion begins; in each flower a contest appears to arise between the sister anthers, as to which can first escape from their confined quarters! The frail, delicate filament, which is elongating more and more, can no longer support the weight of the anther; it tips over, and the others follow, scattering around themselves little dust-clouds of pollen-grains. A strange twisting and quivering seems to agitate them; they become flattened out, right to the apex. More and more anthers emerge, and the pollen clouds become more numerous and larger, until finally the powdering is universal; millions of pollen grains cover the table round about the jar. The whole thing is the work of a few minutes—not only can one literally see the grass grow, but even hear it grow!"

The pollen-clouds mentioned in this description may recur in other grasses. An American botanist[3] in Ohio describes a particular occasion, at the end of June, when exceedingly large numbers of Timothy-grass spikelets (*Phleum pratense* L.) were in bloom. The

[1] Percival, J. (1921).
[2] Zuderell, H. (1909).
[3] Evans, M. W. (1927).

air was very quiet that morning until a little past seven o'clock, when a breeze arose. When gusts of wind moved the stems of Timothy plants growing nearby, the pollen could be distinctly seen, rising in small clouds. It was observed that, as the wind passed over a Timothy meadow, about half a mile away, the clouds of pollen appeared as a haze over the field. Similar phenomena have been observed in the case of Rye.

Under the microscope, the process of anthesis is peculiarly fascinating. One observer[1] describes how he thus watched a spikelet from an ear of Barley that was just ripe. There was a slight tremor, and the anthers began to move upwards, the filaments growing visibly. Near the apex of the most advanced anther, a little slit appeared, and presently the next, and the next, opened. Spurts of tiny bullets were sent dancing from each half-open suture, over the enclosing sides of the lemma and palea, and down upon the spreading stigmatic plumes.

The pollen-grains of the Gramineae are fine and smooth and do not cling together;[2] they are thus readily carried by wind to the feathery stigmas. The ease with which they are conveyed necessitates delicate precautions when grasses are being pollinated artificially. The greenhouse must be draught-proof, not only during the experiment, but for an hour beforehand, to allow any free pollen to settle down.[3] Though wind carriage is the usual method of cross-pollination throughout the family, yet there are certain rare exceptions. It is said, for instance, that small insects may convey pollen into the Wheat flower,[4] while in India the spikelets of *Andropogon monticola* Schult. are much visited by pollen-collecting bees.[5] Another suggestive example has been discovered recently by P. W. Richards, in the course of ecological work in British Guiana. He found a species of *Pariana* growing as a social plant under swampy forest conditions, and bearing inflorescences, though seldom many together. The massed flowers were rendered conspicuous by the numerous brilliant yellow anthers. Several times they were noticed to be sur-

[1] Wilson, A. S. (1874).
[2] Hildebrand, F. (1873).
[3] Jenkin, T. J. (1924).
[4] Percival, J. (1921).
[5] Hole, R. S. (1911).

rounded by clouds of insects, which seemed actually to be alighting on the flowers. On one occasion the insects were Phorid flies, and on another, stingless bees (*Melipona* spp.). The air in the forest was generally so still that there was little chance of pollen conveyance by wind.[1]

When grasses are in flower, the slender yellow or violet anthers hanging from the flexible filaments, and the stigmatic feathers looking like frosted glass, give a delicate air to the inflorescence. It is not difficult to understand the enthusiasm which led Trinius, a century ago, to declare that he would willingly sell his last coat for a new grass.

In England, the way in which flowering may continue far into the winter is a remarkable feature of the Gramineae. For some years I have kept a note of the grasses which I have found in flower (showing anthers or stigmas) in the late autumn and winter, chiefly by the roadside within a few miles of Cambridge. It may be worth while to cite a few of the December and January finds. I have seen *Poa annua* L. in anthesis on Dec. 13, 1927, Dec. 28, 1930, Jan. 6, 1925; *Arrhenatherum avenaceum* Beauv., Dec. 1, 8, 19, 1924, Jan. 22, 1925; *Agropyron repens* Beauv., Dec. 4, 1932, Jan. 9, 1927; *Lolium perenne* L., Dec. 13, 1927; *L. multiflorum* Lam., Jan. 9, 1927; *Brachypodium sylvaticum* Beauv., Dec. 1, 8, 1924, Jan. 23, 1933 (in a severe frost); *Dactylis glomerata* L., Dec. 11, 1926, Jan. 7, 1925; *Phleum pratense* L., Dec. 16, 1931; *Avena sativa* L., Jan. 2, 1927; *Trisetum flavescens* Beauv., Jan. 9, 1927; *Bromus sterilis* L., Jan. 9, 1927.

These dates bear witness to one of the great advantages of grass-hunting as a hobby—the fact that there is almost no close season for the flowers. In the depths of winter, when other plants are mostly quiescent, the devotee of grasses can find what he seeks, even if his search has to be restricted to the borders of the hard high-road.

Now that we have considered the process of anthesis, it remains to touch upon the later history of the gynaeceum.

Cytological details about the reproductive organs fall outside the scope of this book, but it may be noticed that the most unusual

[1] Davis, T. A. W. and Richards, P. W. (1933), and additional information from Mr Richards.

feature of the grass embryo-sac is the tendency to multiplication of the antipodal cells,[1] in the time preceding fertilisation. A botanist, writing in 1926,[2] stated that of the forty-five Gramineae, whose antipodals had been studied at that date, this peculiarity was shown by thirty-six. The highest numbers recorded are fifty to sixty in a species of *Bambusa*[3] and fifty to one hundred in Maize.[4]

The ovule[5] of the Gramineae has two integuments (Fig. 68, E, p. 153), each composed, in general, of two cell layers only. The outer integument always disappears a little time after fertilisation. The inner integument, on the other hand, generally persists to the maturity of the fruit, and forms the testa. The nucellar epidermis may in certain cases survive and take part in the constitution of the integument (e.g. *Bromus* and *Brachypodium*). The inner face of the pericarp (ovary wall) is nearly always partially resorbed. In most of the grasses, the remaining portion enters into intimate fusion with the integument, giving the dry, one-seeded indehiscent fruit, peculiar to the Gramineae, which is technically known as the *caryopsis*,[6] but, in common parlance, the 'seed' or 'grain'.

In certain grasses—*Eleusine*, *Crypsis* and *Sporobolus*—the connection of pericarp and seed-coat is less intimate. The outer integument and almost the whole of the pericarp are resorbed in *Eleusine* after fertilisation. The ripe seed has a well-developed testa, which is enclosed only by a thin pellicle, representing the outermost layers of the pericarp. Such a fruit may be included among achenes. In *Crypsis* and *Sporobolus*, on the other hand, the pericarp at maturity is transformed into mucilage. It has been described[7] that, when inflorescences of the Mediterranean species, *Crypsis aculeata* Ait., are examined in the summer after heavy rains, it is not unusual to find the panicles quite covered with objects resembling the shiny eggs of certain insects. These prove to be the caryopses, which have emerged from between the glumes, but still adhere to their tips. If ripe ovaries,

1 For references, see Schnarf, K. (1929).
2 Shadowsky, A. E. (1926).
3 Schnarf, K. (1926).
4 Weatherwax, P. (1931).
5 On the history of the seed and fruit, see Guérin, P. (1898[2]), (1898[3]), (1899).
6 Richard, A. (1811), p. 235; see also True, R. H. (1893).
7 Duval-Jouve, J. (1866).

which have not yet been extruded in this way, are put into water, it is found that they float, and after two or three minutes surround themselves with a whitish aureole. This mucilaginous swelling of the pericarp more than doubles its diameter. It opens like a bivalve shell and the seed emerges laterally, remaining adherent to the pericarp by a delicate funicle. *Sporobolus pungens* Kunth behaves like *Crypsis*.

In most of the flowering plants that have one-seeded indehiscent fruits, it is the naked ovary in its ripened state which falls to the ground. This happens among the grasses in *Eragrostis*, in which the grains are of dust-like smallness; but such a condition is rare among the wild members of the family.[1] Usually the palea and lemma, at least, are shed with the caryopsis (e.g. *Bromus*). In *Phalaris canariensis* L. these glumes become light brown and shining, and closely invest the grain, which in this condition is known commercially as 'Canary-seed'.[2] Besides the lemma and palea, a segment of the rachilla may be included in the unit which is detached (e.g. *Phragmites*). A further stage is that in which the whole spikelet falls, either without the outer empty glumes (e.g. *Avena barbata* Brot., Fig. 65, p. 148), or complete with them (e.g. *Melica ciliata* L.), or even with a segment of the inflorescence axis in addition (e.g. *Hordeum bulbosum* L.). The ultimate elaboration is reached when several spikelets fall together as a single fruit-complex (e.g. *Anthephora pubescens* Nees, Fig. 72, C 1 and C 2, p. 165, and *Tragus racemosus* (L.) All.).

It may seem astonishing that so many varieties of 'fruit' structure can be produced in a family committed to so simple and uniform a gynaeceum; but the power of presenting innumerable variants—without exceeding the limits of the family type—is one of the inherent traits of the great groups of flowering plants. It may be compared with the capacity for word-production possessed by a limited alphabet.

[1] On the nature of the units detached as 'fruits' or 'seeds', see Hildebrand, F. (1872); for a defence of the use of the word 'seed' in this connection, see Reid, C. (1899); for figures of the 'seeds' of British grasses, and a key to their identification, see Armstrong, S. F. (1917).

[2] Syme, J. T. B. (afterwards Boswell) (1873).

CHAPTER IX

THE REPRODUCTIVE SHOOT IN GRASSES: COMPRESSION AND STERILISATION

IN the preceding chapter, we considered the structure and anthesis of the reproductive shoot in grasses, omitting the more aberrant forms. Even when the field was thus limited, we met with a series of instances in which the divergence from the 'typical' Monocotyledonous floral diagram—or even from the more restricted scheme developed in the Bambuseae—followed the direction of fusion or suppression of parts. Examples which may be recalled are: fusion of the outer empty glumes; absence of the third lodicule; fusion of the remaining pair; absence of one whorl of stamens; loss of two of the three stamens of the remaining whorl; and the uniovular character of the gynaeceum. In this chapter we shall continue to study the subject of reduction, both in typical grasses and in the more peculiar examples in which the factor is specially apparent. It is remarkable that this reduction-trend—though itself, in the last analysis, a negative character—has led to an exceedingly varied series of forms.

Certain modifications in the outer empty glumes may first be discussed. The inflorescences of the curious little grass, *Cornucopiae cucullatum* L., have delicate cupules, each of which encloses one spikelet; they are drawn in Fig. 72, D and E, while the structure is shown in section in F–H. Unlike certain former writers,[1] I interpret the cupule as consisting of the first and second outer empty glumes in a state of fusion. This duplex origin is indicated in Fig. 72, F, in which it is seen that the halves of the cupule, to right and left, do not become free from the axis simultaneously; that to the right is still attached to the rachilla at a level at which that to the left is becoming free. Above the cupule, the rachilla bears a succession of

[1] Neither the description in Bentham, G. and Hooker, J. D. (1883), nor in Goebel, K. von (1884), seems to me to be correct; I think that this is due to the fact that the structure can scarcely be understood without serial sections.

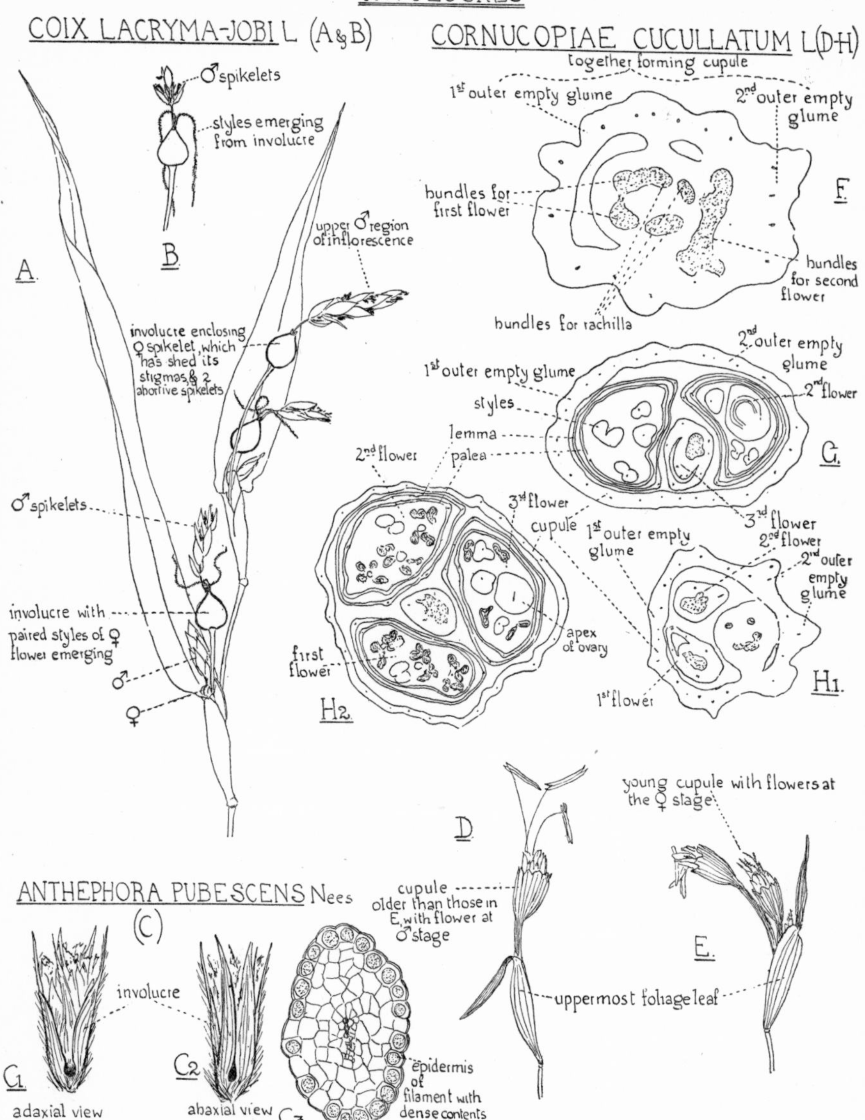

Fig. 72. A and B, *Coix lacryma-Jobi* L., shoot with inflorescences ($\times \frac{1}{2}$), Cambridge Botanic Garden. C, *Anthephora pubescens* Nees. C 1 and C 2, partial inflorescence from Otjitkondo, South-West Africa (enlarged; actual length under 1 cm.). C 3, transverse section of a stamen filament from material from the National Herbarium, Pretoria ($\times$ 193 *circa*). D–H, *Cornucopiae cucullatum* L., Kew Gardens. D and E, inflorescences (enlarged; the actual length of the cupule is about 0·5 cm.). F–H, transverse sections from microtome series of partial inflorescences. F, near the base of the cupule ($\times$ 47). G, higher in a cupule than F, to show three flowers ($\times$ 23). H 1 and H 2, sections ($\times$ 23) at the base and higher up through another cupule, to show departure from distichy in the arrangement of the flowers. [A.A.]

flowers, with each of which a tunicate lemma and palea are associated. Lodicules are absent, and the crowding of the flowers, on the abbreviated rachilla, results in a departure from distichy in their arrangement. We seem here to have a clear case of the connection of crowding and compression with fusion and loss. The departure from distichy can be paralleled in certain abnormal forms of Rye-grass (*Lolium perenne* L., Fig. 74, A–D, p. 168), but there the cause is less obvious.

Another grass in which the outer empty glumes form a cupule-like structure is *Anthephora pubescens* Nees. Fig. 72, C1 and C2, show two views of a partial inflorescence of this grass. I have not been able to examine material young enough to allow of a satisfactory analysis of the shoot. The accepted view is that the partial inflorescence consists of four crowded spikelets, each of which possesses one outer empty glume only; the involucre is held to be formed by the basal fusion of these four glumes.

Lolium perenne L. offers a more familiar instance in which the absence of one of the two outer glumes is generally assumed. It is, however, difficult to prove this contention. The position is that the terminal spikelet of the inflorescence has the normal pair of outer glumes (Fig. 73, C), but in each lateral spikelet there is no empty glume on the adaxial face (B4). One is tempted to associate the 'absence' of the back glume with the pressure of the spikelet against the axis, which is deeply excavated where the spikelet impinges upon it. There is, however, an abnormal 'cristate' form of this species (Fig. 74, E–G), in which the lateral spikelets stand away from the axis of the inflorescence, so that the two ranks of spikelets suggest the open (F) or closed (E) wings of a butterfly. Here each of the two first glumes behaves like a lemma, bearing a flower, normal or abortive, in its axil (e.g. the upper spikelets in G). How these forms are to be related to one another is a puzzle, because, when we find no empty glumes, it is impossible to say whether they are absent in reality,[1] or have developed flowers in their axils and so passed into the category of lemmas. The normal and abnormal[2] spikelet-types

[1] In the labelling of the figures it is assumed that they are absent.
[2] Rouville, P. de (1853) and Bergevin, E. de (1891); see also Fig. 195, p. 387.

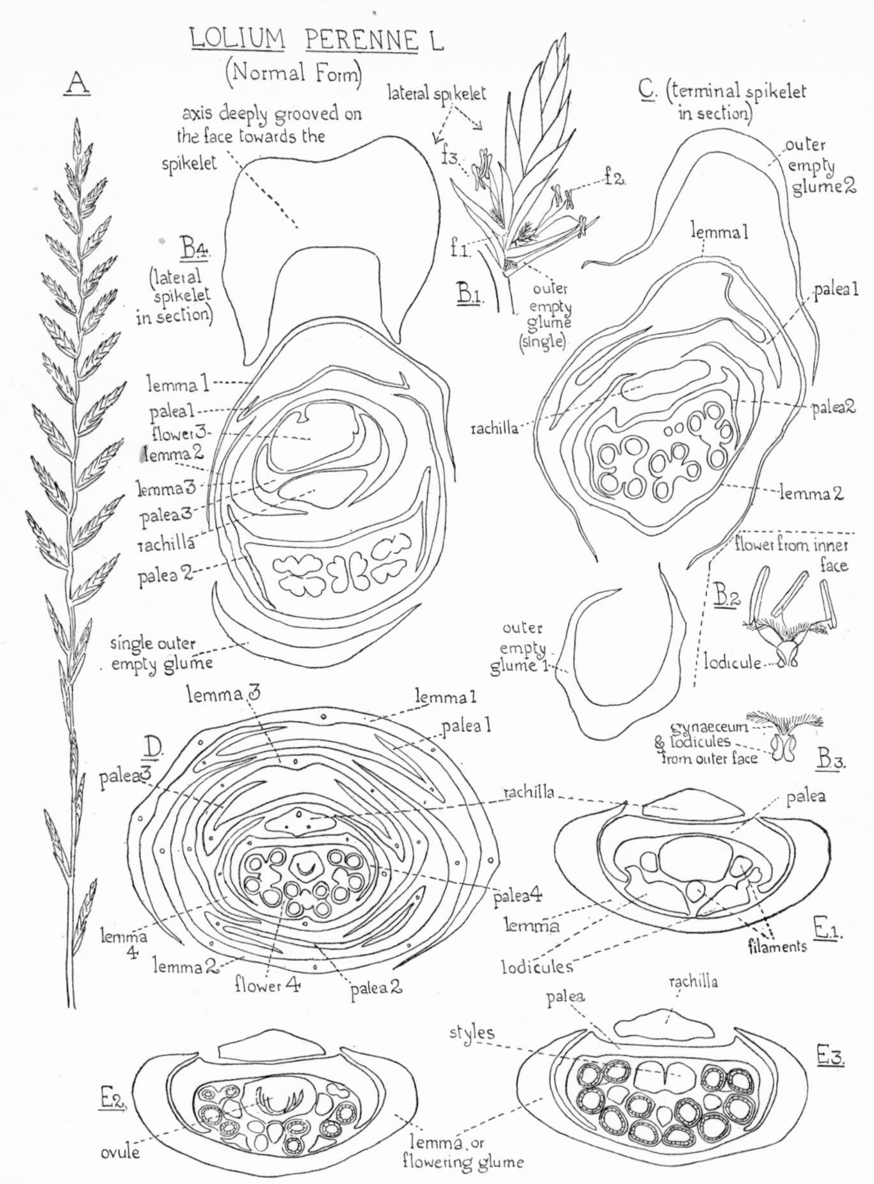

Fig. 73. *Lolium perenne* L. (normal form). A, inflorescence of the slightly awned type (× ½), showing distichous arrangement of spikelets combined with torsion of axis. B 1–B 3, details of spikelet enlarged. B 1, a lateral spikelet at anthesis. In the basal flower, *f* 1, opposite the single outer empty glume, the anthers have fallen and the glumes have closed again. The second and third flowers, *f* 2 and *f* 3, are widely open. B 2, lodicules, stamens and gynaeceum, seen from the side of the palea; B 3, lodicules and gynaeceum seen from the side of the lemma. B 4, lateral spikelet in section, to show the one outer empty glume facing the deeply channelled axis (× 47). C, terminal spikelet in section to show the pair of outer empty glumes (× 47). D, transverse section of part of a spikelet passing through the rachilla and four flowers (× 47). E 1–E 3, sections from a transverse series from below upwards through a flower (× 47). [A.A.]

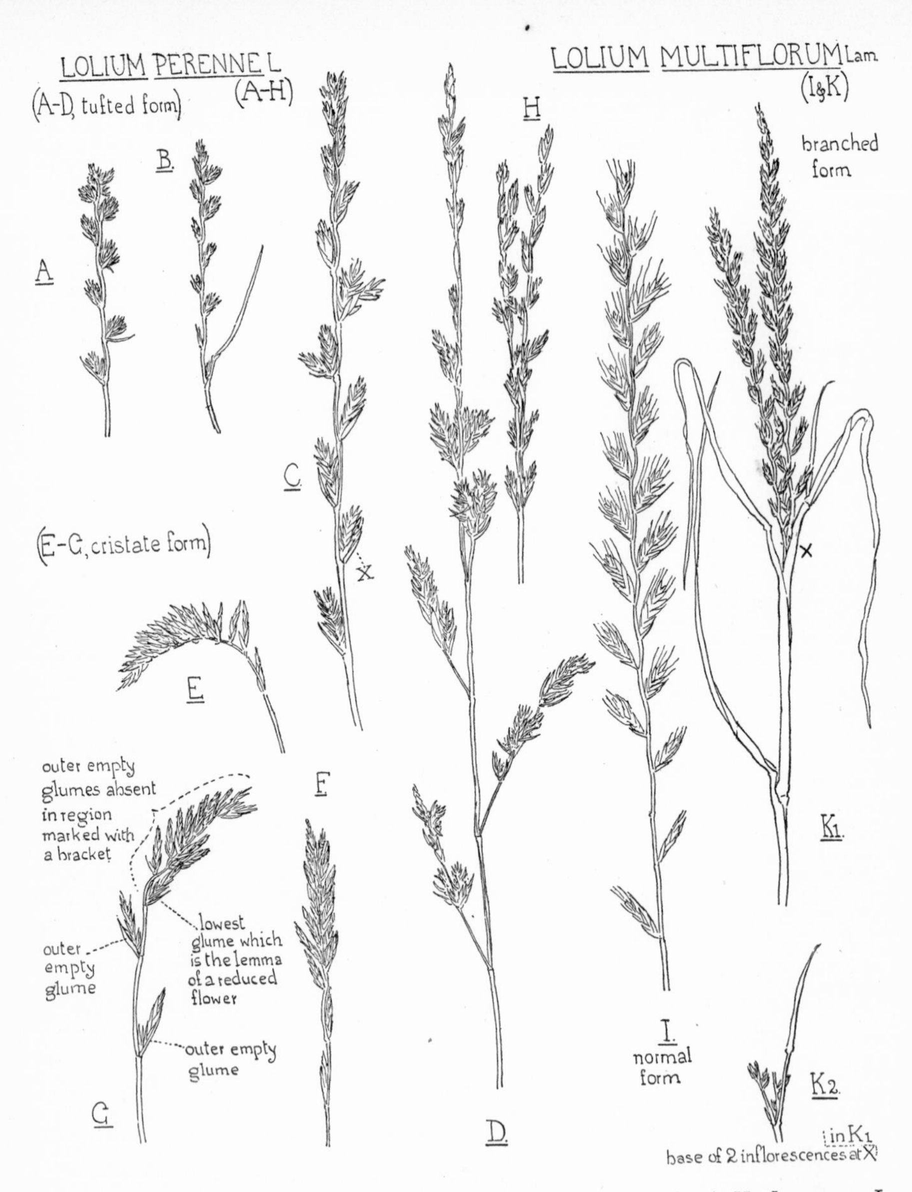

Fig. 74. *Lolium* (normal and abnormal forms). All drawings × ½. A–H, *L. perenne* L. A–D, inflorescences in whose spikelets the flowers form a close tuft, the distichous arrangement being lost. A and B, from the banks of the Cam, near Clayhithe. C, collected by I. H. Burkill, Surrey; the spikelet *X* in C has three orthostichies below and bifurcates above. D, from Kettlewell, Yorks; the individual spikelets are of the tufted form, or show transitions between that and the flat form. E–G, cristate form; E and F, Clayhithe; G, Grantchester. H, bifurcated inflorescence collected by I. H. Burkill, Surrey. I and K, *L. multiflorum* Lam. (*L. italicum* A.Br.) from Whitwell, Isle of Wight. I, normal inflorescence at a stage just after full anthesis when the glumes are held widely open. K, branched inflorescence. [A.A.]

of *Lolium* need a more thorough analysis than they have yet received.

Another example in which the axis is excavated to receive the spikelet is *Nardus stricta* L.[1] (Fig. 75). The reproductive shoot of this grass is much more reduced than that of *Lolium*, and, as in the

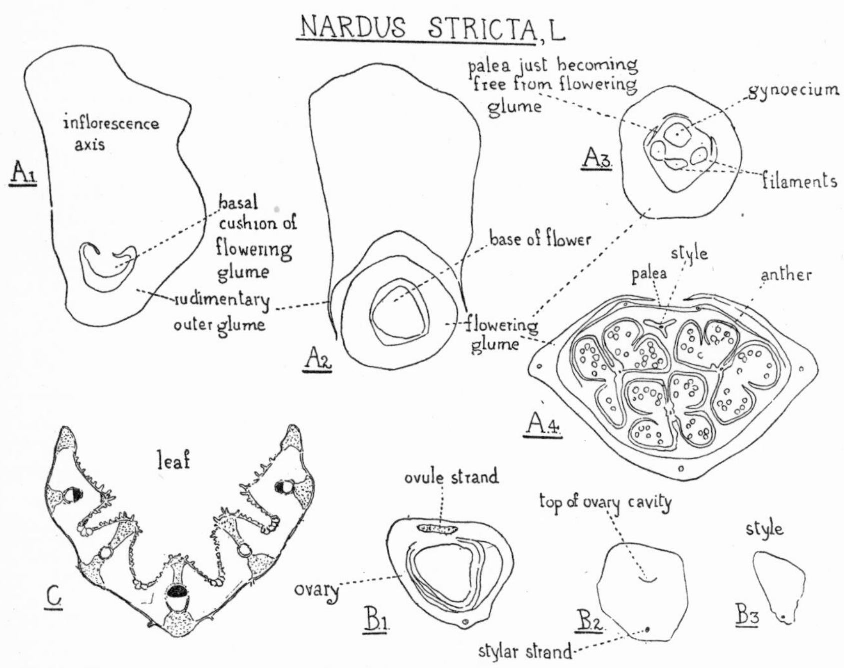

Fig. 75. *Nardus stricta* L. A 1–A 4, transverse sections from a series from below upwards through a spikelet; herbarium material, Switzerland (× 47); A 1 shows an outgrowth (which may represent the rudimentary outer glume or glumes) and the cushion of the lemma; A 2, base of flower enclosed by the ring-like base of the lemma; A 3, palea beginning to separate from lemma, and the gynaeceum and filaments visible; A 4, anthers and stigma. B 1–B 3, transverse sections from a series from below upwards through a gynaeceum, Epping Forest. B 1, ovary; B 2, single stigma (× 77). C, transverse section of the limb of a leaf, Epping Forest (× 47). [Arber, A. (1928[2]).]

next example that we shall consider, the spikelet is one-flowered. The second outer empty glume is absent, and possibly the first also;[2]

[1] On *Nardus*, see Arber, A. (1928[2]), pp. 404–5.

[2] It is possible to regard the structure which I have labelled "rudimentary outer glume" in Fig. 75, A 1, as merely an outgrowth of the inflorescence axis.

there are no lodicules, and the single style is served by one anterior bundle only.

Lepturus filiformis Trin.[1] recalls *Lolium perenne* L. in the difference between the terminal and lateral spikelets. Here, however, we do not get actual loss of one outer glume in the lateral spikelets, but these two glumes are forced to one side by pressure against the inflorescence axis (Fig. 76, A 2), while in the terminal spikelet, where there is no continuation of the axis to displace them, they face one another in the normal manner (C 3).

The example of *Lepturus* shows how the outer glume may be driven into an unusual attitude, and leads us on to consider the Barleys,[2] in which the fact that the glumes both lie to the same side of the spikelet has often been a stumbling-block to interpretation. The unit of the *Hordeum* 'spike' is a triad of spikelets, of which the median member is hermaphrodite, while the laterals are either hermaphrodite (Fig. 77, D, p. 172), male (E), or abortive (F). As will also be seen from Fig. 78, p. 173, the main inflorescence axis, and also the base of the triad, are strongly dorsiventral in form and anatomy. It is this dorsiventrality, in combination with the close backing of the spikelets upon the main axis, which renders the unusual position of the glumes inevitable.

The peculiarities of *Hordeum* are carried still further in the strange South American genus *Pariana*. In *P. campestris* Aubl., the inflorescence, illustrated in Fig. 79, B, p. 174, is made up of a series of units threaded, as it were, on the axis, and each consisting of an apparent whorl of five male spikelets, whose bases are flattened and fused to form a cup-like involucre enclosing the solitary female spikelet. This partial inflorescence is equivalent to a pair of *Hordeum* triads; the correspondence will be understood if Fig. 77, D, p. 172, and Fig. 79, A 6, be compared. In each case, the boundary between the spikelet triads is indicated by a dotted line, but to make the figures strictly comparable, Fig. 79, A 6, should be turned through a right angle. The dorsiventrality of the spikelet bases (involucral segments) is

[1] On *Lepturus*, see Arber, A. (1931[1]), pp. 416–18; *Lepturus filiformis* Trin. is more correctly regarded as *Pholiurus incurvus* A. Camus, sub-sp. *P. filiformis* A. Camus; see Camus, A. (1922).

[2] On *Hordeum* and *Pariana*, see Arber, A. (1929[2]).

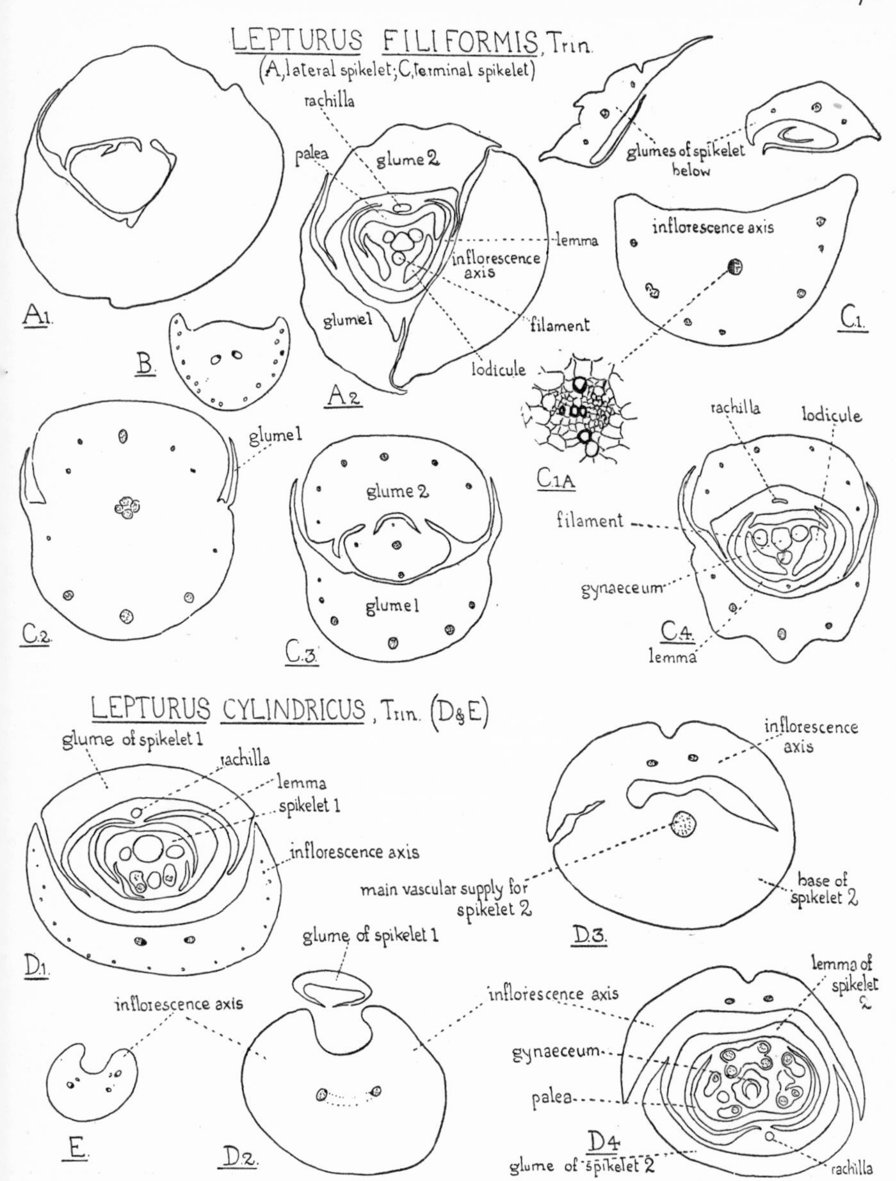

Fig. 76. *Lepturus*. A–C, *L. filiformis* Trin. (more correctly described as *Pholiurus incurvus* A. Camus, sub-sp. *P. filiformis* A. Camus). A 1 and A 2, transverse sections from a series through a lateral spikelet (× 47). B, transverse section through an inflorescence axis below a lateral spikelet (× 23). C 1–C 4, transverse sections through a terminal spikelet (× 47). C 1 A, the bundle in C 1 (× 193). D and E, *L. cylindricus* Trin., transverse sections through a segment of an inflorescence axis bearing two spikelets (× 47). E, transverse section of an inflorescence axis (× 23). [Arber, A. (1931[1]).]

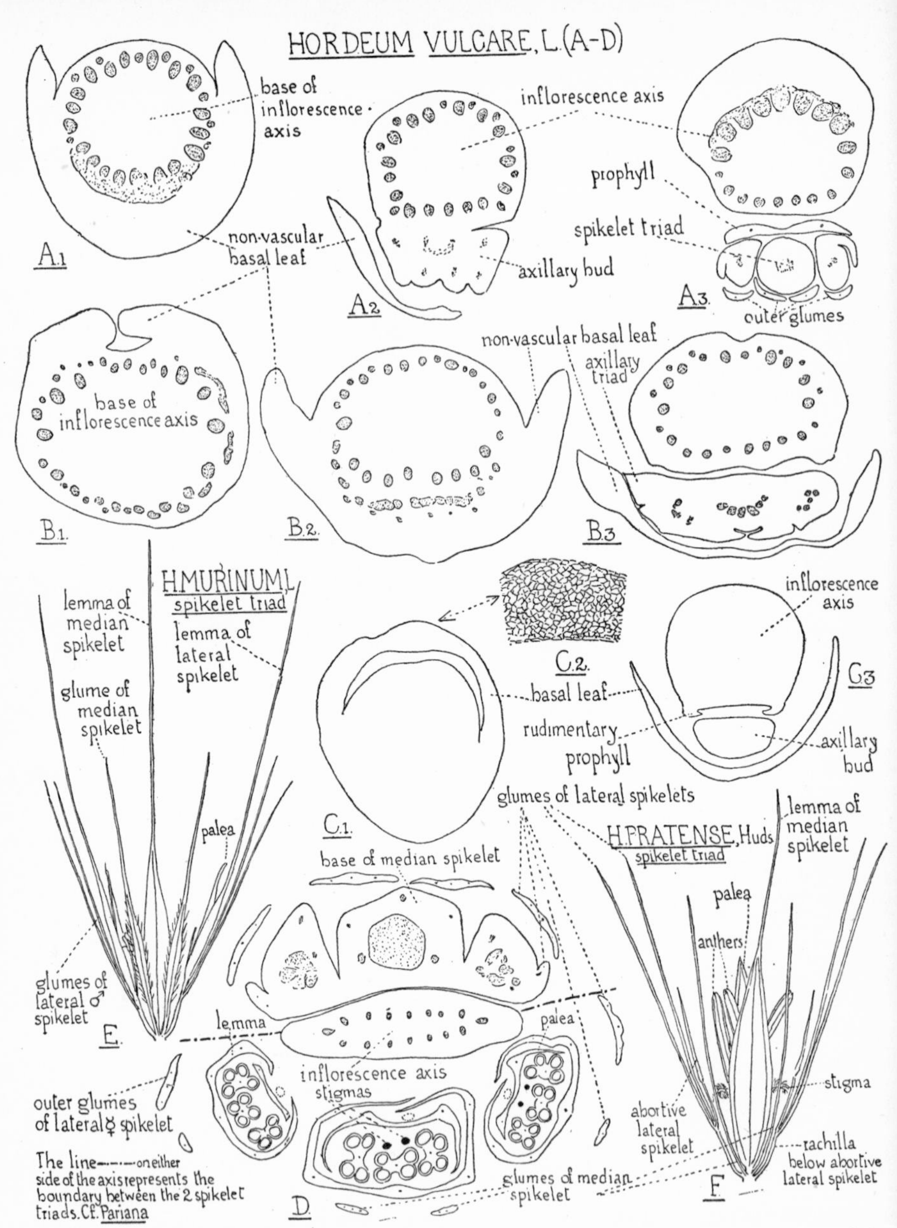

Fig. 77. A, B, D, *Hordeum vulgare* L.; C, *H. distichon* L. (wrongly labelled *H. vulgare* L.). A–C, series of transverse sections from below upwards through bases of inflorescences to show collar-leaf (× 24). C 2, small part of the free region of the leaf in C 1 (× 48). D, transverse section (× 24) to show general structure of six-rowed Barley. The broken line on either side of the axis marks the boundary between the triads, cf. *Pariana*, Fig. 79, A 6, p. 174. E and F, triads of spikelets from abaxial side. E, *H. murinum* L. F, *H. pratense* Huds. (*H. secalinum* Schreb.). [For fuller legend, see Arber, A. (1929[2]).]

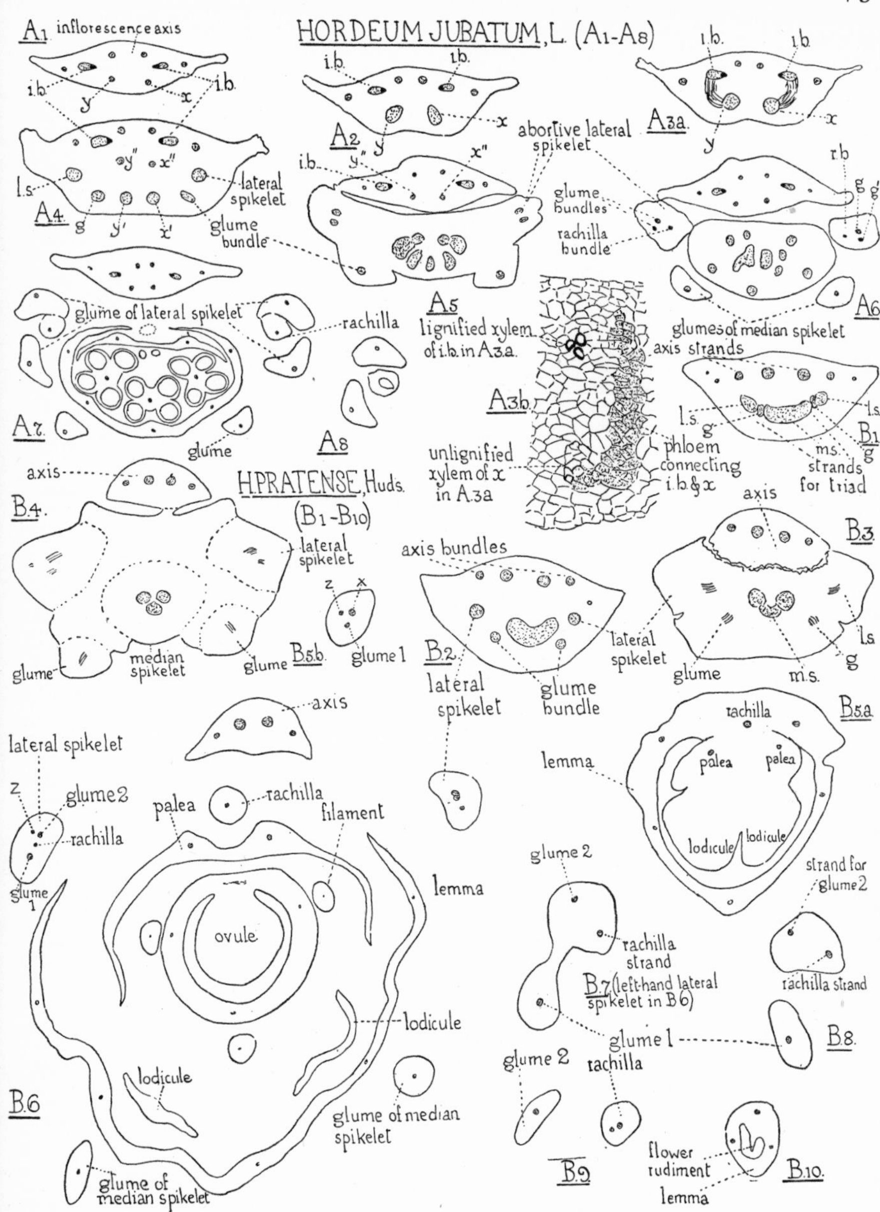

Fig. 78. A1–A8, *Hordeum jubatum* L. Series of transverse sections from below upwards through a segment of a young inflorescence (× 48, except A3*a*, × 199). B1–B10, *H. pratense* Huds. Series of transverse sections from below upwards through a segment of an inflorescence (× 48). [For fuller legend, see Arber, A. (1929²).]

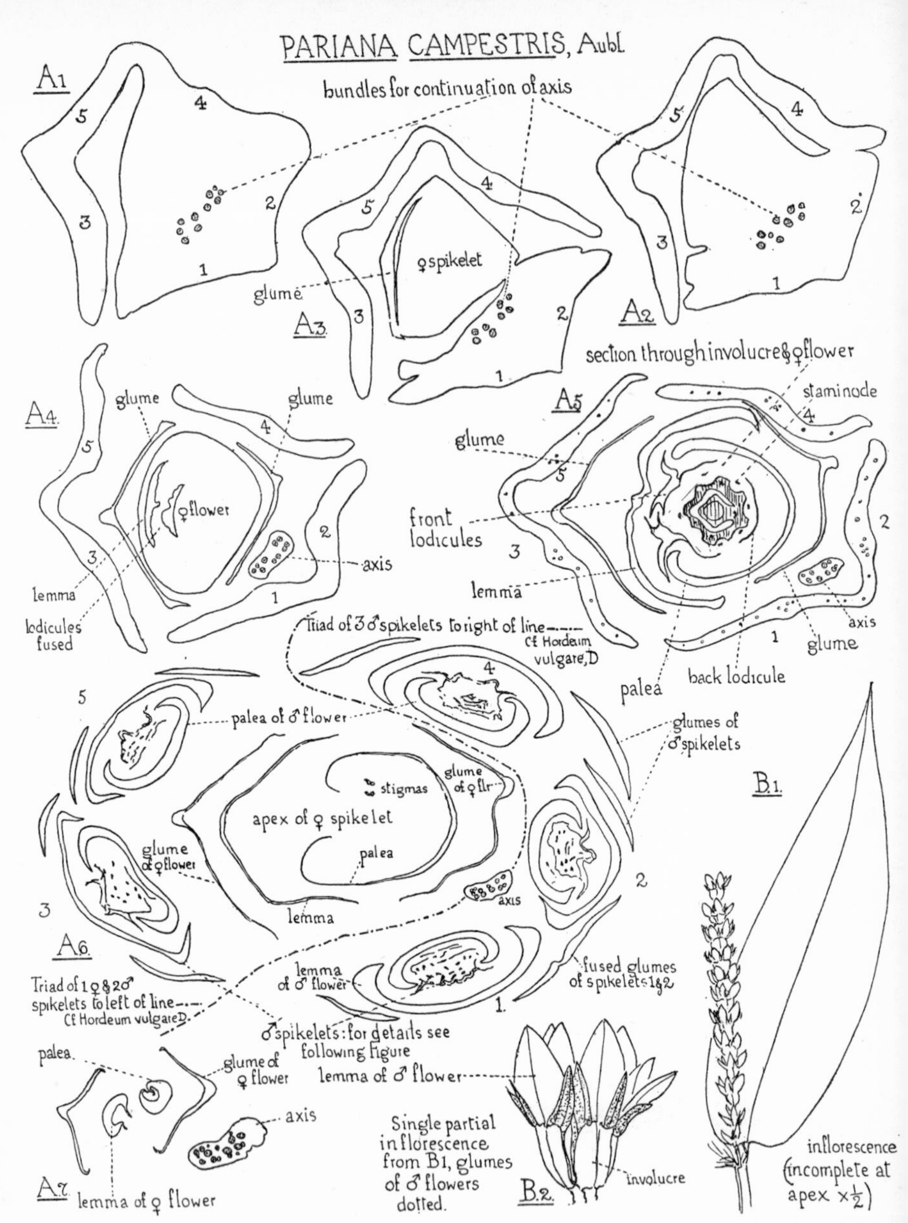

Fig. 79. *Pariana campestris* Aubl., Surinam. A 1–A 7, series of transverse sections from below upwards through a partial inflorescence (× 14). Except in A 5, all bundles omitted except those destined for the continuation of the axis. The male spikelets are numbered in the order in which the bases of the flowers come into view. In A 6, which shows the four male spikelets surrounding the tip of the female spikelet, a broken line indicates the boundary between the two hypothetical spikelet triads, cf. *Hordeum vulgare* L., Fig. 77, D, p. 172. B 1, inflorescence (incomplete) (× ½ *circa*). B 2, single partial inflorescence from B 1, with glumes dotted (enlarged). [Arber, A. (1929[2]).]

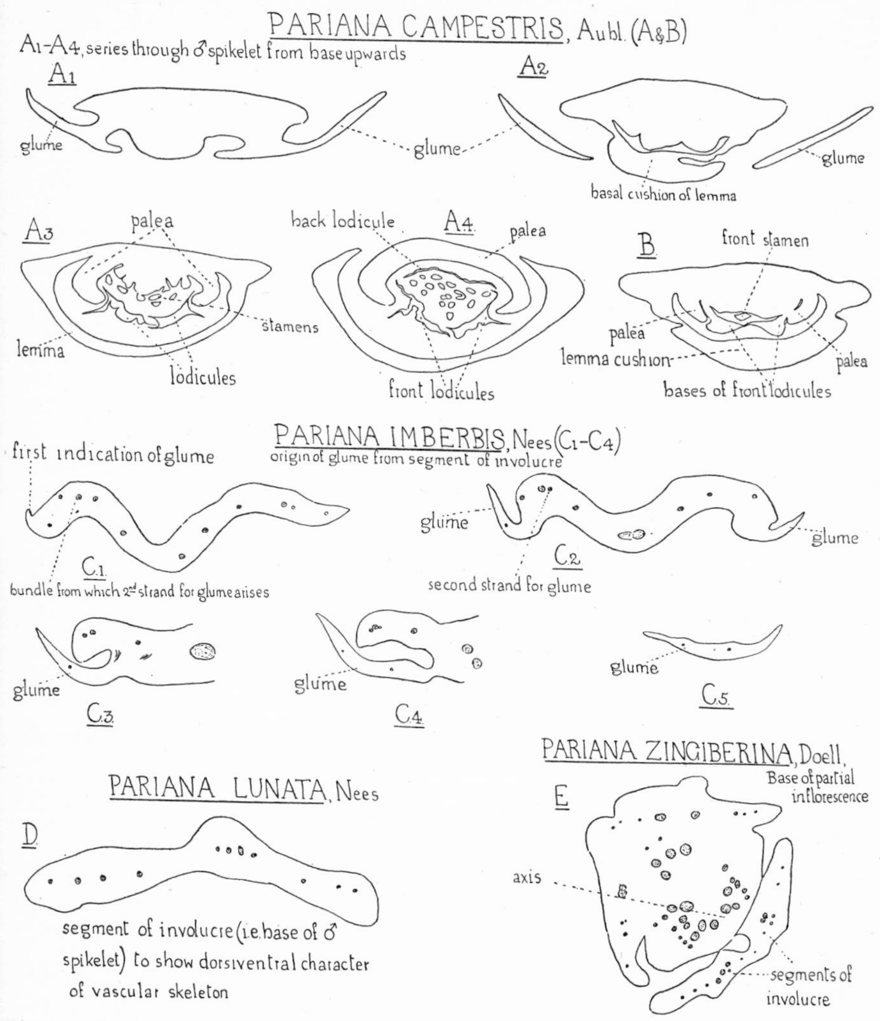

Fig. 80. A and B, *Pariana campestris* Aubl., Surinam. A 1–A 4, series of transverse sections between Fig. 79, A 5 and A 6, p. 174, showing structure of *male spikelet* 1 in greater detail (× 24). C 1–C 5, *P. imberbis* Nees, prope Barra, Prov. Rio Negro, R. Spruce, 1855; series of transverse sections from below upwards (× 24) to show origin of glume from involucral segment. D, *P. lunata* Nees, French Guiana, transverse section of an involucral segment (× 24). E, *P. zingiberina* Doell, British Guiana, transverse section of base of a partial inflorescence (× 14). [Arber, A. (1929²).]

shown in Fig. 80, C 1, D, E, p. 175. The outer empty glumes of the male spikelets are forced out of position exactly as are those of *Hordeum*.

When we leave the outer empty glumes, and consider the main region of the grass spikelet, we find that the tendency to reduction may manifest itself in many ways, one of which is the presence of abortive flowers in addition to those that are normal. These rudiments are commoner than is apparent from taxonomic descriptions; this is not surprising, since they are often indistinguishable without the aid of serial sections. As an example we may take *Melica nutans* L. In the spikelet from which the series of sections in Fig. 81, A 1–A 7 were drawn, there were two perfect flowers at the base, which are seen in A 3. These were succeeded by four abortive flowers, each less developed than the last (A 4–A 7). A further degree of reduction, met with in other grasses, leads to one-flowered spikelets. Above the base of such a spikelet, the rachilla may continue, so that the flower occupies the usual lateral position, e.g. *Phalaris tuberosa* L. (Fig. 82, A 2, p. 178), or the entire apex of the rachilla may be transformed into the flower and its lemma, which are thus actually 'terminal', e.g. *Alopecurus pratensis* L. (Fig. 83, p. 179). Here the outer glumes are united, and the lemma is tubular at the base. The glumes thus form a closed case, which is narrower from side to side than it is in the antero-posterior plane. Hence the flower develops under some lateral pressure, and, with this compression, we may perhaps associate several of the peculiarities of the spikelet: absence of any rudiment of the rachilla above the flower; absence of palea; absence of lodicules; and displacement of one of the lateral stamens into the posterior position.

We touched on the part which pressure may play in modification of the floral shoot when we were considering the palea in the bamboos; the bikeeled form of the palea in grasses is explicable on the same lines. Indeed, the more one studies grass flowers in general, the more conscious does one become of the importance of the restraint which they suffer during development. Its significance was recognised as long ago as 1819 by Turpin,[1] whose

[1] Turpin, P. J. F. (1819).

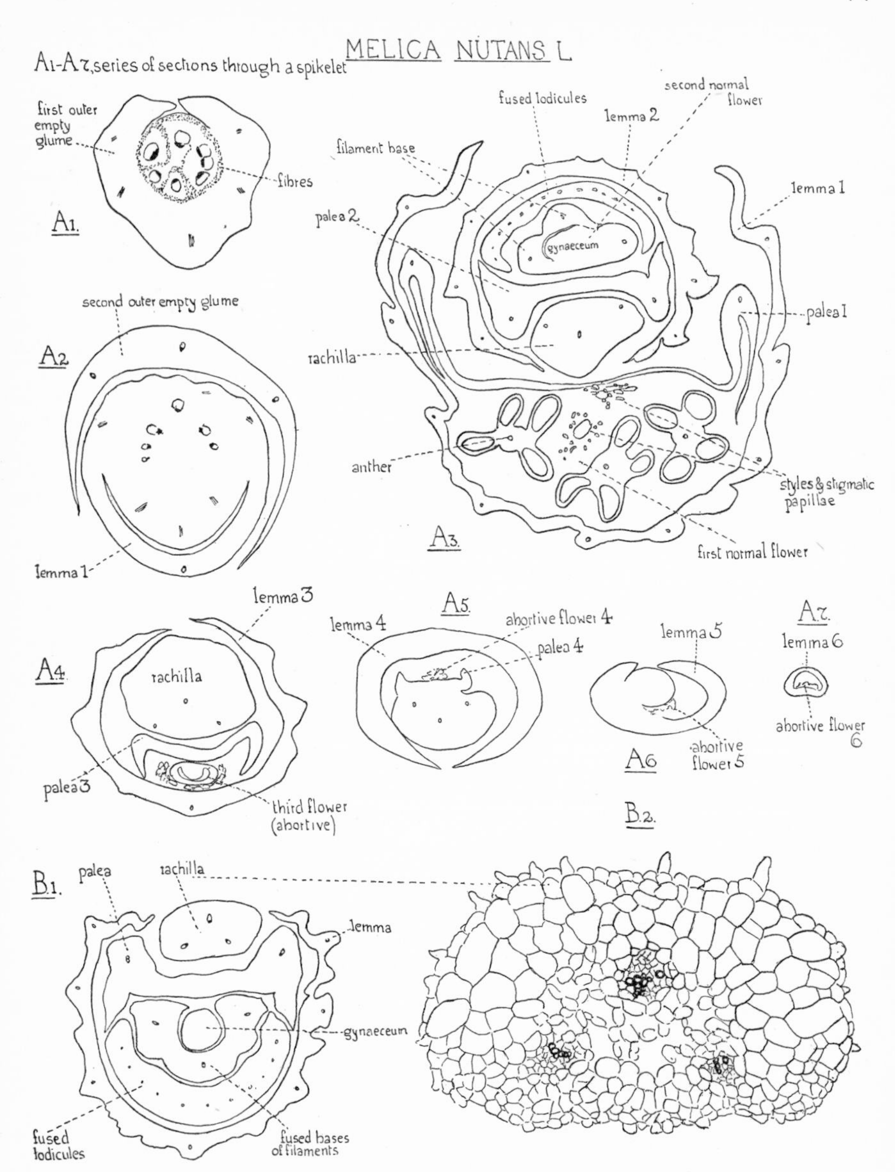

Fig. 81. *Melica nutans* L. Spikelet structure. A 1–A 7, sections from transverse series from below upwards through a spikelet (× 47). B 1, transverse section near the base of a flower from a second spikelet (× 47). B 2, rachilla from a little lower than B 1, at level of Fig. 193, C, p. 384 (× 193 *circa*). [A.A.]

explanation of the frequent absence of the back lodicule in grasses is, I think, worth citing. He notices that in the bamboo, *Guadua angustifolia* Kunth, this posterior lodicule, "placée entre l'ovaire et la *spathelle* [palea], est plus foible que les deux autres qui sont sur le côté extérieur. Cette foiblesse, très-probablement produite par le voisinage de la *spathelle*, qui affame cette écaille, annonce déjà qu'elle doit insensiblement disparoître dans le plus grand nombre des *graminées*, où en effet on ne retrouve plus que les deux écailles

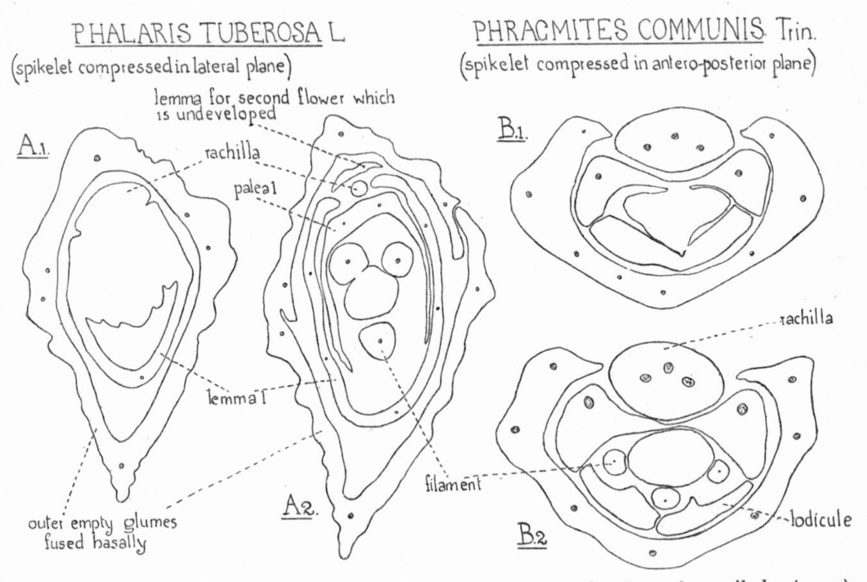

Fig. 82. A, *Phalaris tuberosa* L., sections from a transverse series through a spikelet (× 47); A 1 below, and A 2 above. B, *Phragmites communis* Trin., sections below (B 1) and above (B 2) from two different flowers (× 77 *circa*). [A.A.]

extérieures, c'est-à-dire celles qui s'adossent à la *bractée* [lemma], lorsque celles-ci ne disparoissent pas elles-mêmes entièrement". There is little doubt that Turpin was right, and that it is the situation of the back lodicule—pinched between the essential organs and the palea—which leads to its reduction and ultimate disappearance. Pressure in the antero-posterior plane is more likely to be important in the grasses than in Dicotyledons, because the Gramineae follow the Monocotyledonous plan of a single bracteole (palea) facing the bract (lemma); in Dicotyledons, on the other hand, there are two bracteoles, whose right-and-left position prevents their co-operating with the bract in a back-and-front pressure. Among Monocotyledons in general, there is a distinct tendency to overgrowth on the

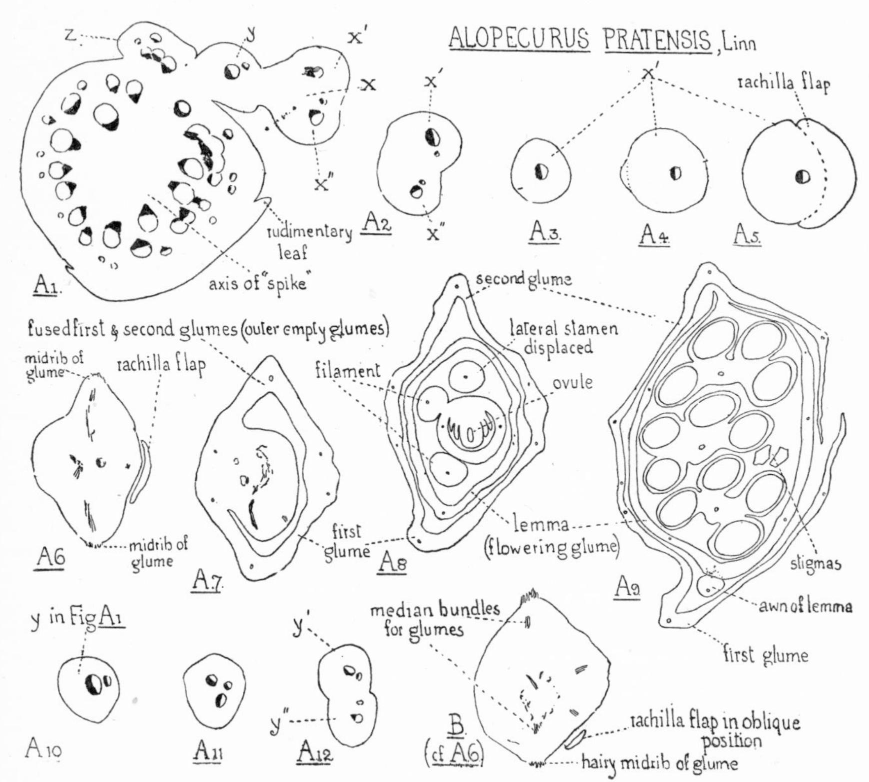

Fig. 83. *Alopecurus pratensis* L. A 1–A 12, sections from a transverse series from below upwards through a segment of a 'spike' (× 47). A 1, axis of 'spike' showing base of a fascicle of three branches, *x*, *y*, *z*. Fig. A 2, branch *x* just about to divide into *x′* and *x″*. Figs. A 3–A 9, the further history of *x′*; A 8 passes through the ovule and A 9 through the anthers. Figs. A 10–A 12, the further history of branch *y*, which is shown attached in Fig. A 1. Fig. B, transverse section of the base of another spikelet (× 47) in which the rachilla flap is placed obliquely. [Arber, A. (1931[1]).]

part of the prophyll, with a correlated reduction in the lateral axis, which lies between it and the axillant leaf.[1] Floral reduction in the Gramineae is an expression of this tendency. In the many-flowered inflorescences of this family, the spikelets are tightly packed in young

[1] Arber, A. (1925), pp. 136–55.

stages, even when they are diffuse at maturity;[1] the compressive action of the prophyll (palea) and axillant leaf (lemma) has thus, in the grasses, a special opportunity of affecting the final form of the axillary shoot (flower).

The interpretation of the grass spikelet on the pressure hypothesis has been expounded chiefly by French writers, more particularly Godron.[1] He points out that at an early period of development of the grass shoot, a series of leaf-sheaths are included within one another, as a result of the shortness of the internodes. These superposed sheaths constrict the haulm, and the rudimentary inflorescence, which they enwrap. The difficulty that is encountered in detaching the leaf-sheaths from a young grass shoot, is an indication of the amount of pressure which they can exert upon the plastic rudiment of the inflorescence; and the effect of the pressure is increased by the fact that this rudiment may be imprisoned for a long time. In Maize, for example, the 'tassel' is already in existence by the time the plant is a few weeks old.[2]

Indirect evidence of the existence of pressure in the antero-posterior plane in the grass flower, is afforded by the fact that the wedge action of the turgescent lodicules is needed to part the lemma and palea. An old observation on the opening of *Lolium temulentum* L.[3]—which perhaps needs modern confirmation—indicates that the force of expansion of the lodicules may counterbalance a tension of 110 grains. Moreover, it has been suggested, that in *Leersia oryzoides* Sw., the cleistogamy[4] is brought about by the mechanical difficulty of separating the closely interlocked lemma and palea.[5]

Important as compression undoubtedly is, it cannot, I think, be made wholly accountable for the reduction effects which are traceable in the reproductive shoots of the Gramineae. There appears to be, in addition, an inherent tendency towards sterilisation, which may possibly be compared with the "advancing sterility" studied by McLean Thompson in the Leguminosae.[6] The joint action of pres-

1 Godron, D. A. (1879); see also Lestiboudois, T. (1819–22).
2 Weatherwax, P. (1931).
3 Wilson, A. S. (1874).
4 See p. 204.
5 Sablon, M. Leclerc du (1900).
6 Thompson, J. McLean (1924) and later papers in this series.

sure, and the sterilisation trend, may be illustrated in our Sweet-vernal-grass, *Anthoxanthum odoratum* L., in which the peculiar features of the spikelet are most easily interpreted as due to reduction from a fuller type. In each spikelet, the solitary terminal flower is preceded by six glumes, of which the third and fourth are awned; these can be seen in Figs. 71, p. 158, 84, 85, p. 182. The flower has no lodicules and only two stamens placed in the antero-posterior plane. This reduction may be correlated with the extra number of glumes which imprison the flower. The difficulty of harmonising the spikelet

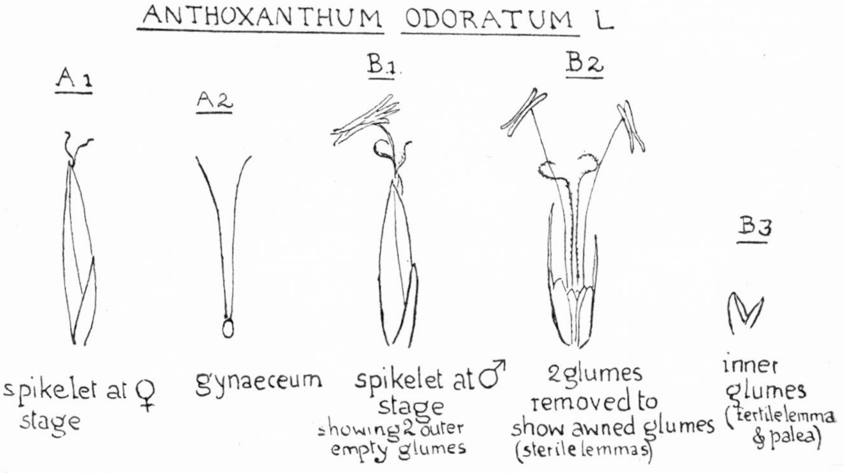

Fig. 84. *Anthoxanthum odoratum* L. Details of spikelet (enlarged). A 1 and A 2, at the stage when the styles protrude, but the stamens have not yet elongated. B 1–B 3, spikelet and dissection at the later male stage. [A.A.]

of *Anthoxanthum* with that of the other Gramineae, lies in the presence of the middle glumes. The two first glumes may reasonably be treated as outer empty glumes, and the two adjacent to the flower as lemma and palea; but we are left with the two intermediate awned glumes to interpret. The solution is to be found in the comparison with the related genus *Hierochloë*;[1] a section of the spikelet of *H. redolens* R.Br. is shown in Fig. 86, p. 183. It has a terminal flower below which there are, as in *Anthoxanthum*, six glumes; but *Hierochloë* also has a male flower, with a palea, in the axil of the third glume and

[1] This comparison was instituted by Doell, J. C. (1868).

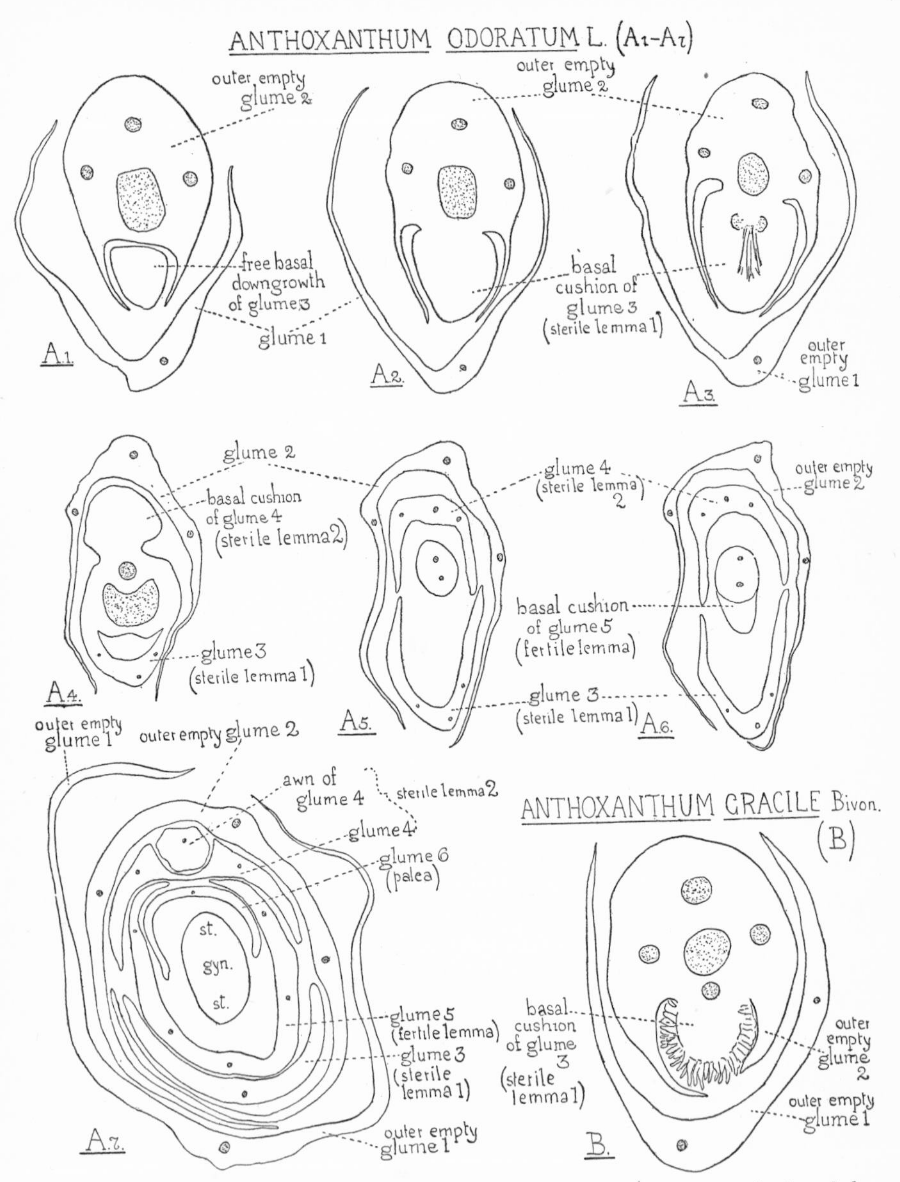

Fig. 85. *Anthoxanthum*. A 1–A 7, *A. odoratum* L. Transverse sections from a series from below upwards through a spikelet (× 77 *circa*); vascular tissue dotted. B, *A. gracile* Bivon, transverse section of the base of a spikelet (× 77 *circa*). [Modified from Arber, A. (1927[2]).]

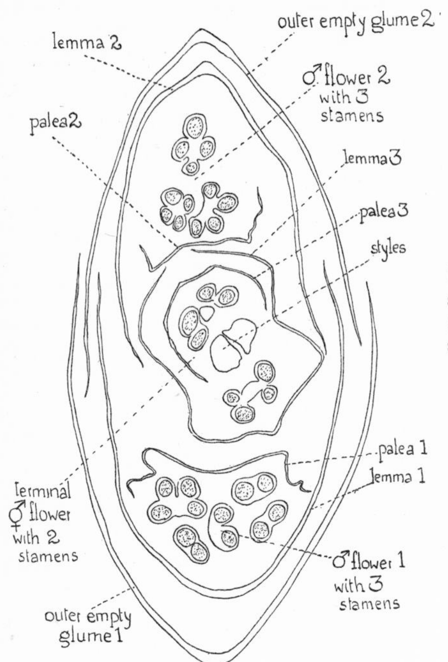

Fig. 86. *Hierochloë redolens* R.Br. Transverse section of a spikelet, herbarium material (× 47); outer empty glumes broken in the sections and somewhat reconstructed. [A.A.]

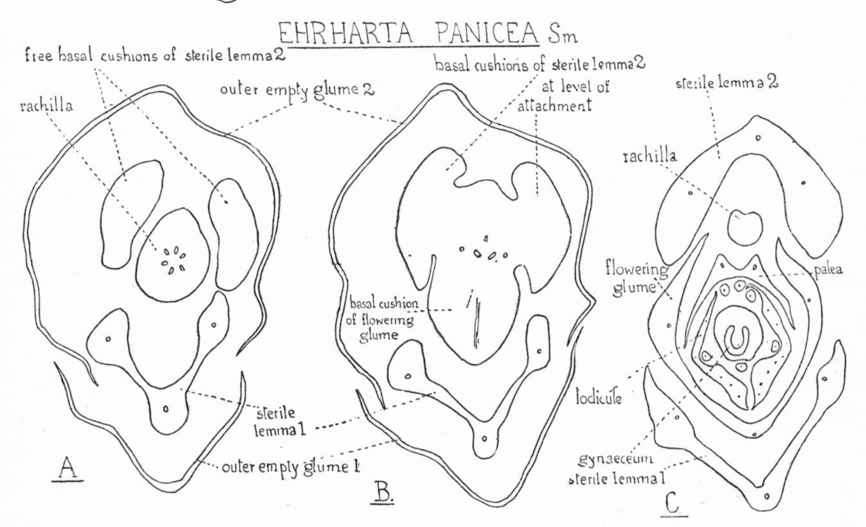

Fig. 87. *Ehrharta panicea* Sm. (*E. erecta* Lam.). A–C, sections from a transverse series from below upwards through a spikelet from Port Elizabeth, Cape Colony, Cambridge Botany School Herbarium; two outer glumes omitted in C (× 47). [Modified from Arber, A. (1927²).]

of the fourth. The empty awned glumes of *Anthoxanthum*, which occupy the positions of the lemmas of the lateral male flowers in *Hierochloë*, may thus be regarded as the lemmas of absent flowers.

In *Ehrharta panicea* Sm. (Fig. 87) there are also four glumes below the fertile lemma. These are generally described as outer empty glumes,[1] but there is little doubt that, as in *Anthoxanthum*, the third and fourth are sterile lemmas; the higher of the two has a pair of remarkable basal wings. Sterile lemmas sometimes show unusual features. In certain species of *Panicum* they bear "crateriform glands",[2] which might provide an interesting study for someone who was able to obtain stages showing their development.

The reduction series beginning with *Hierochloë* and passing through such forms as *Anthoxanthum* and *Ehrharta*, throws light upon the morphology of Rice. In the one-flowered spikelets of this cereal (Fig. 88, A 2), we find a lemma and palea, preceded by two glumes, which have often been interpreted as outer empty glumes; below them are two small structures, which have been regarded by some authors as merely cupular outgrowths. Serial sections through the young spikelet (Fig. 88, D 1–D 5) show, however, that these basal outgrowths are, in reality, the outer empty glumes, though in an extremely reduced form.[3] The two leaves succeeding them should hence be called sterile lemmas (by analogy with those of *Anthoxanthum*) and not outer empty glumes. Vestigial, non-vascular outer glumes are found in another swamp grass related to Rice, *Luziola Spruceana* Benth.[4] (Fig. 168, I 2, p. 319, and Fig. 171, C 7 and C 8, p. 323). A member of the Gramineae in which the outer empty glumes are entirely absent is *Lygeum Spartum* Loefl.[5] (Fig. 63, p. 146).

In a number of the genera which we have hitherto considered, the effect both of development under pressure, and of an inherent trend towards sterilisation, may be invoked to account for the present condition. In others, pressure effects are less obvious, and the

[1] In Arber, A. (1927[2]), these glumes are misinterpreted as outer empty glumes.
[2] Hitchcock, A. S. and Chase, A. (1910).
[3] These sections thus confirm the view expressed by Stapf, O. (1898).
[4] On the glumes of other Oryzeae, see Weatherwax, P. (1928[2]).
[5] On *Lygeum*, see Arber, A. (1928[2]).

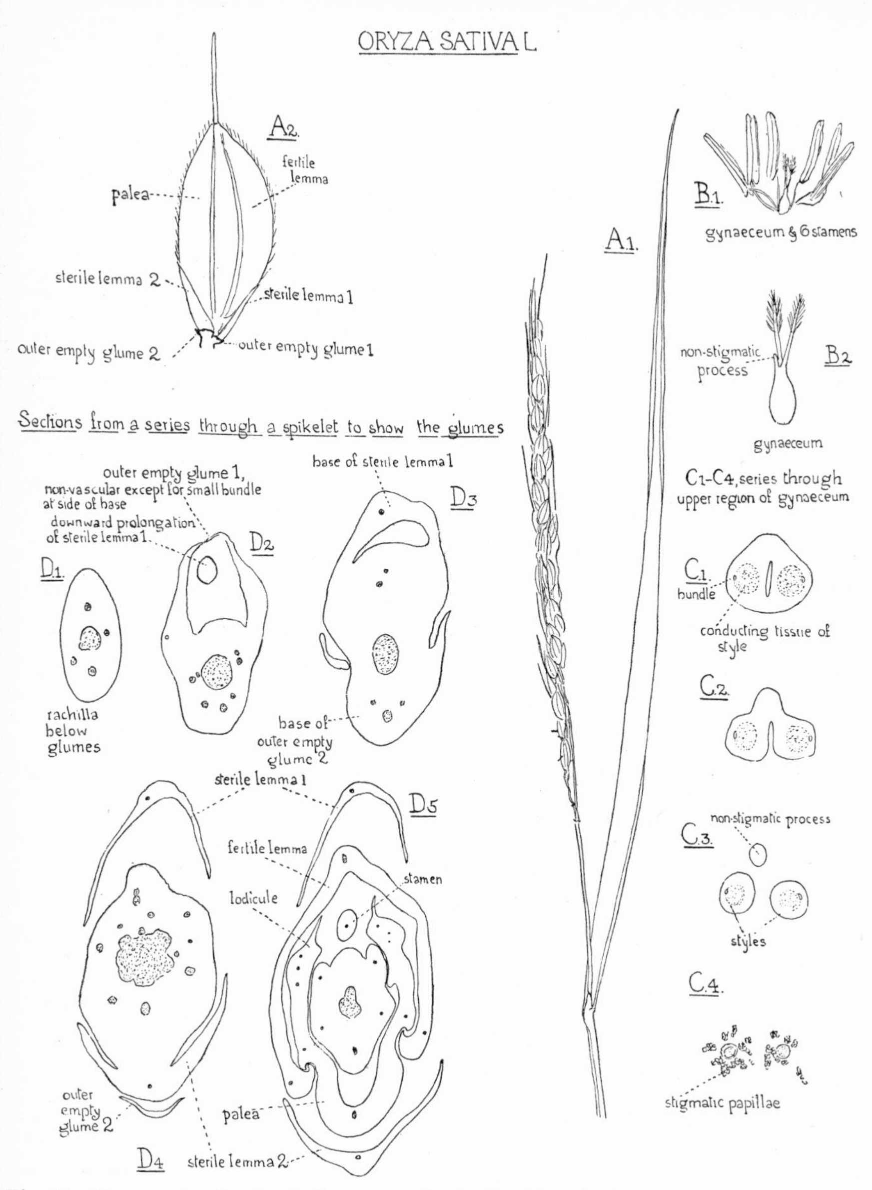

Fig. 88. *Oryza sativa* L. A 1, inflorescence, Cambridge Botanic Garden (× ½). A 2, spikelet (× 3 *circa*). B 1, gynaeceum and six stamens. B 2, gynaeceum more enlarged. C 1–C 4, series of transverse sections from the base of the styles upwards (× 47). D 1–D 5, sections from a transverse series from below upwards (× 47) through a young spikelet to show the succession of glumes. [A.A.]

tendency to sterilisation seems to be at work alone. This tendency may be expressed in a sterile region at the apex of the inflorescence; in *Anthochloa colusana* (Davy) Scribn.,[1] for instance, the 'spike' ends in a tassel of narrow empty bracts, differing in shape and texture from the flower-enclosing glumes below. In other grasses, the reduction of reproductive activity takes another path—certain spikelets developing as sterile shoots, with an increased number of leaves replacing the glumes. Among the bamboos, this vegetative replacement occurs sporadically (cf. Figs. 26, p. 74, 33, p. 109, 37, p. 113, 106, p. 221), while in certain grasses it is a regular and more differentiated feature. When, for instance, the little tufts or fascicles which make up the 'spike' of the Dog's-tail-grass, *Cynosurus cristatus* L.,[2] are examined with a lens, it is seen that, among the fertile spikelets, there are delicate, doubly pectinate structures, looking at first sight like tiny pinnate leaves (Fig. 170, p. 322). Despite their minute size, however, they are not equivalent to single leaves, but to complete shoots, for they each consist of a slender axis bearing a series of small distichous appendages. A related Mediterranean grass, *Lamarkia aurea* Moench,[2] Golden-top, has sterile spikelets corresponding to those of *Cynosurus*. In Golden-top, however, the tendency to reduced fertility is stronger than in Dog's-tail-grass, for whereas the fertile spikelet of *Cynosurus* contains several flowers, that of *Lamarkia* contains one normal flower and one abortive flower only; moreover, the tendency to sterilisation seems to be at work even on this already simplified spikelet, for the fascicles frequently contain reduced spikelets, including glumes, but no flowers. There are thus three types of spikelet in *Lamarkia*, which can be seen in Fig. 172, A 1, p. 324: fertile spikelets; spikelets belonging to the fertile type, but reduced to glumes only; and sterile leafy spikelets.

More striking examples of sterilisation in the inflorescence than those we have as yet considered, are found in the Bristle-grasses—*Pennisetum*, *Setaria*,[3] etc.—in which the spikelets are intermingled

[1] Davy, J. Burtt (1898); this species was described under the name of *Stapfia colusana* Davy.

[2] On *Cynosurus* and *Lamarkia*, see Arber, A. (1928[1]) and Thoenes, H. (1929).

[3] On *Pennisetum*, *Setaria*, and *Cenchrus*, see Arber, A. (1931[1]); for a study of the North American species of *Pennisetum*, see Chase, A. (1921[2]).

with various forms of seta. The 'spikes' of a few of these grasses are illustrated, at half their natural size, in Fig. 89. *Pennisetum* may be taken as an example. The species drawn in Figs. 90 and 91 have

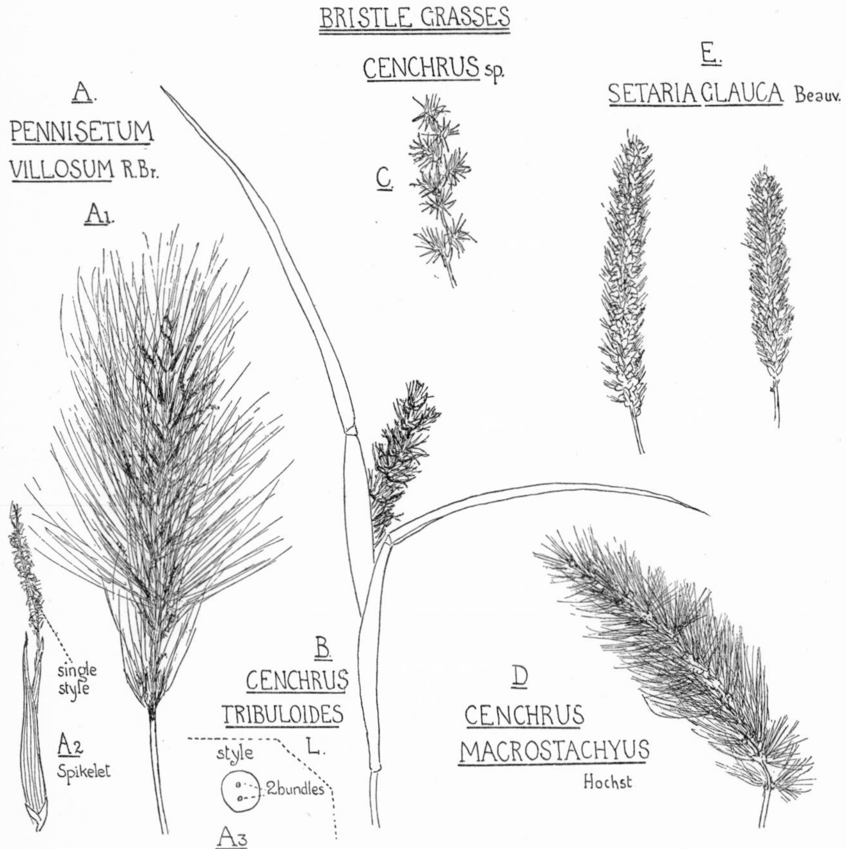

Fig. 89. Bristle-grasses. A, *Pennisetum villosum* R.Br. (*P. longistylum* of gardens). A 1, inflorescence ($\times \frac{1}{2}$) to show feathery styles. A 2, single spikelet enlarged to show the single style which may be inconspicuously bifurcated at the apex (bristles removed). A 3, transverse section of the style to show the two bundles (enlarged). B, *Cenchrus tribuloides* L., inflorescence ($\times \frac{1}{2}$). C, *Cenchrus* sp., New Mexico; D, *C. macrostachyus* Hochst., Abyssinia; inflorescences ($\times \frac{1}{2}$), Cambridge Botany School Herbarium. E, *Setaria glauca* Beauv., inflorescence ($\times \frac{1}{2}$), Cambridge Botany School Herbarium. [A.A.]

fascicles of one, two or three fertile spikelets associated with bristles. One of these bristles is longer and stronger than the rest, and represents the termination of the fascicle axis. The others are associated into tufts, each tuft being, on my view, equivalent to a spikelet. The

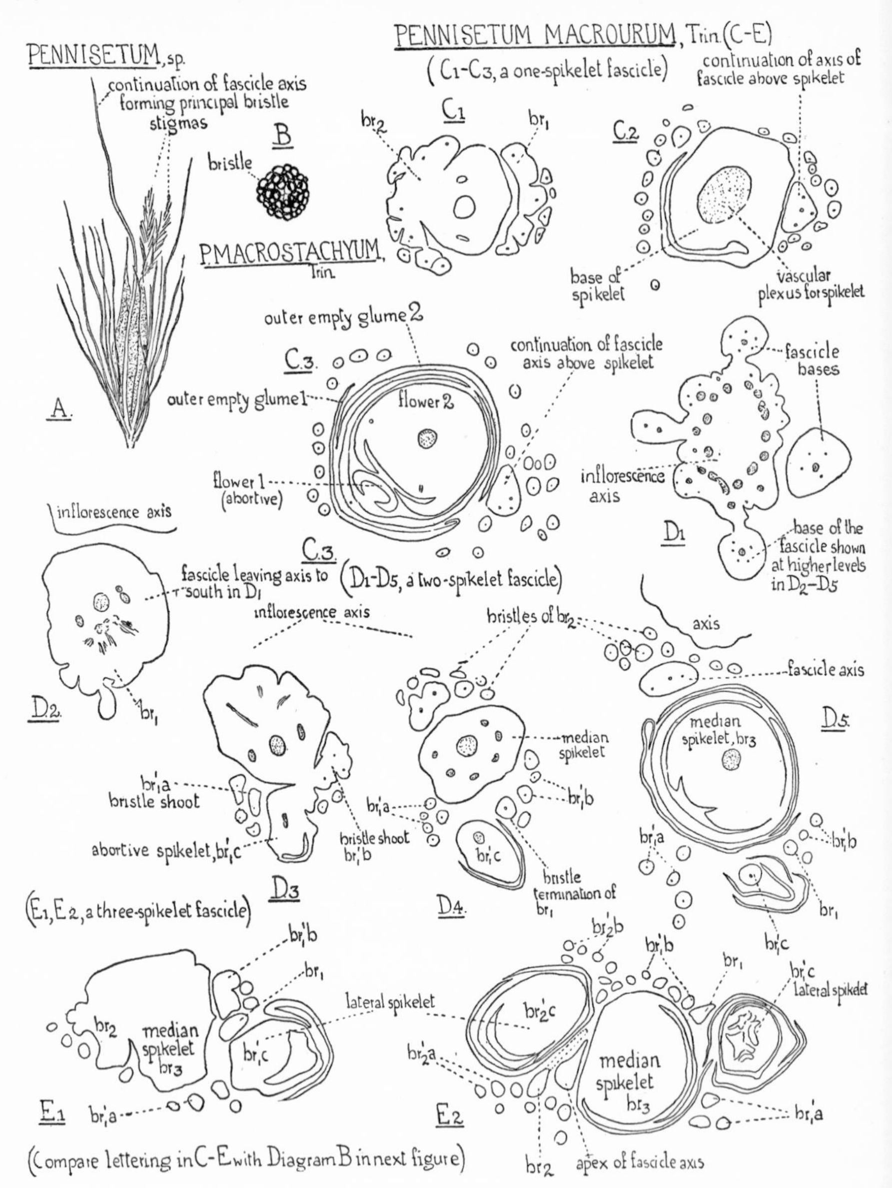

Fig. 90. *Pennisetum*. A, *P.* sp., Buitenzorg Garden. Single fascicle (× $3\frac{1}{2}$ *circa*). B, *P. macrostachyum* Trin., Buitenzorg Garden. Transverse section of a bristle (× 193). C–E, *P. macrourum* Trin., Kew Gardens. Sections of the inflorescence; for explanation of the lettering see diagram B, in Fig. 91, p. 189. C 1–C 3, a one-spikelet fascicle (× 47). D 1–D 5, a two-spikelet fascicle; D 1 (× 23); D 2–D 5 (× 47); E 1 and E 2, a three-spikelet fascicle (× 47). [For fuller legend, see Arber, A. (1931[1]).]

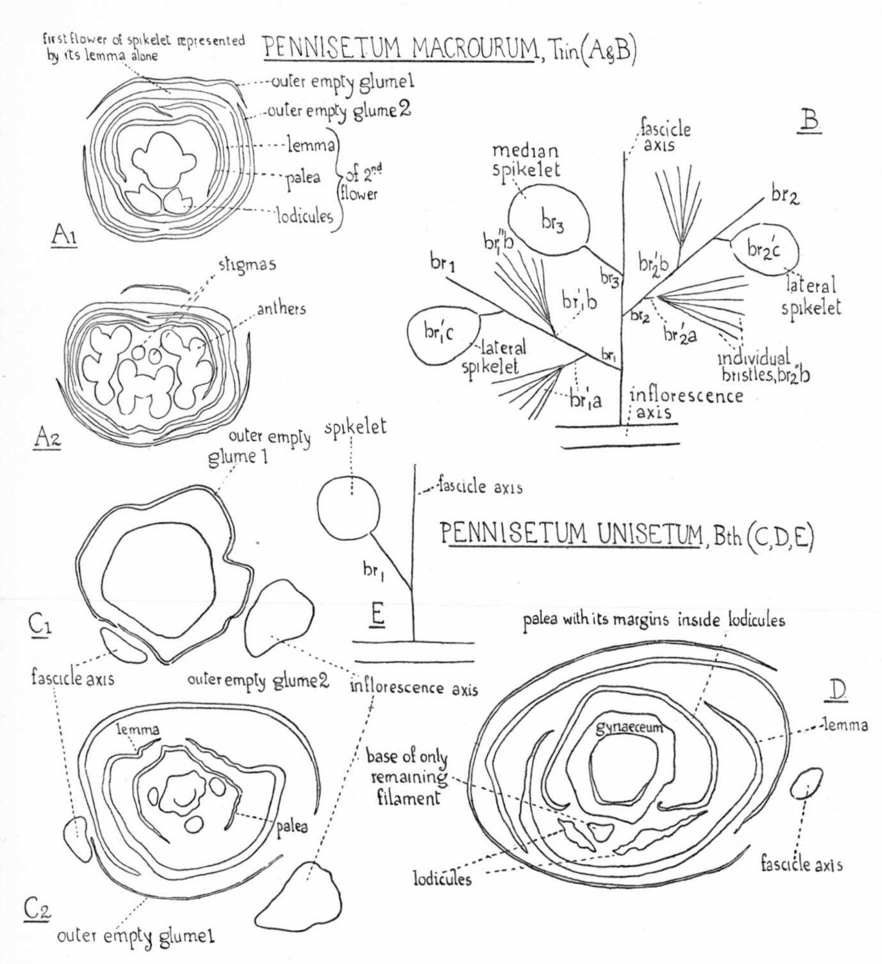

Fig. 91. *Pennisetum*. A and B, *P. macrourum* Trin., Kew Gardens. A 1 and A 2, sections through a young spikelet (× 47). The first flower is represented by its lemma alone. B, diagram of a three-spikelet fascicle such as that shown in section in Fig. 90, E 1 and E 2, p. 188. C–E, *P. unisetum* Bth. C 1 and C 2, sections from a transverse series through a fascicle from Tanganyika Territory, showing the inflorescence axis, and the bristle representing the fascicle axis, to which the spikelet is lateral (× 77 *circa*). D, transverse section of a spikelet from Nyasaland (in which the flower is faded) to show margins of palea lying *inside* lodicules (× 77 *circa*). E, diagram of fascicle. [Arber, A. (1931[1]).]

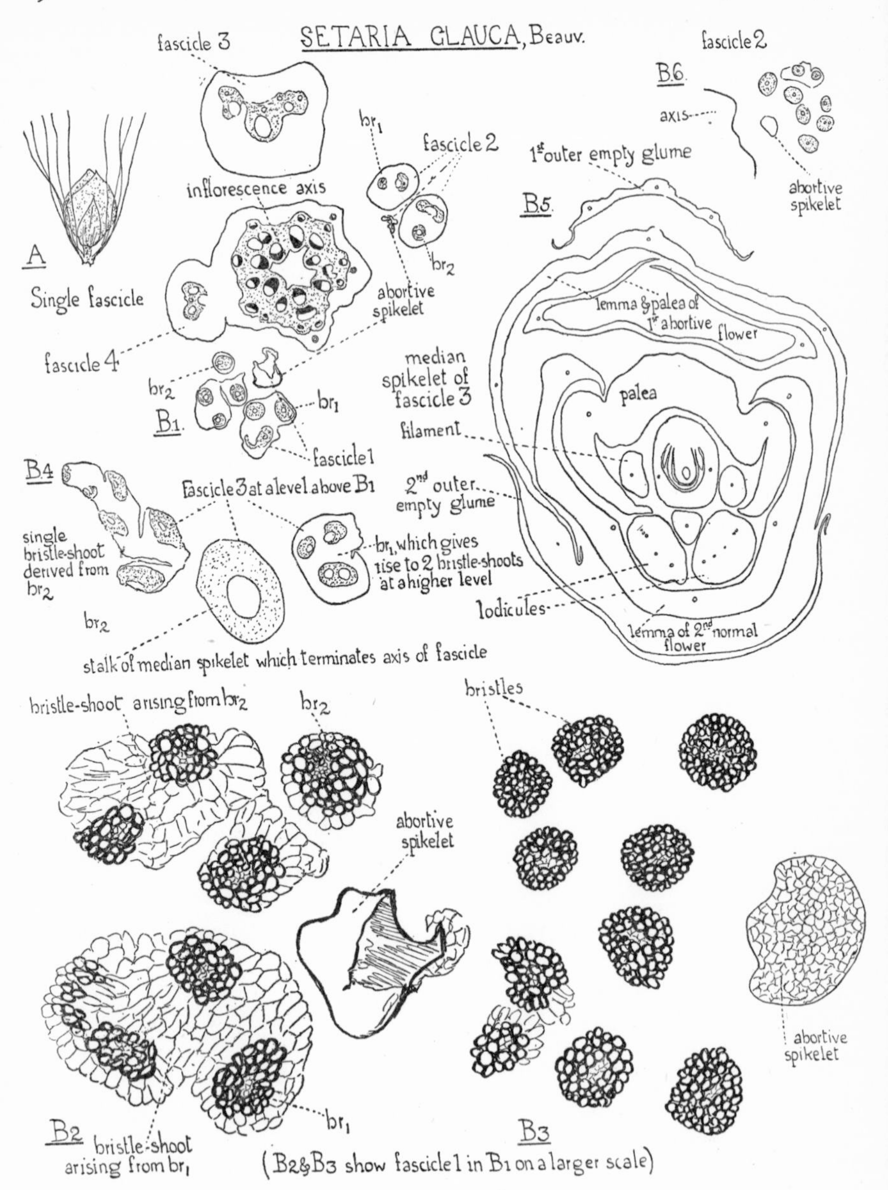

Fig. 92. *Setaria glauca* Beauv., Cambridge Botanic Garden. A, fascicle consisting of a one-flowered spikelet with accompanying bristles (× nearly 3). B 1–B 6, sections from a transverse series through an inflorescence. B 1, axis with the bases of *fascicles* 1–4 (× 47). B 2 and B 3, *fascicle* 1 at level of B 1, and at a higher level (× 193 *circa*); B 4 and B 5, sections through *fascicle* 3 at levels above B 1 (× 47). In B 5, only the median spikelet is shown; B 6, *fascicle* 2 at a higher level (× 47). [Arber, A. (1931[1]).]

fascicle scheme is analysed diagrammatically in Fig. 91, B, p. 189. In *Setaria* the nature of the bristles is essentially similar; the differences will be recognised from the diagram, Fig. 93, B, p. 192.

The inflorescence axis of *Cenchrus* resembles that of *Pennisetum* and *Setaria* in bearing a series of fascicles, but the spikelets of which they are composed, instead of being associated with discrete bristles, are enclosed in an involucre, consisting of the two bristle-bearing branch systems in a state of more or less complete union. The structure is analysed in Figs. 94 and 95. A link between *Pennisetum* and *Cenchrus* is furnished by *Pennisetum ciliare* (L.) Link (*P. cenchroides* Rich.), in which there is basal fusion of the involucral branches (Fig. 94, C 1 and C 2, p. 193).

The bristly fascicles of *Cenchrus* form burr fruits, which attach themselves readily to animals and clothing and are thus dispersed. In *Cenchrus australis* R.Br.[1] they catch in the manes and tails of horses and the wool of sheep. This grass has a long, slender, erect 'spike' of about eighty fascicles, each of which is surrounded by a fringe of some fifty bristles, each bristle being studded with at least 500 retrorse spines. These prickly burrs form a death-trap for insects. A large and powerful beetle has been seen caught by the hind-legs in a head of this grass, while ladybirds, grasshoppers, flies and small Hymenoptera may also become entangled. This insect-catching habit serves no 'purpose' in the plant's economy; it is merely a mechanical outcome of the particular structure of the spines. There is a legend[2] that an Indian *Cenchrus* called Bhurat (*C. catharticus* Del.) once upon a time was pitted against an army, and won. An Emperor of Delhi was on his way to attack Bikaner, when a burr of Bhurat stuck in his arm. He picked it off, and it stuck in his finger; he then tried to bite it off, and it stuck in his lip and gave him intenser pain than ever. When told that the country was full of these things, he ventured no further, and Bikaner was left in peace.

In the examples of sterilisation which we have hitherto considered, entire flowers, or even entire spikelets, lose their fertility. A slighter manifestation of the same tendency is the suppression of either the androeceum or the gynaeceum, but not of both—a suppression which

[1] Lea, A. M. (1915). [2] Coldstream, W. (1889).

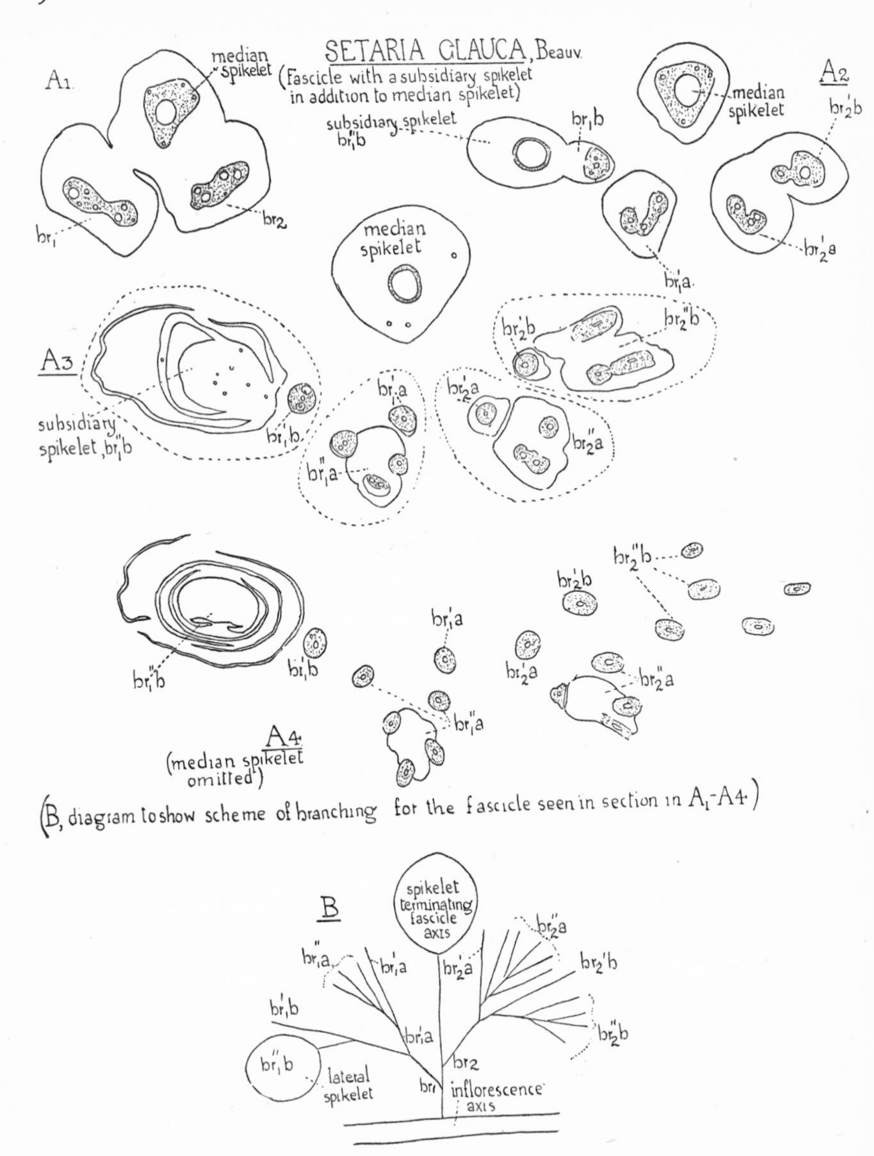

Fig. 93. *Setaria glauca* Beauv. A 1–A 4, series of transverse sections (× 47) through a fascicle with a subsidiary lateral spikelet in addition to the central spikelet. B, diagrammatic scheme of the mode of branching of the fascicle illustrated in A 1–A 4. [Arber, A. (1931[1]).]

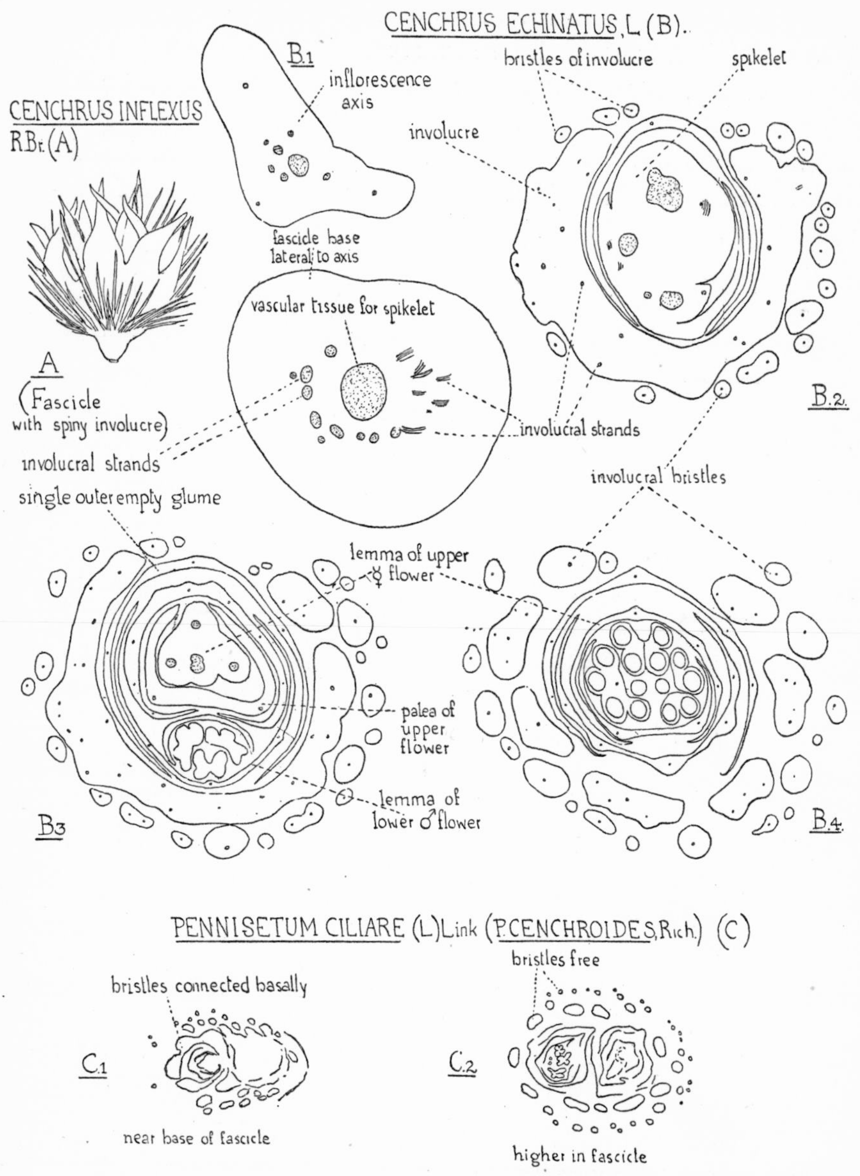

Fig. 94. *Cenchrus*. A, *C. inflexus* R.Br., Buitenzorg Garden. Single fascicle viewed from the abaxial side to show the involucre (× 3½ *circa*). B, *C. echinatus* L., Cambridge Botanic Garden. B 1–B 4, transverse sections from series upwards from below through a young one-spikelet fascicle (× 47). C, *Pennisetum ciliare* (L.) Link (*P. cenchroides* Rich.) from Angola. Transverse sections, C 1, below, and C 2, above, through a fascicle to show the bristles partially fused basally in C 1, and free in C 2 (× 14). [Arber, A. (1931[1]).]

produces unisexual flowers.[1] Various examples of such flowers have been mentioned incidentally, but we still have to consider the general question of the distribution of the sexes in those grasses in which the flowers are not all hermaphrodite.

The grasses show every grade[2] between 'dioecious' species, which have their male and female spikelets on different plants, and are thus

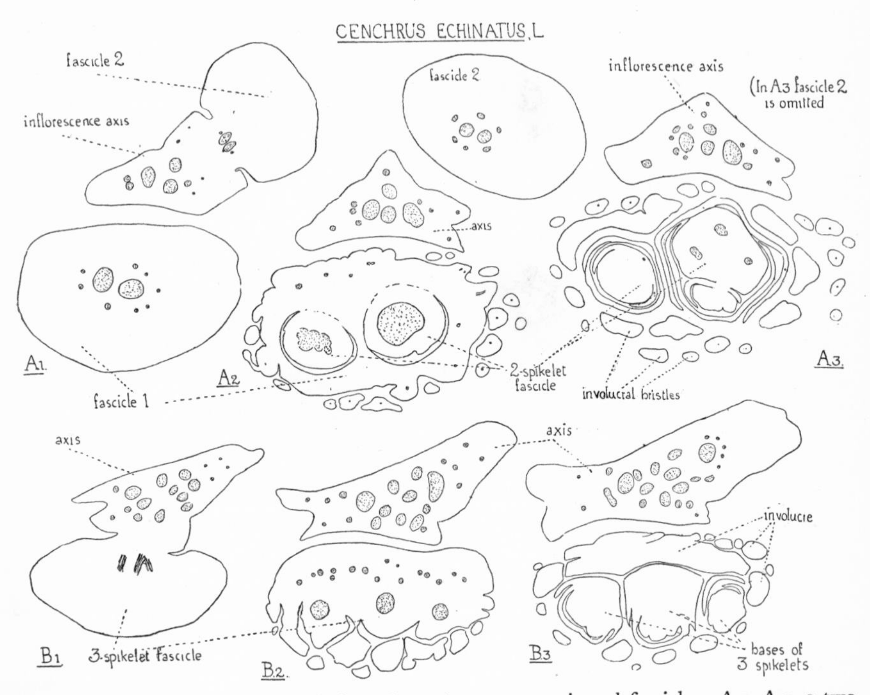

Fig. 95. *Cenchrus echinatus* L., sections through a young axis and fascicles. A 1–A 3, a two-spikelet fascicle (*fascicle* 1) (× 36 *circa*). In A 1, *fascicle* 1 is detached from the axis; another (*fascicle* 2), which is cut at the level of detachment, has two bundles at the base. B 1–B 3, a three-spikelet fascicle; B 1, base; B 3, higher level, passing through three spikelets. [Arber, A. (1931[1]).]

wholly dependent upon crossing between different individuals, and species prevailingly 'cleistogamic', in which the flowers are, as a rule, self-pollinated without opening. As examples of complete dioecism we may mention two grasses of the New World, *Monantochloë littoralis* Engelm. and *Jouvea pilosa* (Presl) Scribn. The male

[1] On monoecism and dioecism, see Pilger, R. (1905).
[2] Hildebrand, F. (1873).

flower of *Jouvea*, a genus which suggests an impoverished and dioecious *Lepturus*,[1] is shown in Fig. 66, C1 and C2, p. 150. In another of the American Gramineae, Buffalo-grass, *Buchloë dactyloides* Engelm., the male and female plants are so unlike that they were originally allotted to two different genera; the staminate form was called *Sesleria dactyloides* Nutt., and the pistillate form, *Anthephora axilliflora* Steud. There is some controversy, however, as to the degree of dioecism in this species. According to certain observers,[2] one plant may bear both male and female spikelets. This monoecism is described as occurring at an early stage in the life-history, but when stolons are put forth, the sexes become to some extent separated, as the branches belong to one or other sex exclusively. The staminate stolons are much stronger than the pistillate ones, and they are liable to outgrow and overrun the female part of the stock. Another writer[3] maintains that the Buffalo-grass is fully dioecious. He raised sixteen seedlings, each in a separate pot, and brought them to the flowering stage. He obtained eight carpellate and eight staminate plants, and found no sign of monoecism. Such a species, in which both dioecism and monoecism may apparently occur, in different phases of the life-history, and under different conditions, would well repay experimental and ecological investigation. Light on the general problem of sex-distribution in the flowering plants, might also, I think, be gained by a comparative study of closely related grasses with differential sex-behaviour. There are two species of *Eragrostis*, for instance, which were at one time confused, but it has now been shown[4] that one of them, *E. hypnoides* (Lam.) B.S.P., is always hermaphrodite, while the other, *E. reptans* (Michx.) Nees, is dioecious. There is a marked difference between the stamens of the two species. In the hermaphrodite grass, the anthers are minute—only 0·2 mm. long—while in the species with separate sexes, the anthers are ten times as long, and probably two hundred times as great in mass.

The most familiar of the grasses in which separate male and female inflorescences are borne regularly on the same plant, is Indian-corn,

[1] Fournier, E. (1876).
[2] Engelmann, G. (1859); Plank, E. N. (1892); Hitchcock, A. S. (1895).
[3] Schaffner, J. H. (1920).
[4] Hitchcock, A. S. (1926).

Zea Mays L., in which the male inflorescence or 'tassel' is terminal, while the 'cobs' bearing the grain are lateral (Fig. 182, p. 356). In the genus *Luziola*, also, there are species in which the male and female inflorescences are distinct. In one of these species, *L. Spruceana* Benth., the female flower has six stamen rudiments, while the male flower includes an abortive gynaeceum (Fig. 168, B, p. 319, and Fig. 171, D 1 and D 2, p. 323). Other species of the genus have the spikelets of the two sexes associated in the same inflorescence. In the related genus *Zizania*, the reproductive shoot bears female spikelets above and branches with male spikelets below (Fig. 96, C 1 and C 2). In Job's-tears, *Coix lacryma-Jobi* L., on the other hand, the male spikelets are borne above the female. Typical inflorescences of this species are seen in Figs. 16, p. 35 and 72, A, p. 165. Except when the two stigmas are protruding, the female spikelet is hidden within the involucre, while a slender axis, emerging from the small aperture at the tip, bears a series of male spikelets. Gerard in 1597 wrote of the "graie shining seed or graine...bored through the middle like a bead and out of which cometh a small idle or barren chaffee eare like unto that of Darnell". The structure[1] of the parts concealed within the involucre can best be understood by means of serial sections cut while the involucre is still so young that it does not resist the razor. Such sections reveal the fact that the 'tear' is the sheath of the leaf axillant to an abbreviated shoot bearing a triad of spikelets. This triad consists of one fertile and two abortive spikelets (Fig. 97, D 1 and D 2). The abortive spikelets are reduced to little but stalks; their tips can often be seen emerging from the apex of the 'tear' (Fig. 97, C). The structure of the fertile spikelet is shown in Fig. 97, E 1–E 3. It has two outer empty glumes, followed by a lemma axillant to a flower-rudiment, and then by the lemma of the fertile flower. The fertile flower has two lodicules, three abortive stamens and a gynaeceum. Normally, the continuation of the axis above the solitary involucre bears male spikelets alone (Fig. 72); but, in the plants from which Fig. 97 was drawn, there was a tendency to repetition of spikelet triads—with or without involucres—in which

[1] Weatherwax, P. (1926); in this paper the first detailed description of *Coix* spikelets will be found.

the median flower was female, hermaphrodite or abortive. Such inflorescences had no purely male region; their female trend may have been due to lack of sunlight during development.[1]

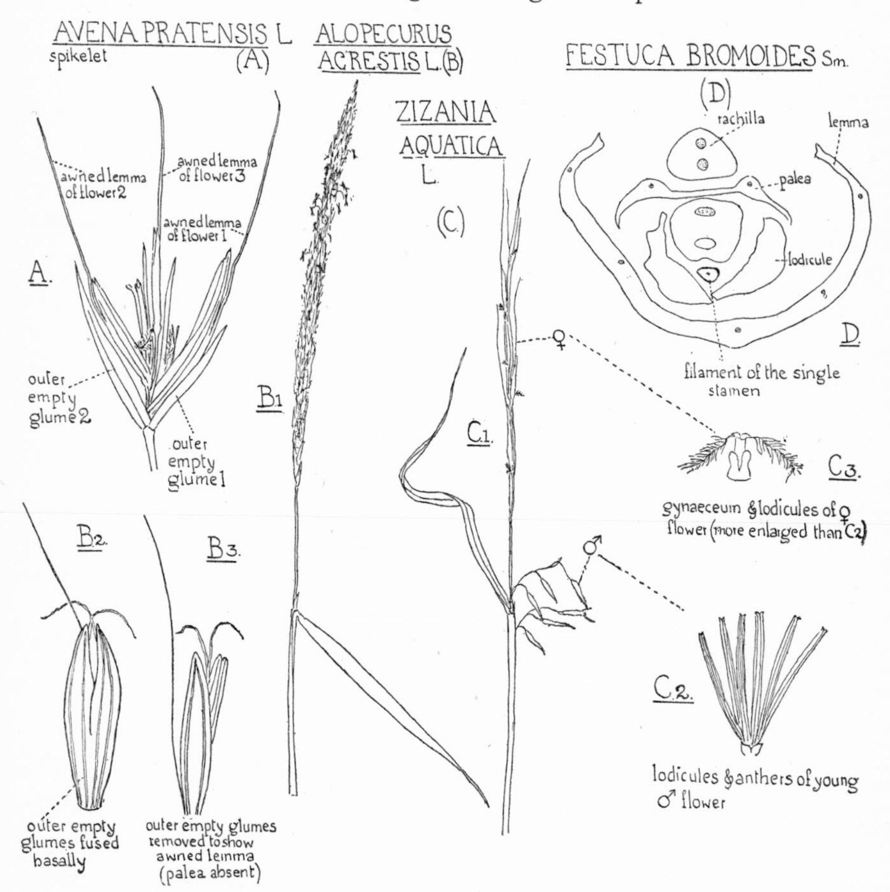

Fig. 96. A, *Avena pratensis* L., spikelet enlarged (long hairs on rachilla omitted). B 1–B 3, *Alopecurus agrestis* L.; B 1, inflorescence (× ½); B 2 and B 3, details of spikelet (enlarged). C 1–C 3, *Zizania aquatica* L. C 1, a small inflorescence (× ½); C 2 and C 3, details of the ♂ and ♀ flowers (enlarged). D, *Festuca bromoides* Sm. (*F. myuros* L., sub-sp. *sciuroides* Roth). Transverse section of flower (× 77 *circa*) to show single stamen. Section broken and slightly reconstructed. [A.A.]

A parallel for the spikelet triads within the *Coix* involucre can be found in the genus *Hordeum*, where the median member of each trio is hermaphrodite, while the laterals may be male or

[1] Weatherwax, P. (1926).

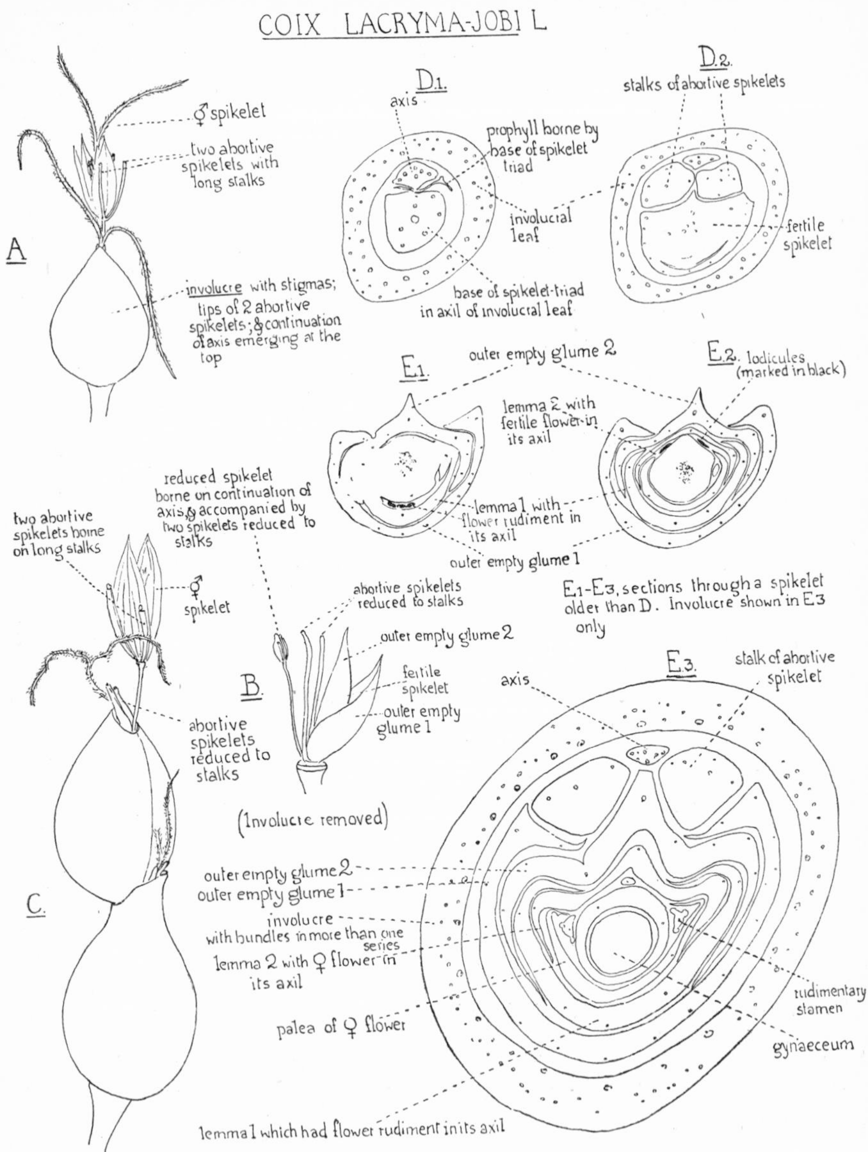

Fig. 97. *Coix lacryma-Jobi* L. Material grown in a hothouse at the Cambridge Botanic Garden. A and C, habit drawings of reproductive shoots (enlarged). B, top of a reproductive shoot with involucre removed (enlarged). D 1 and D 2, transverse section of an involucre and the parts enclosed (× 23). E 1–E 3, sections from a transverse series through a spikelet older than that shown in D; E 1 and E 2 (× 14); E 3 (× 23). [A.A.]

abortive; the reduced spikelets may be stalked, which increases their resemblance to those of *Coix* (Fig. 77, E and F, p. 172). Among the Barleys it has been shown that there is a correlation between length and attenuation in the pedicels, and feebleness of development in the spikelets in which they terminate.[1]

A similar relation between the pedicellate character and sterility may be traced in *Ischaemum rugosum* Salisb.[2] Here the spikelets are arranged in pairs—a sessile fertile spikelet, and a stalked spikelet, which is variable in character. In the lower part of each inflorescence, the pedicel of the stalked member of each pair is short, and the stalked spikelet is well developed and usually fertile. About the middle of the inflorescence, the pedicel is longer and the spikelet is smaller and often male, while, at the apex, the pedicel is still longer, and the spikelet is often barren and much reduced. Fig. 98, p. 200, illustrates the structure of a spikelet pair, in which the pedicelled spikelet is abortive. The tendency to sterilisation is shown not only in the pedicelled spikelet, but in the fertile spikelet itself, which includes two flowers, the first of which is abortive. The tendency to fusion of parts, which so often goes with reduction, manifests itself in the union of the outer glume of the fertile spikelet with the axis on one side, and with the abortive spikelet on the other. The specimens of *Ischaemum rugosum*, from which the sketches were made, reached me in the interstices of a package from Burma. It is worth while to examine the wrappings of parcels from distant countries, since they often include hay, which may provide agreeable treasure trove.

So far we have considered those genera in which either *individuals*, or *inflorescences*, or *spikelets* within the same inflorescence, differ in sex. It is only the ultimate term of this series which we have still to add—that in which the individual spikelets include both hermaphrodite and unisexual *flowers*. As British examples we may take *Holcus lanatus* L. (Fig. 99, p. 201) and *Arrhenatherum avenaceum* Beauv. (Fig. 55, p. 137); in both grasses each spikelet has two flowers—the lower *male* and the upper hermaphrodite. The occurrence of *female* and hermaphrodite flowers in the same spikelet

[1] Weideman, M. (1927).
[2] I am indebted to Mr C. E. Hubbard for this information.

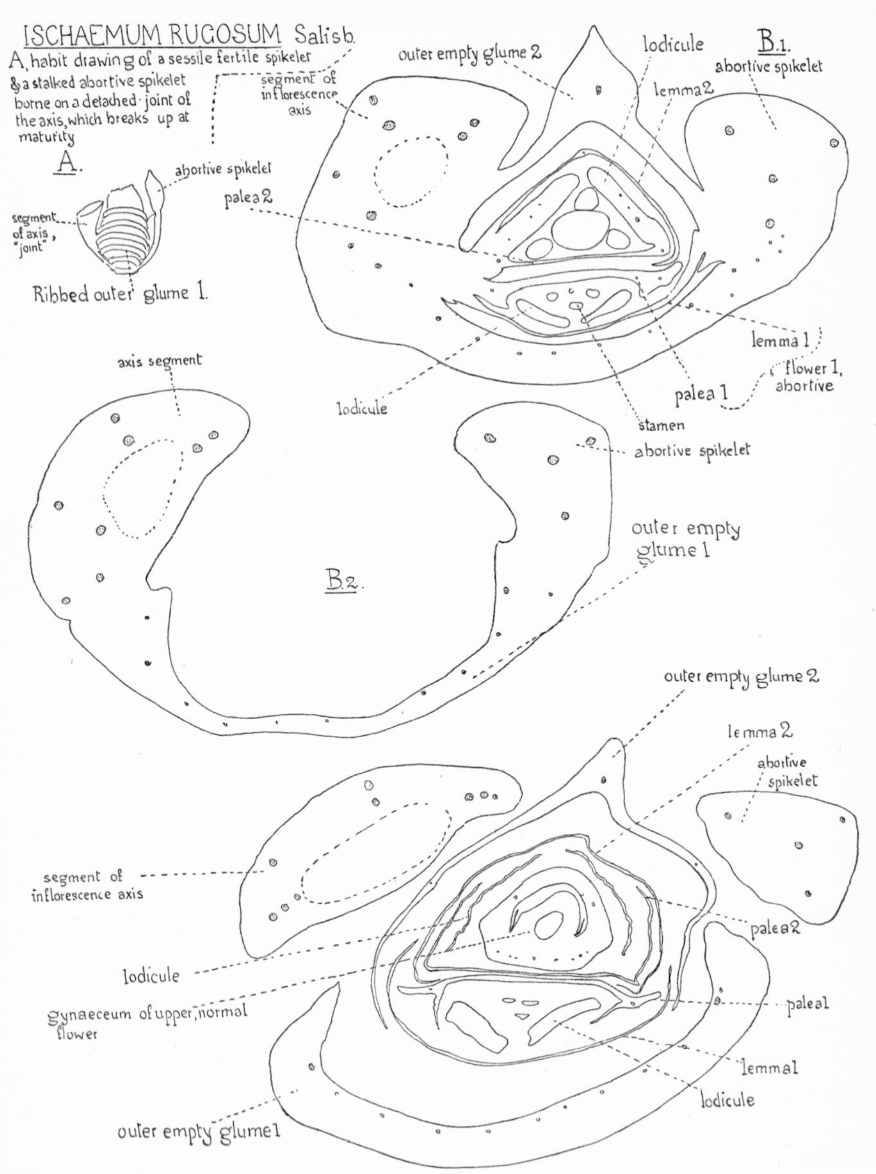

Fig. 98. *Ischaemum rugosum* Salisb., Burma. A, a pair of spikelets, one sessile and fertile, and one pedicelled and abortive, borne on one of the segments into which the axis disarticulates (enlarged). The outer empty glume is the only part of the fertile spikelet shown; hairs omitted. B 1–B 3, sections from a transverse series through a two-flowered fertile spikelet, with a sterile spikelet on one side and an axis segment on the other (× 47). In B 2, the only part of the fertile spikelet included is the outer empty glume. (In the dried material used, the bundles in the outermost glume were difficult to follow, so they may not be shown correctly.) [A.A.]

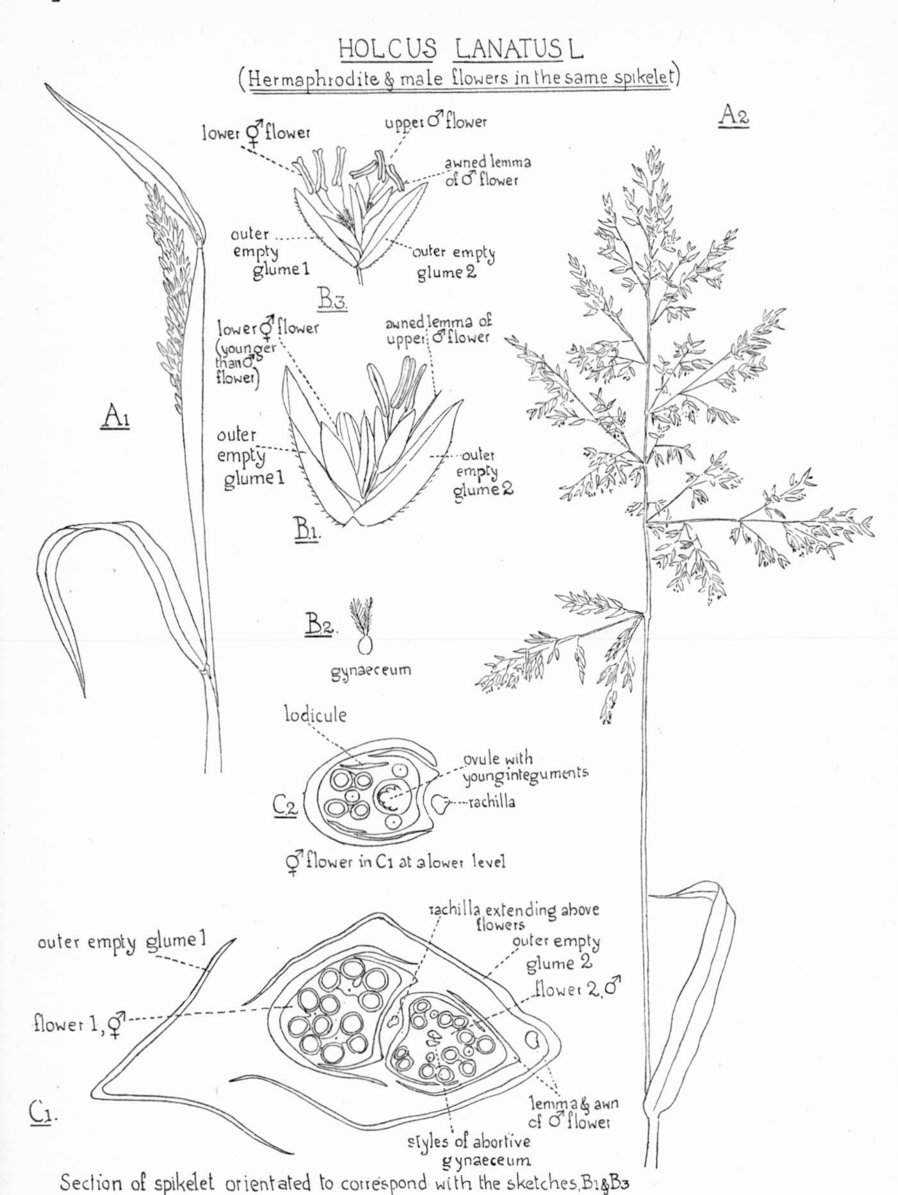

Fig. 99. *Holcus lanatus* L. A, inflorescence ($\times \frac{1}{2}$); A 1, in bud; A 2, in anthesis. B 1, young spikelet; B 2, gynaeceum; B 3, an older spikelet on a smaller scale. C 1 and C 2, sections from a transverse series through a spikelet ($\times$ 47). [A.A.]

is much rarer, but it happens that the commonest of our British grasses, *Poa annua* L., shows this peculiarity[1] (Fig. 100).

In the majority of grasses, the flowers are all hermaphrodite; nevertheless cross-fertilisation may be the rule, and certain species may be, to a large extent, self-sterile (e.g. Rye, p. 12). On the other hand, self-fertilisation may occur, either occasionally or regularly. It

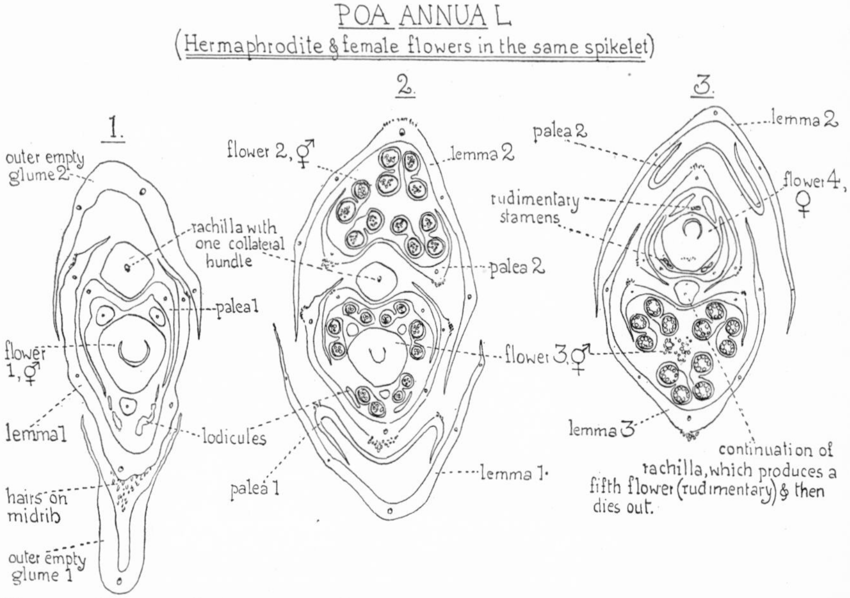

Fig. 100. *Poa annua* L. (gynomonoecism). *1, 2, 3*, sections from a series from below upwards through a spikelet from the same clump as that illustrated in Fig. 103, p. 213 (× 77 *circa*). The spikelet includes three hermaphrodite flowers, one female and one rudimentary. (In *2*, the outer empty glumes are omitted). [A.A.]

has been shown that, in those Gramineae which are habitually self-fertilised, the pollen is particularly difficult to germinate. Two authors,[2] who succeeded in germinating Barley pollen, have concluded that the difficulty is due to the extreme delicacy of the water adjustment of the pollen-grains. If they are exposed to dry air for two or three minutes, the walls collapse, through loss of moisture, while, if they are placed in a saturated atmosphere for an even shorter

[1] Kirchner, O. von, Loew, E. and Schroeter, C. (1908, etc.).
[2] Anthony, S. and Harlan, H. V. (1920).

time, they imbibe water so fast that they burst: to breathe upon them is fatal to them. Pollen-grains of such susceptibility are indeed ill-adapted to the exigencies of crossing.

In considering the distribution of sexes in the grasses, we began with dioecism. The condition, which forms the actual antithesis to this state, is cleistogamy, in which the lemma and palea do not separate, and the anthers pollinate the adjacent stigmas in the unopened flower. Unfortunately the term cleistogamy has been employed in various senses by different writers. Those, who use it most rigidly, would confine it to species in which the pollen-grains germinate within the unopened anthers, so that the tubes pass directly from the anther cavity to the stigma;[1] but the term is more frequently applied to any grass in which pollination takes place while the flower is still completely enclosed in the lemma and palea, even if the anthers dehisce. Cleistogamy, in this sense, is widespread among grasses, and recurs in various cycles of affinity; there is an impression that it is specially prevalent in the New World,[2] but this may be due merely to the fact that it has been searched for in America more intelligently and more consistently than elsewhere. Statistics about the occurrence of cleistogamy have to be scrutinised with some care, because so many of the examples cited have not been observed in the living state. Fairly good evidence can however be obtained from dried material;[3] serial sections of the spikelets may, indeed, clinch the matter. When cleistogamic forms are examined in the fresh condition, it is found that the anthers do not rise above the styles, but become surrounded by the stigmatic branches, on to whose papillae the pollen-grains are shed directly. In the developing, and even in the ripe fruit, one finds the empty anthers embedded in the entanglement of stylar branches, so that it is difficult to free them. It seems tolerably safe to conclude that, if herbarium material shows this peculiarity, the plant was cleistogamic.

There are some species which are known only in the cleistogamic

[1] Sablon, M. Leclerc du (1900).

[2] See Chase, A. (1908), (1918), (1924); Parodi, L. R. (1924); Hitchcock, A. S. (1925); Weatherwax, P. (1928[1]), (1929).

[3] Hackel, E. (1906).

form; certain Barleys may be of this type. Other grasses exist in two forms, in one of which the flowers are chasmogamic (opening normally), while in the other they are cleistogamic. *Festuca microstachys* Nutt., for instance, has a chasmogamic form with normal stamens, whose anthers are 2·0–2·5 mm. long, and lodicules, 1 mm. long; it also has a cleistogamic form, in which the short anthers (only 0·3 mm. long) are closely entangled with the stigmas, and the lodicules are scarcely 0·3 mm. long.[1]

Cleistogamic inflorescences sometimes take the minimal form of 'cleistogenes'—solitary, sessile, single flowers, with lemma and palea, but without the usual outer empty glumes. Agnes Chase,[2] to whom we owe much of our knowledge of these structures, first observed them in *Triplasis purpurea* Chapm., where they occurred in the lower leaf-sheaths of the flowering culms. Her eyes having been opened by this discovery, she looked for these reduced spikelets elsewhere, and found them in about twenty other species, twelve of which belonged to the genus *Danthonia*. The cleistogenes are often so markedly unlike the normal spikelets, that, if their source were unknown, they would not be attributed to plants of the same tribe. They are poorly adapted for dispersal, since they are often deeply enclosed; it appears that, as a rule, they are freed from the parent plant by the disjointing of the culms.

The British grass flora is not rich in cleistogamic species, but there is one example in which the character has long been recognised. This is the rare Cut-grass, or Rice-grass (*Leersia oryzoides* Sw.), which grows in swampy ground in a few of our southern counties. This grass has two forms; in one, the inflorescences remain enclosed in the leaf-sheaths, and the flowers are cleistogamic and completely fertile. The other form produces elongated haulms; the inflorescences emerge and the flowers open, but the fertility is low.[3] Another cleistogamous British grass, *Sieglingia decumbens* Bernh. (*Triodia decumbens* Beauv.),[4] belongs to a genus related to *Danthonia*. *Sieglingia decumbens* possesses aerial panicles, whose flowers are fertilised while they are still enclosed in the leaf-sheath. Sessile cleistogenes

[1] Hackel, E. (1906).
[2] Chase, A. (1918).
[3] Koernicke, F. (1885).
[4] Beddows, A. R. (1931).

also occur in the axils of the basal leaf-sheaths of the flowering stem. In addition a chasmogamic form is known, but it is rare.

An unusual variant on the cleistogamic type is that in which the

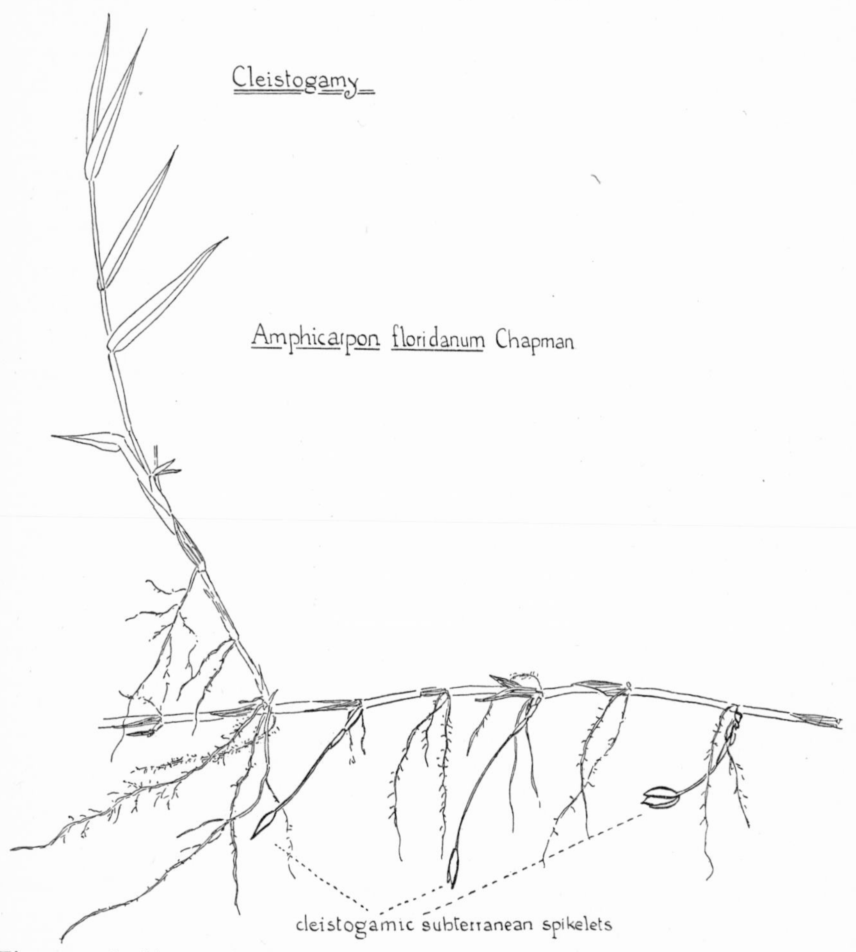

Fig. 101. *Amphicarpon floridanum* Chapm. Specimen collected by Paul Weatherwax from wet sandy soil just west of Groveland, Florida, and identified by Agnes Chase. (× ½.) [A.A.]

closed spikelets are borne on underground shoots. Fig. 101 illustrates this feature in an American grass, *Amphicarpon floridanum* Chapm.; occasionally, though rarely, this species also produces fertile aerial spikelets.[1]

[1] Chase, A. (1908).

It has been suggested for *Danthonia*[1]—and it is probably true for other genera—that the key to the peculiarities of cleistogamic spikelets lies in the reduced condition of their lodicules. These members tend to be rudimentary and poorly supplied with vascular tissue, or they may even be absent; in either case, the flower suffers from failure of its opening mechanism.

It is possible that cleistogamy may play a part in species production, owing to the isolation which it involves. It has been suggested, for instance, that *Poa Chapmaniana* Scribn., a cleistogamous grass which inhabits the United States, may be an offshoot from the chasmogamic *P. annua* L.[2]

A review of the argument in this and the preceding chapter, leads us to the conclusion that the special features of both the more normal and more aberrant grass flowers seem to be due, in the main, to reduction processes, modifying a type which can be seen in a less restricted form in the bamboos. This reduction can often be related to the effects of crowding and compression during ontogeny. The many-flowered inflorescence, in its rudimentary stages, is squeezed within the leaf-sheaths; the individual spikelets are imprisoned inside the outer empty glumes; and each young flower is entrapped between the lemma and palea, as between the two valves of an oyster shell. However, besides those losses and fusions whose correlation with pressure is obvious, we can recognise a number of forms of spikelet sterility, and of unisexuality or floral abortion, for which pressure can be, at the most, only partially responsible. It seems clear that the Gramineae, in addition to their liability to pressure effects, suffer from a definite inherent trend towards some degree of sterilisation of the reproductive shoot.

1 Weatherwax, P. (1928[1]). 2 Weatherwax, P. (1929).

CHAPTER X

INDIVIDUALITY AND LIFE-PHASES IN BAMBOO AND GRASS

IT would be difficult to find a parallel in any other corresponding group of flowering plants for the range in individual size met with in the Gramineae. At one end of the scale are the bamboos, which, as we have shown, may exceed 120 ft. in height. Although the bamboos arrive at greater statures than the other Gramineae, the examples of tall forms among grasses outside this tribe are more numerous than is generally realised by those accustomed only to the European flora. Spruce,[1] in his account of the forests of Chimborazo, describes the Uva, *Gynerium saccharoides* H. et B.[2] (Festuceae, related to *Arundo*), as attaining "its maximum of development on stony springy declivities, at an elevation of about 1500 feet above the sea, where a forest of Arrow-cane [Uva], with its tall slender stems of 30 to 40 feet, each supporting a fan-shaped coma of distichous leaves, and a long-stalked thyrse of rose and silver flowers waving in the wind, is truly a grand sight". Spruce goes on to say—"The longest stem I ever measured was one I met a man carrying on his shoulder at Tarapoto. From that stem had been cut away the leaves and peduncle, and the base of the stem, which is generally beset with stout-arched exserted roots (serving as buttresses), to a height of 1 to 3 feet; yet the residue was 37 feet long, so that the entire length must have been at least 45 feet". Other grasses which may reach 20 ft. are the South African Elephant-grass, *Pennisetum purpureum* Schum.;[3] the Giant-millet, *Andropogon Sorghum* (L.) Brot.;[4] and *Arundo Donax* L.[5] The size of the last of these grasses has been

[1] Spruce, R. (1908).

[2] According to Hitchcock, A. S. (1920), the name should be *G. sagittatum* (Aubl.) Beauv. *G. argenteum* Nees (*Cortaderia argentea* Stapf) is the Pampas-grass of gardens; see Stapf, O. (1897).

[3] Bews, J. W. (1925).

[4] Koernicke, F. (1885).

[5] Hitchcock, A. S. (1920).

turned to account in the south of France, by growing it from cuttings to form "larges rideaux contre les vents du nord".[1]

Reference has already been made to the great lengths reached by the stems of liane bamboos (p. 61); their elongated axes can be paralleled among the climbing Gramineae[2] of other tribes. Spruce describes rampant species of *Panicum*, which occur in the Red-bark forests, and "thread among adjacent branches to a height of 15 feet or more".[3] These climbers are, however, outclassed in their measurements by certain swamp and water grasses. The aquatic form of *Coix lacryma-Jobi* L., for instance, is said to reach a length of 100 ft.[4] From a floating grass island on the upper Amazon, consisting exclusively of *Paspalum pyramidale* Nees (*P. repens* Berg), Spruce[3] succeeded in drawing up an entire stem, which measured 45 ft. in length and possessed seventy-eight nodes. Even in our cold climate, we can find a parallel for this elongation. There is a singular form of the Reed, *Phragmites communis* Trin. var. *repens* Meyer, which has been described from "the slipped banks of wet and almost semifluid clay, skirting the southern shores" of the Isle of Wight.[5] In this form the culms are said to depend like long and slender ropes on the steep sides of the landslips, or to trail in a straight or serpentine direction on the shingly beach, or smooth and level sand, without rooting at the joints, to the length of 20 to 40, or even 50 ft. These elongated axes are apparently always sterile, and their leaves are very much reduced. Occasionally a few rootlets arise from the joints, but in general the plant lies quite prostrate and entirely unconnected with the soil, so that it may be wound about any object like a cord.

The barren shoots of various grasses, especially *Agrostis palustris* Huds. and species of *Alopecurus*, may attain great lengths when they grow under swamp conditions. The classic instance is the Ebbesbourne or Orcheston Long-grass, which figures in *The Natural History of Wiltshire*, compiled by John Aubrey[6] in the seventeenth century. Aubrey writes that "At the east end of Ebbesbourne Wake

[1] Raspail, F. V. (1824).
[2] For a list of the climbing grasses of South Africa, see Bews, J. W. (1925).
[3] Spruce, R. (1908).
[4] Watt, G. (1904).
[5] Bromfield, W. A. (1844).
[6] Aubrey, J. (1847).

is a meadowe called Ebbesbourne, that beareth grasse eighteen foot long. I myself have seen it of thirteen foot long; it is watered with the washing of the village. Upon a wager in King James the First's time, with washing it more than usuall, the grasse was eighteen foot long. It is so sweet that pigges will eate it; it growes no higher than other grasse, but with knotts and harles like a skeen of silke (or setts together). They cannot mowe it with a sythe, but they cutt it with such a hooke as they bagge pease with".

"At Orston [Orcheston] St. Maries is a meadowe of the nature of that at Ebbesbourne aforesayd, which beares a sort of very long grasse. Of this grasse there was presented to King James the First some that were seventeen foot long;.... In common yeares it is 12 or 13 foot long."

This Long-grass of Wiltshire, "one of the most singular vegetable products of this country", has also been discussed by later writers.[1] The general conclusion seems to be that about half-a-dozen species of Gramineae, all more or less elongated, grew in the Orcheston meadow, so that the name, Orcheston Long-grass, is a comprehensive term; but the plant to which the name is most usually assigned is Fiorin-grass, *Agrostis palustris* Huds.

At the other end of the scale from these species which have an indefinite capacity for shoot elongation, we meet with certain dwarf grasses, in which the whole plant, with its inflorescence, is comprised within a few inches, or even a single inch. *Phippsia algida* R.Br., *Coleanthus subtilis* Seid. and *Mibora verna* Beauv. (*Chamagrostis minima* Borkh.) are said to be among the smallest of flowering plants.[2] The total height of *Mibora verna* (Fig. 102, E, p. 210) is from $\frac{1}{2}$ to 3 in. Within the sub-tribe of the Agrostideae, known as the Phleinae, to which *Mibora* belongs, nanism shows a curious type of incidence.[3] This group consists of *Alopecurus* with fifty species, *Phleum* with ten species, *Heleochloa* with eight species, and eight other genera, which are all monotypic and dwarf. One cannot but feel that this peculiar systematic distribution—if only it could be interpreted—

[1] Stillingfleet, B. (1811), and references in Preston, T. A. (1888), pp. 393–4.
[2] Schumann, K. (1895).
[3] Attention has been drawn to these facts by Bews, J. W. (1929).

might perhaps reveal the clue to some of the evolutionary relations of the dwarf habit.

Besides species which are normally on a small scale, there are

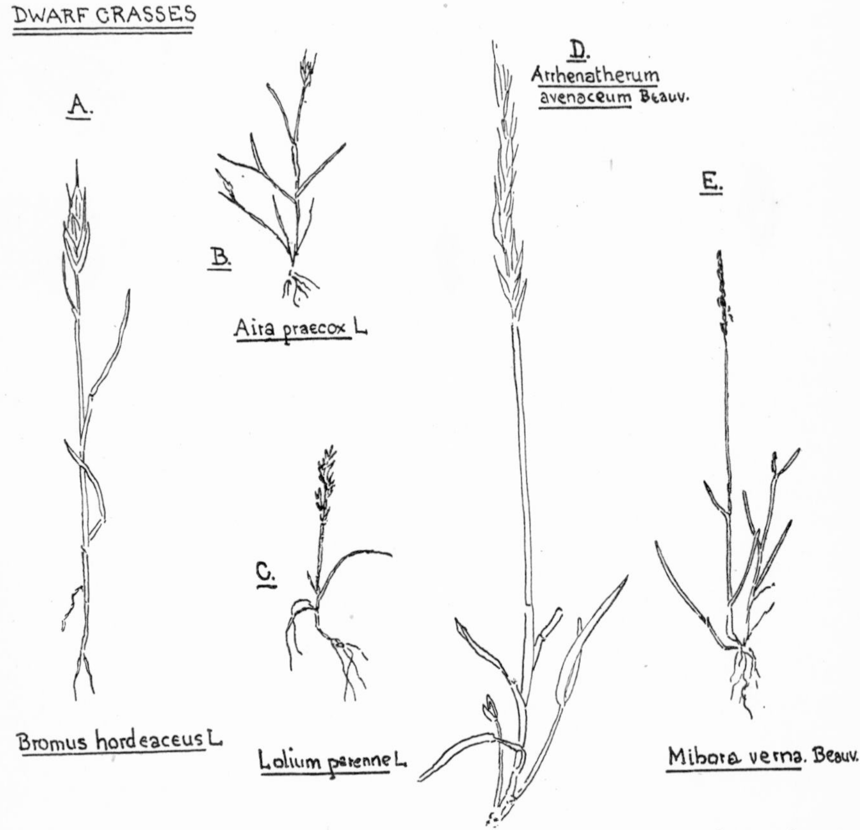

Fig. 102. Depauperate Grasses; all drawings natural size. (All dwarfed specimens of grasses which normally grow to a larger size, except *Mibora verna* Beauv., which is typically small.) A, *Bromus hordeaceus* L., downs between Whitwell and Niton, Isle of Wight, August 13, 1924. B, *Aira praecox* L., wood at Sandy, Beds, June 1932; total height, 3 cm. C, *Lolium perenne* L., from a rock surface, Paignton, Devon, collected by I. H. Burkill, May 17, 1929. D, *Arrhenatherum avenaceum* Beauv., downs between Whitwell and Niton, Isle of Wight, August 13, 1924. E, *Mibora verna* Beauv., Cambridge Botanic Garden, February, 1932. [A.A.]

other dwarf grasses, which are starveling forms, whose minute size is due to the influence of their environment. The tiny Rye-grass (*Lolium perenne* L.) drawn in Fig. 102, C, for instance, grew on a rock surface in Devonshire; the total length of the shoot was less than $1\frac{1}{5}$ in. (3 cm.), whereas the usual height to which this species attains is 1–2 ft. A writer,[1] who figured a depauperate example of Rye-grass 130 years ago, called it "a pitiful plant". Fig. 103, C, p. 213, shows another example of precocious reproduction—a flowering specimen of Causeway-grass, *Poa annua* L., whose height above ground was 16 mm. (less than $\frac{3}{4}$ in.). Flowering has also been described in miniature specimens of *Setaria* and *Panicum* germinated in sand.[2]

As soon as we begin to compare the large and small members of the Gramineae, it becomes clear that a correlation can sometimes be recognised between small size and a brief life-history on the one hand, and between great stature and deferred maturity on the other, a contrast which is also familiar among animals. In many of the smaller, as well as some of the larger grasses, the life-history is relatively so short that the fruiting stage is completed while the remains of the seed from which the plant sprang are still recognisable at the base of the axis (Fig. 108, p. 230). On the other hand, the vast shoot-development of the bamboos, and of such forms as the Orcheston Long-grass and the creeping variety of the Reed, represents a prolonged sterile phase intervening between the seedling stage and the reproductive period. In certain bamboos, as we have seen, this vegetative stage may persist for thirty years or more. A brief hint of such a preliminary sterile phase can be detected in many grasses. Even the 'summer' cereals, which are harvested in the year of sowing, show a pause in their obvious development, while vegetative activity—rooting and preparation for tillering—takes place underground; and in 'winter' cereals this deferring of the fertile period is much more pronounced.

There is remarkable variation among different grasses in the degree to which they show periodicity in their life-history.[3] At one end of the series come those in which the time-rhythm appears to be fixed and inescapable; among these are such winter annuals as

[1] Knapp, J. L. (1804); this book is worth examining for the sake of the faithful and delicate coloured engravings of British grasses.

[2] Hitchcock, A. S. (1892).

[3] Kirchner, O. von, Loew, E. and Schroeter, C. (1908, etc.).

Mibora verna Beauv., which refuse to flower in their first season, even if sown in the spring. These grasses, being unable to change their habits, are apt in severe winters to fall martyrs to their own conservatism. At the opposite pole are such forms as *Poa annua* L. (Fig. 103), which show no fixed periodicity in their growth phenomena, and may either run through several generations in a year, or may perennate. Shoots which presumably had lived through the winter are shown in Fig. 103, D–G. D and E, for instance, represent stolons found lying on the surface of close-cropped meadow grass near Lyme Regis in April. In the examples drawn, the axis of the stolon was thickly clothed with roots, most of which lay parallel to it, forming a dense coat. The lower leaves were dead, dry and whitish, thus contrasting sharply with the vivid green of the closely folded young leaves near the shoot tips. In F, which was found in September on a roadside in Cambridge, the length of the stolon was 14 cm.

Species differentiation may sometimes come about by a change in the life periodicity affecting the flowering time. For instance, an American form closely related to *Hierochloë borealis* Roem. et Schult., which has been given specific rank, flowers from July to September, while Holy-grass itself flowers from mid-April to June.[1]

The relation between the fertile and the sterile phases in the life-history of grasses is affected by an environmental factor which has only been distinguished in comparatively modern times—the number of hours intervening between sunrise and sunset.[2] This factor—*length of day*—has been found to possess a remarkable influence on plant growth; it exercises the casting vote, as it were, between the purely vegetative and the sexually reproductive forms of development. This seems surprising when we recall that *light intensity*—within the range from full sunlight to one-third or one-quarter of the normal, or even less—is *not* a factor of importance in controlling the achievement of the reproductive phase. Nevertheless it has now become clear that any given species can attain to the flowering and fruiting stages only when the *length of day* falls within certain limits.

[1] Bicknell, E. P. (1898). This grass is called *Savastana Nashii* Bicknell; on the synonymy of Holy-grass see footnote, p. 55. On seasonal dimorphism, see Arber, A. (1925), p. 216.

[2] Garner, W. W. and Allard, H. A. (1920).

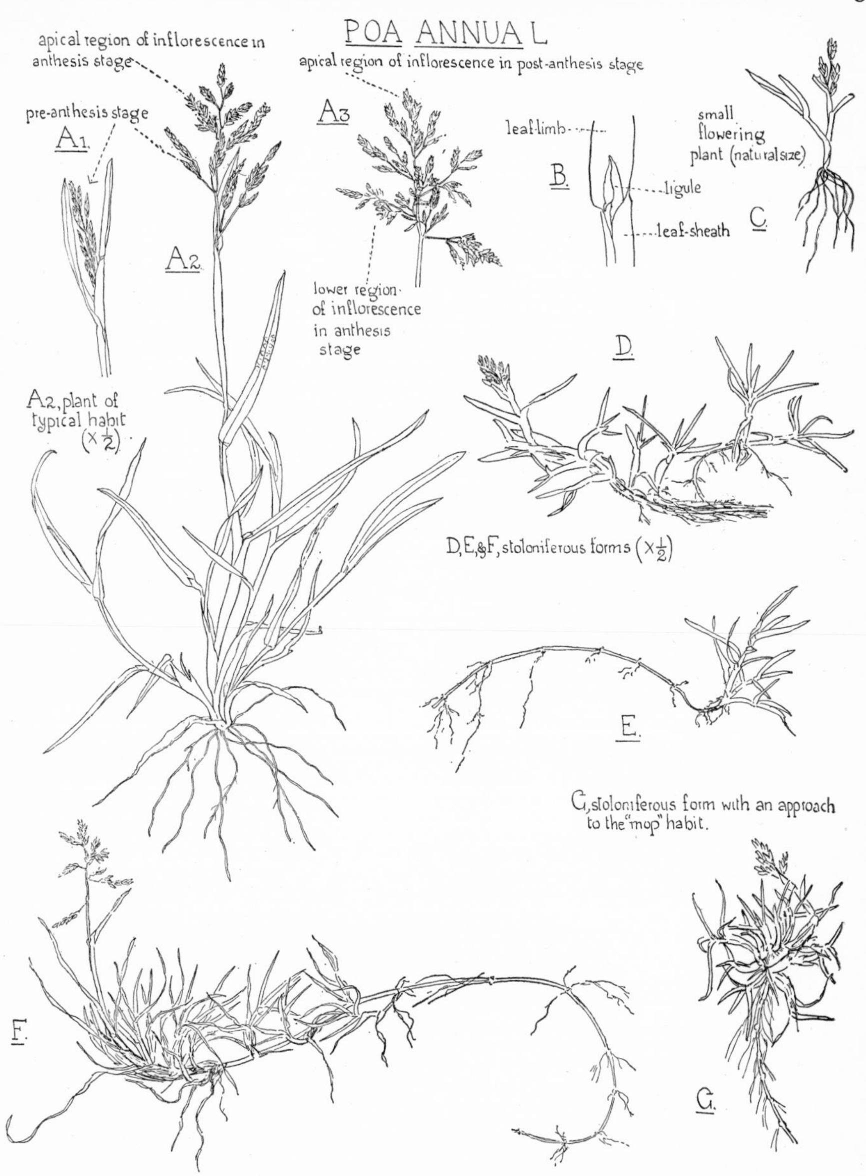

Fig. 103. *Poa annua* L. A 1–A 3, drawings (× ½) of a flowering plant, Cambridge, April 2, 1924. A 1, inflorescence in bud stage; A 2, plant with slightly older inflorescence; A 3, inflorescence older than A 2. B, ligule (enlarged). C, a plant which flowered at the height of 16 mm. (nat. size). D–G, stoloniferous forms (all × ½). For spikelet structure see Fig. 100, p. 202; leaf structure, Fig. 157, C, p. 299, and Fig. 159, B, p. 302. [A.A.]

Some species, 'long-day plants', respond to relatively long days, and 'short-day plants' to short days. Length of day seems, indeed, to be unique among environmental factors in its action on sexual reproduction. It has been tentatively suggested that not only plants, but birds also, may be sensitive to changes in length of day, and that it may be this factor which affords the stimulus to periodic migrations.

In the Gramineae[1] it has been found that experimental shortening of the hours of daylight retards flowering. It makes early varieties late, and prevents late varieties from flowering: it tends to prolong the tillering period, so that more tillers are produced. In Oats and in an early type of Timothy, *Phleum pratense* L., but not in other grasses examined, it was found that the plant subjected to the shorter days, though unable to flower normally, made an effort in this direction; it left the rosette tillering stage, and shot up into an erect form, as if about to flower.

Some experiments upon *Phleum pratense* L. may be cited to illustrate the behaviour of a 'long-day plant'.[2] If in the autumn, when the cold weather is about to put a stop to growth, plants of Timothy are transferred to a greenhouse, where conditions for growth are favourable, but where the sun supplies the only illumination, they continue their vegetative growth throughout the winter; but they do not produce inflorescences until about the same time as the plants left in the field, which have been practically dormant for weeks or even months. If, however, at any time during the winter, the plants are illuminated from dusk to midnight each evening by means of a 200-watt electric light, placed about 3 ft. above them, the length of day is increased approximately to that of early summer; and the result is that inflorescences are soon produced.

The daily period of illumination not only affects the reproductive process in general, but it appears that it may influence sex. In investigations on a certain type of *Zea Mays* L., it was found that 'short-day' conditions induced the development of gynaecea in the male tassels.[3] For instance, in Indian-corn sown in a greenhouse on

[1] Tincker, M. A. H. (1925). [2] Evans, M. W. (1931).
[3] Schaffner, J. H. (1927); see also Schaffner, J. H. (1930).

November 1, all the individuals showed some degree of femaleness in the tassel, while that planted in spring or summer produced pure staminate tassels.

Where the vegetative phase in grasses is of some length, it tends to become composite, owing to the capacity of the nodes for forming fresh centres of growth, a capacity which finds expression in the tillering of cereals. This focusing of activity at one or more nodes, and the production of complex branch clusters at these points, leads to the 'pompon' growth of liane bamboos (p. 76); to corresponding forms of the inflorescence in this tribe; and also to a curious mop-like habit, which many of our grasses tend to assume at certain stages of development. Examples of 'mops' from the commonest British grasses are shown in Fig. 104, p. 216. The 'bird's-nest' specimen of Cock's-foot-grass (D) consisted of a group of three shoots at one node, followed by a fan-like cluster of eight shoots, all in one plane, and appearing to arise from the stem apex. *Holcus lanatus* L., Yorkshire-fog, is particularly liable to grow in mop fashion (A and B). In April 1925, I found it at Lyme Regis, under a clipped Hawthorn hedge, and scrambling up through it. On the flat hedge-top, where the shoots emerged into light and air, pompons of leaves were developed. The effect was very odd; the horizontal upper surface of the hedge-row was carpeted with bird's-nest-like green tufts of Yorkshire-fog, matted together into a miniature meadow. When pulled out, the plants looked like cobweb-mops borne on long, slender, flexible canes. The apicalness of the pompons was only apparent, since a segment of dry, dead axis was generally to be seen beyond the base of each of them. I have found an example in which the stem bearing the pompon (though incomplete) consisted of nine internodes and was 79 cm. long; elongated axes, such as this, were no doubt the growth of the previous year. It is clear that, in such hedgerows as that described, many of the mops must inevitably die; in the season in which I observed them, a large proportion of them showed decadence before the end of April. They were entirely dependent for their water supply upon the old elongated axes; though the mops of this and other species produce roots very readily if put into water, there is no chance of rooting in a hedgerow.

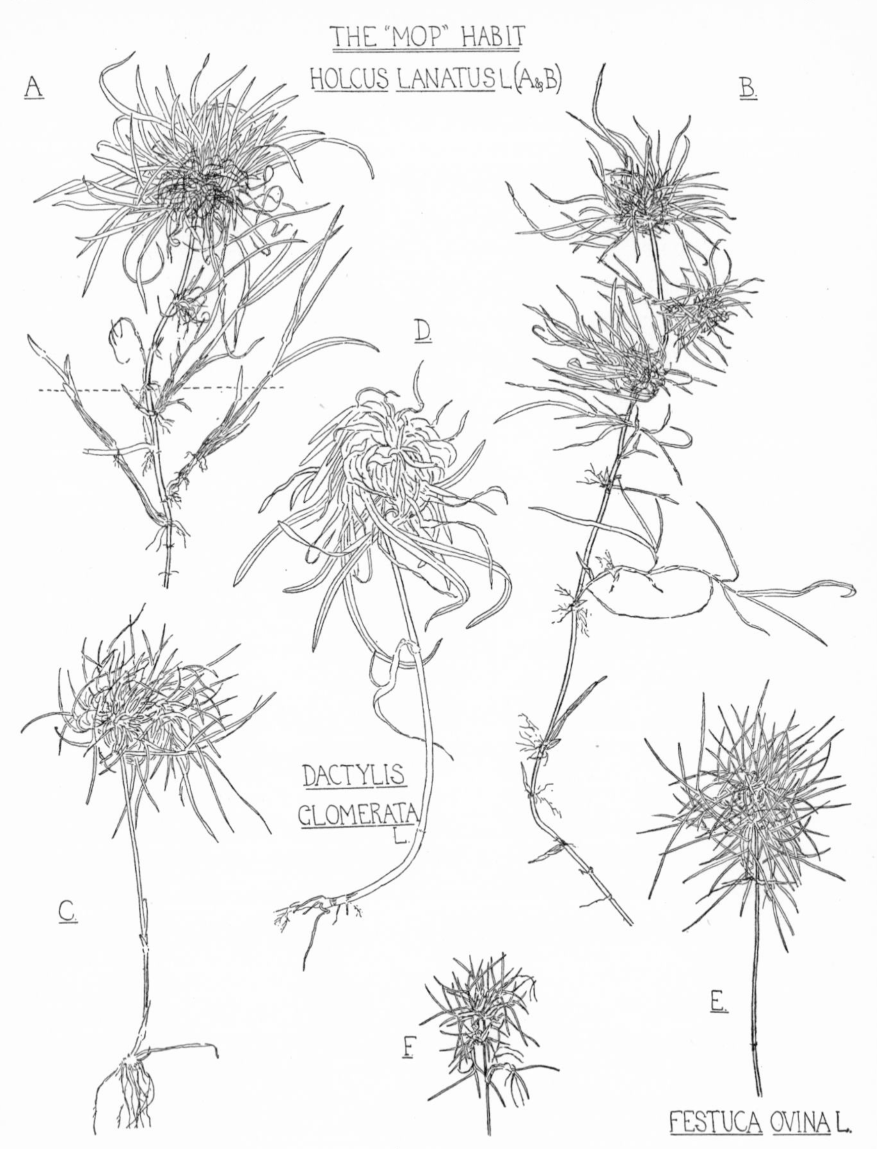

Fig. 104. The 'mop' habit in grasses; all drawings × ⅓. A–E, from neighbourhood of Lyme Regis, Dorset, April 1925. A and B, *Holcus lanatus* L.; the dotted line in A gives the approximate ground level. C, probably *Agrostis palustris* Huds. D, *Dactylis glomerata* L. E, *Festuca ovina* L. F, *Glyceria maritima* Mert. et Koch. Collected by G. Lister, Axmouth, Devon, April 1925. [A.A.]

The tendency to great vegetative activity at the nodes, which culminates in 'mop' development, may easily lead to multiplication of the plant soma, since rooting readily occurs from the nodes, and the intervening segments of hollow internode are liable to perish. This is an everyday occurrence, but it may be worth citing, as a striking instance, that a stolon of *Ischaemum muticum* L., 5 m. in length, found upon a beach in Western Java, bore no less than 122 young plants.[1] 'Vegetative reproduction' of this type suggests the question as to where we should set the limit that bounds the individual. It makes for clearness of thought on this difficult problem if we distinguish between 'major' and 'minor' individuals as plant units.[2] The 'major plant unit', which seems to be equivalent to the 'clone' of the ecologist, may be defined as the total vegetative output which one egg-cell is capable of initiating, while 'minor units' are those into which the major unit dissociates. Such vegetative multiplication into 'minor individuals' may be repeated in generation after generation, and may cover long periods of time. Indeed, as far as age is concerned, the most ancient Oak may have to give precedence to certain inconspicuous grasses, in which the 'major individual' has a prolonged existence. It has been pointed out[3] that much of the Buffalo-grass (*Buchloë dactyloides* Engelm.), which today forms a continuous turf on the North American prairies, may represent the actual plants that took possession of the plains after the retreat of the glaciers. The writers to whom we owe this suggestion, also note that the mode of development of the 'bunch' grasses—which grow outwards from the centre, and eventually assume the form of 'fairy rings'—makes it possible to some extent to trace the time-element in their history. "Eventually the ring, which sometimes becomes as much as a hundred feet in diameter, breaks up into segments..., which finally form the beginnings of new rings. The increase in diameter of a bunch may be only a fraction of an inch each year, hence a large fairy ring may represent the growth of hundreds or even thousands of years."

Phragmites communis Trin., the Reed (Fig. 105), is a plant in

[1] Backer, C. A. (n.d. [1930]). [2] Pallis, M. (1916).
[3] Hitchcock, A. S. and Chase, A. (1931).

which the conception of the 'major individual' has been fully worked out.[1] From this somewhat speculative study the general conclusion has been reached that the *length of life* of the major individual

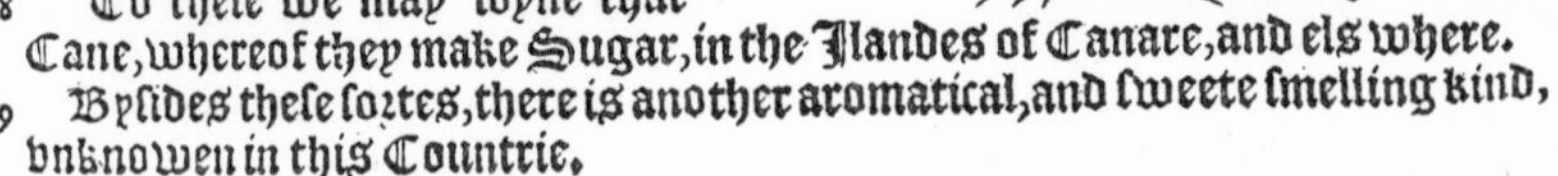

Of Pole Reede/ or Canes. Chap. liiij.

The Kindes.

THERE are diuers kindes of Reedes, as Dioscorides and Plinie do write, whereof the sixth kinde is very common and well knowen in this Countrie.

The Description

6 THE common Reede or Cane hath a long stalke or strawe full of knottie ioyntes, whereuppon grow many long rough blades or leaues, and at the top large tufts, or eares spread abrode, the whiche do change into a fine downe or cotton, and is carried away with the winde, almost like the eares of Mill or Millet, but farre bigger. The roote is long & white, growing outwardly in the bottome of the water.

7 The Cane of Inde, or ẏ Indian Cane, is of the kind of Reedes, very high, long, great, and strong, the which is vsed in temples & Churches to put out ẏ light of candels, whiche they vse to burne before their Images.

8 To these we may ioyne that Cane, whereof they make Sugar, in the Ilandes of Canare, and els where.

9 Besides these sortes, there is another aromatical, and sweete smelling kind, vnknowen in this Countrie.

Fig. 105. "Pole Reede"[2] (*Phragmites communis* Trin.). [From the *Nievve Herball* of Rembert Dodoens, translated by Henry Lyte, 1578.]

—the product of the seed—is not in itself fixed, but that the *total mass* of vegetable substance which can be developed from such a seed has a definite limit, though this limit can be reached either at a

[1] Pallis, M. (1916).

[2] Lyte calls *Scirpus lacustris* L., "the pole Rushe, or bull Rushe"; it is probable that in both cases "pole" stands for "pool".

rapid or at a leisurely pace. The earlier rhizomes[1] of the Reed are the thickest, and, as the individual increases in age, the successive shoot generations become slenderer and slenderer, until senescence and death are reached. This final stage may be slow in arriving, for the life of the major individual in the Reed is much longer than that of man, but the end comes at last. Viewing the life-history of the Reed as a whole, we must regard the thickest rhizomes as being, morphologically, the basal members of a vast branch system. They may be compared with the trunk and principal branches of a forest tree, but in visualising this analogy, it must be borne in mind that in the Reed the 'trunk' and the ultimate twigs do not coexist in time.

The variation in size of the axial members, which is found in the Reed at different periods of its life-history, recurs in the bamboos. In this tribe, when the individual has a long life, the growth of the culms is not uniform throughout, but the stems which succeed one another become larger year by year until the adult stage is reached.[2] In contrast with this crescendo up to the normal size limit, there is a diminuendo when clumps of bamboo are weakened by overcutting. The culms become gradually shorter and thinner, until finally the rhizome is capable of producing whip-like branches only.[3]

In many grasses and bamboos the reproductive shoot is a continuation of one which has been previously vegetative. There is one bamboo, however, which produces three types of shoot—vegetative, reproductive, and sterile reproductive. This is *Glaziophyton mirabile* Franchet,[4] which is found near Rio de Janeiro. The long rhizomes of this bamboo creep almost at the surface of the soil, producing at intervals nodal swellings clothed with broad shining scales, whence arise erect stems, 3 or 4 m. high. These stems, which are leafless, stiff, and gradually tapering to the apex, hardly attain the thickness of a little finger. Their bases, for a height of 40–50 cm., are covered with sheathing scales. Above this they are nodeless and naked until the apex is reached, where there is a terminal swelling, marking the arrest of development, if—as is generally the case—

[1] More information is needed about the development of the Reed shoot from the seedling phase up to that of the full-sized rhizome.

[2] Rivière, A. and C. (1878).

[3] Brandis, D. (1899).

[4] Franchet, A. (1889).

these shoots are sterile. In this state the plant recalls the Bulrush, *Scirpus lacustris* L., in the non-flowering condition. For years *Glaziophyton mirabile* was known only in this leafless phase, but, after a fire in which the junciform stems were destroyed almost to the level of the soil, they were replaced by other shoots with fascicles of leaf-bearing branches. On one occasion, also, flowering panicles were seen to occur on the junciform stems. It thus became clear that these junciform stems were reproductive shoots, but in an inhibited, sterile condition, while the leaf-bearing shoots were a distinct development. These inhibited fertile branches are an example of the tendency towards sterilisation of the reproductive shoots[1] which shows itself not infrequently among perennial grasses. It has been recorded, for instance, that in *Avena pratensis* L. and *Koeleria cristata* Pers., it is possible to examine a great many inflorescences without finding fruits.[2] We have already spoken of the replacement of spikelets by vegetative buds in the bamboos[3] (Fig. 106, A). In the Reed, *Phragmites communis* Trin., detached fragments of the stem, floating on the surface of the water, develop their axillary buds and form roots, and the stolons also root at the nodes. In correlation with this activity of vegetative reproduction, it is found that the flowers are rarely fertile.[4] Moreover, certain leafy shoots, which seem externally to be merely vegetative, reveal a rudimentary inflorescence when they are dissected, so that they come into the category of perverted fertile shoots. These abortive inflorescences are also to be found in *Arrhenatherum avenaceum* Beauv.—another plant with a peculiarly active vegetative life.[5] A marked inhibition of fertility also occurs in certain grasses cultivated for their vegetative organs, e.g. Sugar-cane (p. 53), Kikuyu-grass (p. 47) and some oil-grasses (p. 56). The opposite condition is reached in certain bamboos in which consummation of fertility involves the exhaustion and death of the sterile parts. The Gramineae would seem, indeed, to offer a suitable field in which to study the relation—or, rather, the antagonism—which prevails between the vegetative and reproductive phases.

[1] The tendency to sterilisation within the inflorescence has been discussed in Chapter IX.
[2] Hackel, E. (1882). [3] See pp. 73–4. [4] Royer, C. (1883).
[5] Kirchner, O. von, Loew, E. and Schroeter, C. (1908, etc.).

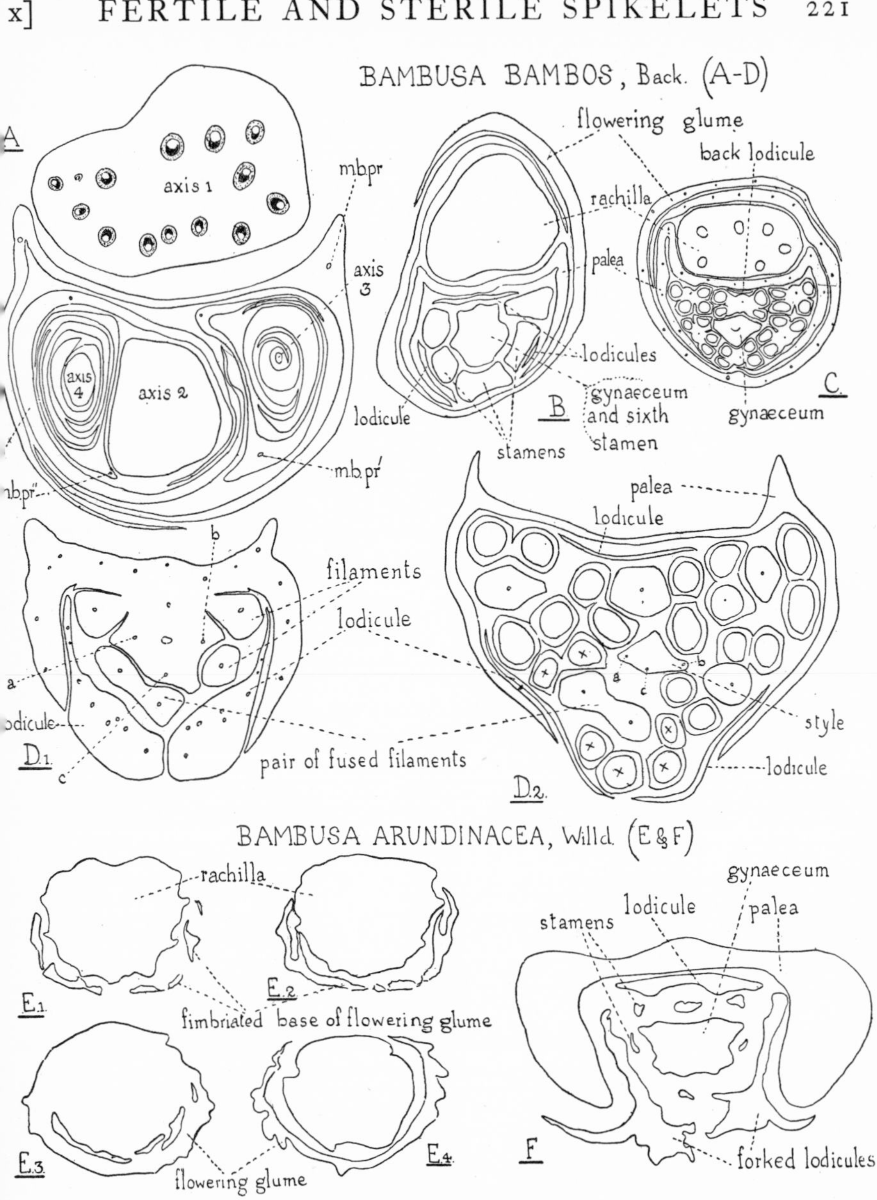

Fig. 106. *Bambusa*. A–D, *B. Bambos* Back. A, transverse section showing two vegetative spikelets, whose axes are marked *axis* 3 and *axis* 4 (× 47); *m.b.pr.*, median bundle of prophyll borne by *axis* 2; *m.b.pr.′*, median bundle of prophyll borne by *axis* 3, which arises in the axil of the prophyll of *axis* 2; *m.b.pr.″*, median bundle of the prophyll borne by *axis* 4, which arises in the axil of leaf *l*, which succeeds the prophyll of *axis* 2. B–D, transverse sections of flowers; B (× 77), C and D (× 47); in D, two stamens are united. E and F, *B. arundinacea* Willd.; E (× 23), F (× 47). [For fuller legend, see Arber, A. (1926).]

CHAPTER XI

THE GRASS EMBRYO AND SEEDLING

IT is a matter of common observation that no family of flowering plants includes so great a number of individuals as the Gramineae —a fact which seems in part accounted for by their extraordinarily prolific and effective seeding. There are a number of detailed records confirming this general impression. As long ago as 1660, Sir Kenelm Digby related[1] that the "Fathers of the Christian Doctrine at *Paris* doe still keep by them for a Monument (and indeed it is an admirable one) a Plant of Barley consisting of 249. stalkes, Springing from one Root or Grain of Barley, in which they counted above 18000. Grains or seeds of Barley". Later in the seventeenth century, Thomas Everard[2] recorded that from one grain of Wheat he had obtained eighty ears and above 4000 grains. Seeding on this scale may lead to a rapidity of increase which seems almost miraculous. A few years ago the chief kind of Wheat grown in Canada was that known as 'Marquis'. One grain of Marquis was planted in an experimental plot at Ottawa in the spring of 1903, and in 1918, from the progeny of this grain after fifteen years, 300,000,000 bushels of Wheat were produced in Canada and the United States.[3]

In the grasses there is normally a period of seed-rest before germination. The whole subject of seed-rest, and the related problem of delayed germination, seem to need a thorough comparative study, for our present knowledge of these questions is only complete enough to reveal much obscurity. It is a striking fact that seeds harvested in full ripeness seem to be in a condition of profounder rest, and to germinate with more difficulty than those which are gathered when not quite ripe. The need for a longer or shorter seed-rest may be merely one symptom of a deep-seated difference in constitution, for a species may exhibit parallel behaviour in vegetative rest and

[1] Digby, K. (1661). [2] Everard, T. (1692 and 1693/4).

[3] Buller, A. H. R. (1928); on the genic constitution of 'Marquis', see Hurst, C. C. (1933).

seed-rest. When *Oryza sativa* L. perennates, it is said to pass through a period of complete rest, and it also has a very marked seed-rest;[1] the grains of wild Rice are shed while still green, and though they drop into water or mud, they do not germinate until the succeeding monsoon.[2] *Poa pratensis* L., again, with its decided vegetative rest in winter, germinates less easily than the definitely evergreen *P. trivialis* L.[1] It seems that the normal seed-rest of some grasses may, on occasion, be dispensed with, for certain experiments of which the results were recorded more than eighty years ago,[3] and which perhaps need confirmation, showed that the grains of various cereals possessed the power of germination long before their maturity, when the endosperm was still "presque en lait". More recently it has been stated that most races of Maize will also germinate when quite unripe—"in the milk".[4] At the opposite end of the scale, there is the tradition that the seeds of Gramineae may remain living, but dormant, for thousands of years. Time after time, there have been reports of the germination of grain of immemorial antiquity from Egyptian tombs—the so-called 'Mummy-wheat'—and recently similar claims have been made for Wheat said to have been found in an ancient tomb in India.[5] There is, however, no case on record in which the evidence for such survival has withstood critical examination; indeed experimental work has made it clear that the life duration of the seeds of the Gramineae is strictly limited. Some results from the United States[6] showed that, whereas practically all grains of Wheat retained their vitality for the first five years, more than 75 per cent. had lost their power of germination before they were fifteen years old. A few kept their vitality for seventeen years, but scarcely any lived to eighteen years and none to nineteen. Germination of Wheat, harvested twenty-five years before, has been obtained, however, in some experiments in England.[7] The longevity of Oats, in the American trials, proved to be greater than that of Wheat; at the age of nineteen years, 41 per cent. were still alive. In some

[1] Kirchner, O. von, Loew, E. and Schroeter, C. (1908, etc.).
[2] Graham, R. J. D. (1927).
[3] Duchartre, P. (1852).
[4] Sturtevant, E. L. (1894).
[5] *The Times*, August 4, 1933.
[6] Sifton, H. B. (1920).
[7] Percival, J. (1921).

earlier experiments carried out in this country,[1] the survival times were shorter; none of the seeds of Wheat or Barley proved to be alive in the tenth year, and no Oats survived beyond the sixteenth year. All the pasture grasses which were tested, died between the eighth and thirteenth years.

Grass 'seed' is of a somewhat complex structure, as will be seen from the grain of Maize (*Zea Mays* L.) cut in half radially, which is shown in Fig. 107. The external surface is formed by the pericarp. The embryo is seen lying against one face of the endosperm, with which it makes contact by means of the scutellum—a cushion-like body, attached to, and partly enwrapping, the embryonic axis. The plumular bud does not arise at the level of attachment of the scutellum, but is separated from it by a stalk, the mesocotyl. The primary root, with its root-cap, is enclosed in a special thimble-like sheath, known as the coleorhiza, which in Wheat, Rice, Timothy-grass, etc., produces 'root'-hairs.[2] The plumule is also enclosed in a sheath—the coleoptile. In Maize, two adventitious roots (Fig. 107, *r.'*) arise at an early stage from the junction region of scutellum and axis. The Maize embryo is not wholly typical for the Gramineae, because it does not include an epiblast. This little non-vascular outgrowth, which arises from the face of the axis opposed to the scutellum, can be seen in Fig. 117, A and C, p. 244 (*Avena sativa* L.), and Fig. 108, D2, p. 230 (*A. barbata* Brot.).

It will be recognised from this brief description that the embryo of the Gramineae is decidedly different from that of Monocotyledons in general; it thus becomes necessary to treat its parts in some detail. Before turning to the embryo itself, we will consider the endosperm—that curious food tissue, derived from the triple fusion of the second male nucleus and the two polar nuclei of the embryo-sac, one of which is the sister nucleus of the egg.[3] In one sense, the endosperm may be regarded as a 'spoiled' second embryo; it is unique among living things in having, in a sense, three 'parents', and never

[1] Carruthers, W. (1911).

[2] Percival, J. (1921); Nishimura, M. (1922); Graham, R. J. D. (1927); Howarth, W. O. (1927).

[3] Sargant, E. (1900); Vries, H. de (1900).

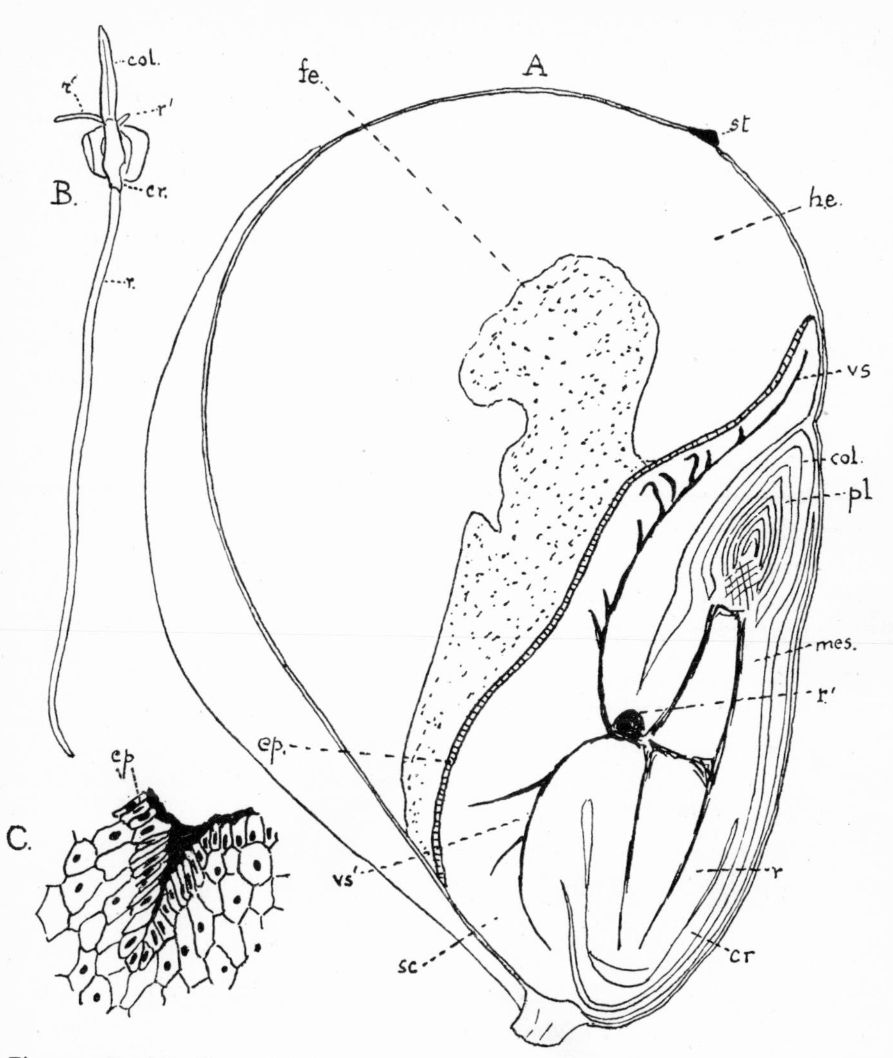

Fig. 107. *Zea Mays* L. A, diagram of the grain in radial longitudinal section, modified from the figure in Sachs, J. von (1882). *h.e.* and *f.e.*, horny and floury parts of endosperm; *st.*, remains of base of style; *sc.*, scutellum; *ep.*, epithelium; *v.s.*, main vascular trunk of upper part of scutellum; *v.s.'*, one of the vascular strands which ramify in the lower part of the scutellum; *r.*, radicle; *r.'*, vascular tissue for first adventitious roots; *col.*, coleoptile; *pl.*, plumule; *cr.*, coleorhiza; *mes.*, mesocotyl. B, seedling six days old (nat. size). C, transverse section of part of margin of scutellum from a dry seed, showing glandular infolding of secretory epithelium (× 137). [Arber, A. (1925), adapted from Sargant, E. and Robertson (Arber), A. (1905).]

leaving any descendants.[1] The earlier cells originating from the triple fusion form a single layer lining the sac. This lining layer produces, from its inner surface, a series of thin-walled elements, which become packed with starch; when it ceases to divide, its cells become filled with aleurone grains, and their walls increase in thickness. This aleurone layer, clothing the endosperm, may thus be regarded as a 'resting cambium'.[2] The history of the main part of the endosperm is the history of the degradation of a living tissue to the point at which it becomes a mere mass of food. While the starch is being laid down in the endosperm, the nuclei are at first in full activity and show their nucleoli clearly; but they begin to change at a stage at which all the endosperm cells have received some portion of their starch. The nucleoli disappear, and the nuclei become solider and denser in appearance, staining very deeply with haematoxylin. They develop irregularities of shape, become transformed into a coarse network, and are at last completely disorganised, though when the grain is ripe, their relics are still in evidence. These changes seem to be a direct result of the overwhelming pressure exerted by the expanding starch-grains.[3]

Both simple and highly compound starch-grains are to be found in the endosperm of Wheat, Rye and Barley. In Wheat and Rye the compound grains may contain as many as twenty-five component members, and, in Barley, as many as twenty.[4] Although starch is the principal seed-reserve in grasses, it is not the only one. In the translucent endosperm of 'sugary' Maize, only a very small amount of fine-grained starch is present; the greater part of the starch is replaced by an amorphous sugary substance, which, in the completely ripe and dry condition, shines like gum arabic.[5] In the endosperm of 'waxy' Maize, and in some other cereal varieties, a carbohydrate is present, which has been identified as erythrodextrin,[6] while in the Maize embryo, fat occurs.[7] These examples will show

[1] Weatherwax, P. (1930), (1931).

[2] McLennan, E. I. (1920); Gordon, M. (1922); another view is that it is not the aleurone layer, but the layer within it, which is cambial.

[3] Brenchley, W. E. (1912).

[4] Peter, A. (1900).

[5] Koernicke, F. (1885).

[6] Weatherwax, P. (1922[2]).

[7] Toole, E. H. (1924).

the complexity of the chemistry of the seed; it is too technical a subject for consideration in the present book. We must, however, make a brief mention of the protein constituent, which, so far as it is obtained from the endosperm, seems to be the product of the residual cytoplasm left between the starch-grains; possibly the relics of the nuclei also contribute something to it. This substance, which is called *gluten*, can be separated without difficulty by tying up a quantity of Wheat flour in fine muslin, and kneading it under a tap. The starchy part of the flour is thus washed away, leaving the gluten as a coherent mass.[1] Gluten is of great practical importance, for upon its quality and quantity depends the 'strength' which retains the gas liberated during fermentation, and thus produces a well-formed and well-risen loaf.[2] This baking capacity is closely bound up with the physico-chemical properties of the gel structure which gluten forms with water. The higher the protein content, the easier the dough is to handle, and the more water does it hold. This water-holding power is an advantage to the baker, but not always to the consumer, who does not invariably share the baker's admiration for a flour which "will make water stand upright".

The aleurone layer secretes diastase, and thus plays some part in rendering the store of food in the endosperm available for the embryo. In a seedling of Rye,[3] two to four days old, the starchy part of the endosperm has degraded into a white pulp. At this stage the cells of the aleurone layer are rich in cytoplasm and look granular; they are papillose on the inner face. The experiment has been tried of isolating fragments of aleurone layer, placing them on damp filter paper, and covering them with a mixture of starch and water. At the same time, as a control, a similar mixture of starch and water was laid upon damp filter paper with no aleurone layer. After twenty-four hours the starch-grains of the control experiment were still intact, while those on the aleurone layer were corroded and many of them had fallen to pieces.

The aleurone layer of the Gramineae is generally one cell thick,

[1] Biffen, R. H. and Engledow, F. L. (1926).
[2] Hunter, H. (1931); Hunter, H. and Leake, H. Martin (1933).
[3] Haberlandt, G. (1890).

but in many cultivated Barleys, the cells of the external face of the endosperm elongate radially and divide by two or three tangential walls, so as to form a layer three or four elements deep. It seems probable that this thickness of the digestive layer may explain the preference given to Barley over other cereals in malting.[1]

In the grass family, it is the endosperm which is the principal source of human food; under cultivation, it seems that man has encouraged hypertrophy of the endosperm, just as he has aimed at a milk yield in cows which is altogether excessive from the point of view of the calf. Experimental work confirms this general impression, for it has been shown that the supply of endosperm produced in some of the cereals definitely exceeds the amount needed by the developing seedling.[2] For instance, in a six-rowed Barley, in which the lateral grains were much smaller than the median ones, it was found that as high a proportion of the small as of the large grains germinated, and that the progeny of the small grains was not inferior to that of the large grains in vegetative or reproductive vigour. In the Oat spikelet, again, the lower grains are larger than the upper ones. When the large and small grains were collected and sown separately, it was found that the smaller were in every respect as good as the larger; the plants derived from the larger grains became self-supporting long before the whole supply of endosperm had been absorbed, thus showing that this supply was excessive. In the wild species, *Avena fatua* L., on the other hand, practically the whole of the endosperm was utilised by the seedling.

In certain grasses the development of the embryo from the fertilised ovum has been studied.[3] In *Poa annua* L., for instance, the apical region of the segmented egg gives the scutellum and part of the coleoptile; the next zone—the hypocotyl, plumule, and the remainder of the coleoptile; and the zone nearest the suspensor—the root-cap, coleorhiza, and (as an epidermal outgrowth) the epiblast. The question is, how these facts are to be interpreted. In the nineteenth

[1] Tieghem, P. van (1897).
[2] Drabble, E. (1906).
[3] Souèges, R. (1924); see also Noerner, C. (1881) and Schnarf, K. (1929).

century, it was assumed that the 'regional planning' of the embryo was of high significance morphologically; but it now seems doubtful whether the first cleavage planes have, in truth, much more meaning than the cell divisions of the mature plant, which show little regard for morphological boundaries. We need, indeed, a fresh approach to the problems of early embryology—we have been contented for too long with the avenue marked out in the eighteen-seventies. What the new approach may be cannot be foretold, but we may at least hope that the study of early embryology in the plant will some day include the use of biochemical methods, corresponding to those which have already been applied to the animal egg.

In the ripe seed, the most noticeable member of the embryo is the sucking organ, or scutellum (*sc.*, Fig. 107, p. 225). This organ is not generally green, but, in *Spartina Townsendi* Groves, it is exceptional in containing chlorophyll when the seed is mature.[1] The scutellum continues to be conspicuous in the seedlings of various grasses long after germination. It can be seen in Fig. 110, B4 and B5, p. 234 (*Coix lacryma-Jobi* L.); Fig. 108, D1 and D2, p. 230 (*Avena barbata* Brot.); and Fig. 117, A and C, p. 244 (*A. sativa* L.). The scutellum does not lightly sever its connection with the caryopsis, so that the survival of remains of the last generation is often a striking feature. In Fig. 108, B, a plant of *Avena sterilis* L. is seen in fruit, but its base is even yet enclosed by the residue of the parent spikelet; the basal spikelet, the vegetative plant, and the embryo in the ripe seed thus represent three successive sporophyte generations remaining in union.

The face of the scutellum which is in contact with the endosperm shares with the aleurone layer the capacity to secrete diastase. Its cells elongate into finger-like processes (Fig. 117, B1–B3, p. 244); in *Briza minor* L. they increase, when the seed germinates, to ten times their original length.[2] In Wheat the epithelium cells in the resting stage are 30–40μ long. As germination advances, they elongate to 80–90μ, while their tips become swollen and club-shaped and project

[1] Oliver, F. W. (1925).
[2] Kirchner, O. von, Loew, E. and Schroeter, C. (908, etc.).

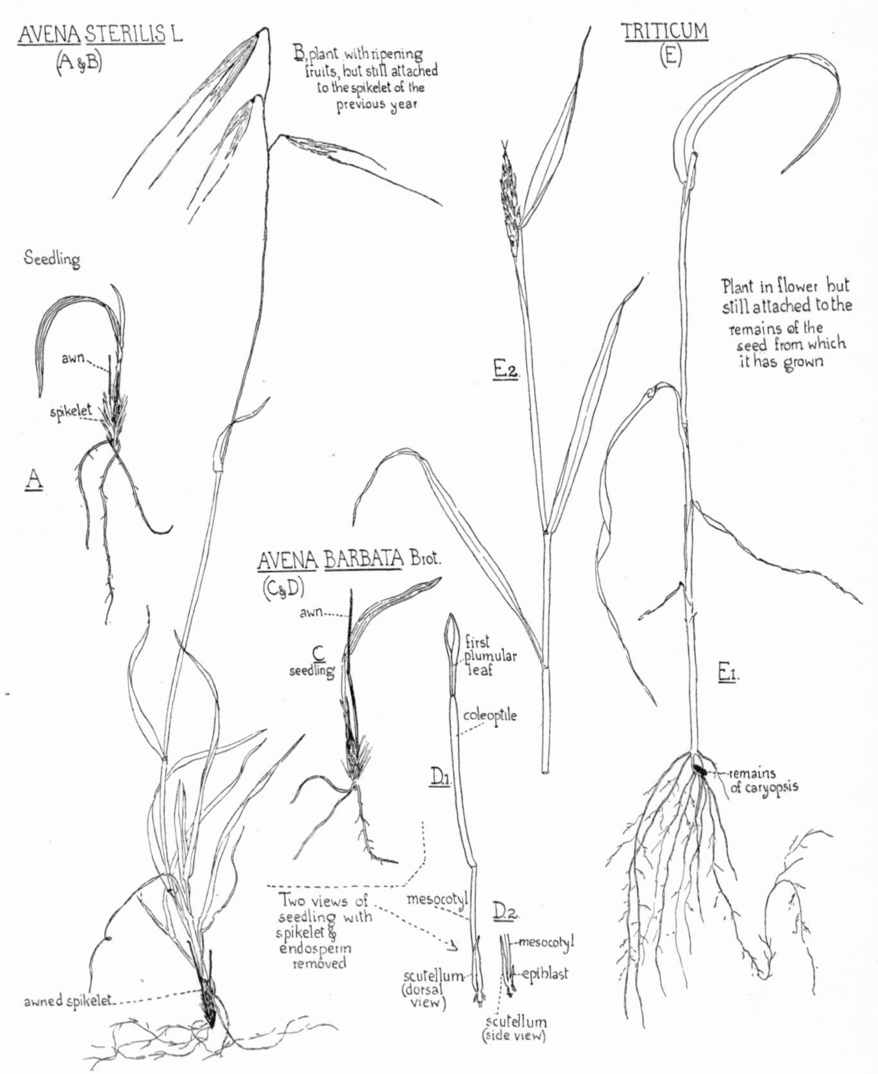

Fig. 108. A and B, *Avena sterilis* L. A, seedling; B, small plant in infructescence stage. (Both A and B × $\frac{2}{5}$ *circa*.) C and D, *Avena barbata* Brot. C, seedling (× $\frac{2}{5}$). D 1 and D 2, seedling with parent spikelet and endosperm removed to show scutellum (× $\frac{4}{5}$ *circa*). E 1 and E 2, *Triticum* sp. (Wheat), small flowering plant cut in two (× $\frac{2}{5}$ *circa*). [A.A.]

freely into the endosperm.[1] In *Zea Mays* L., the epithelium shows numerous infoldings, which form definite glands.[2] One of these clefts is seen in section in Fig. 107, C, p. 225. The number of glands is variable, and it is difficult to reckon them with any exactness, but as many as thirty-eight have been reported from a single scutellum. Glands of the same type, but less well-developed, are found in *Coix lacryma-Jobi* L.,[2] and they have also been described in certain Wheats.[1]

The upper part of the scutellum of *Zea Mays* L. is served by a single main bundle (*v.s.* in Fig. 107, A), but a number of subordinate bundles, of which *v.s.′* is one, enter the lower region. The main bundle bears numerous slender branches, of which the longer commonly arise from the lateral faces of the bundle-trunk, and extend into the wings of the scutellum, giving off short branches towards its secretory surface. From the dorsal face of the bundle-trunk, on the other hand, short branches spread out towards the dorsal epithelium. As the scutellum narrows towards the apex, these short branches become more numerous, and a fairly thick radial section through the scutellum shows the main bundle feathered on its dorsal face by a close crop of vascular branchlets all bending outwards (Fig. 107, A). The bundle-trunk terminates in a tuft of such branchlets, which reach almost to the very tip of the scutellum.

A few days after germination, the main bundle of the scutellum has become completely lignified. Just above its origin from the axis, it is circular or oval in section; the ventral xylem group is small, but the phloem is massive (Fig. 109, A, p. 232). Higher up, this collateral bundle shows a tendency to become amphivasal—the xylem creeping round the phloem; this tendency is shared even by some of the minor bundles (Fig. 109, A2 and A3). Amphivasal bundles of a corresponding type are found in the scutellum of Job's-tears (Fig. 109, B) and of Barley.[3]

[1] Percival, J. (1921).
[2] This account of the scutellum of *Zea Mays* L. is taken from Sargant, E. and Robertson (Arber), A. (1905), in which further details will be found.
[3] Arber, A. (1930[1]).

It seems generally to be agreed among botanists that the scutellum represents either the whole, or the terminal region only, of a seed-

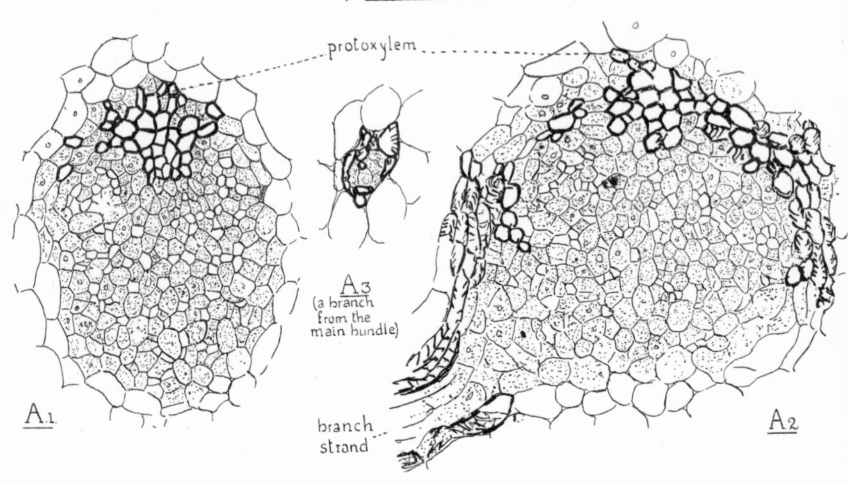

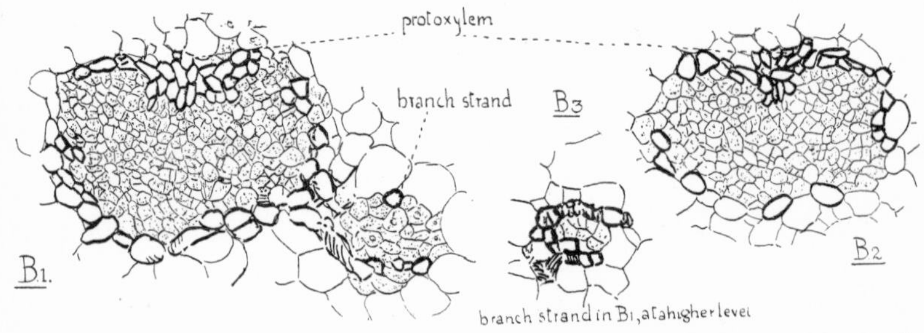

Fig. 109. Amphivasal bundles of the grass scutellum. Transverse sections from microtome series from below upwards (× 193). A 1–A 3, *Zea Mays* L. A 1 shows the collateral structure of the base of the bundle; in A 2 an amphivasal structure is approached; an amphivasal branch-bundle is being given off. The phloem and conjunctive tissue of the bundle are dotted, but on the phloem side it is doubtful where the limit of this should be set. A 3 shows an amphivasal branch strand. B 1–B 3, *Coix lacryma-Jobi* L. B 1, showing amphivasal character of main trunk near base; B 2, main trunk, higher; B 3, the branch in B 1 at a higher level. [Arber, A. (1930[1]).]

leaf. These alternatives will be considered further, when the other parts of the embryo have been discussed.

A feature in which grass seedlings differ from those of Monocotyledons in general, is the frequent presence of a segment, which appears to be axial, inserted between the cotyledon-sucker, and the sheath which encloses the plumular bud. This intercalated segment, the mesocotyl,[1] which we have already noticed in Maize, is shown for Job's-tears in Fig. 110, A and B, and for the Manchurian-water-rice in Fig. 116, A 1 and A 2, p. 242, where it is indicated in solid black. It may elongate greatly if the seed is planted deep; a length of 28 cm. has been recorded for Teosinte and 36 cm. for a variety of Maize cultivated from time immemorial by the American Indians in regions where the drought is too severe for strains which need the ordinary shallow planting.[2] Variation in mesocotyl length plays a part also in pasture grasses. In Timothy, *Phleum pratense* L., the rooting system of the seedling tends to become established near the surface of the soil, irrespective of the depth of sowing. This adjustment is due to the mesocotyl;[3] it is probably lack of light which induces it to elongate, for it has been shown that in *Coix lacryma-Jobi* L. the mesocotyl may reach only 5 mm. in continuous light, but 20 cm. in darkness.[4] The Oat, *Avena sativa* L., is a plant in which the anatomy of the mesocotyl can easily be studied. Figs. 117, A, C, p. 244, show the structure of the embryo within the seed, and of the seedling after germination. It will be recognised that the mesocotyl scarcely exists in A, but that the strand supplying the scutellum gives a slight downward dip in passing outwards. This downward dip is the first indication of a process which will go much further in the seedling. In C, the intercalary growth, which has lifted the plumular bud and its sheath upward, away from the exsertion of the scutellum, has also drawn up the proximal part of the scutellum-trace, so that now the scutellum bundle is carried for the full length of the mesocotyl parallel to the mesocotylar stele, instead of connecting with the main cylinder at the level of the scutellar attachment.[5] Apart from the details of its stelar anatomy, the mesocotyl

[1] This name was suggested by Čelakovský, L. J. (1897).
[2] Collins, G. N. (1914). [3] Evans, M. W. (1927).
[4] Prat, H. (1932); see also Hamada, H. (1933).
[5] This process is clearly shown in Avery, G. S. (1930), Figs. 25–7, p. 11.

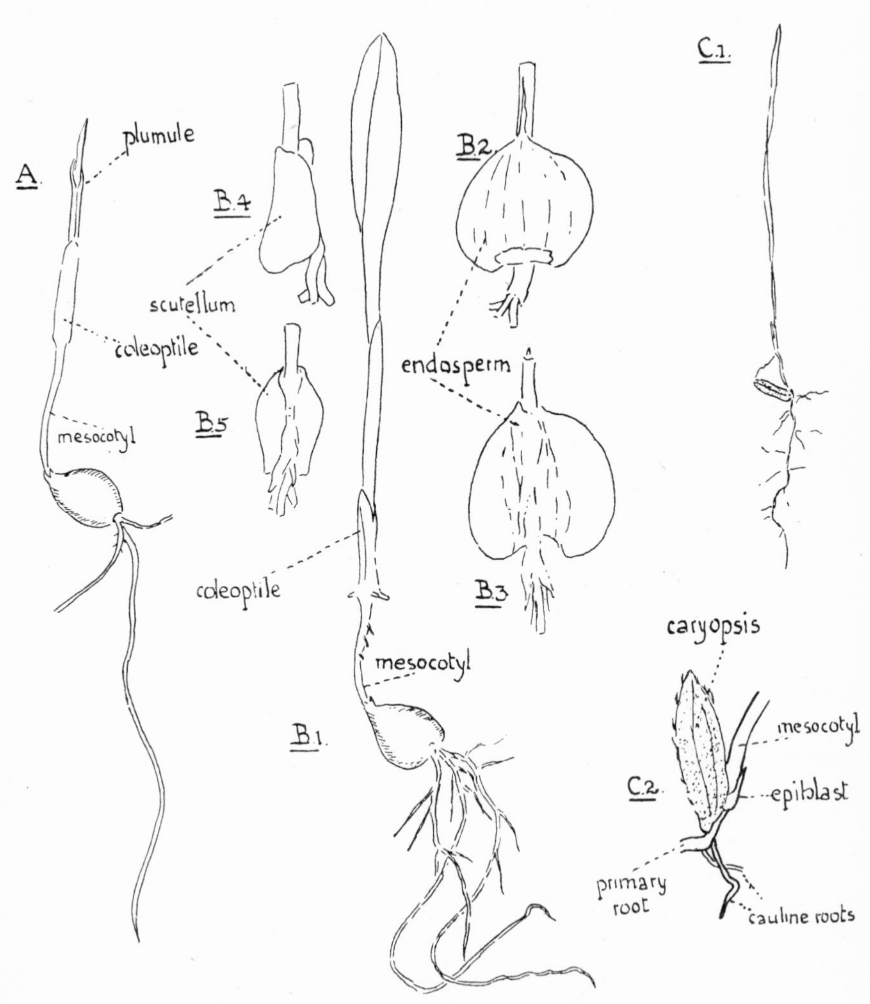

Fig. 110. A and B, *Coix lacryma-Jobi* L. A, young seedling (× 1); B 1, older seedling (× 1); B 2 and B 3, endosperm from back and front (× 2); B 4 and B 5, scutellum from side and front (× 2). C 1 and C 2, *Leersia oryzoides* Sw.; C 1, seedling (× 1); C 2, caryopsis and base of seedling to show leaf-like epiblast (× 5). [E.S. and A.A.]

of *Avena* thus differs in two respects from an ordinary grass internode—in the presence of the inverted scutellum-trace outside the stele,[1] and in the fact that the region of intercalary growth is apical, instead of basal.[2] So it seems best to interpret the mesocotyl,[3] not as an internode, but as the cotyledonary node, telescoped out into an elongated form. Possibly its apical growth may be regarded as due to a meristematic 'infection' from the basal growing region of the succeeding internode.

At the top of the mesocotyl we come to the sheath known as the coleoptile, enclosing the plumular bud. It is a delicate scale with no midrib, but with two bundles placed approximately opposite to one another, in a plane at right angles to the plane of symmetry of the scutellum. In Oats, Maize, Barley and Wheat, these bundles, which are collateral below, become concentric, with external xylem, in passing up, while at the apex the xylem forms a closed cone of spiral tracheids capping the bundle. Large, conspicuous stomates occur at the coleoptile apex in association with the bundle terminations; in a moist atmosphere drops of water exude from these pores. This exudation of water was noticed early in the eighteenth century by the agriculturalist, Edward Lisle[4]. "I had often observed", he wrote, "in the spring-time, when the blades of barley first begin to shoot out of the ground, dewy drops standing every morning on the points of the blades." He proceeded to carry out an experiment, which was remarkable for that early date, in order to test his idea that this water-production was due to root-pressure. "I took a pot of fine garden-mold", he wrote, "and placed it in my study; the earth was

[1] The inverted scutellum-trace may in some grasses be included within the limits of the stele instead of remaining outside, e.g. *Zea Mays* L. and *Andropogon Sorghum* (L.) Brot. This difference is, however, non-essential; see Sargant, E. and Arber, A. (1915).

[2] Avery, G. S. (1930); Prat, H. (1932).

[3] I do not propose to enter into detail about the history of the controversy relating to the morphology of the grass seedling; I will only say that I have abandoned certain theoretical views expressed in Sargant, E. and Arber, A. (1915), where references will be found to earlier papers. The interpretation offered in the present chapter is based on further work, and on a consideration of later papers: Worsdell, W. C. (1916); Weatherwax, P. (1920); Bugnon, P. (1921); Howarth, W. O. (1927); Avery, G. S. (1930); Prat, H. (1932); Boyd, L. (1931) and (1932).

[4] Lisle, E. (1757); see p. 48, note 1.

but moderately moist, and I put into it a handfull of barley; when the barley shot up about half an inch or an inch, at the ends of the points appeared the said pearly drops; I wiped them all off, and carefully took up half a dozen of the blades of barley by the roots, then with a pair of scissars cut off the roots close to the grains of corn, and covered them in the same earth again; the next day I looked on the blades, and found the pearly drops of water settled on the blades as before; but on the tops of those blades, whose fibrous roots I had cut off, not the least moisture appeared, tho' the blades continued in good verdure through the moisture of the earth they were put in; this shews plainly, those watery globules are not collected from the moisture of the outward air, but from the juices drawn upwards from the roots."

Though the coleoptile is typically a two-bundled structure, occasionally a higher number of strands may occur. As many as six have been recorded in varieties of *Triticum dicoccum* Schübl.,[1] and three to five in certain strains of Maize;[2] even in these unusual examples, however, we do not find symmetry about a median strand.

That the coleoptile and scutellum are parts of one member, is suggested by Barley seedlings, in which—in the absence of a mesocotyl—these two parts are closely associated. We can find analogies for such a subdivision of the seed-leaf in other Monocotyledons in which there is differentiation into a sucking region and an upstanding sheath. If we adopt the view that the scutellum and coleoptile of Barley are components of one seed-leaf, we must assume that the basal sheathing region of this seed-leaf is suppressed, and that the ligular region[3] has, in correlation, suffered an increased development, forming the coleoptile. The comparison with an awned flowering glume seems to favour the ligular interpretation; here the scutellum would be equivalent to the awn, and the coleoptile to the region above the departure of the awn. Since the main bundle passes into the awn, the part of the glume above the exit of the awn is left midribless like the coleoptile (Fig. 111). The glume differs from the cotyledon in the fact that the basal sheathing region is not entirely suppressed.

[1] Percival, J. (1927). [2] Avery, G. S. (1928). [3] See p. 247, note 1.

By taking the Barley cotyledon as an example, we have eliminated the complication introduced by the mesocotyl. *Triticum* provides a link between the Barley seedling and the more usual type, for, in the Wheat, the vertical distance between the entry of the scutellum bundle into the axis, and its fusion with the coleoptile-traces, is perhaps 0·05 mm., so that there is a mesocotyl, though it is minimal. From such intermediate forms, we pass to the more typical members

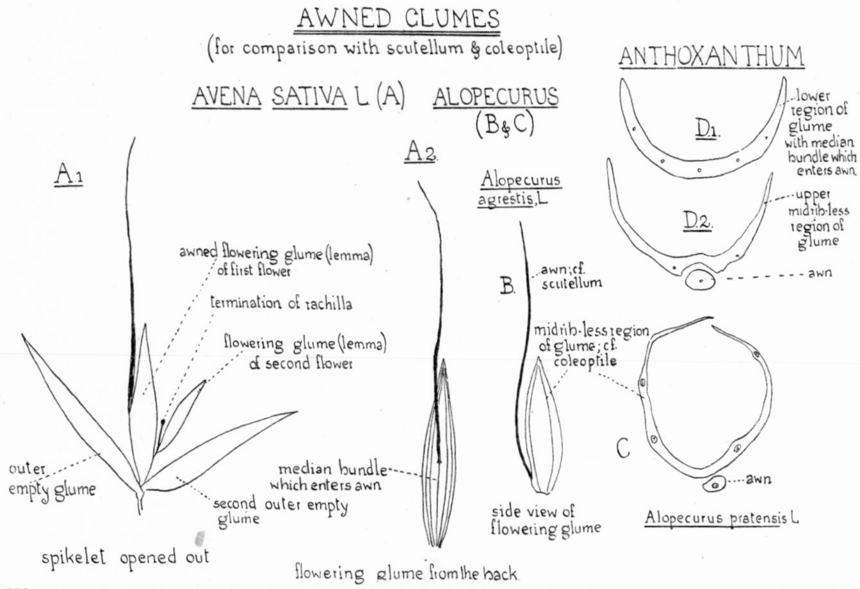

Fig. 111. A, *Avena sativa* L. A1, spikelet opened out, slightly magnified. A2, flowering glume from the back. B, *Alopecurus agrestis* L., awned lemma (enlarged). C, *A. pratensis* L., transverse section of awned lemma at level above detachment of awn (× 47). D1 and D2, *Anthoxanthum odoratum* L., transverse sections of an awned glume, below and above the detachment of the awn (× 77 *circa*). [A.A.]

of the family, in which the mesocotyl is a conspicuous feature. Now even if the view that the coleoptile is the ligular region of the cotyledon be accepted for Barley, the question still remains whether it can be accepted for those grasses in which the coleoptile and scutellum are separated by a mesocotyl. It seems to me that it can, for we have the analogy of the Cyperaceae to show that such a separation may occur between parts of what is undoubtedly one seed-leaf.[1] In

[1] Tieghem, P. van (1897).

Fig. 112, the seedling structure of *Carex glauca* Scop. is illustrated. A single cotyledonary strand leaves the stele (B 2). It runs upwards into the sheath of the cotyledon, and when it reaches the apex of the sheath, it turns downwards, doubling so closely on itself that the phloems of the upward and downward limbs, while within the sheath,

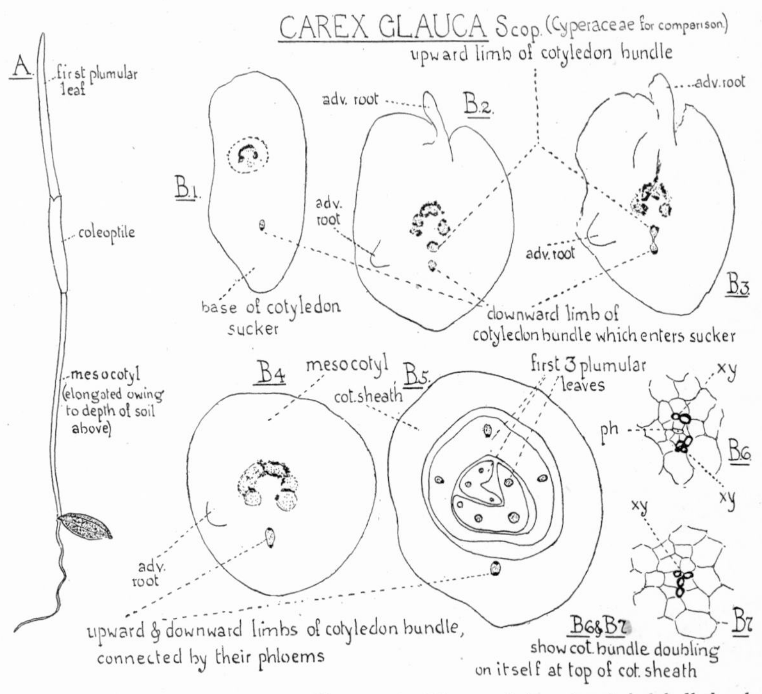

Fig. 112. *Carex glauca* Scop. A, seedling (× 2). The cotyledon sheath is labelled coleoptile. This seedling was buried under a layer of soil after germination had taken place, with the result that the mesocotyl elongated to 1·5 cm.; in another example it reached 2 cm., while in seeds sown at the surface, it was so short that it was not recognisable externally. B 1–B 7, sections from a transverse series passing upwards through a seedling; B 1–B 5 (× 47); B6 and B 7 (× 193 *circa*). [A.A.]

form one group (B 6 and B 7).[1] At a lower level the downward limb detaches itself from the upward limb (B 3) and continues its course towards the base of the mesocotyl (B 2), finally turning outwards into the cotyledon sucker (B 1). The example of the Cyperaceae thus removes any *a priori* doubt that might be felt as to the possibility of

1 This doubling of the bundle on itself has been figured by Goebel, K. von (1922), Fig. 1215, p. 1244.

the separation of the sucking and foliar parts of a single cotyledon by an elongation of the cotyledonary node.

The mesocotyl and coleoptile are not the only parts of the grass seedling whose homologies are not obvious; the epiblast and the coleorhiza are also puzzling. The epiblast, which is shown in Figs. 117, A, C, p. 244; 116, A 1 and B 2, p. 242; 108, D 2, p. 230; and 113, has been interpreted by some as a second cotyledon. This possibility cannot, indeed, be excluded, since reduced leaves, which are wholly non-vascular, are known to occur among the grasses. (A number of examples of such leaves are figured in this book; it will suffice to

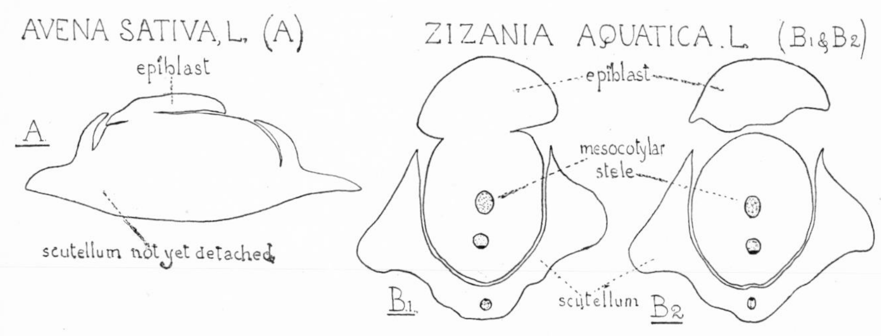

Fig. 113. Epiblasts. A, *Avena sativa* L. Transverse sections of the embryo from a seed, cut at a level at which the scutellum is still attached, and the epiblast just freeing itself (× 23); vascular tissue omitted. B 1 and B 2, *Zizania aquatica* L. Two transverse sections from a series from below upwards through the attachment of scutellum and epiblast in a seedling (× 23); in B 1 the scutellum is free and the epiblast still attached, while in B 2 both are free. [Arber, A. (1927[2]).]

recall the 'collar-leaf' which occurs just below the ear in *Avena barbata* Brot., Fig. 114, p. 240, and in other Gramineae.) There is, however, another possibility which must be taken into account. It is that an analogy for the epiblast may be found in certain non-vascular outgrowths of the reproductive axis,[1] rather than in any form of reduced leaf. In Fig. 115, p. 241, I have illustrated the structure which I have called the 'rachilla-flap' of *Cephalostachyum virgatum* Kurz. This outgrowth is the lip of an oblique cupule, forming the upward continuation of an internode of the rachilla; it faces the leaf (lemma) borne at the node which is the upper boundary of this internode. The 'articulations' of the rachilla of certain grasses, and the "appen-

[1] Arber, A. (1927[2]).

dages of the joints" described as reaching a length of more than 4 mm. in a grass from Tropical Africa, *Urelytrum squarrosum* Hack.,[1] probably belong to the same category as the 'rachilla-flap'. It will be seen from Fig. 113, that the epiblast faces the scutellum, as the rachilla-flap faces the succeeding glume in Fig. 115, A4 and A6. A tendency to the formation of these non-vascular excrescences seems to be characteristic of the Gramineae; it finds expression, not only in the cases cited, but also in the development of pulvini at the junction of inflorescence branches with the axis; basal cushions in connection

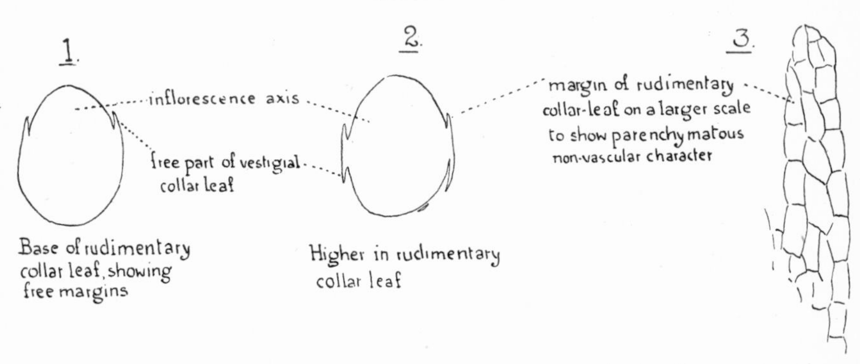

Fig. 114. *Avena barbata* Brot. *1* and *2*, transverse sections (× 19 *circa*), *1*, lower, and *2*, higher, through the 'collar-leaf' at the base of an inflorescence axis. *3*, free marginal region of this vestigial collar-leaf; the parenchymatous non-vascular character is maintained throughout the whole of the leaf, including what would be the midrib region. [A.A.]

with various glumes; and the coleorhiza which ensheathes the radicle in the embryo.

We must now consider, briefly, the vascular scheme of the seedling as a whole, since this scheme should be kept in view in trying to arrive at a judgment about the relation of the parts. *Lolium multiflorum* Lam. may be taken as an example, because it has a slender seedling, which presents the vascular system in a simpler form than that found in the solider plantlets of the cereals. Fig. 116, B1, is cut through the base of the scutellum, and shows the pentarch stele of the primary root. In B2, two of the poles of the pentarch xylem have united and the

[1] Stapf, O. (1898).

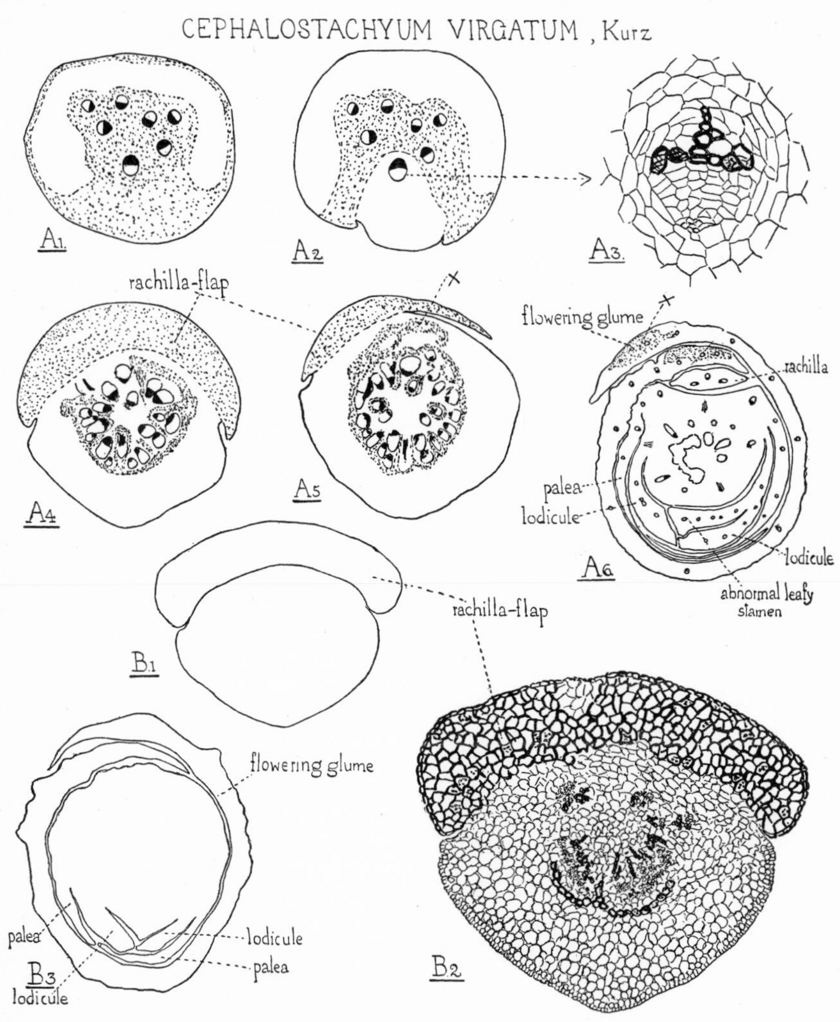

Fig. 115. *Cephalostachyum virgatum* Kurz (from the Calcutta Botanic Garden). A 1–A 6, transverse sections from a series from below upwards through a spikelet (× 47 except A 3, representing the largest bundle in A 2, which is × 318). B 1–B 3, transverse sections between the two flowers of another spikelet (B 1 and B 3, × 47; B 2, × 77). The axis and rachilla-flap are shown in B 1, and on a larger scale in B 2; the flowering glume and base of the flower are reached at a slightly higher level in B 3. [Arber, A. (1927[2]).]

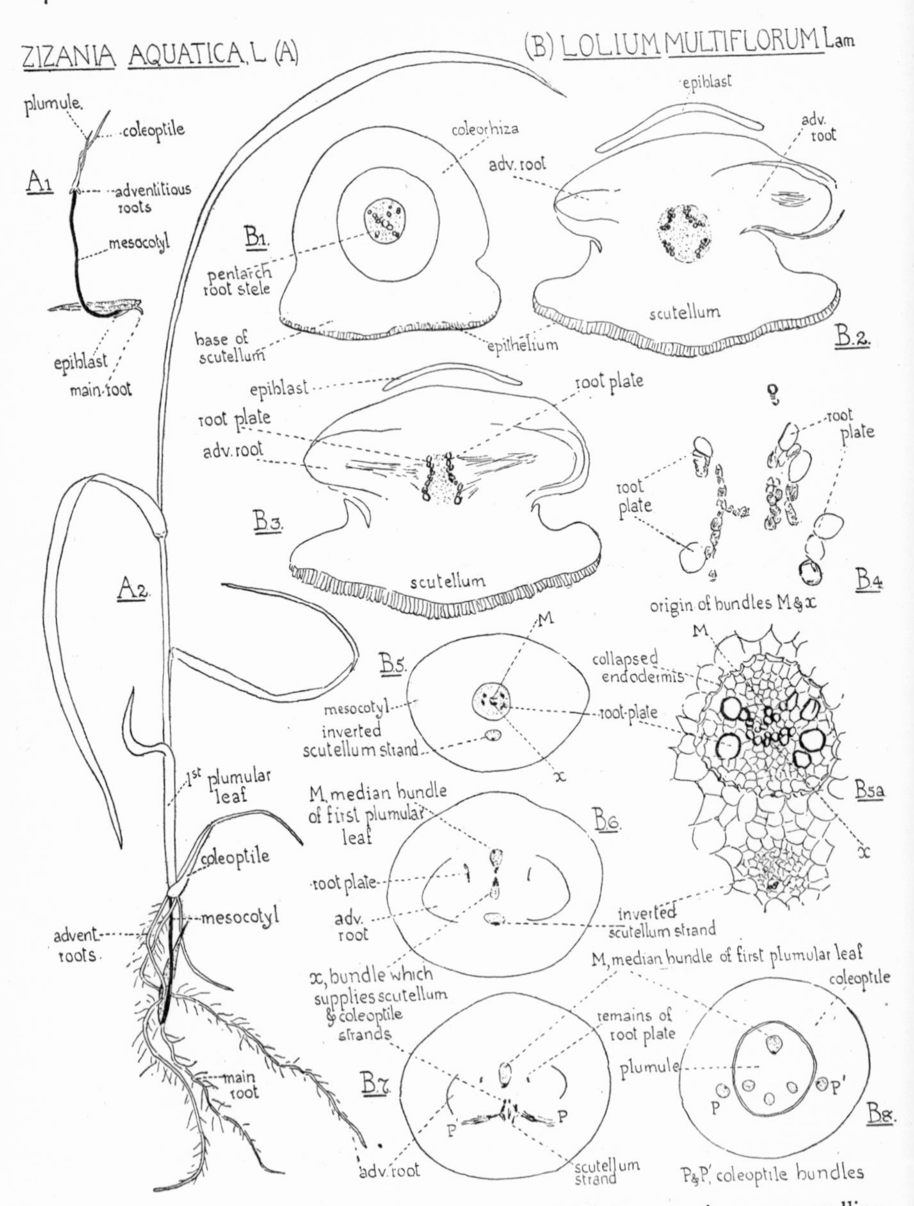

Fig. 116. A, *Zizania aquatica* L., from drawings by Ethel Sargant. A 1, young seedling; A 2, older seedling (× ½); mesocotyls indicated in solid black. B 1–B 8, *Lolium multiflorum* Lam. Sections from a transverse series passing through a seedling from the upper region of the main root (B 1) to the base of the coleoptile (B 8). B 1–B 3, B 5–B 8 (× 47); B 4 and B 5 *a* (× 193 *circa*). Only the lignified bundles are represented. [E.S. and A.A.]

xylem has arranged itself as four crescents with their concavities outwards. At this level the epiblast is cut facing the scutellum. In B 3, in which the scutellum is still cut below the level of the bundle, the four crescents have united into two plates, which are destined in part to connect with the adventitious roots to right and left. In B 4 it has become clear that these bands consist of two distinct cell-types—large-lumened root-plate tissue, and smaller elements, which, in B 5, have arranged themselves to form the xylem of two endarch bundles to north and south, which we may call *M* and *x*. At this level the inverted scutellum bundle is seen outside the main stele. The relation of parts can be recognised more clearly in B 5 *a*, which is on a larger scale. In B 6, the dwindling of the root-plates reveals *M* and *x* more distinctly; *M* will be the median bundle of the first plumular leaf, while *x* is the cotyledonary bundle. B 7 shows *x* turning outwards and dipping down into the mesocotyl (*en route* for the scutellum) and, at the bend, giving off two branches, *P* and *P'*, which will supply the coleoptile. In B 8, the coleoptile is seen detached from, and surrounding, the plumular bud. No attempt is made in these diagrams to show the relation of the plumular bundles, other than *M*, to the scutellar bundles; they are embryonic at this stage, and it is not possible to follow them with any exactness. The fact that the three bundles, *P*, *P'* and *M*, are the only lignified strands in B 8, makes the essentials of the story show more clearly than if the seedling were older and the vascular system more mature.

A less simplified example will be found in Fig. 117, C, p. 244, in which a seedling of *Avena sativa* L., cut lengthways, is diagrammatically reconstructed, so as to show the course of the vascular system. In Fig. 118, a few detailed transverse sections are given, to indicate the type of evidence on which the diagrams are based. The reconstruction (Fig. 117, C) extends from just below the epiblast and scutellum to the base of the plumule. The exact method of transition from the stele of the main root to that of the mesocotyl cannot be described, as it is almost completely masked by root insertions. In one seedling in which the main root could be followed downwards, it passed from the diarch to the heptarch state. Five of the protoxylems arose by

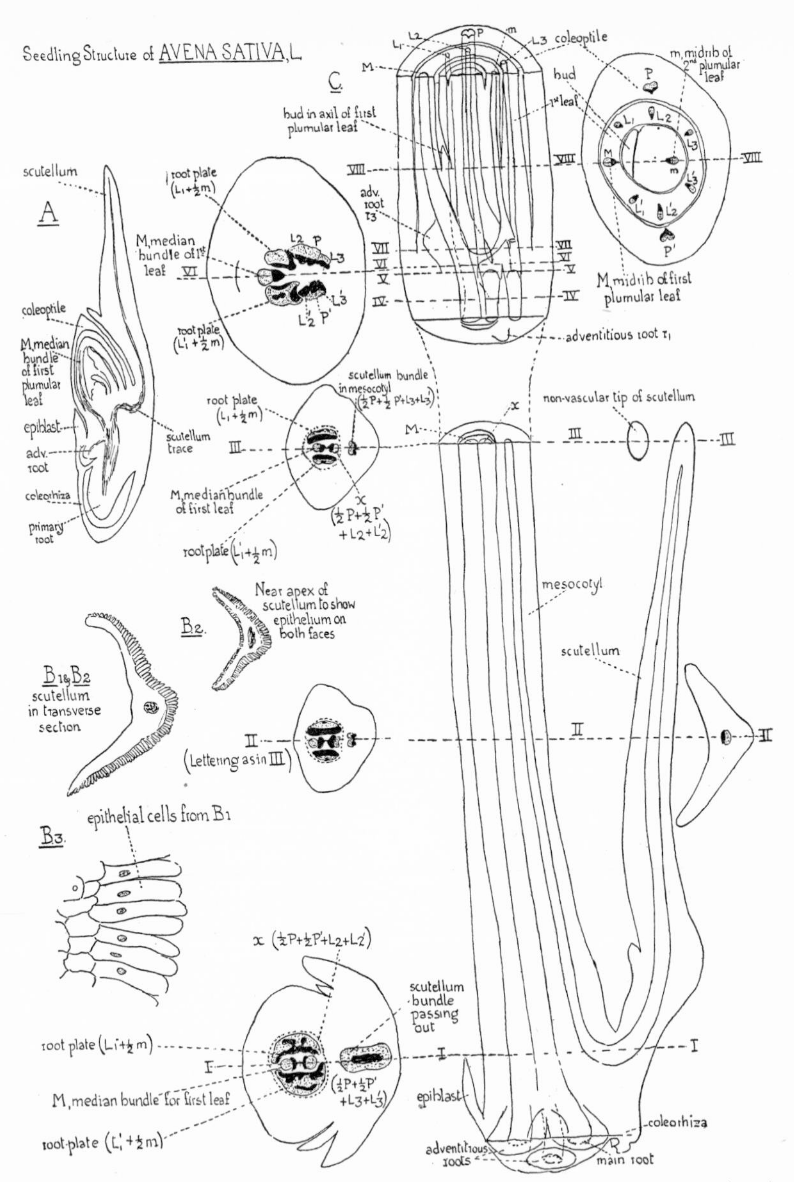

Fig. 117. *Avena sativa* L. A, embryo within seed in median longitudinal section (× 12). B 1 and B 2, transverse sections of scutellum of seedling younger than C (× 20 *circa*). B 3, a few cells of the epithelium from B 1 (× 165 *circa*). C, diagram (× 20) of a seedling cut in half longitudinally, showing the region between the insertion of the epiblast and scutellum and the insertion of the second plumular leaf. The structure at the levels marked with Roman numerals is explained in diagrammatic transverse sections, and, in Fig. 118, p. 245, in more detailed sections. *P* and *P′*, coleoptile bundles; *M*, *L* 1, *L* 1′, *L* 2, *L* 2′, *L* 3, *L* 3′, bundles of first plumular leaf; *m*, median bundle of second plumular leaf. [A.A.]

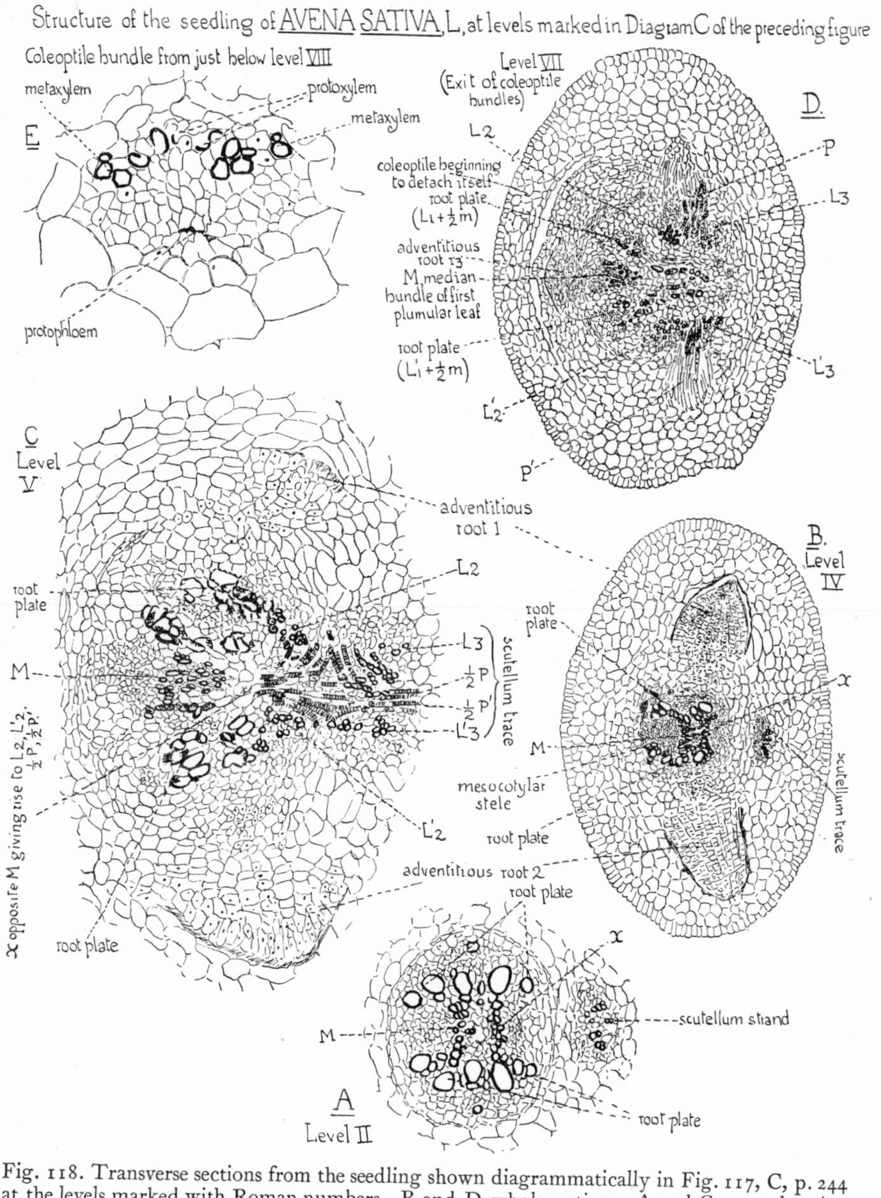

Fig. 118. Transverse sections from the seedling shown diagrammatically in Fig. 117, C, p. 244 at the levels marked with Roman numbers. B and D, whole sections; A and C, central region only. A (× 75); B and D (× 32); C (× 77 *circa*); E (× 318). From the same preparations as some of those illustrated in Sargant, E. and Arber, A. (1915). [A.A.]

division of the poles of the diarch root, while two were derived from the root-plates.

As in *Lolium multiflorum* Lam., so in *Avena sativa* L., the strand for the scutellum, *x*, runs up to the level of the base of the coleoptile within the stele. At this level it emerges from the stele, gives off two branches, *P* and *P'*, for the coleoptile, and itself turns downwards, running through the mesocotyl parallel to the stele until it reaches the level of the scutellum, where it passes out. In all the grasses which have been examined anatomically, this close and characteristic connection between the scutellum and coleoptile bundles is found, despite the intercalation of the mesocotyl. This anatomical connection confirms the view that the scutellum and coleoptile are components of one organ, the seed-leaf. Attention must, however, be paid to a certain complication, which is shown in Fig. 117, but with which we were not concerned in the *Lolium* seedling figured. The scutellum bundle not only gives off branches which form the two bundles of the coleoptile, but it is also connected with four of the lateral strands of the first plumular leaf—*L* 2, *L'* 2, *L* 3, *L'* 3. It has been claimed[1] that connections such as this undermine the anatomical argument for regarding the scutellum and coleoptile as parts of one leaf; but I do not feel convinced that they invalidate it. It must not be forgotten that the vascular skeleton of the plant is necessarily a linked system, and that the attachment of the later to the earlier members of this system depends upon where they can find a place. One would not, for instance, draw any morphological conclusions from the fact that the root-plates, seen in various sections of the mesocotyl of *Avena sativa* L. drawn in Fig. 117, C, connect with two of the laterals of the first plumular leaf, and the midrib of the second. How casual and apparently fortuitous such connections may be, is realised when one studies the attachments of bud-bundles to the vascular system of the parent shoot (p. 267). The relation between the scutellum and coleoptile strands exists from a very early stage, and it is of a peculiarly orderly and symmetrical type; I think it is possible to believe that it falls into a somewhat different category from the various vascular

[1] Avery, G. S. (1930).

connections established at a later period. I fully recognise, however, that there is no finality in such a view.

The conclusions regarding the morphology of the grass seedling reached in the present chapter, may be recapitulated as follows, with the qualification that they are all tentative. The scutellum represents the distal sucking region of the cotyledon, and is equivalent to the blade of the foliage leaf, while the coleoptile is equivalent to the ligule[1] of the foliage leaf, the basal sheath being suppressed. The mesocotyl is the cotyledonary node, elongated by intercalary growth in its upper region. The epiblast and the coleorhiza are non-vascular outgrowths, which may perhaps be compared with such non-foliar excrescences as the 'rachilla-flap' of certain Gramineae.

Though so much laboratory work has been done upon the grass seedling, we have comparatively little outdoor information about the early germination stages of most of our native species. A happy exception is *Molinia caerulea* Moench, the Purple-heath-grass, whose life-history has been studied by Jefferies.[2] He tells us that this grass makes use of sun-cracks as germination beds. On the moors about Huddersfield, where he had it under observation, little peat hollows, sometimes only about two square feet in area, are common; these become pools in wet weather, but are robbed of their surface water by a few days' drought. Bright sunshine and a strong wind playing on the unprotected surface of such a hollow, cause the formation of a network of cracks. Into these the seeds of *Molinia* are blown, and are sheltered there from the wind. When the water has disappeared in the early summer, lines of crowded seedlings appear, marking out the meshwork of last season's cracks like miniature green hedgerows. Another common seed-anchorage for *Molinia* is the tangle of dead inflorescences, stalks and leaves, which is often found in the peat hollows—each tangle becoming a bed of seedling plants. Field studies of the germination habits and the first phases of the life-history in other grasses, are much needed.

[1] In this connection I use the term 'ligule' in a broad sense for the whole development of the leaf-sheath above the level of detachment of the petiole or limb; see Arber, A. (1925), p. 98, Fig. lxxiv.

[2] Jefferies, T. A. (1915).

CHAPTER XII

THE VEGETATIVE PHASE IN GRASSES: ROOT AND SHOOT

OUR knowledge of the root systems of grasses falls far short of that of their visible parts—not because there is less to know, but because it is so troublesome to get at facts when they are hidden under layers of earth. There is no short cut to this knowledge; merely uprooting the plants is of little use, for it partially destroys some of the underground organs and displaces the rest. An American botanist, J. E. Weaver, however, has worked out a laborious and thorough technique, by the aid of which he has arrived at a much clearer picture of root systems than had been obtained before. The method he employed was to dig a trench by the side of the plant to be examined, about 5 ft. in depth and of convenient width. This afforded an open face which could be explored with a hand-pick. The original trench often had to be deepened to 8 or 10 ft. and sometimes more. To ensure certainty as to the maximum depth of the root endings, the soil was usually undercut for about a foot below the deepest of them, and was carefully examined as it was removed. It was found that when the root system thus exposed was photographed, many of the finer branches and root ends were always obscured; experience showed that a more accurate record of the extent, position and minute branching of the root system could be obtained by means of a drawing to scale on a large sheet of squared paper. For Weaver's results, his beautifully illustrated monographs[1] must be consulted. They reveal that root development is, as a rule, far more extensive and elaborate than everyday methods of observation would lead one to imagine. He concludes that the "general characters of the root systems of species are often as marked and distinctive as are those of the aerial vegetative parts". It is true that few botanists could be trusted to determine

[1] Weaver, J. E. (1919), (1920).

species from a collection of root systems; but this is merely due to lack of the knowledge born of familiarity.

The root-hairs of the grasses are particularly well developed, and also tend to be persistent, instead of being confined to the younger parts. In Wheat a maximum of more than a thousand root-hairs to a millimetre length of root has been recorded for the regions of most active absorption.[1]

The tenacious hold upon the soil which is often characteristic of the roots of the Gramineae, may make it almost impossible to pull up the tufts. The Purple-heath-grass (*Molinia caerulea* Moench)[2] is particularly notable for the firmness of its root system. The individual roots commonly remain functional through three seasons; they are so numerous that, close to the soil, they form an almost solid, tangled mass. In some grasses this firmness is of service, incidentally, to man. *Epicampes macroura* Benth., a native of Mexico, has strong roots which are turned to account in the manufacture of scrubbing brushes;[3] while *Eragrostis plana* Nees, of South Africa, is so deeply and rigidly rooted, that the early colonists, while on trek, used to look for a tuft of this grass to which they might fasten the fore yoke, when they wished to tether their teams of oxen.[4]

The root systems of desert grasses have a special interest. As an example we may take *Aristida pungens* Desf.,[5] of North Africa, which, when seen from a distance, resembles Marram-grass. The roots do not plunge vertically downwards, but run near the surface, reaching lengths for which 20 m. is probably far too low an estimate. They are thus independent of the subsoil, and they are able to make use of every available drop of rain water, for they are clothed with root-hairs throughout their length. As in other desert grasses, the cells of the piliferous layer secrete mucilage in the region near the root-tip, and thus cement the neighbouring sand-grains together, forming a sheath of agglutinated particles, which has been compared to a caddis-worm case. This sheath is illustrated for a second species, *A. plumosa* L., in Fig. 119, p. 250.

[1] Percival, J. (1921).
[2] Jefferies, T. A. (1916).
[3] Hitchcock, A. S. (1920).
[4] Bews, J. W. (1918).
[5] Massart, J. (1898); see also Price, S. R. (1911).

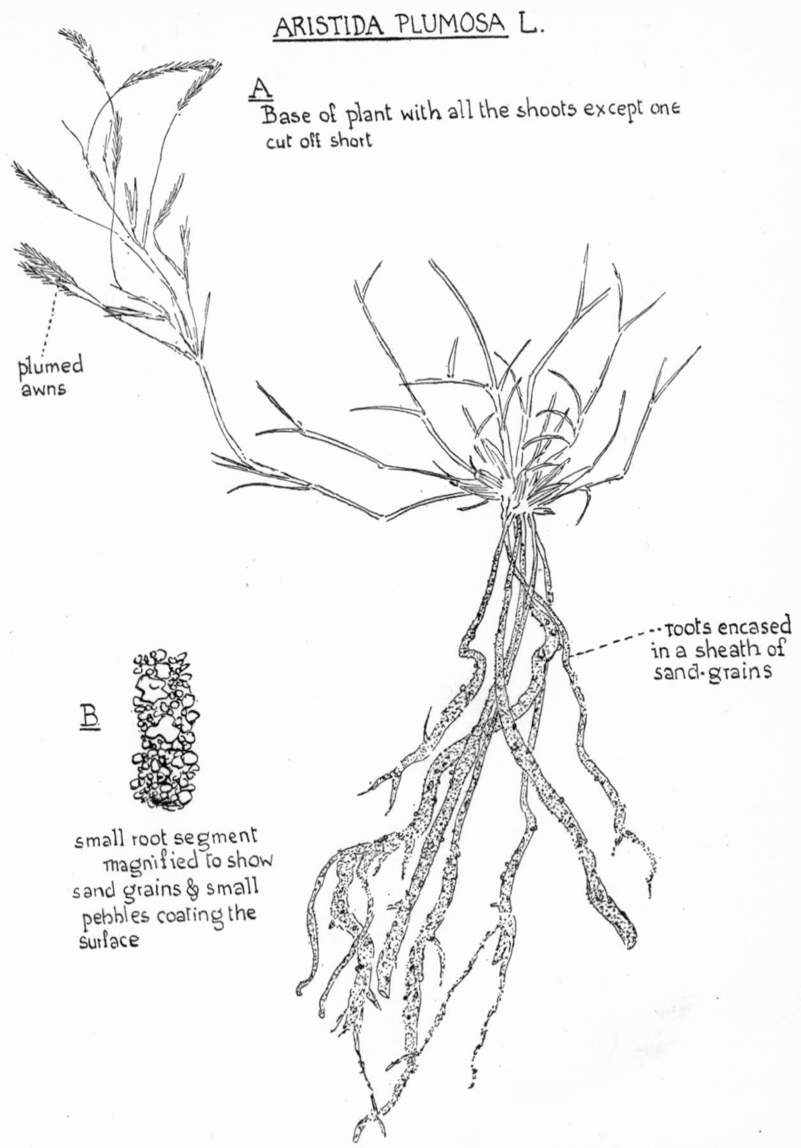

Fig. 119. *Aristida plumosa* L. A, base of a plant from the Egyptian desert (Babington's herbarium, Cambridge Botany School). All but one of the numerous inflorescences omitted ($\times \frac{2}{3}$). B, small segment of root enlarged to show irregular casing of sand grains and tiny pebbles. [A.A.]

Many grasses show a distinction between long unbranched and shorter branched roots (e.g. *Phalaris arundinacea* L.,[1] Fig. 120, p. 252). This dimorphism has been studied in the Reed (*Phragmites communis* Trin.),[2] where the parts of the rhizome buried in mud bear soft, white, unbranched roots, sometimes attaining a length of 3·5 m. (about 11 ft. 6 in.) and a thickness of 6 mm. (close on ¼ in.). The parts of the rhizome in the water, on the other hand, bear water-roots, which only reach a length of about 15 cm. (about 6 in.), and are branched to the third degree; they are hardish, brown, and almost threadlike. The boundary regions between the mud- and water-zones of the rhizome bear roots of an intermediate type. A similar dimorphism has been described in the roots of Townsend's-grass,[3] while the cereals also show some degree of distinction between branched and unbranched roots.[4] An attempt has been made to relate these root-types to variations in the permeability of the endodermis,[5] but at present this suggestion does not seem to have been tested critically.

It is remarkable that we have, apparently, no record of the existence of contractile roots among the Gramineae,[6] though such roots are a marked feature in other Monocotyledonous families.[7] Mycorrhizal mycelium occurs in connection with roots of various grasses.[8]

Our knowledge of the anatomy of grass roots[9] is in the most unsatisfactory state. A great many individual observations have been made, but a section cut here and there is not enough, and we still await a broad and thorough study of the structural relations, not only of roots of comparable ages and orders belonging to different species, but also of the different regions of the same root. Such a study would be well worth making, for, even from the disjointed information at present available, it is clear that there is great variety in the skeletal scheme. The number of protoxylems may, for instance, range from a few up to such numbers as 40;[10] and they may be in direct contact

[1] The two forms of root in this species were noticed by Duval-Jouve, J. (1875).
[2] Pallis, M. (1916).
[3] Oliver, F. W. (1925).
[4] Jackson, V. G. (1922); Kokkonen, P. (1931).
[5] Priestley, J. H. and Radcliffe, F. M. (1924).
[6] Rimbach, A. (1922).
[7] See Arber, A. (1925), pp. 16–21.
[8] For references, see p. 69.
[9] On bamboo root-anatomy, see pp. 68–9.
[10] Price, S. R. (1911).

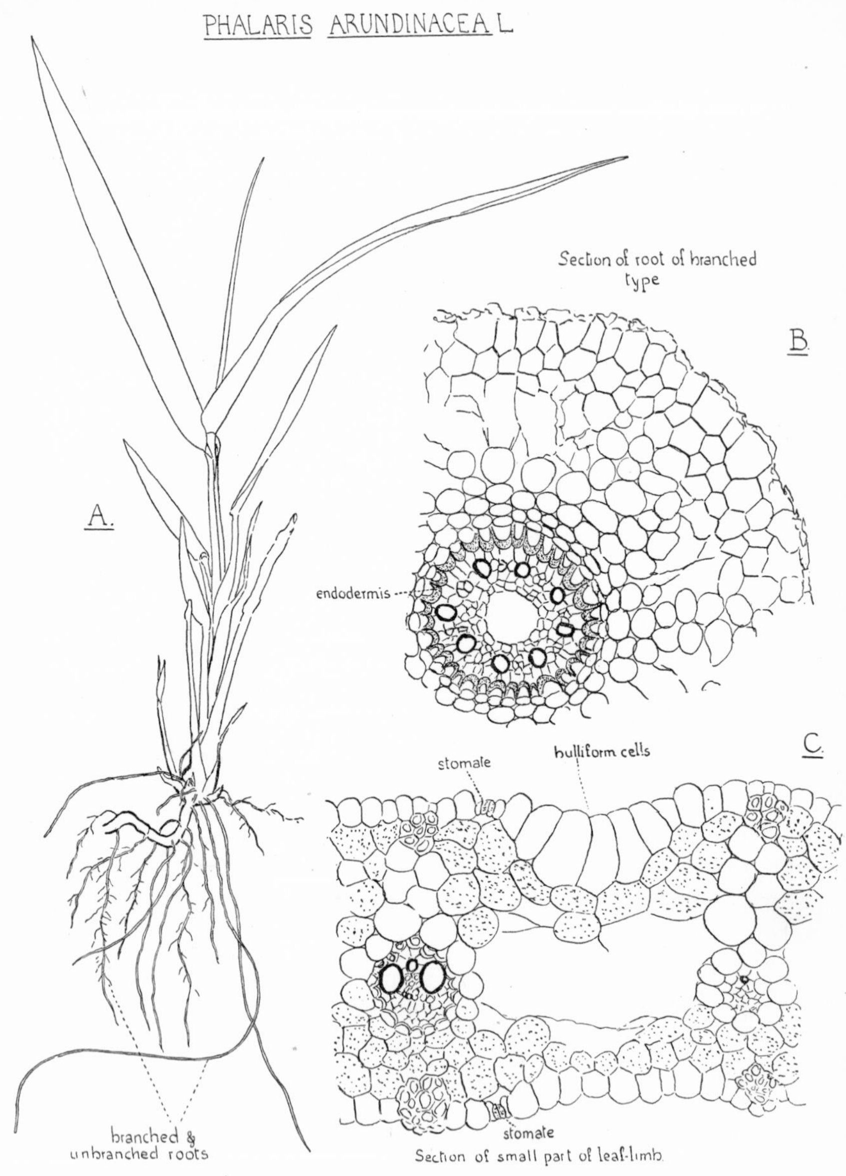

Fig. 120. *Phalaris arundinacea* L. A, part of a plant from a small roadside stream, Cannington Viaduct, April (× ½). B, part of a transverse section of a branched root (× 193 *circa*). C, transverse section of a small part of a leaf-limb (× 193 *circa*); the large air-spaces between the bundles contain some derelict remains of cells. [A.A.]

with the endodermis, or separated from it by a pericycle. The behaviour of the phloem and the degree of fibrosis of the stele vary greatly, while the thickenings on the endodermal walls, also, deserve

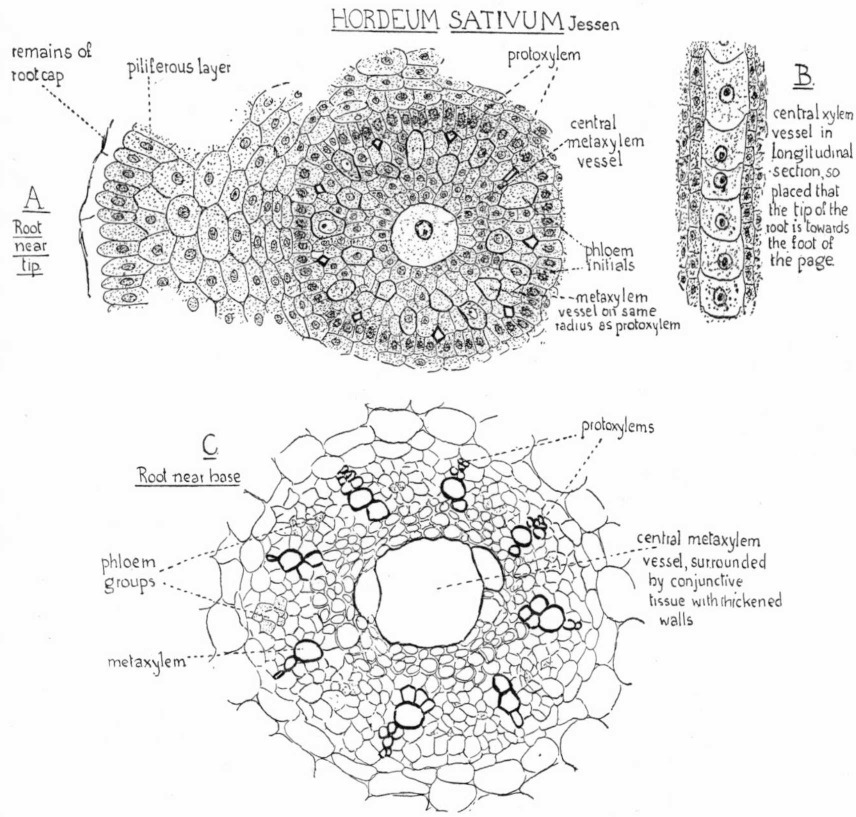

Fig. 121. *Hordeum sativum* Jessen. A, part of a transverse section of a young root from a seedling, so near the tip that it is still enclosed in the remains of the proximal region of the root-cap (× 193 *circa*). For clearness, the outlines of metaxylem elements and phloem initials are slightly darkened. B, part of a median longitudinal section near the root-tip of another root from the same seedling, to show the elements which will fuse to form the central metaxylem vessel (× 193 *circa*). C, a root from the same seedling cut in transverse section near its base (× 193 *circa*). [A.A.]

special study. In certain species of *Andropogon*, bodies interpreted as silica cystoliths have been found embedded in these thickenings; they look, in section, like screws attaching the endodermis to the central cylinder.[1] In Fig. 121, a few points in the anatomy of the Barley

[1] Borissow, G. (1924).

root are indicated—for instance, the ontogeny of the large central metaxylem vessel.

With rare exceptions (e.g. *Phalaris arundinacea* L., Fig. 120, A, p. 252), all the roots of the Gramineae which succeed the radicle of the seedling, are borne on the nodes of the stem. Their vascular supply is often closely connected with the 'nodal plexus',[1] and in the plumular axis there may be a 'girdle' of vascular tissue to which they

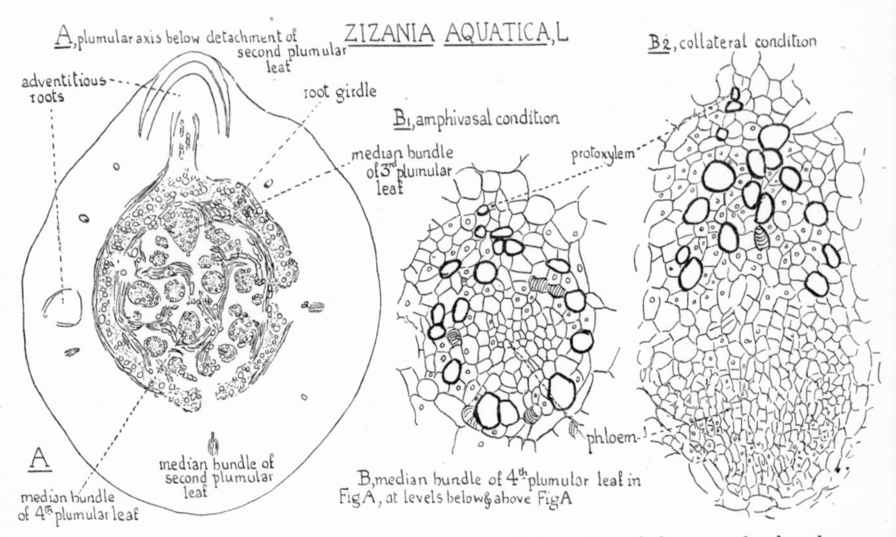

Fig. 122. *Zizania aquatica* L. A, transverse section of plumular axis between the detachment of first leaf and second leaf. Semi-diagrammatic: only the lignified xylem elements, and those approaching lignification, indicated individually (× 23). In the region of A, some of the bundles have a definitely amphivasal structure. One of these, the median bundle of the fourth plumular leaf, is shown in B 1, at a level a little below A; B 2, the same bundle above A; there is less than a millimetre between B 1 and B 2. B 1 and B 2 (× 193). [Arber, A. (1930[1]).]

are attached (Fig. 122). The horizontal rooting axes show much variety in their behaviour.[2] They may be of the nature of subterranean rhizomes, producing a single aerial shoot or tuft of shoots (e.g. *Poa pratensis* L.), or a succession of aerial shoots (e.g. *Agropyron repens* Beauv.); or they may be above-ground stolons ending in a single flowering shoot (e.g. *Poa annua* L., Fig. 103, F, p. 213), or a succession of aerial shoots (e.g. *Cynodon dactylon* Pers.).

The sharp distinction between nodes and internodes, which is so

[1] See p. 259, et seq. [2] Oakley, R. A. and Evans, M. W. (1921).

marked a feature of every bamboo cane, is also characteristic of grass axes in general; but since the leafy investment is apt to hide it, this type of structure is best studied in rhizomes. Fig. 123, B, is a sketch

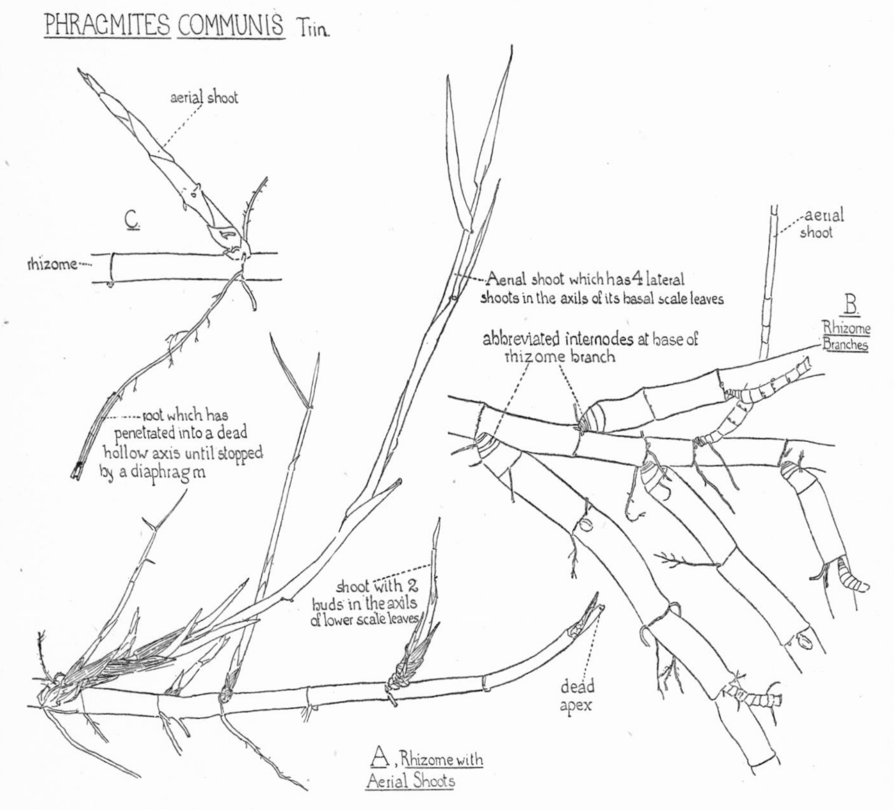

Fig. 123. *Phragmites communis* Trin. (All sketches × ⅓ *circa*). A, Ware Cliffs, Lyme Regis. Axis lying horizontally on the mud, dead at the tip, to show aerial shoots with lateral branches in the axils of their lower scale-leaves. B, part of a dead, dry rhizome from the shore at Lyme Regis. C, part of a shoot to show that one of the roots has penetrated for more than an inch into the hollow of a dead axis, which fitted it so closely that it was impossible to withdraw it. The dead axis is represented cut in half. [A.A.]

of a dead and dry rhizome[1] of the Reed, *Phragmites communis* Trin., at one-third the natural size. This specimen was found on the shore at Lyme Regis, where the crumbling and falling of the Lias cliffs, whose slopes are clothed with the Reed, carries down masses of dead

[1] In some specimens the epidermis and outer cortex had shelled off, leaving a ridged pseudo-surface, suggestive of certain types of preservation in fossil axes.

rhizomes on to the beach. It will be noticed that each rhizome branch is conical at the base; the earlier internodes are both short and narrow, and only gradually attain to their full length and breadth. Our knowledge of the internodal development in the Gramineae is scanty, but certain suggestive observations have been made in India,[1] where the rapid growth of the larger grasses favours this study. In *Saccharum spontaneum* L. (a grass belonging to the same section of this genus as the Sugar-cane) an average flowering culm shows about nine very short internodes at the base, aggregating about an inch in length, and then eleven long internodes, in addition to the final segment which terminates in the panicle. The leaves of the lowest six of the short internodes have no green laminae, so there are fourteen internodes with green leaves. It is found that this series of internodes corresponds to fourteen months of vegetative activity, and it seems safe in this species to regard each of the long internodes as, in general, the work of a month. The series of long internodes may, however, be interrupted, for, if the green leaves are destroyed by grazing or by fire, a group of short internodes is intercalated, to be succeeded again by long internodes, when the plant retrieves its assimilating apparatus.

In the individual internodes of grasses, the growth is basal. When children amuse themselves by sucking the joints of Wheat or Oats, they choose the basal part of each internode, since they have learned by experience that it is white, tender and sugary, even when the tip of the same joint has become green, hard, and tasteless.[2] In various meadow grasses, the base of the uppermost internode is still completely unlignified long after flowering.[3] Not only do the basal regions keep the power of growth longer than the rest of the internode, but they have the capacity for either initiating or submitting to curvature, and are thus able to restore the part of the shoot above them to the vertical position, if it gets accidentally 'laid'. In many grasses this curvature is carried out by changes in the turgour of swellings or pulvini at the internode base,[4] which may be formed from the axis alone (e.g. *Phragmites communis* Trin., Fig. 124, C), or

[1] Hole, R. S. (1911).
[2] Duval-Jouve, J. (1871).
[3] Frohmenger, M. (1914).
[4] Lehmann, E. (1906).

the leaf-base alone (e.g. *Agropyron repens* Beauv., Fig. 124, B) or both (e.g. *Zea Mays* L.). These pulvini have no stomates, no large intercellular spaces, and little or no chlorophyll in the parenchyma. The large lateral xylem vessels in each bundle, which in the sheath and the main part of the internode are reticulate or pitted, in the pulvinus have ring or spiral thickenings; they are thus capable of elongating and bending without injury.

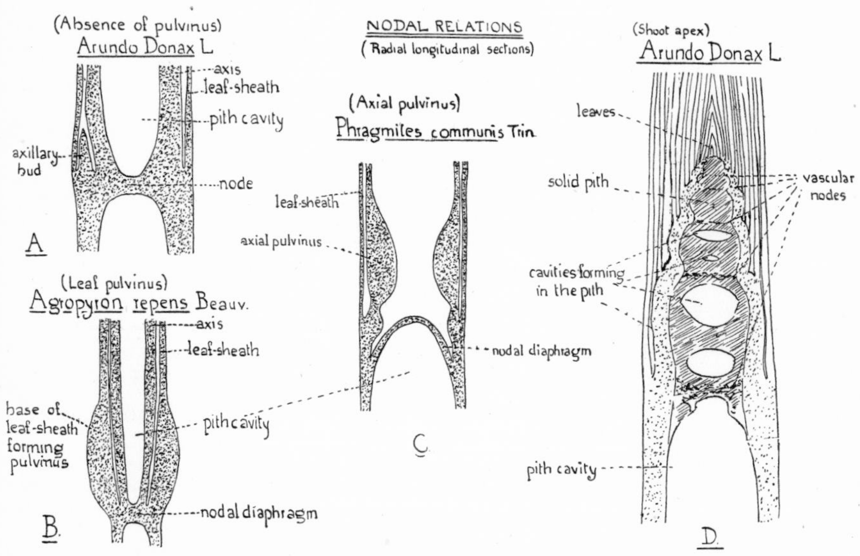

Fig. 124. Nodal relations. A–C, diagrammatic longitudinal sections through nodes. A, *Arundo Donax* L. (no pulvinus); B, *Agropyron repens* Beauv. (leaf pulvinus); C, *Phragmites communis* Trin. (stem pulvinus); D, diagrammatic radial longitudinal section through the shoot apex of *Arundo Donax* L., to show the transition from the solid pith of the youngest internodes, to the pith cavity of the oldest internode drawn. The pith (which is white in nature) is obliquely shaded, and the rest of the haulm, which is green, is dotted. [A.A.]

Silica often occurs in the epidermal cells of the grasses. Its distribution in the stem—like that of growth, and the capacity to bend—is not uniform, but is related to the distance from the base of the internode. In Timothy (*Phleum pratense* L.) it has been found that an internode, 325 mm. long, showed no epidermal silica cells at the base, but 50–100 per sq. mm. in the median region, and 500–600 per sq. mm. in the upper part. The transition from the middle region to the strongly siliceous upper zone was a sudden one.[1] The total quantity of silica present in the tissues of the Gramineae is so great that their ashes may give rise to a basic glass. This explains the source of the vitreous lumps often collected on the sites of mills or grain storehouses which have been burnt down. In certain parts of France these masses used to be called thunderbolts, and were regarded as the cause—not the result—of the disaster.[2] The firmness and persistence of the glumes enclosing the grass flower may be connected with the siliceous character of the tissues in this family.

The pith of grass haulms is generally resorbed, leaving a hollow cavity. For *Arundo Donax* L. this process can be followed in Fig. 124, D, p. 257. Here the pith (obliquely shaded) is solid in the youngest internodes, but, as elongation takes place, splitting occurs, opening up cavities which finally coalesce. In *Zizania aquatica* L., the resorption of the pith is partial, leaving transverse septa; in the Sugarcane, that "pleasant and profitable Reede", the stem "is not hollow as the other Canes or Reedes are; but full, and stuffed with a spungious substance in taste exceeding sweete".[3]

In transverse sections of the internodes of grasses, great variety of pattern is met with. This is due to the distribution of the bundles and fibres, which differs from species to species and also in successive parts of the same haulm. In the last century, these arrangements were interpreted teleologically, but more recent work lends no support to this view.[4] The details of the mechanical system follow a regular cycle of development from the lower to the higher internodes, and this cycle bears no obvious relation to the needs of the plant.

In the shoots of many grasses, mechanical tissue is highly conspicuous. Owing to its prevalence in the Purple-heath-grass, *Molinia caerulea* Moench, this plant is made, locally, into brooms,[5] while thousands of tons of fibrous Esparto-grass (*Lygeum*, *Stipa*, etc.) from Spain and North Africa are used every year for paper and cordage.[6]

[1] Frohmenger, M. (1914). [2] Vélain, C. (1878). [3] Gerard, J. (1597).

[4] Cf. Schwendener, S. (1874) with Roelants, H. W. M. (1922); Schwendener's work deserves consideration even today, for he was feeling his way towards an explanation of plant structure on physical lines.

[5] Ward, H. Marshall (1901). [6] Hitchcock, A. S. (1929).

In the internodes of grass haulms, the course of the bundles is vertical, but, at the level of the nodes, horizontal strands are met with, which connect the vertical strands, and anastomose into an

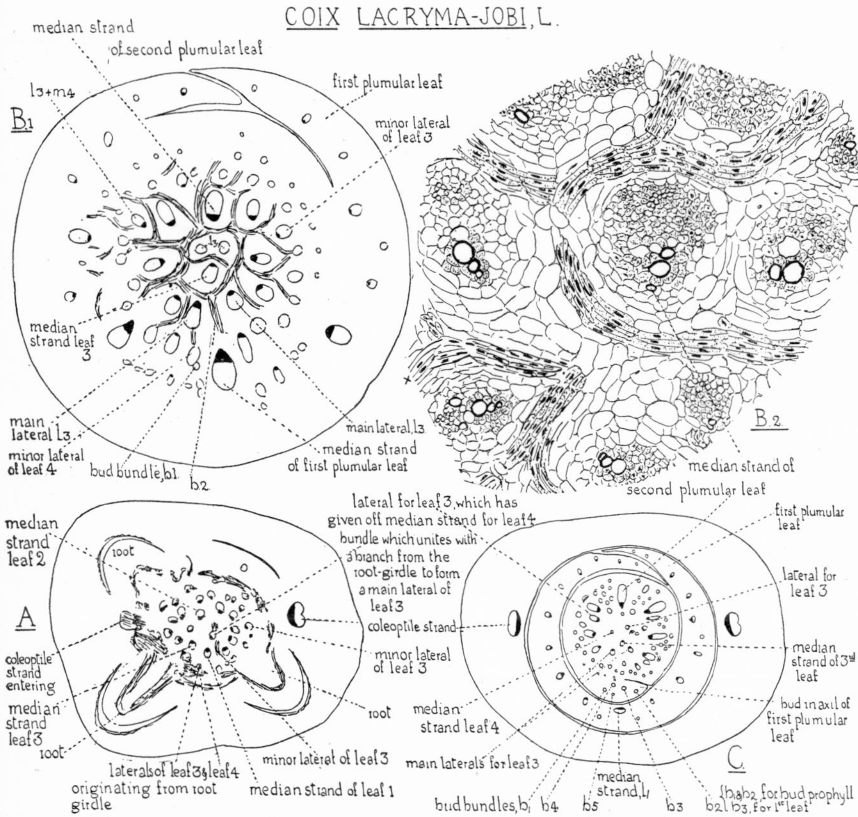

Fig. 125. *Coix lacryma-Jobi* L. Sections from a transverse series from below upwards through a seedling. A, at level of exsertion of coleoptile, and showing attachment of three adventitious roots (× 19). B 1, at the level at which the first plumular leaf is beginning to detach (× 39). The nodal plexus is sketched from about six sections. B 2, a small part of a section slightly lower than that principally used in B 1, to show the nodal plexus in detail in the neighbourhood of the median strand of the second plumular leaf (× 163). Strand × contains a little lignified xylem; all the other horizontal strands are procambial. C, transverse section at a level between the detachment of the first and second plumular leaves, showing the bud in the axil of the first plumular leaf, not yet fully detached (× 19). [Arber, A. (1930[1]).]

elaborate plexus. To some extent the horizontal strands link up the bud- and root-bundles with the vertical system. The nodal network can be seen in its simplest form in the slender axes of seedlings, such as those whose structure is illustrated in Figs. 125, 126 and 127.

The network is also shown for the young node of a mature plant in Figs. 128 and 129. It is not worth while to discuss all the anatomical complexities in these figures;[1] the general conclusion to be drawn

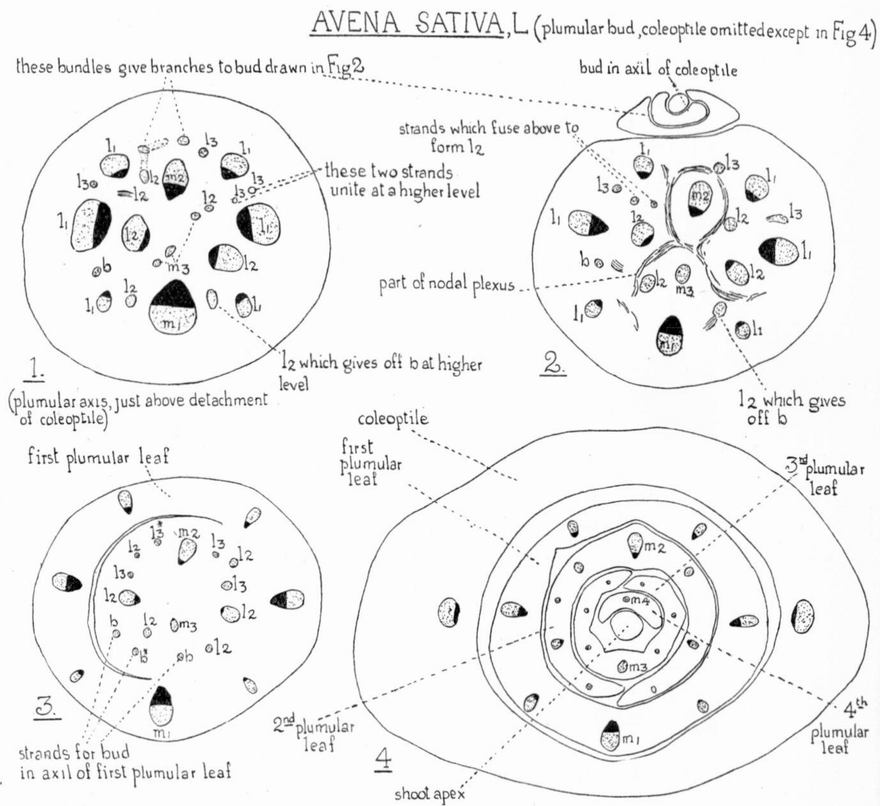

Fig. 126. *Avena sativa* L. Transverse sections from a series from below upwards through the plumular bud (× 47), from *1*, plumular axis just above detachment of coleoptile, to *4*, through shoot apex, traversing four plumular leaves. Coleoptile shown in *4* only; m_1, m_2, m_3, m_4, median strands, and l_1, l_2, l_3, laterals, for the first, second, third and fourth plumular leaves. Fig. *2*, from more than one section, showing part of the nodal plexus of horizontal strands; *b*, bundles of bud in axil of first plumular leaf. The bundles b^* and l_3^* in *3* arise as branches from the horizontal plexus. [Arber, A. (1930[1]).]

from them is that the nodal plexus arises through an outburst of meristematic activity, affecting most of the ground tissue between the bundles, especially the region flanking the cambium (Fig. 129).

[1] For details, see Arber, A. (1930[1]).

This cell-liveliness may be regarded as a 'second wind' of that anatomical activity which, in its first phases, initiates the leaf-traces.

The facies of the grass plant depends largely upon the behaviour

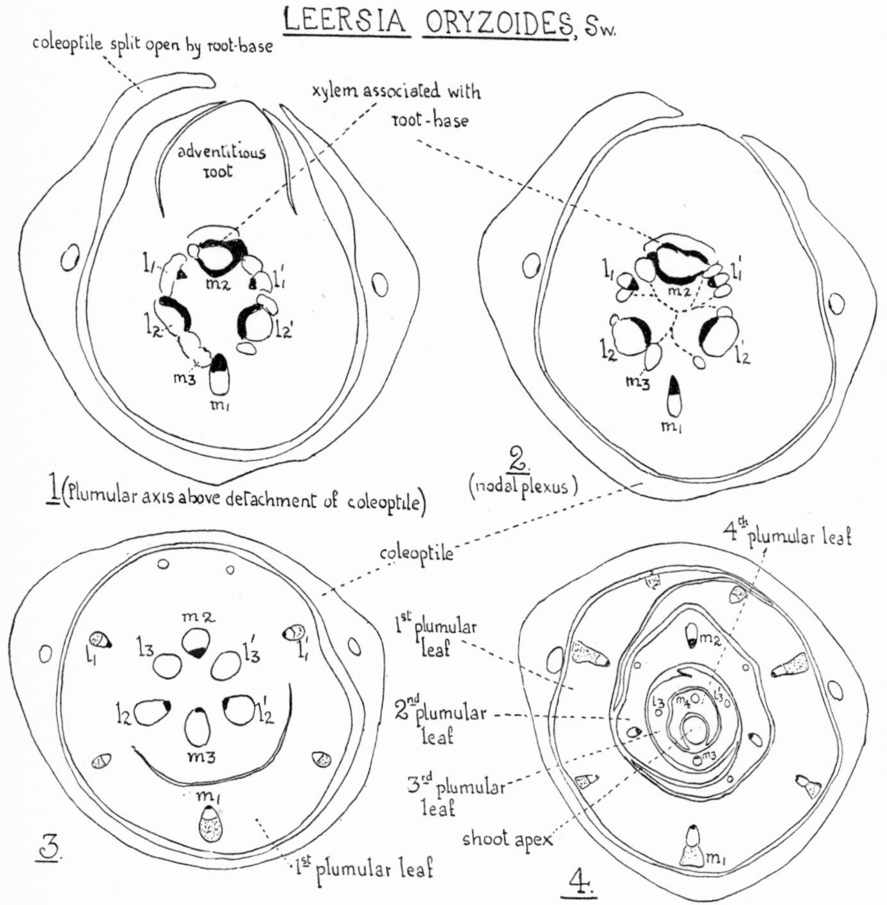

Fig. 127. *Leersia oryzoides* Sw. Transverse sections (× 103) from a series from below upwards through the plumular axis of a seedling from below the exsertion of the first plumular leaf to the apex of the plumular bud. Median strands of successive plumular leaves are marked m_1, m_2, m_3, m_4, and the lateral bundles l_1 and l'_1, l_2 and l'_2, l_3 and l'_3. [Arber, A. (1930[1]).]

of the lateral shoots. The basal branching, which often converts the seedling into an extensive tuft, has already been mentioned in connection with pasture grasses (p. 47). The process of 'tillering', or 'stooling', has a great deal in common with the mode of branching of the panicle, in which axes of successive orders are often crowded together almost indistinguishably (p. 134). In Rice,[1] for instance, primary tillers arise out of one or two of the lower subterranean leaf axils of the mother culm. A secondary tiller then arises from the lowermost leaf axil of the primary tiller; and so on. The tillering has

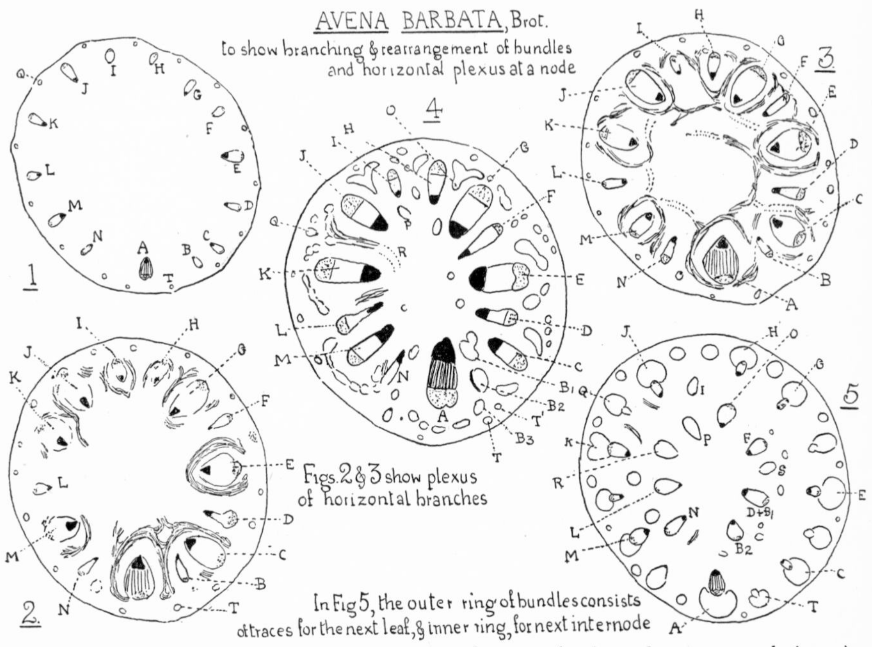

Fig. 128. *Avena barbata* Brot. Transverse sections from a series through a young node (× 23). The median bundle, A, of the leaf exserted at the node is shaded. *1*, internode below node studied. *2*, appearance of horizontal branches from the ring of bundles. For mode of attachment of vertical and horizontal branches see Fig. 129, p. 263, in which the history of bundle E is followed. *3*, slightly higher than *2*, to show horizontal plexus; the branches not visible exactly at this level are dotted. *4*, further history above the plexus. *5*, just below detachment of leaf; the inner bundles form a ring for the next internode. [Arber, A. (1930[1]).]

thus a certain sympodial tendency. In nearly all grasses the plane of symmetry of each branch is at right angles to that of the parent axis; that is to say, the planes of symmetry alternate in the branches of successive orders. The lateral leafy shoots of St-Augustine's-grass, *Stenotaphrum secundatum* Kuntze (Fig. 130) and the spikelets of Rye-grass, *Lolium perenne* L. (Fig. 73, p. 167) are exceptions, for here

[1] Bhalerao, S. G. (1926).

the leaves of the axillary bud are developed in the same plane as those of the parent shoot.

Hackel, in his monograph of *Festuca*,[1] points out that the branch system has a varying aspect, according to whether the growth of the lateral shoots is upward (apogeotropic), horizontal (diageotropic) or downward (positively geotropic); this classification can be

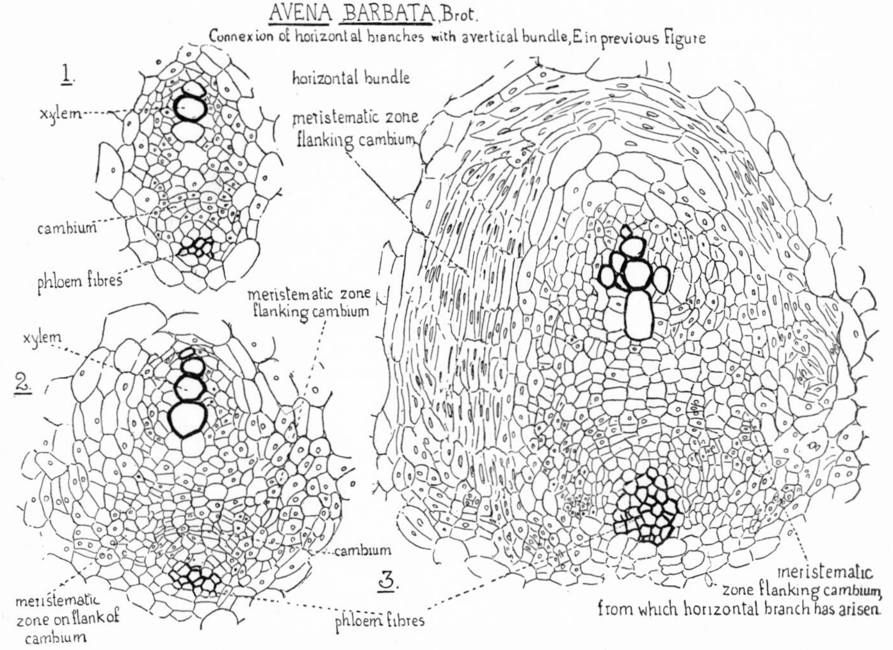

Fig. 129. *Avena barbata* Brot. Transverse sections from a series through the bundle E in Fig. 128, p. 262, to show the connection of the horizontal branches with the meristem flanking the cambium (× 163). (Orientation not uniform with Fig. 128.) *1*, bundle E in Fig. 128, *1*; *2*, bundle E between *1* and *2*, Fig. 128; *3*, bundle E in Fig. 128, *2*. [Arber, A. (1930[1]).]

applied also outside the Fescues. If the branches are apogeotropic, they grow up closely appressed to the mother axis, and remain with their bases enclosed within the sheaths of the axillant leaves. This type of branching, which is called intra-vaginal, is illustrated for *Stenotaphrum secundatum* Kuntze in Fig. 130, p. 264, and for *Phleum pratense* L. in Fig. 140, B 1 and C, p. 276. If the bud axis is, on the other hand, more or less diageotropic, it grows out horizontally or

[1] Hackel, E. (1882).

obliquely, breaking through the base of the axillant leaf; it is then called extra-vaginal. Young shoots bursting through the leaf-base of *Glyceria fluitans* R.Br. can be seen in Fig. 131, B 1 and B 2, while in A, which shows a shoot of *Phalaris arundinacea* L., the lateral branch to

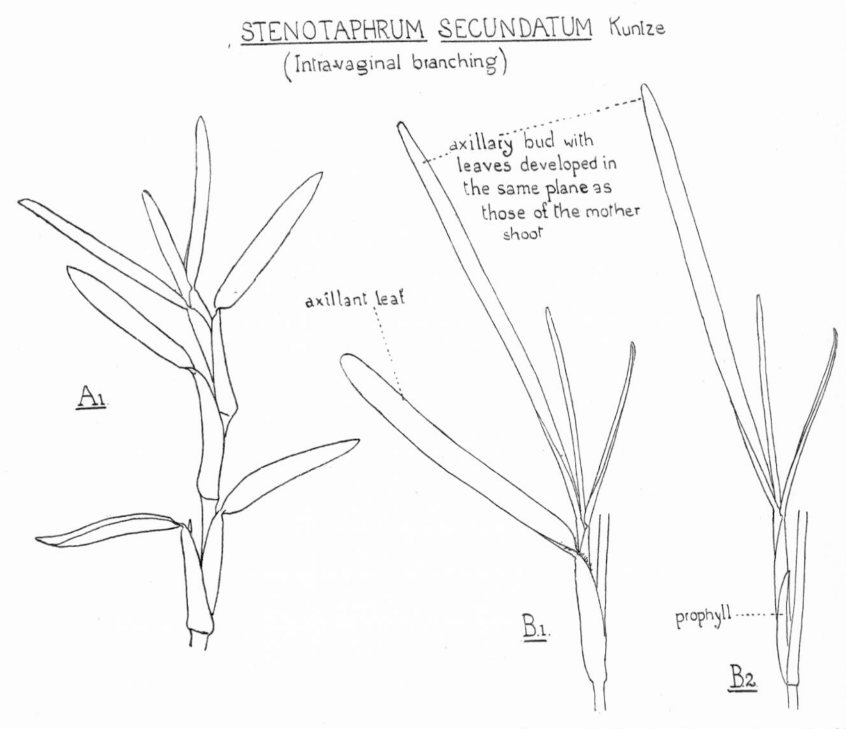

Fig. 130. *Stenotaphrum secundatum* Kuntze. (Cambridge Botanic Garden.) A 1, shoot (× ½), hairs omitted. Leaf-sheath folded very flat, so that above the level at which it diverges from the axis, the two halves are in contact. The young limbs are also folded flat, but expand later. The fold at the top of the sheath remains a permanent feature, as the ligule and its region of attachment conform to this shape, so that the leaf cannot be opened out without tearing. B 1 and B 2 (× ½), an axillary bud, to show that the leaves are developed in the same plane as those of the mother shoot. In B 2 the axillant leaf is removed. [A.A.]

the left has cut its way through the base of leaf 1, while, in that to the right, the leaf-sheath remains external, though split. The bursting of the leaf-sheaths may also be caused by lateral branches growing vertically from a horizontal axis (Fig. 132). The process of extra-vaginal branching was long ago described in vivid terms by Godron for *Arundo Donax* L.[1] At Nancy, where he studied it, lateral buds

[1] Godron, D. A. (1880).

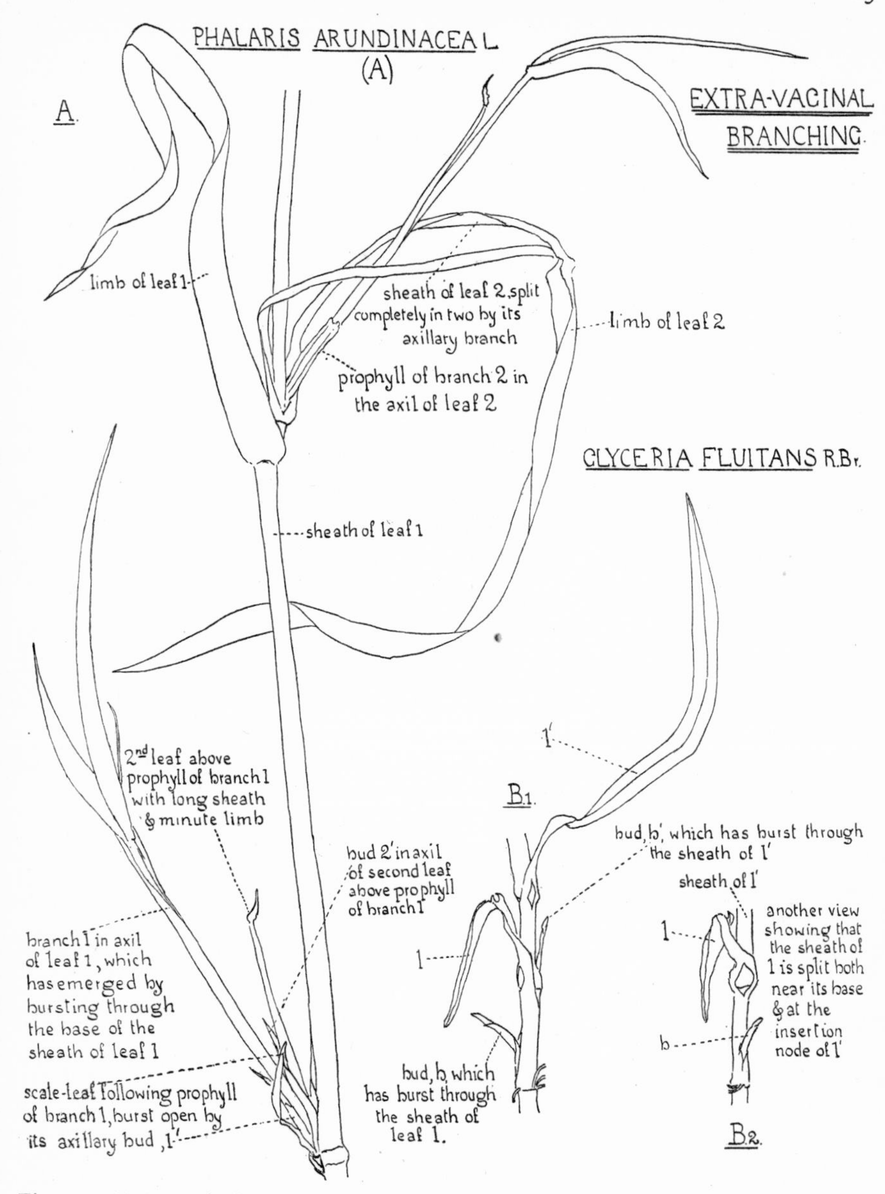

Fig. 131. Extra-vaginal branching. A, *Phalaris arundinacea* L. Part of a shoot, Clayhithe, June (× ½), to show axillary branches which burst the sheaths of their axillant leaves, but do not always emerge through the split. The bursting of the sheath has been accomplished by branches of two orders; branches 1 and 2 of the first order, in the axils of leaves 1 and 2, and bud 1′ of the second order, in the axil of the scale-leaf succeeding the prophyll of branch 1. B 1 and B 2, *Glyceria fluitans* R.Br., Holmbush Fields, Lyme Regis, April. Small segment of shoot (× ½). [A.A.]

are produced at the beginning of the summer, when the sheaths of the axillant leaves are still green, and tightly enwrap the parent stem. The base of the bud below the first leaf is 2–3 mm. long, and adheres obliquely to the stem; both this naked base and the upper part, encased in the prophyll, are much compressed. As the bud grows, it

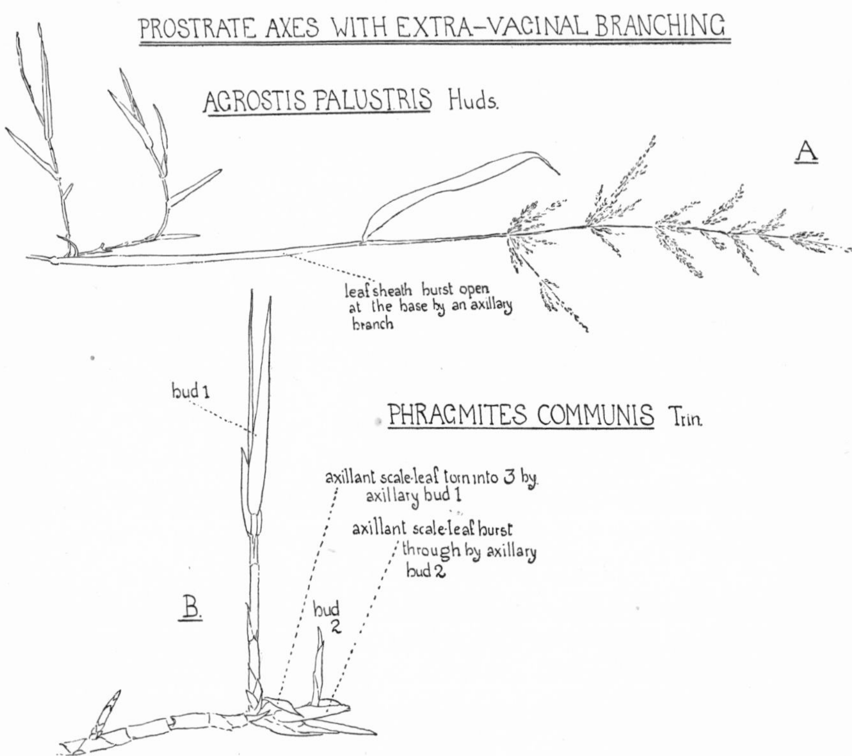

Fig. 132. Prostrate axes with extra-vaginal branching. A, *Agrostis palustris* Huds. Dead faded inflorescence lying horizontally, with axillary branching at base (× ½). B, *Phragmites communis* Trin., Lias Cliff, Lyme Regis (× ½). [A.A.]

becomes convex on its external face, and almost flat on its internal face, where it lightly depresses the stem. Its prophyll, which is bi-keeled, enwraps the younger part of the bud like swaddling-clothes. As growth proceeds, the succeeding leaves, folded and flattened, force the margins of the prophyll-envelope apart. The pressure of their acute, spiny apices splits the base of the foliar sheath lengthways.

The pointed bud emerges into the daylight; the breach enlarges; the branch grows out, and bears foliage leaves.

It is sometimes said that the grasses do not branch freely, but a glance at the sketches in Fig. 133, p. 268, which are all from common British species, should dispel this idea; in fact branching is often the most conspicuous feature of the plants. I have seen a last year's stem of *Agrostis palustris* Huds., over 150 cm. long, and with twenty-four nodes, looped up in a roadside bramble like a climber, and, though this main axis was dry, yellowish, and moribund in appearance, it bore eighteen vigorous lateral branches, one of which was 41 cm. long.

The mode in which the vascular system of a lateral branch is connected with that of the parent axis, can be followed in Figs. 134–6, pp. 269–71, which are all drawn from a single immature node of the Sword-grass, *Phalaris arundinacea* L. This relation of bud to axis is shown also in Fig. 125, p. 259, for Job's-tears, and in Fig. 126, p. 260, for the Oat. Details of these examples will be found elsewhere;[1] it will suffice here to point out that the relation of the vascular supply of the bud to that of the parent axis is curiously irregular. The young bud of *Coix*, drawn in Fig. 125, p. 259, showed no connection with the bundles of its own axillant leaf, or the next leaf, but one of the five traces was attached to a bundle belonging to the third leaf. In *Avena sativa* L. (Fig. 126, p. 260), again, there was no link with the axillant leaf, but one trace came from a lateral bundle belonging to a succeeding leaf. In the older bud of *Phalaris arundinacea* L. (Figs. 134 and 136), though there were a number of connections with the axillant leaf, there were also links with the two succeeding leaves. It seems that the bud-bundles have to connect up with the vascular system of the parent stem where they can; but the mechanism, for all its fortuitous air, appears to be highly efficient.

The branches which we have considered hitherto have been either vertical or more or less horizontal. It remains to touch upon the third and much rarer category, in which the laterals are positively geotropic, and burrow in the soil. Such shoots are found in *Festuca spadicea* L. and *F. caerulescens* Desf. (Fig. 137, p. 272). They are

[1] Arber, A. (1930[1]).

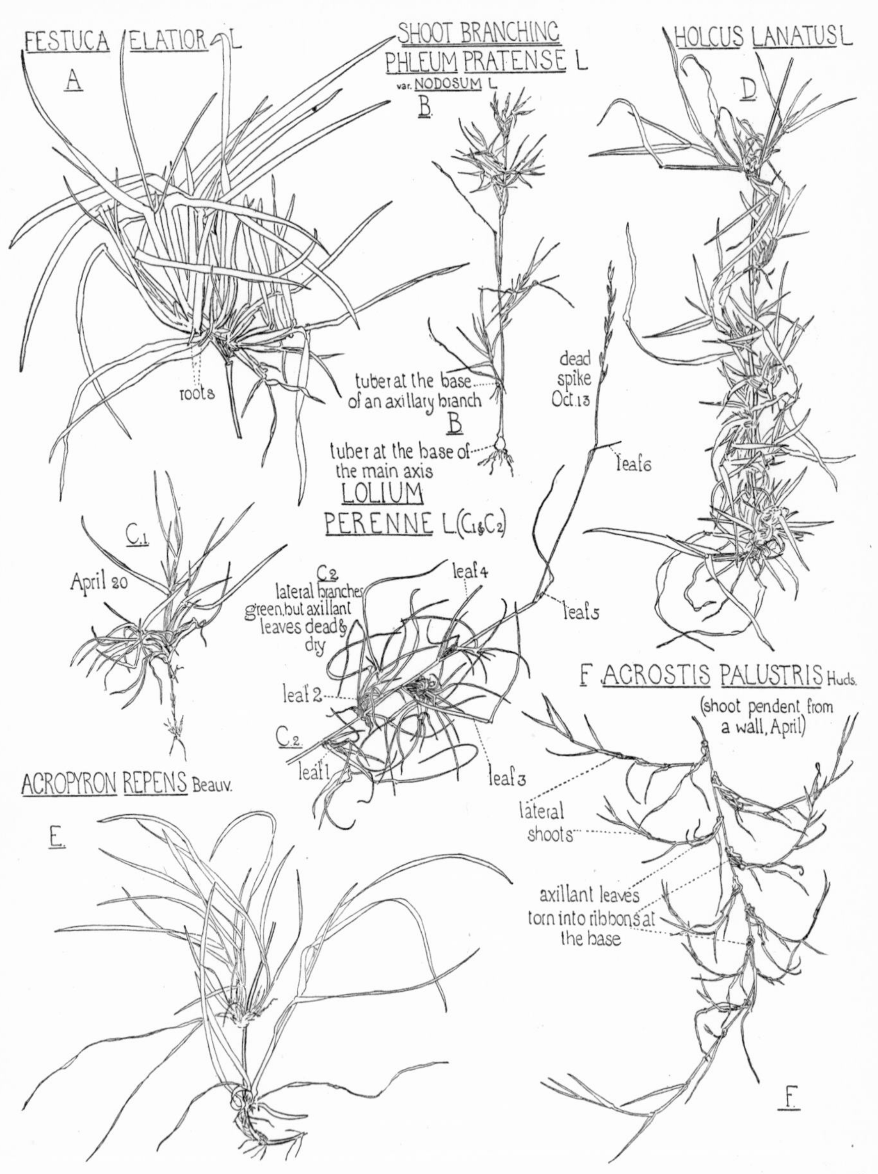

Fig. 133. Grass shoots to show axillary branching (all × $\frac{1}{4}$). All from Lyme Regis, April, except B, which is from near Cambridge in March. [A.A.]

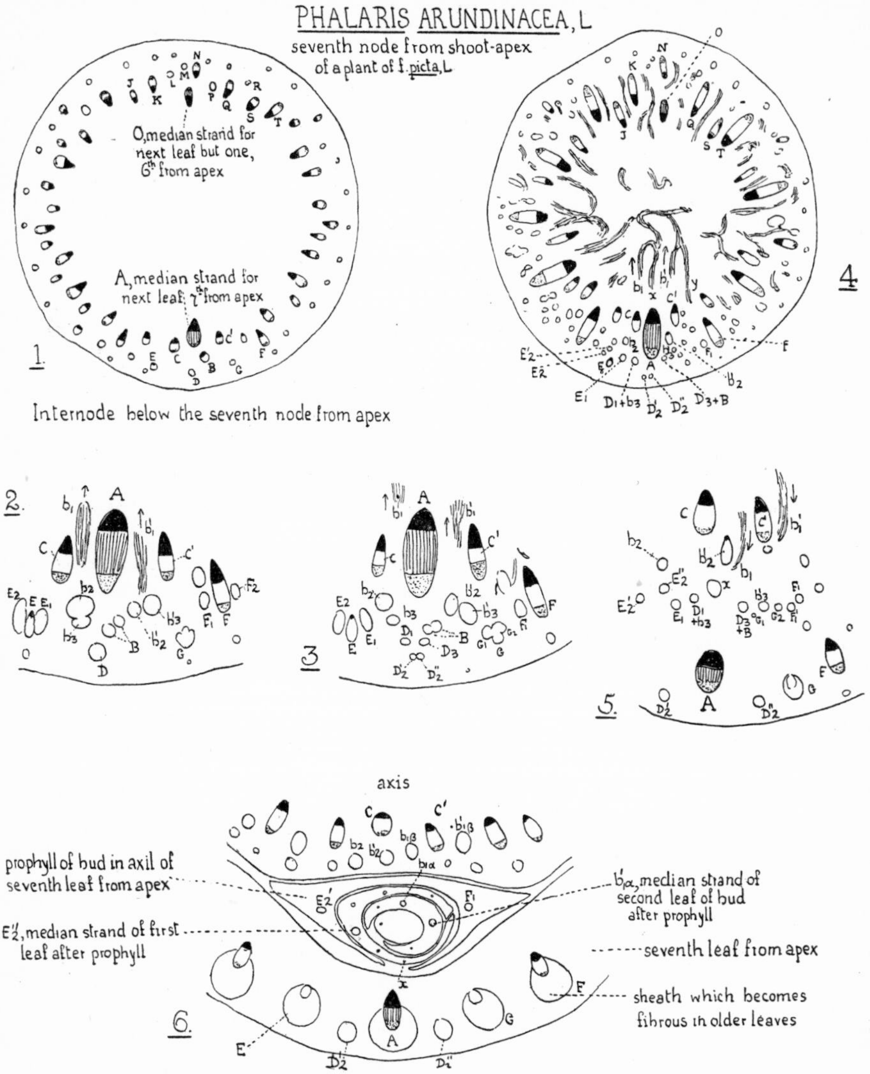

PHALARIS ARUNDINACEA, L
seventh node from shoot-apex
of a plant of f. picta, L

O, median strand for next leaf but one, 6th from apex

A, median strand for next leaf; 7th from apex

Internode below the seventh node from apex

axis

prophyll of bud in axil of seventh leaf from apex

E2', median strand of first leaf after prophyll

b'α, median strand of second leaf of bud after prophyll

seventh leaf from apex

sheath which becomes fibrous in older leaves

Fig. 134. *Phalaris arundinacea* L. f. *picta* L. (Cambridge Botanic Garden). All the diagrams in Figs. 134, 135 and 136, are drawn from one immature node (the seventh from the apex of a shoot) and the region immediately above and below it. *1* and *4* (× 23); *2*, *3*, *5*, *6* (× 47). *4* shows part of the plexus of horizontal strands. *6* shows the bundle-relations of a leaf, its axillary bud, and the next internode. [For fuller legend, see Arber, A. (1930[1]).]

intra-vaginal and are exposed only by the rotting of the axillant leaves. The leaf-sheaths are much thickened below, so that these positively geotropic branches serve as reserve shoots.[1]

Another species with reserve shoots in which the leaf-bases become storage organs, is *Poa bulbosa* L. (Fig. 138, p. 273). In a plant under cultivation, which was dug up in April, the leaf-sheaths of the non-flowering shoots were thickened, succulent and white, forming

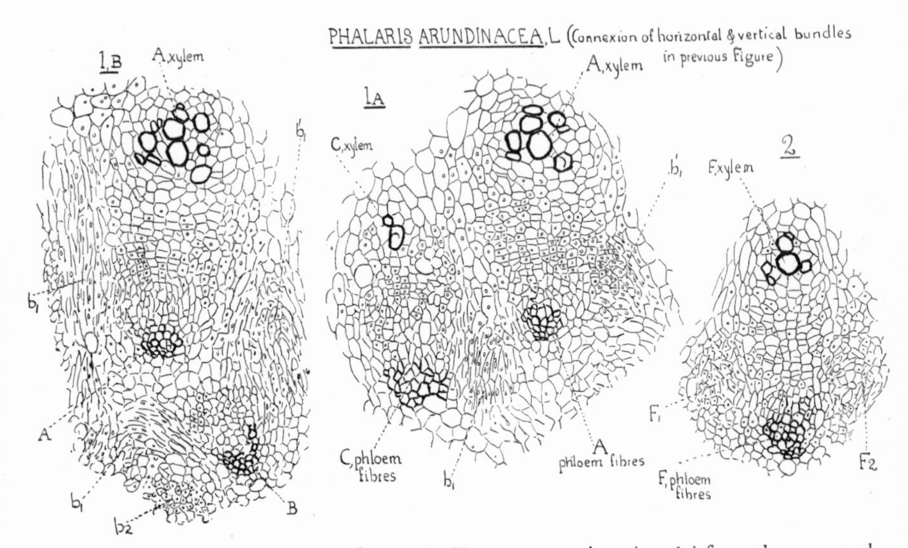

Fig. 135. *Phalaris arundinacea* L. f. *picta* L. Transverse sections (× 163) from the same node as Fig. 134, p. 269, and Fig. 136, p. 271, to illustrate the origin of the horizontal branch bundles in greater detail than in Fig. 134. *1*A (between Figs. 134, *1* and *2*), attachment of b'_1 to bundle *A* seen to the right; b_1 was attached to the cambium of *A* to the left, but at a lower level. *1*B, a slightly higher level in which b_1 and b'_1 have produced outwardly directed branches fusing with *B*. *2* (a little below Fig. 134, *2*) shows bundle *F* giving off strands *F*1 and *F*2 on either side from the cambial region. [Arber, A. (1930[1]).]

the so-called bulbs. Fig. 138, B, shows two of the swollen leaf-bases in section; when the individual cells were examined under a higher power, it was seen that there was little thickening of the walls (Fig. 138, C), but starch-grains were present. In September, however, after the summer's vegetation, deposits of reserve cellulose covered the walls (D); the starch had disappeared, but the cytoplasm included granules and droplets (unidentified). In another member of

[1] For a general study of reserve shoots in grasses, see Hackel, E. (1890[1]).

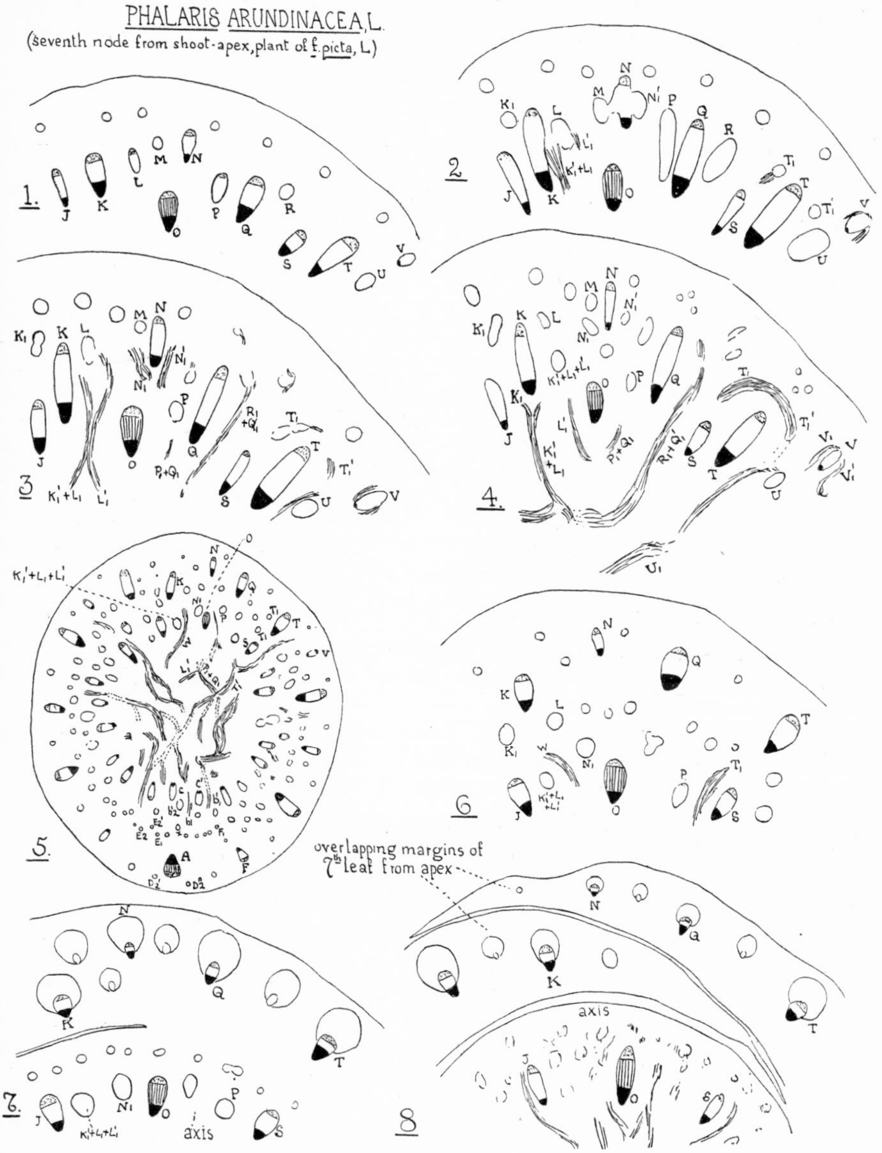

PHALARIS ARUNDINACEA, L.
(seventh node from shoot-apex, plant of f. picta, L.)

Fig. 136. *Phalaris arundinacea* L. f. *picta* L. The same node as that illustrated in Fig. 134, p. 269, and Fig. 135, p. 270. All drawings from transverse serial sections × 47, except 5, which is × 23. The diagrams illustrate the history of the group of bundles to the north in Fig. 134, *1*, which are associated with *O*, the median strand for the next leaf but one. [For detailed legend, see Arber, A. (1930[1]).]

the same genus, *P. flabellata* Hook.,[1] which grows as a tussock-grass in the Falkland Islands, the bases of the stems and the sheaths of the inner leaves are closely compressed, "white and soft, agreeably flavoured, somewhat resembling Filberds, and very wholesome". The young shoots are boiled in water and eaten like asparagus; their taste has been compared to that of mountain-cabbage (the heart of a palm).

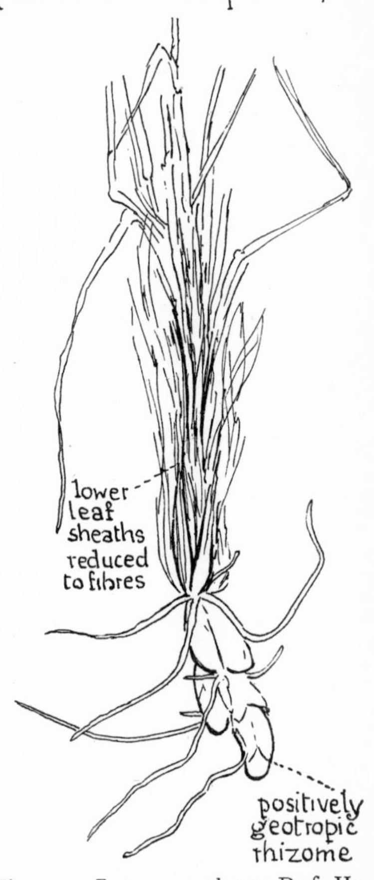

Fig. 137. *Festuca caerulescens* Desf. Herbarium material from North Africa. Base of plant (nat. size) to show downwardly directed rhizome branches. [A.A.]

In grasses the deposition of reserve materials in rhizomes, or the bases of aerial stems, is commoner than the bulb form. Long ago Duval-Jouve[2] followed the history of the rhizome reserves of *Panicum vaginatum* Godr. et Gren. (*P. digitaria* Lat.), a tropical grass naturalised in France. He noticed in October that the cortical parenchyma had thickened walls. About the middle of December, wishing to show these curious cell-walls to a friend, he collected fresh rhizomes, but, to his surprise, sections did not show the same aspect at all; the thickness of the wall was reduced to half, and the inner surface was shagreened, wrinkled, and folded. At first he thought that the individual plants must be diseased, but the same result was obtained on testing others. He tried again in the middle of February, and found that the walls had, by that date, lost all thickening. He then planted these rhizomes, with their emaciated cell-

[1] Hooker, W. J. (1843); see also pp. 337–8. [2] Duval-Jouve, J. (1869).

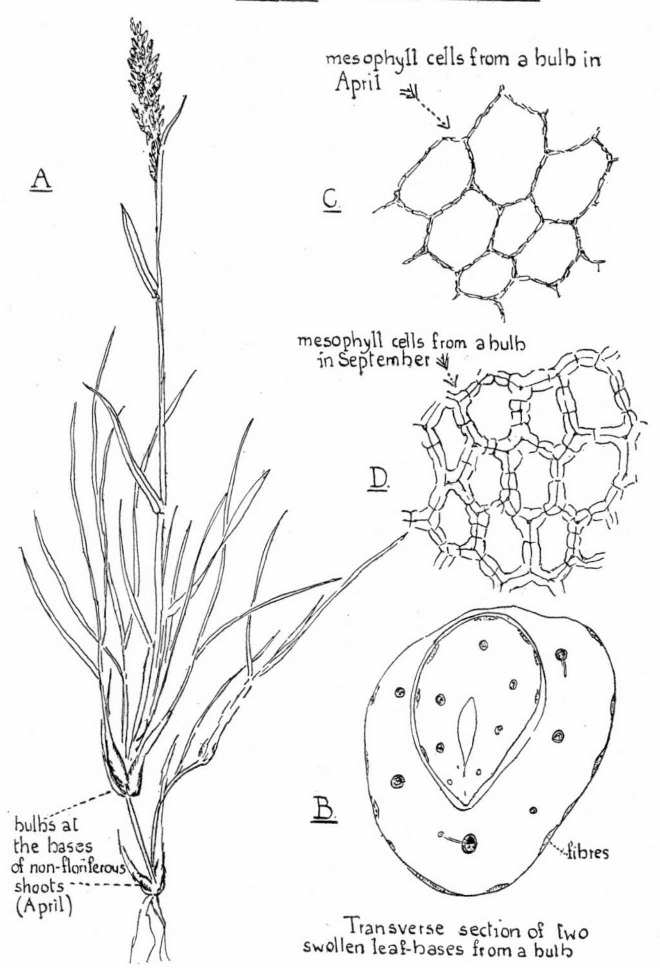

Fig. 138. *Poa bulbosa* L. (Cambridge Botanic Garden). A, plant at end of April (× $\frac{2}{3}$). B, transverse section of two swollen leaf-sheaths from a bulb (× 18 *circa*). C and D, mesophyll cells of a leaf-base from sections such as B, to show the condition of the walls in April and September; cell contents omitted in both (× 258). [A.A.]

walls, in good moist earth in a warm room, and at the end of three weeks buds had come forth, and the walls of the rhizome parenchyma had already recovered part of their thickness.

Among British grasses, *Arrhenatherum avenaceum* Beauv. var. *bulbosum* Lindl. (Fig. 139) offers the most striking type of reserve

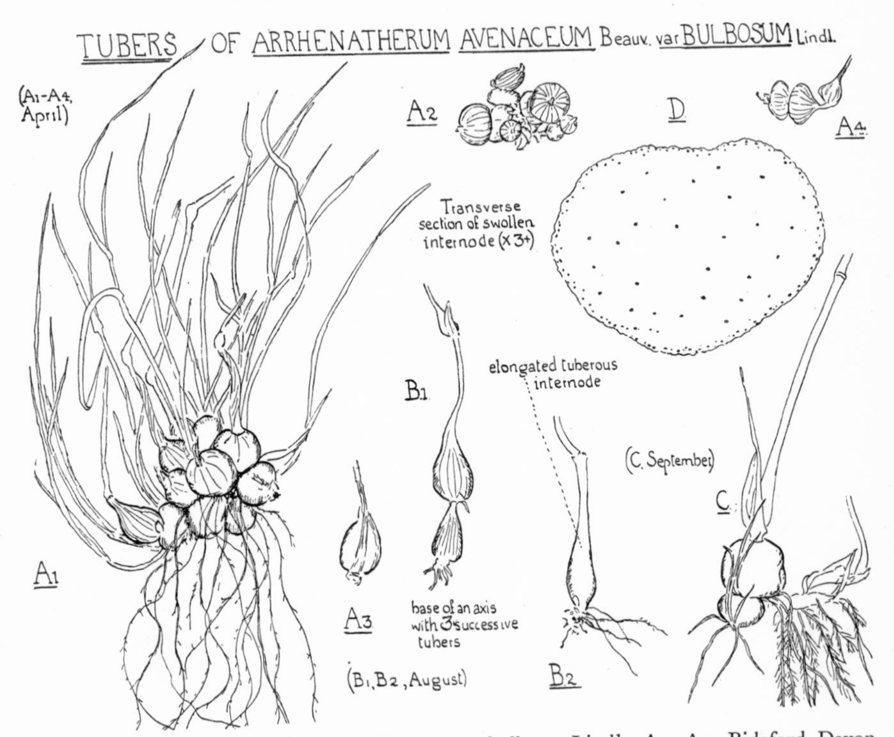

Fig. 139. *Arrhenatherum avenaceum* Beauv. var. *bulbosum* Lindl. A 1–A 4, Bideford, Devon, April (× ½). B 1 and B 2, Whitwell, Isle of Wight, August (× ½). C, sandy soil, Haslemere, Surrey, September (× ½). D, transverse section of a swollen internode (× rather more than 3). [A.A.]

shoot. Often several of the basal internodes are swollen into a bead-like form, from which this species gets its names of Onion-couch and Button-grass. The largest tuber[1] I have found occurred on sandy soil; it measured 1·6 cm. in diameter (Fig. 139, C).

In another of our common grasses, *Phleum pratense* L. var. *nodosum*

[1] I use the word 'tuber' for this and similar organs, for the sake of not multiplying terms; it is, however, open to objection (Evans, M. W. (1927), pp. 29–30).

L.,[1] the thickening is often almost confined to a single basal internode of the main and lateral axes (Fig. 140, A and B, p. 276). In A4, a few cells are shown from a transverse section of a swollen internode. The section was examined in dilute glycerine, and stained in methyl green, which darkened the walls. It will be seen that the cells are filled with large sphaerocrystals; these resist solution in water, but the glycerine was beginning to dissolve them when the drawing was made. Since Timothy tubers were the source from which *phlein* ($C_{90}H_{150}O_{75}$) was originally described,[2] it is probable that these crystals consist of that carbohydrate.

A plant with storage tubers, to whose life-history a good deal of attention has been paid, is Purple-heath-grass, *Molinia caerulea* Moench[3] (Fig. 141, p. 277). In this species the aerial stem bears a succession of leaves in two groups—the lower given off immediately above the rhizome, and the upper, about 5 cm. higher up the axis. The leaves in each series are closely crowded. The internode between the two series, when mature, resembles an Indian club in shape; it may approach 10 cm. in height in tall woodland forms, but 5 cm. is more usual. In a flowering plant this swollen internode is found at the bottom of the culm, with from one to three buds at its base, which will be next year's flowering shoots, while, beside it, a shrunken Indian club, the lower end of last year's inflorescence axis, is often traceable (Fig. 141, A2). In B1 and B2, the structure of the storage internode is indicated. During the summer, the walls of the central ground tissue are thin, but in the autumn they become thickened and pitted. The process is that the cells first fill with a mixture of minute starch-grains and protein granules, in which starch predominates; the next stage is that the starch-grains are dissolved, and their carbohydrate is laid down as reserve cellulose. When the cellulose walls are at their thickest, the tissue is as hard as a date stone. In spring, the reserve cellulose is dissolved, until only the middle lamella of

[1] For figures of the developing stool and tuber in this variety, see Nishimura, M. (1922).

[2] For references, see Kirchner, O. von, Loew, E. and Schroeter, C. (1908, etc.), p. 16.

[3] Schellenberg, H. C. (1897) and Jefferies, T. A. (1916).

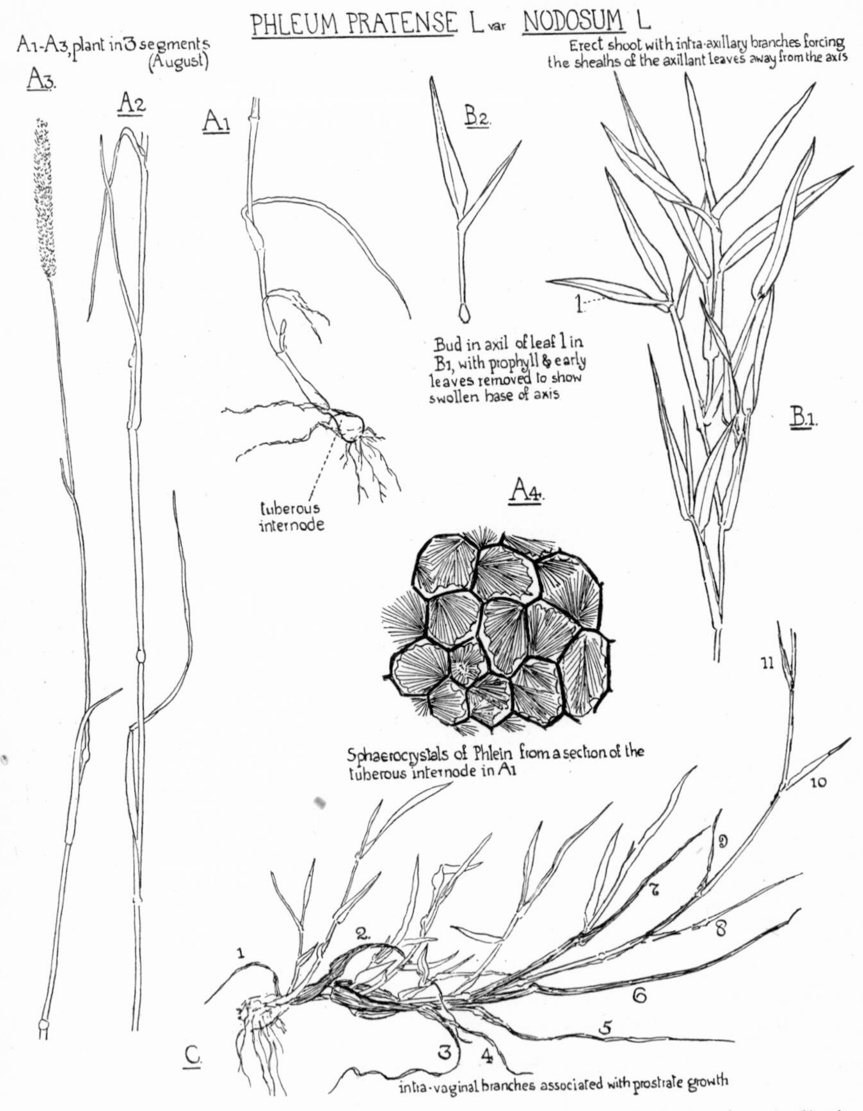

Fig. 140. *Phleum pratense* L. var. *nodosum* L. A 1–A 3, B 1, B 2, C, habit drawings (× ½). A 4, cells with sphaerocrystals of phlein from a transverse section of a tuber (× 77). [A.A.]

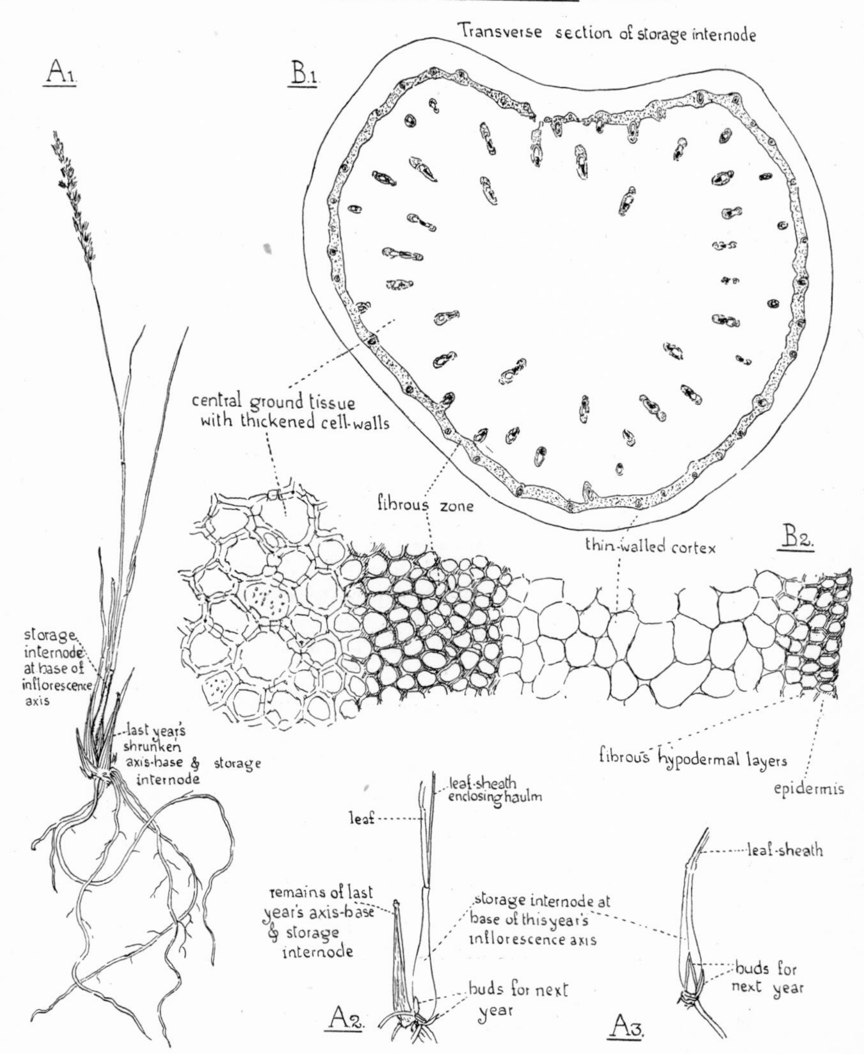

Fig. 141. *Molinia caerulea* Moench. A 1 and A 2, Wharfedale, August; A 3, B 1 and B 2, Kew Gardens. A 1, plant with inflorescence (× ½); the base of the same plant with leaf-bases removed to show the relation of the axes for last year, this year and next year (× rather more than ½). A 3, base of another plant with leaves removed to show the buds for next year at a slightly more advanced stage (× ½). B 1 and B 2, transverse sections of a storage internode; B 1 (× 14); B 2 (× 193 *circa*). [A.A.]

the cell-wall is left, while starch-grains are again formed in the cell, but, during vegetation of the plant, these also are brought into solution. As the buds develop into new shoots, the storage internodes become depleted and shrivelled. When the shoots are attacked by a gall midge, the deposit of reserve cellulose is deflected to the leaf-sheaths around the nests of the larvae, and to the associated roots.

The tuberous internodes of some of the grasses are remarkable for the amount of water which they hold. An experiment has been recorded[1] in which a tuber of *Hordeum bulbosum* L. was left in a dry room until it had shrivelled completely. After fourteen days, when its weight was reduced to little over 0·4 g., it was put into water, and in twenty-four hours it had recovered its original form and weighed 1·6 g.; that is to say, it had taken up nearly three times its weight of water.

Some writers have treated the storage of food and water in the bulbs and tubers of the Gramineae as an "adaptation", but I know of no positive evidence for such an opinion. The accumulation of food reserves and water in certain tissues occurs widely in the flowering plants, and we shall not find an explanation of it by considering one family alone, as we are doing now. All we can say is that the particular *form* which storage takes in the Gramineae depends upon the special structure of the shoot, with its jointed axis and highly developed leaf-sheaths. Indeed, when our general knowledge of root and shoot in the grasses comes under review, it is the high significance of node and internode which is the most vivid impression left on the mind. The bud-bearing node, with its capacity for rooting, and the internode—whether telescoped, in the tillering grass, or elongated, in the slender stalk bearing the ear of Wheat—are the foundations of the vegetative scheme in the Gramineae.

1 Schellenberg, H. C. (1897).

CHAPTER XIII

THE VEGETATIVE PHASE IN GRASSES: THE LEAF

IN the branch systems of grasses, each lateral shoot bears a bikeeled first leaf (prophyll) facing the axillant leaf and addorsed[1] to the main axis (Fig. 182, B, p. 356). Opinion has been divided, in the past, as to whether this bikeeling indicates a derivation from two leaves, but the more recent evidence[2] seems to me to favour the view that the prophyll is a single leaf, whose characters are those of a leaf-sheath. It probably owes its curious form to the pressure due to space conditions in the bud. Although the prophyll has two principal bundles, its symmetry is not really duplex, for one of the bundles is, as a rule, earlier in development, and larger, than the other. Moreover the bud axillary to the prophyll tends to occur opposite to this larger strand, which may thus, on all counts, be interpreted as the median bundle. These points are illustrated in Fig. 142, p. 280, and Fig. 143, A 2 *a*–A 2 *c*, p. 281. In one or more of the earliest leaves succeeding the prophyll, the sheath is apt to predominate, while the limb is absent or reduced (cf. Fig. 131, A, p. 265).

The mode of origin of the leaf members from the stem apex in grasses is a matter of some interest. A few years ago, it was reported that in Wheat the leaves develop from the dermatogen[3] alone, whereas, in other families, not only the dermatogen, but deeper layers as well, play a part in leaf production.[4] The section drawn in

[1] The need for a descriptive term corresponding, in English, to 'adossiert', 'adossé' or 'ge-adosseerd', has been pointed out to me by Dr A. M. Hartsema. I propose to take the word 'addorsed' from heraldry, because this seems to be the nearest equivalent to the French 'adossé', which is employed in describing either a man with his back to the wall, or two heraldic beasts placed back to back, and which will be found in the quotation from Turpin on p. 178.

[2] Bugnon, P. (1921); for full citations of the literature, see Guillaud, M. (1924) and for arguments in favour of the bifoliar view, Collins, G. N. (1924).

[3] The outermost cell-layer of the growing apex.

[4] Roesler, P. (1928); but see also Bugnon, P. (1924[1]) for a more cautiously qualified statement.

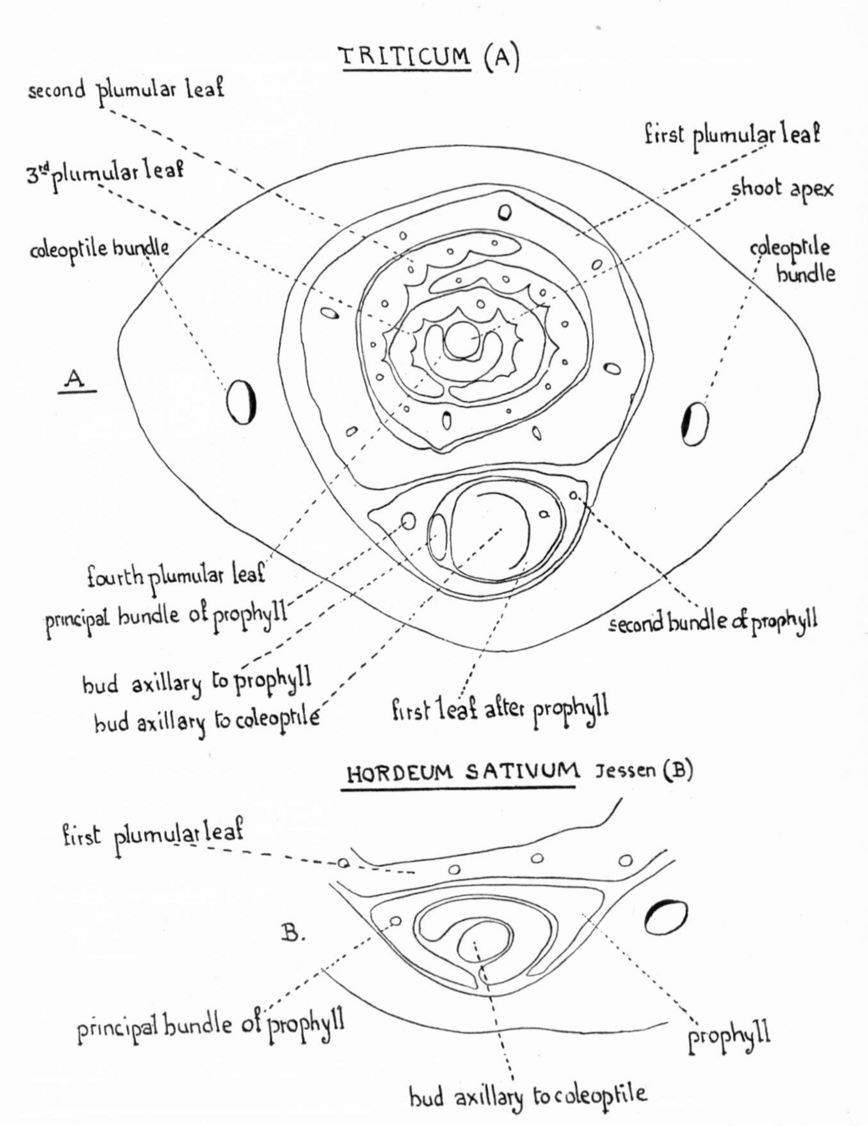

Fig. 142. Buds in the axil of the coleoptile. A, *Triticum* sp. Transverse section (× 47) through a seedling just above the first node to show, in the axil of the coleoptile, a bud whose prophyll itself has a bud in its axil. B, *Hordeum sativum* Jessen. Transverse section (× 77) of a bud in the axil of the coleoptile. The bundle marked "principal bundle" is the only one yet differentiated in the bud. [Adapted from Arber, A. (1923).]

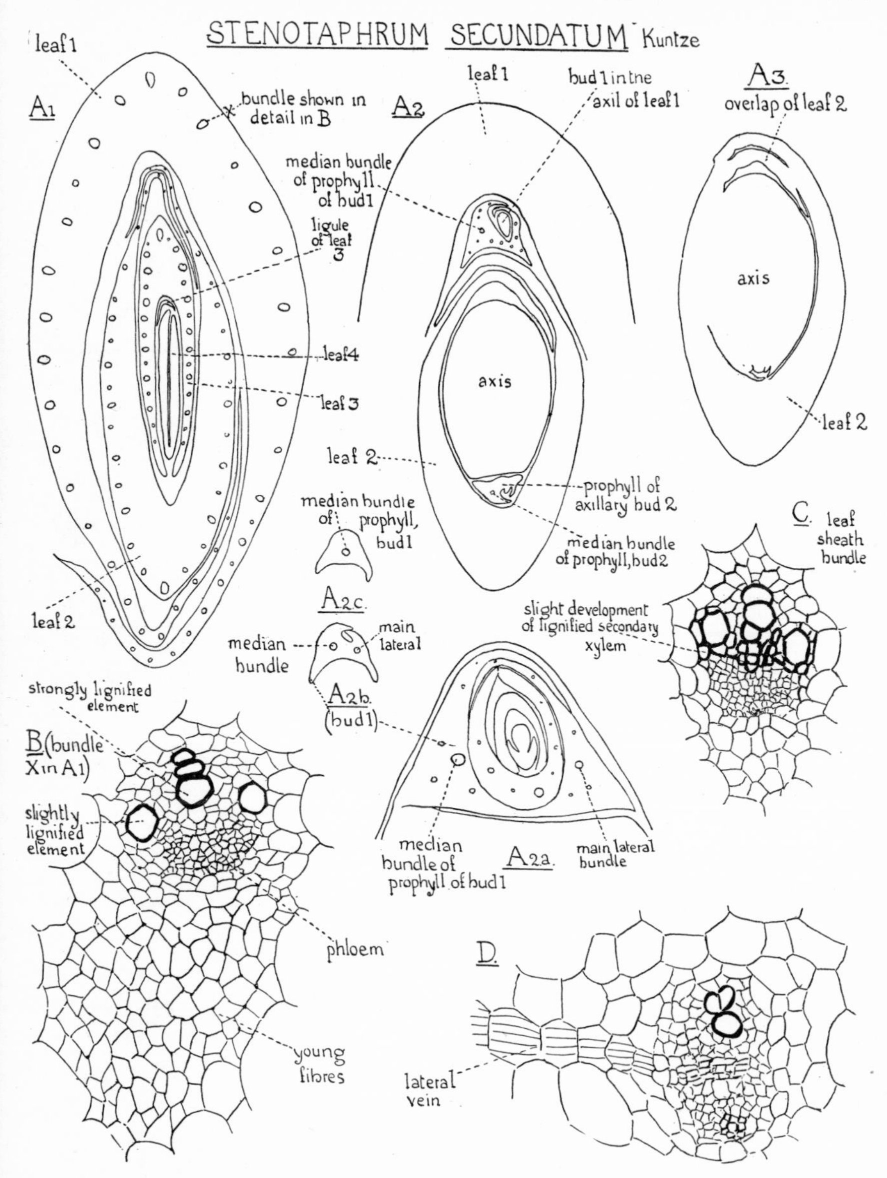

Fig. 143. *Stenotaphrum secundatum* Kuntze (Cambridge Botanic Garden). A 1–A 3, sections from a transverse series from above downwards through a shoot apex (× 23). A 2 *a*–A 2 *c*, sections of *bud* 1, below and above A 2 (× 47). A 2 *a*, close to the base of the bud; A 2, A 2 *b* and A 2 *c*, at successively higher levels. B, bundle *X* in the sheath of *leaf* 1 in A 1, but cut at level of A 2 (× 318). C, a smaller bundle from the sheath of *leaf* 1 (× 318). D, a bundle of leaf 2 at a higher level than A 1, to show a branch bundle (× 318); the tissues were embryonic and the drawing is slightly diagrammatised. [A.A.]

Fig. 144 does not conflict with this account, though it is by no means conclusive. It shows a radial section of a plumular bud; for clearness the rudiment of the third plumular leaf is dotted, though there is, in reality, no difference between its cell-contents and those of the surrounding elements. The word 'dermatogen' becomes etymologically absurd when it is used for plants in which this layer gives rise, not to the epidermis alone, but to the leaf as a whole; the alternative 'tunica', which has been suggested, certainly seems more reasonable.

The distichous (two-ranked) arrangement of the leaves, seen for

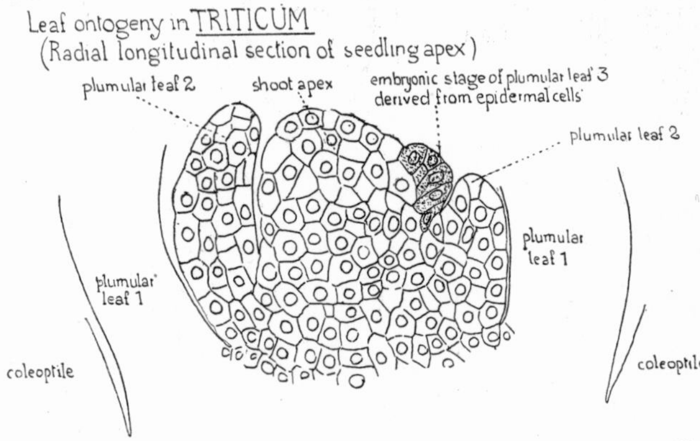

Fig. 144. Radial longitudinal section (× 193 *circa*) through the apex of a seedling of *Triticum* to show the central region of the plumular bud. The coleoptile and the first and second plumular leaves are cut on either side of the section, as their insertion encloses the apical bud. Owing to the extreme youth of the third plumular leaf, it is, at this stage, visible as a hump on one side of the shoot apex only. [A.A.]

instance in Fig. 143, A 1, p. 281, is usual for grasses; that it long ago caught the eye of man, is shown by the Millet in the Nineveh bas-relief reproduced in Fig. 13, p. 24. This characteristic appearance is, however, masked in some species by the twisting of the axis, while in the Reed, *Phragmites communis* Trin., it ceases to be obvious owing to the special qualities of the leaf-sheath. The lower parts of the shoots in this grass are usually sheltered by its gregarious growth, and in this region the leaves stand out to right and left; in the upper, more exposed parts, on the other hand, they are all blown to the lee side of the stem. This is due to the fact that the inner surfaces of the long leaf-sheaths are smooth and polished, so that the leaves readily slide round the axis.[1]

Fig. 143, A 3, p. 281, illustrates a feature characteristic of many grasses—the overlap at the base of the leaf-sheath, which is so wide that it more than encircles the stem. Leaf-sheaths[2] in the grasses are usually open, but sometimes (e.g. *Glyceria aquatica* Wahl., Fig. 148, B 1, p. 287) they form closed hollow cylinders. Among twenty-seven species of *Festuca* examined from this point of view, the sheaths were open in eighteen, while in eight they were closed, and in one they varied in character in different sub-species.[3]

In India it has been noticed that there is separate disarticulation of sheath and limb in *Aristida cyanantha* Steud. ex Trin. The lamina is shed, leaving the culms covered with the straw-coloured persistent sheaths. After an interval, these also fall, and the green culms are left naked from base to apex.[4] A British example, in which the leaf-sheaths are more persistent than the limb, is the Couch-grass, *Agropyron repens* Beauv., Fig. 145, B, p. 284. In the same species the leaf-sheaths may become relatively enlarged as the result of the attacks of a Chalcid-fly, *Isosoma graminicola* Giraud[5] (Fig. 146, p. 285). An exaggeration of the size of the leaf-sheath, and a corresponding reduction in the limb, occur as part of the normal development in the uppermost foliage leaf enclosing the inflorescence in many grasses, such as *Phalaris canariensis* L. (Fig. 145, E) and *Alopecurus pratensis* L. (Fig. 145, D). In the xerophytic South African grass, *Ehrharta aphylla* Schrad. (Fig. 145, C 1 and C 2), this type of change has gone so far that the lamina is reduced to a mere point.

In perennial grasses, the lowest internodes of each haulm, and those of the basal lateral shoots, remain short, so that there is a series of leaf-sheaths inside one another. In the grasses of dry places, these leaf-sheath bases are liable to persist, and their dead fibres, becoming matted together, form a 'thatch-tunic' round the younger

[1] Yapp, R. H. (1908).

[2] On the morphology of the grass leaf, see Bugnon, P. (1921), (1924[1]), (1924[2]), and Arber, A. (1923).

[3] Hackel, E. (1882).

[4] Hole, R. S. (1911).

[5] Houard, C. (1904); Dr G. Salt kindly tells me that the correct generic name is *Harmolita* (Motschulsky) which has replaced *Isosoma* (Walker).

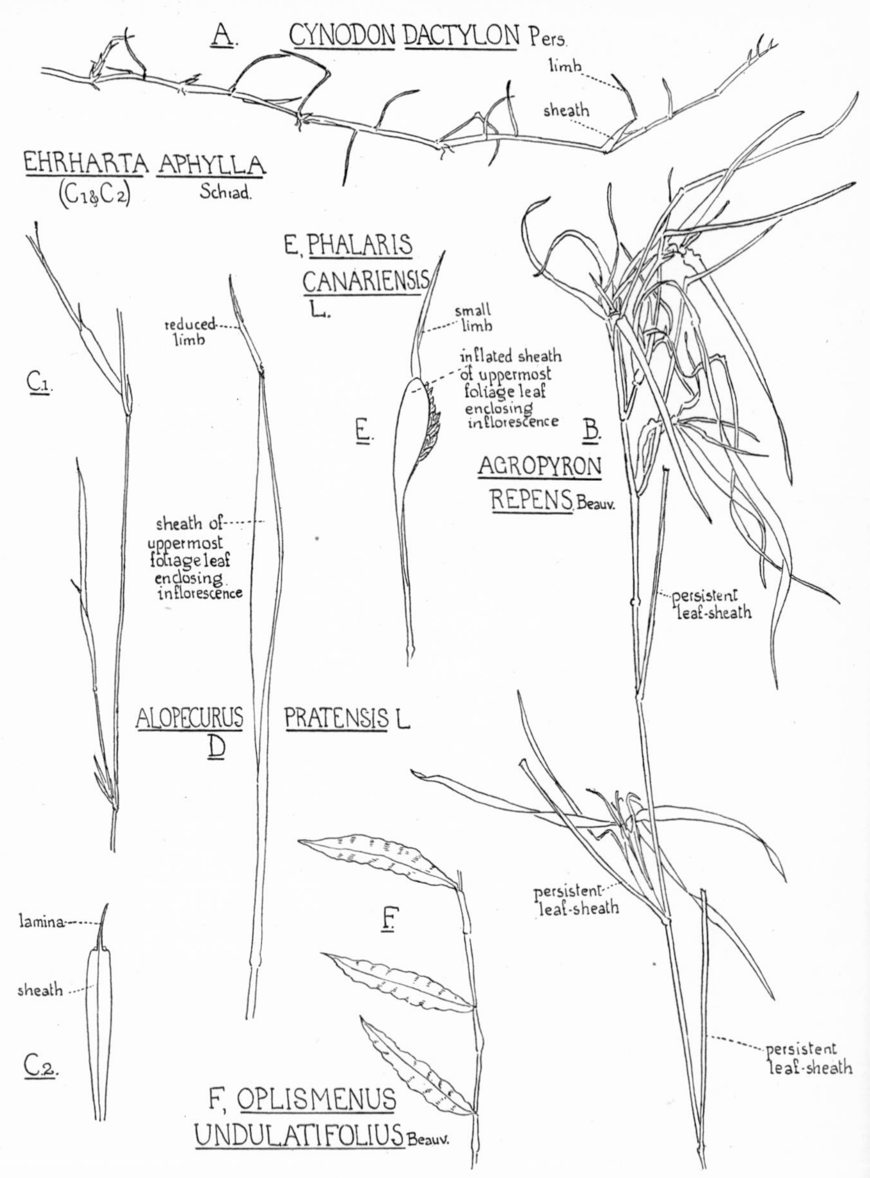

Fig. 145. Leaf morphology. A, *Cynodon dactylon* Pers., Mer de Sable, Carteret, Normandy, August. Apex of creeping shoot (× ½). B, *Agropyron repens* Beauv., shoot (× ½). C 1 and C 2, *Ehrharta aphylla* Schrad., from South African Museum, Cape Town. C 1, part of shoot (× ½); C 2, single leaf (nat. size). D, *Alopecurus pratensis* L. (× ½). E, *Phalaris canariensis* L. (× ½). F, *Oplismenus undulatifolius* Beauv. (Welwitsch, Iter Angolense) (× ½). In some of the dried material examined the leaves show a slight crisping (cf. Arber, A. (1925), p. 66). [A.A.]

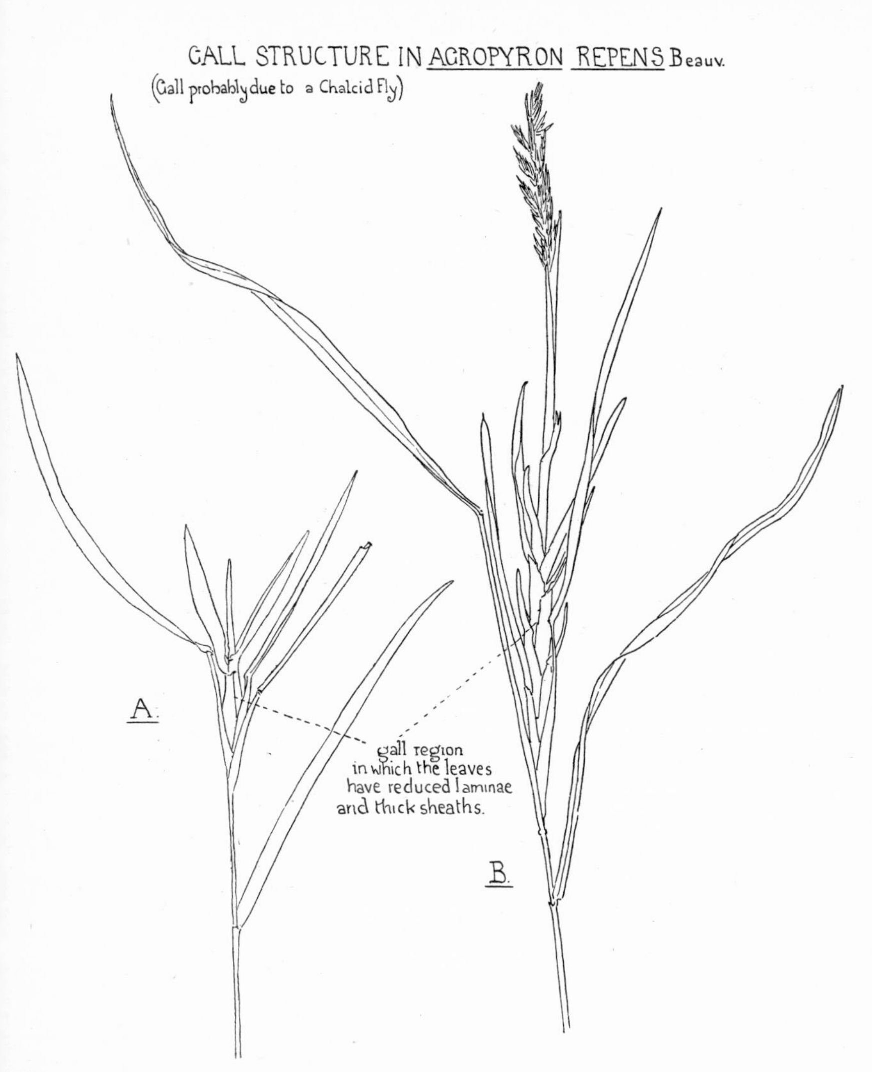

Fig. 146. *Agropyron repens* Beauv. Examples of shoot galls found near Cambridge in the autumn of 1924, when they were specially prevalent. A, September 1; B, October 13 (both × ½). These galls are probably due to *Isosoma* (*Harmolita*) *graminicola* Giraud. [A.A.]

leaves.[1] *Festuca arundinacea* Schreb., sketched in Fig. 177, p. 341, is a tunicate Fescue from our own flora, but more striking examples may

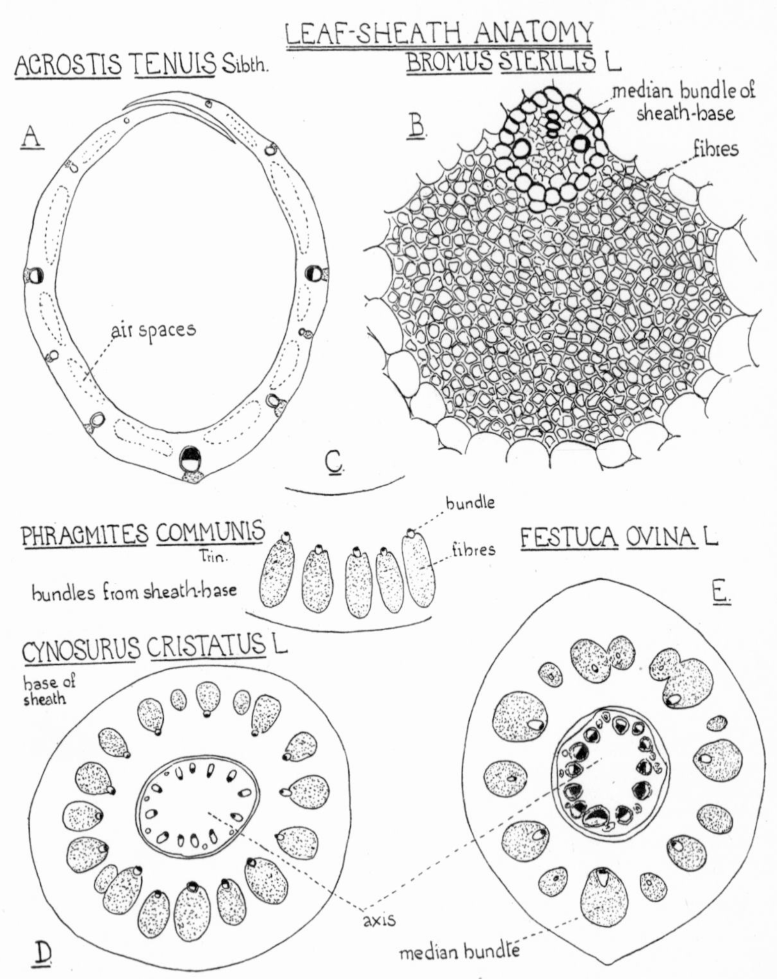

Fig. 147. Leaf-sheath anatomy. A, *Agrostis tenuis* Sibth.; transverse section of leaf-sheath (× 47). B, *Bromus sterilis* L., median bundle of thickened region of sheath-base (× 193 *circa*). C, *Phragmites communis* Trin. Bundles from sheath-base (× 14). D, *Cynosurus cristatus* L.; transverse section of sheath-base just above node (× 23). E, *Festuca ovina* L.; transverse section of sheath near base, below 'wrap-over' (× 23). [A.A.]

be found elsewhere. For instance, in the Australian grass, *Eragrostis eriopoda* Benth., the base of the slender haulm appears to be

[1] Hackel, E. (1890[1]); Brockmann-Jerosch, H. (1913).

swollen, but the thickening is due exclusively to the felted sheaths, which are clothed with long, finely curled, closely interwoven hairs. *Pollinia phaeothrix* Hack., from the Nilgiri and Ceylon, has dark purplish-brown wool on the tunic. A glance at the anatomy of leaf-

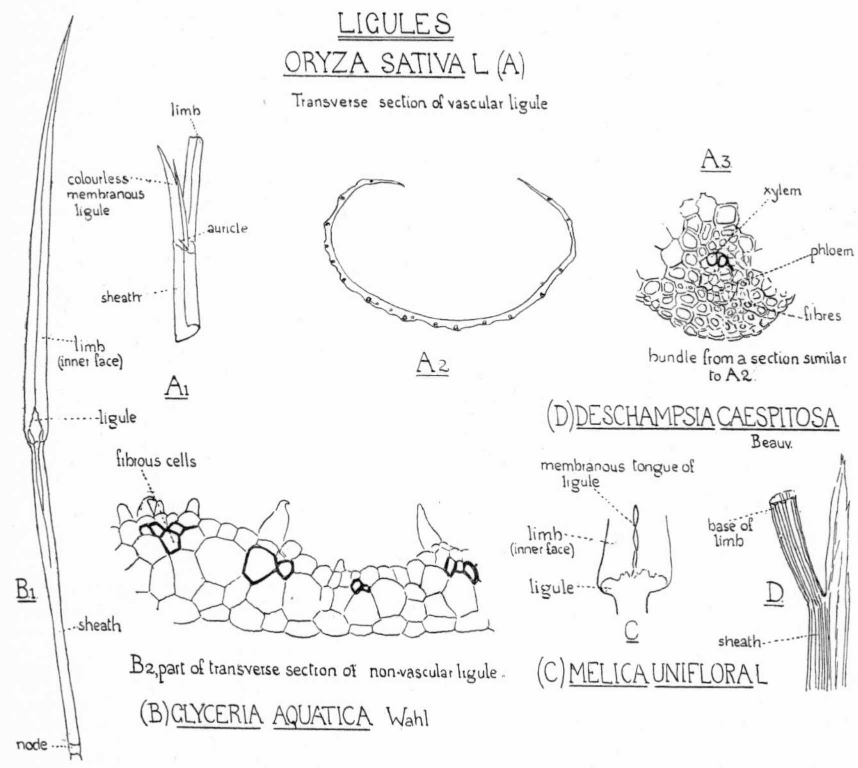

Fig. 148. Ligules. A, *Oryza sativa* L. A 1, base of limb and top of sheath with ligule (× ½). A 2, transverse section of ligule (× 14) to show vascular bundles. A 3, a bundle from another section, similar to A 2 (× 193 *circa*). B, *Glyceria aquatica* Wahl. B 1, a leaf and the node from which it arises (× ½). B 2, part of transverse section of ligule (× 193 *circa*). C, *Melica uniflora* L. Base of lamina, seen from front, enlarged, to show ligular peg (herbarium material). D, *Deschampsia caespitosa* Beauv., base of limb with ligule, enlarged. [A.A.]

sheaths explains how their skeletons can survive in a tunicate form. Fig. 147 illustrates their structure in a few common British grasses, and it will be seen that in several of them (B–E) each bundle is accompanied by a mass of fibres, altogether disproportionate to it in size.

An outgrowth, the *ligule*, occurs at the distal inner margin of the leaf-sheath (Figs. 148 and 149). In *Dactylis glomerata* L., in which its development has been traced, it has been found to originate from the epidermis alone.[1] In most grasses it is thin and membranous, and sometimes it is fimbriated; in *Melica uniflora* L., it has a peculiar tongue-like projection in the median line (Fig. 148, C). Usually it contains no bundles, but in the diaphanous, silvery ligule of *Glyceria aquatica* Wahl., which may exceed a centimetre in length, it

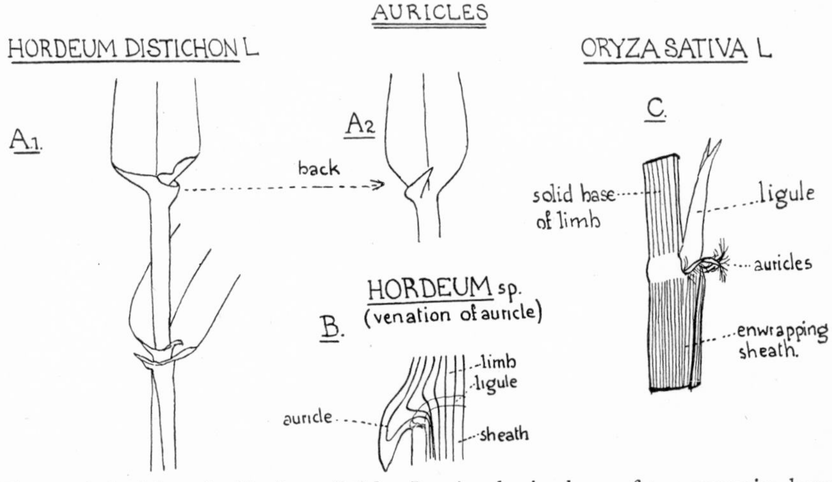

Fig. 149. Auricles. A, *Hordeum distichon* L.; A 1, lamina-bases of two successive leaves; A 2, back of lamina-base of upper leaf (about nat. size). B, *Hordeum* sp. Diagrammatic sketch of one auricle with part of the junction of sheath and limb to show vascular supply of auricle (enlarged). C, *Oryza sativa* L., top of sheath and base of limb; the auricles arise from the whitish band at the extreme base of the limb at each side of the ligule. [A.A.]

includes fibrous elements (Fig. 148, B), while in a few species it is fully vascular, e.g. *Oryza sativa* L. (A 1–A 3) and *Psamma arenaria* Roem. et Schult. Genetical experiments upon Wheat have shown that in this cereal the absence of a ligule is a character recessive to the normal.[2] A wild grass from California, *Anthochloa colusana* (Davy) Scribn., has been described as having liguleless foliage leaves, in which there is no distinction of sheath and limb.[3]

Just above the ligule, we sometimes find a pair of outgrowths

[1] Bugnon, P. (1921). [2] Vavilov, N. I. and Bukinich, D. D. (1929).
[3] Davy, J. Burtt (1898).

known as *auricles*. In certain grasses these outgrowths are obviously equivalent to the basal lobes of sagittate limbs, e.g. *Hordeum* (Fig. 149, A and B), while in others they have, in appearance, little connection with the leaf-limb, e.g. *Oryza* (C).

Dr Johnson once said, "a blade of grass is always a blade of grass, whether in one country or another"; it is sad that we cannot know what his reaction would have been if some of the broad-leaved forest grasses of tropical countries[1] had been forced upon his attention. A few of these wide leaf-limbs are illustrated in Figs. 150, p. 290, 151, p. 291, and 211, p. 406; they may narrow to a distinct petiole (*Orthoclada laxa* Beauv., Fig. 211, B), which in *Panicum sagittaefolium* Hochst., from Abyssinia, may be as much as 5 in. in length.[2] In this species, and in another petiolate form, *Phyllorachis sagittata* Trim.,[3] the leaf-limb is definitely arrow-shaped (Fig. 211, E). These broad leaf-limbs are often asymmetrical about the midrib (e.g. *Oplismenus loliaceus* Lam., Fig. 150, A). In asymmetric leaves it is found that it is that flank of the blade whose margin was internal in the bud which is the wider; this is shown for St-Augustine's-grass, *Stenotaphrum secundatum* Kuntze, in Fig. 143, A 1, leaf 1 and leaf 2, p. 281. Asymmetry may also occur in our narrow-leaved meadow grasses, e.g. *Poa pratensis* L.,[4] but it is naturally less conspicuous here than in types with wide leaves.

Some of the broad-limbed South American grasses produce long, many-leaved shoots, which often grow horizontally; the blades, also, are brought into the horizontal plane by torsions in the stem or leaf-stalk. In one of these grasses, *Olyra guineensis* Steud. (*Strephium guianense* Ad. Br.) sleep movements have been described.[5] The leaf-sheath is separated from the limb by a cushion forming a very short petiole, recalling the pulvinus at the base of the leaflets in *Mimosa pudica* L. The shoot with its distichous leaves is spread out like a pinnate leaf with alternating leaflets; by a slight twist of the pulvinus the leaves are brought into one plane, with their upper surfaces

[1] On the form and the torsions of broad leaf-limbs of the S. American forests, see Lindman, C. A. M. (1899); on *Planotia nobilis* Mun., a broad-leaved bamboo, see p. 71.
[2] Hochstetter, C. F. (1847).
[3] Trimen, H. (1879).
[4] Duval-Jouve, J. (1875).
[5] Brongniart, A. (1860).

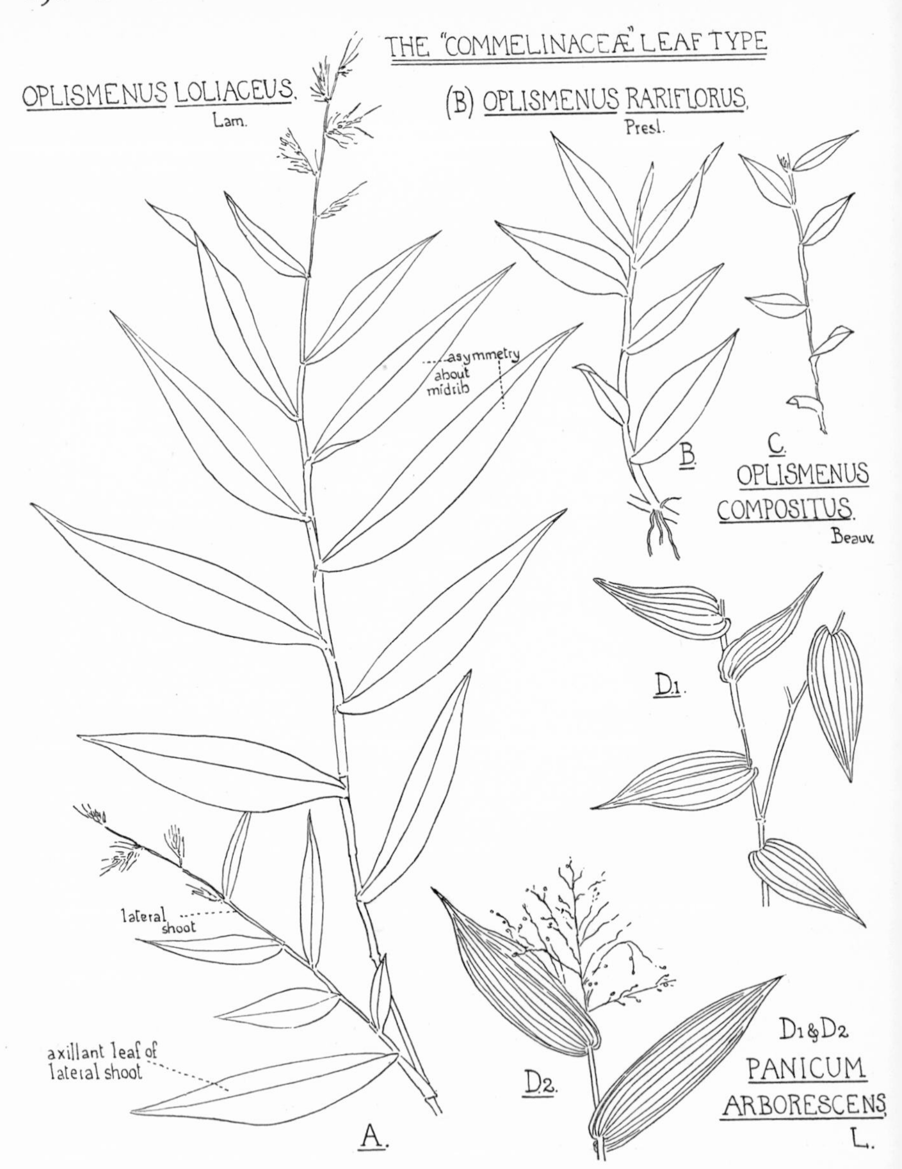

Fig. 150. Sketches of grass leaves (all × ½) in the herbarium of the British Museum (Nat. Hist.) to illustrate the approximation to the Commelinaceous leaf type in certain genera. A, *Oplismenus loliaceus* Lam., wooded rocky hillside, Cuba. B, *O. rariflorus* Presl, moist banks, Mexico. C, *O. compositus* Beauv., damp shaded places, Queensland, Australia. D 1 and D 2, *Panicum arborescens* L.; D 1, dampish woods, Welwitsch, Iter Angolense; D 2, Lake Nyassa. [A.A.]

towards the sky. The first effect of darkness is to undo the torsion, so that the leaves return to their original position; by a subsequent

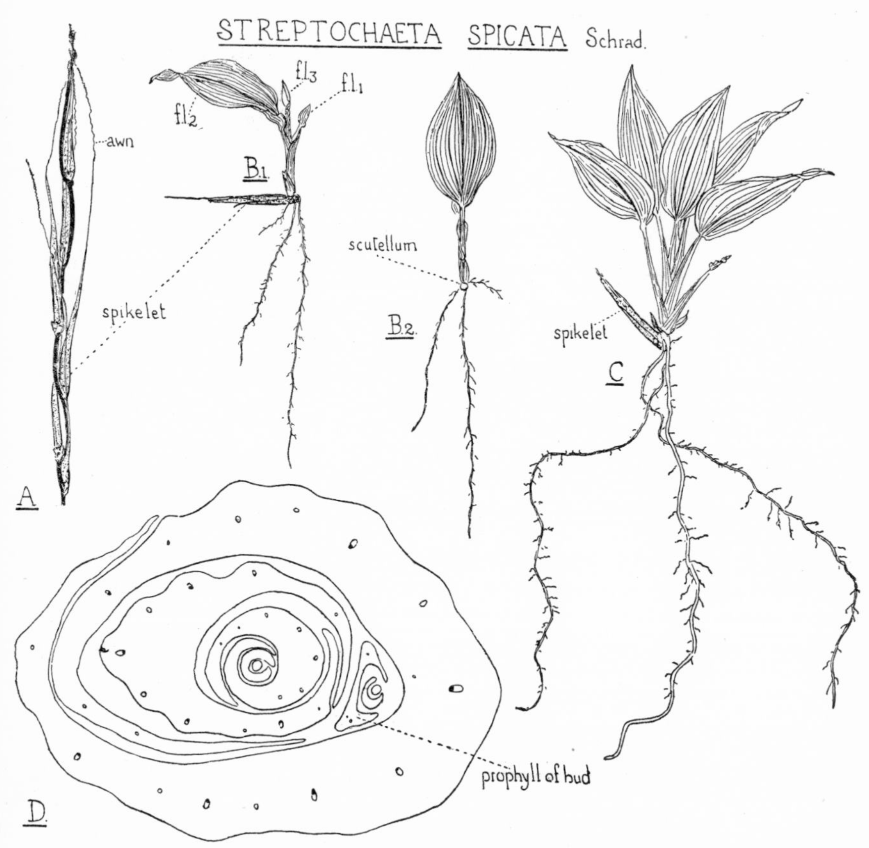

Fig. 151. *Streptochaeta spicata* Schrad. A, an inflorescence gathered January, 1927, near Rio de Janeiro, removed from its enclosing leaf-sheaths (× ½ *circa*). It shows six spikelets arranged in a ⅖ spiral; five of the awns are twisted together at the apex. The successive spikelets are dotted, and the main axis indicated in black. B and C, seedlings from seed collected near Rio de Janeiro, January 10, and sown at the Cambridge Botanic Garden, February 21, 1927. B 1 and B 2, drawn on July 6, 1927 (× ½ *circa*): B 1, $f.l._1$, $f.l._2$ and $f.l._3$, foliage leaves. B 2, view at right angles to B 1, to show small scutellum; the remains of the spikelet, caryopsis and endosperm have been dissected off. C, plantlet drawn September 27, 1927 (× slightly more than B). D, transverse section (× 45) of an apical bud of a seedling of the same age as B to show that the leaves originate on two orthostichies as in typical grasses. [Arber, A. (1929[1]).]

upward movement they come to embrace the part of the shoot above them.

A different kind of torsion is found in *Pharus glaber* H.B.K., in which the shoot has short internodes and long leaf-sheaths; between the sheath and limb there is an unusually long stiff petiole. Throughout development the leaves remain folded inside one another, so that the upper surface is never exposed to the light. The habit thus initiated seems too powerful to be overcome, for, in the final stage, before the leaves open out, each leaf-stalk twists through 180°, so that the morphologically lower surface always retains the upper position which it has usurped.[1] The anatomy does not seem to have been examined in this species, but in *P. latifolius* L. (Fig. 211, D 1, p. 406), in which the same twist occurs, it has been shown that there is assimilating tissue under each epidermis, and stomates on both surfaces, and that the assimilating tissue develops as a palisade on the face exposed to the light.[2]

Resupination may be seen in some of our British grasses. When the development of the radical leaves of *Brachypodium sylvaticum* Beauv. is observed in early spring, it is found that the first leaf emerges and unrolls without a trace of torsion or resupination, but, when it attains 6–7 cm. in length, it becomes horizontal, and begins to twist towards the middle of the limb. Older and larger leaves are completely inverted by a twist at the base of the limb.[3] In addition to those species in which the leaves are definitely inverted, there are others, such as *Briza maxima* L., illustrated in Fig. 152, which show one or more turns in the leaf-limb. In Wheat, the blade is twisted in the form of a loose right-handed screw. With rare exceptions, this direction of the screw is the same for all the leaf-blades; it is independent of the alternating, right-over-left and left-over-right rolling of the young leaves.[4] The problem of the extent to which torsion and resupination are autonomous movements, and of the degree to which they are influenced by external conditions, such as light and gravity, has not been solved with any completeness. It is a direction in which further work is needed; the spiral coiling of grass leaves,

[1] Lindman, C. A. M. (1899).
[2] Kugler, H. (1928); on leaf movements, see also Goebel, K. von (1926).
[3] Porta, N. H. (1928).
[4] Percival, J. (1921); see also Macloskie, G. (1895[1]).

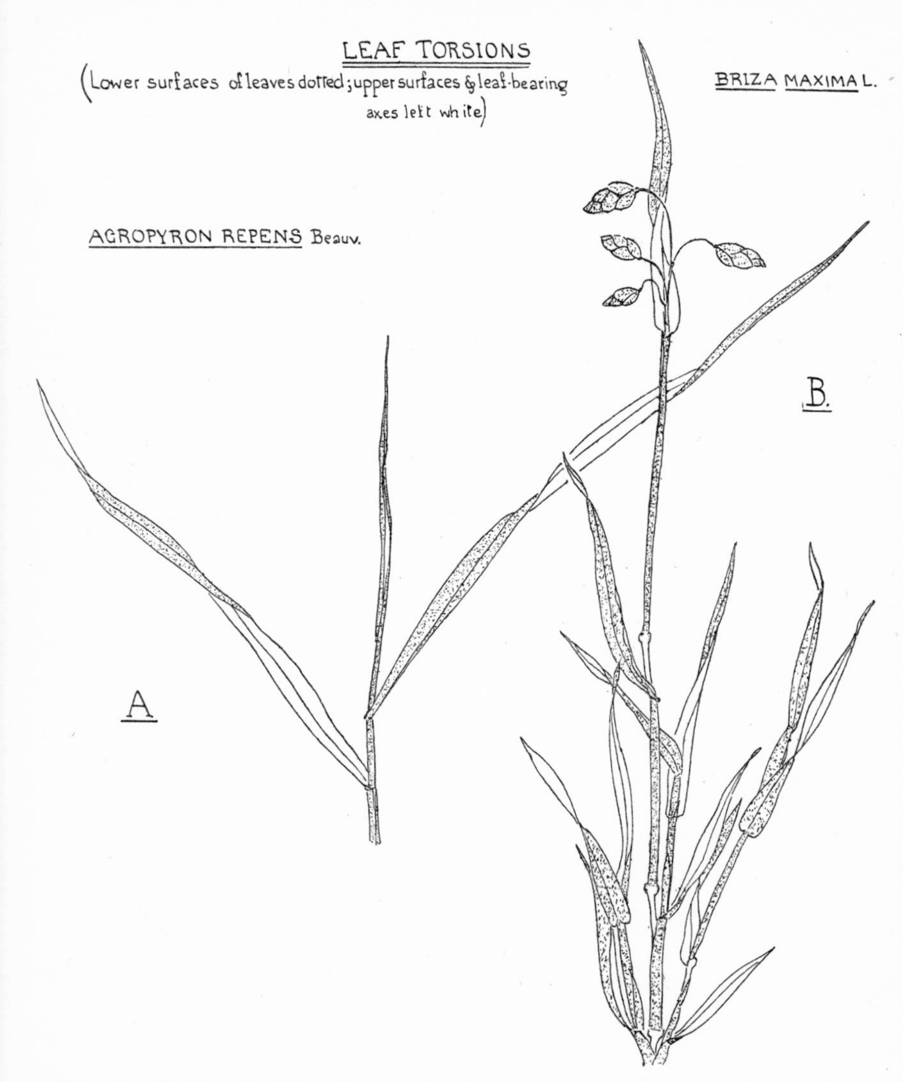

Fig. 152. Torsions in foliage leaves. A, *Agropyron repens* Beauv. Apex of young shoot (reduced). B, *Briza maxima* L. Part of a young plant (reduced). [A.A.]

sometimes met with as an abnormality in individual specimens,[1] also wants explanation.

The narrow or grooved leaf-limbs, some of which are markedly fibrous, shown in Figs. 153 and 154, form a sharp contrast to those of the broad-leaved shade types, with their tendency to torsion and resupination. As examples of the extremes, we may take, on the one hand, the fibrous leaf of *Sporobolus rigens* (Trin.) Desv.,[2] which approaches radial symmetry in its apical region (Fig. 154, A4) and, on the other hand, the large, flat limb of *Olyra latifolia* L. (Fig. 211, A, p. 406). Indeed, the study of the leaves of the Gramineae[3] awakens continual surprise at the range of variation which they present. They show so many constant anatomical differences that transverse sections often give critical help in cases of doubtful identification.[4]

Many grasses have a limb which in section is seen to be ridged and furrowed; the furrows may form deep, narrow grooves. How such a form originates is not obvious at first glance, but some light is thrown on the process by a study of serial sections through the transition region between sheath and limb in the Marram-grass (Fig. 155, D and E, p. 297). These sections suggest that the ridging is due to growth in connection with each conducting strand; this growth takes place ventrally because this is the direction in which the leaf is not confronted by a boundary wall formed by the sheaths of its predecessors. The leaf of *Deschampsia caespitosa* Beauv. (*Aira caespitosa* L.) illustrated in Fig. 156, p. 298 is another example of this deep grooving. This grass is often left uneaten in pastures, forming rough tussocks known as "Bull-hassocks" or "Bullpates".[5] The reason why cattle avoid it becomes clear on looking at the transverse section of the leaf; it is sharply ridged and each ridge is crowned by a knife-edge of fibres.

1 I have seen this in *Avena sativa* L., *Agropyron repens* Beauv., and *Arrhenatherum avenaceum* Beauv.

2 Parodi, L. R. (1928), Fig. C2, p. 119.

3 On nervation, see Ettingshausen, C. Ritter von (1866).

4 On grass leaf anatomy, see Duval-Jouve, J. (1875); Pée-Laby, E. (1898); Ward, H. Marshall (1901); Lewton-Brain, L. (1904); Burr, S. and Turner, D. M. (1933); etc.

5 Britten, J. and Holland, R. (1886).

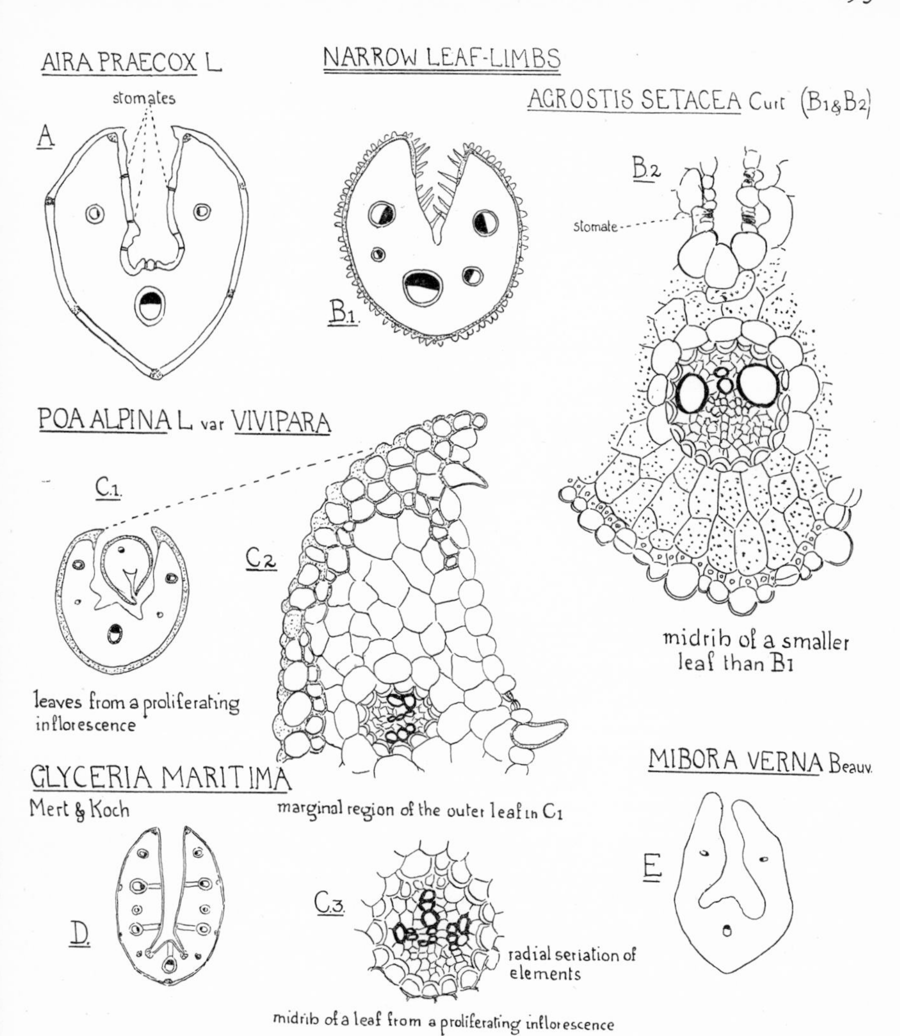

Fig. 153. A, *Aira praecox* L., transverse section of leaf-limb (× 77 *circa*). B, *Agrostis setacea* Curt B 1, transverse section of limb (× 77 *circa*); B 2, midrib of a smaller leaf than B 1 (× 318). C, *Poa alpina* L. f. *vivipara* L. Transverse section of leaves from a proliferating inflorescence; C 1 (× 47); C 2, marginal region of outer leaf in C 1 (× 318); C 3, midrib of another leaf to show evidence of cambial activity (× 318). D, *Glyceria maritima* Mert. et Koch, Axmouth transverse section of leaf-limb (× 14). E, *Mibora verna* Beauv., transverse section of leaf-limb (× 77 *circa*). [A.A.]

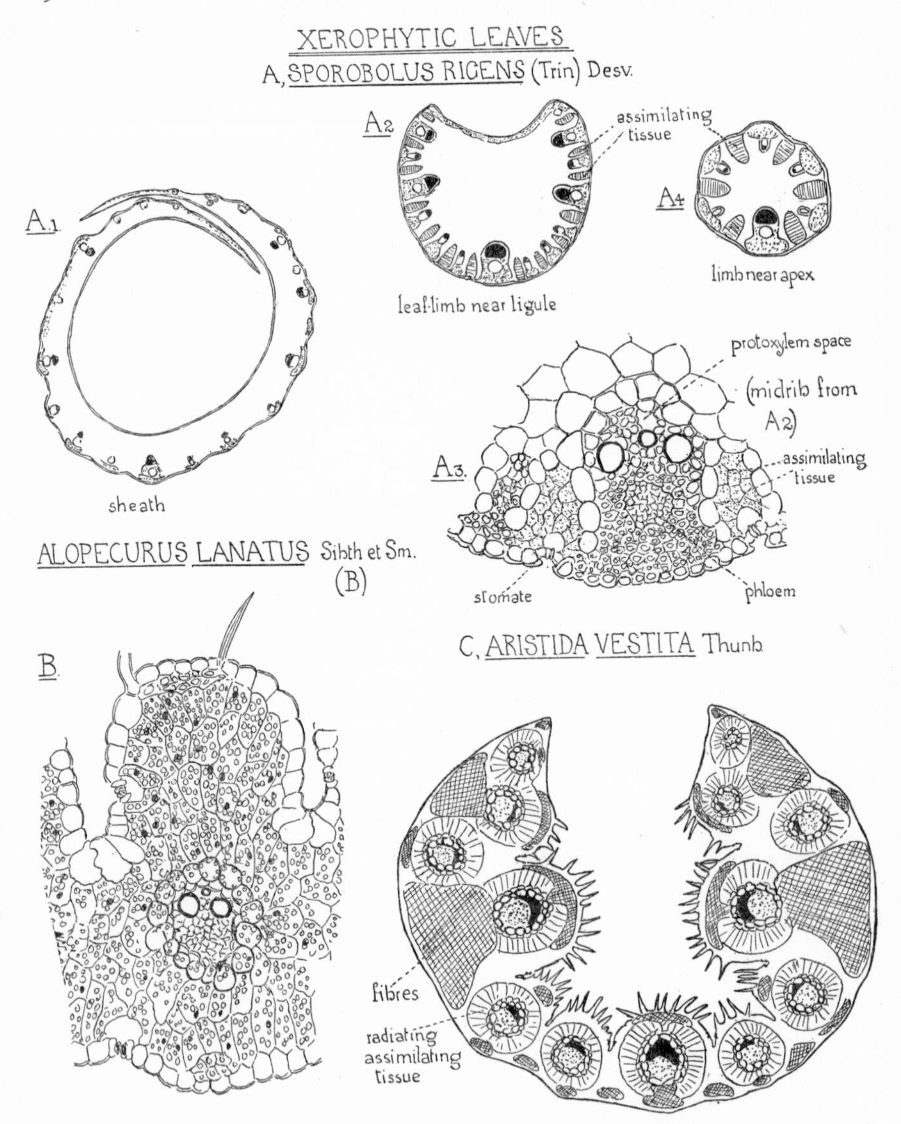

Fig. 154. Xerophytic leaves. A, transverse sections of the leaf of *Sporobolus rigens* (Trin.) Desv., grown in Cambridge Botanic Garden from seed sent from Buenos Aires by Professor Parodi; A 1, leaf-sheath (× 47); A 2, limb near ligule (× 47); A 3, midrib of A 2 (× 193 *circa*); A 4, limb near apex (× 77 *circa*). B, *Alopecurus lanatus* Sibth. et Sm., Cambridge Botanic Garden; a single rib of a leaf in transverse section (× 193 *circa*); for the leaf as a whole, see Fig. 159, p. 302. The mesophyll of the hair-clothed leaf is poor in air spaces. C, *Aristida vestita* Thunb. Transverse section of a leaf, herbarium material from the Transvaal (× 77 *circa*); an extreme type of fibrous xerophytic leaf (fibres cross-hatched). [A.A.]

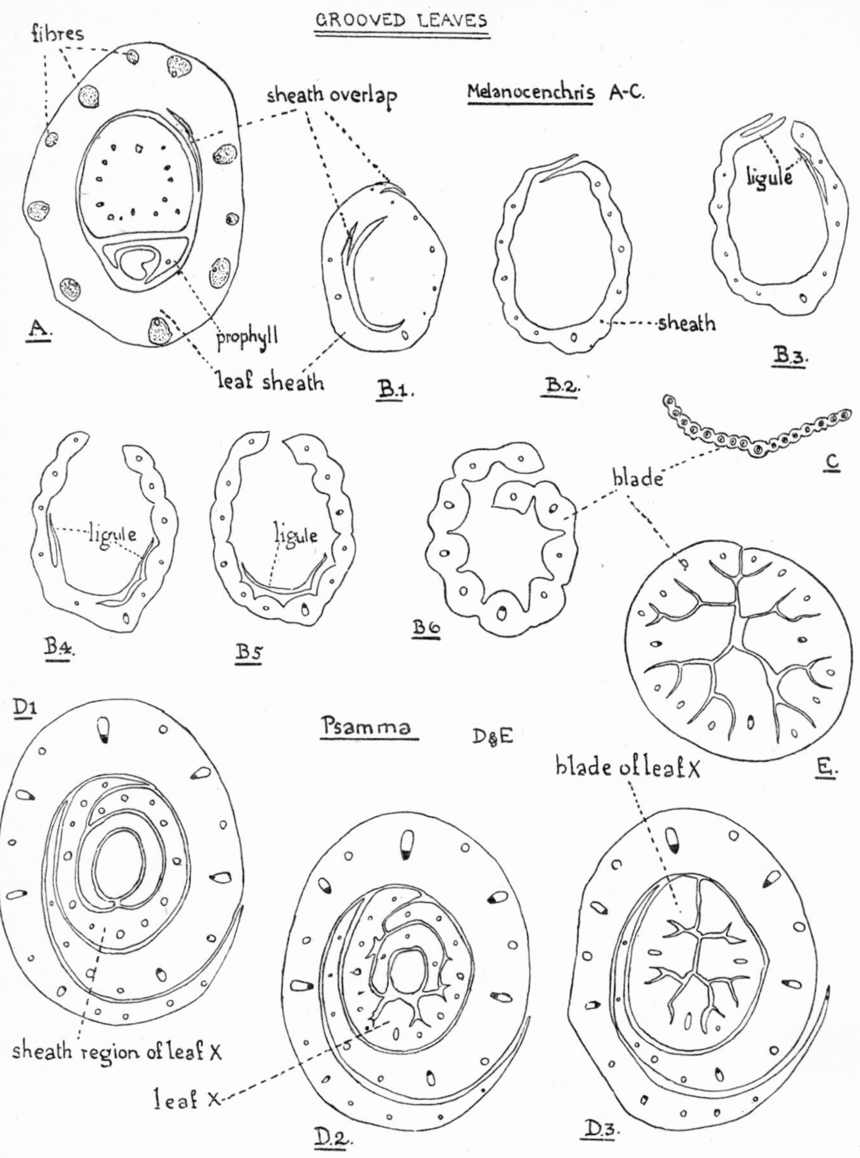

Fig. 155. A–C, *Melanocenchris Royleana* Nees (*Gracilea Royleana* Hook. f.). A, transverse section through apical bud (× 77 *circa*); for details see Fig. 192, p. 383. B 1–B 6, series of transverse sections (× 77 *circa*) through a young leaf from the base upwards. C, transverse section of the limb of a mature leaf (× 14). D and E, *Psamma arenaria* Roem. et Schult. D 1–D 3, transverse sections from series through apical buds to show origin of invaginations in the limb (× 47). E, transverse section of a young leaf from another bud (× 47). [Adapted from Arber, A. (1923).]

In section, ridged leaves generally show a group of large epidermal cells—'bulliform cells'—at the base of each groove. Occasionally they are little developed, e.g. *Bromus hordeaceus* L. (Fig. 157, A 1

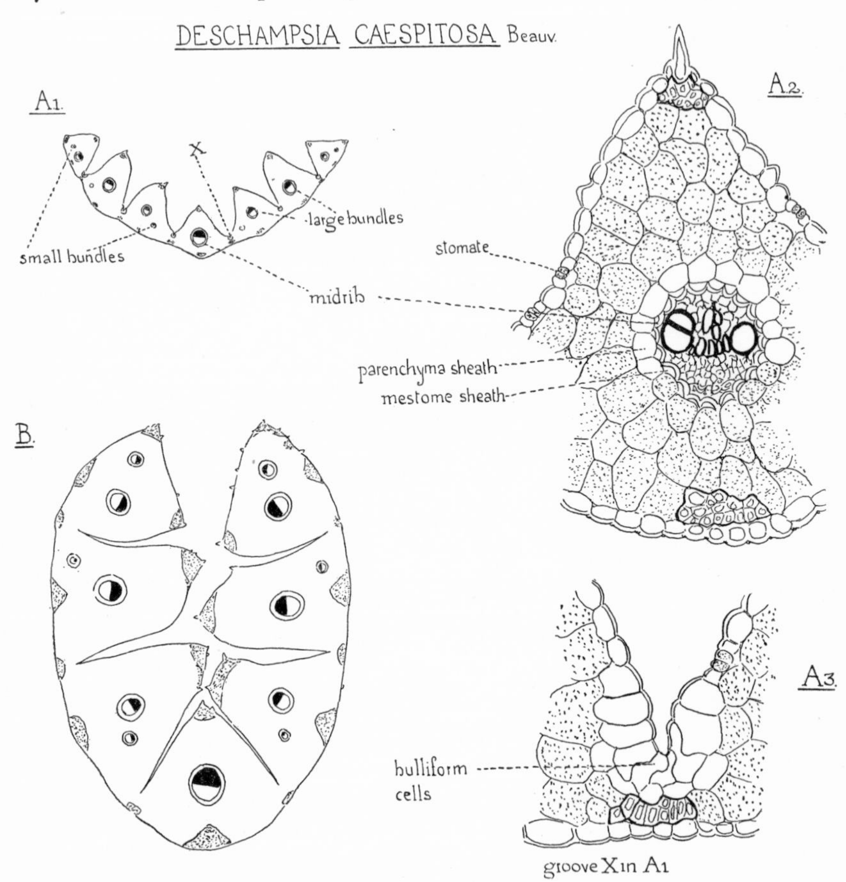

Fig. 156. *Deschampsia caespitosa* Beauv. (*Aira caespitosa* L.) A 1, transverse section of open limb (× 23). A 2, midrib of A 1 (× 193 *circa*). A 3, groove *X* in A 1 (× 318). B, transverse section of closed limb, nearer the base than A 1 (× 47). [A.A.]

and A 2). In the species in which they are well differentiated, their distribution is highly variable. In *Dactylis glomerata* L. there is a single group over the midrib (Fig. 157, B 1); a pair are found in *Poa annua* L. (C 1); while numerous groups occur in other grasses, such as *Festuca elatior* L. (Fig. 158).[1] When the individual cells

[1] See also Fig. 212, A, p. 407.

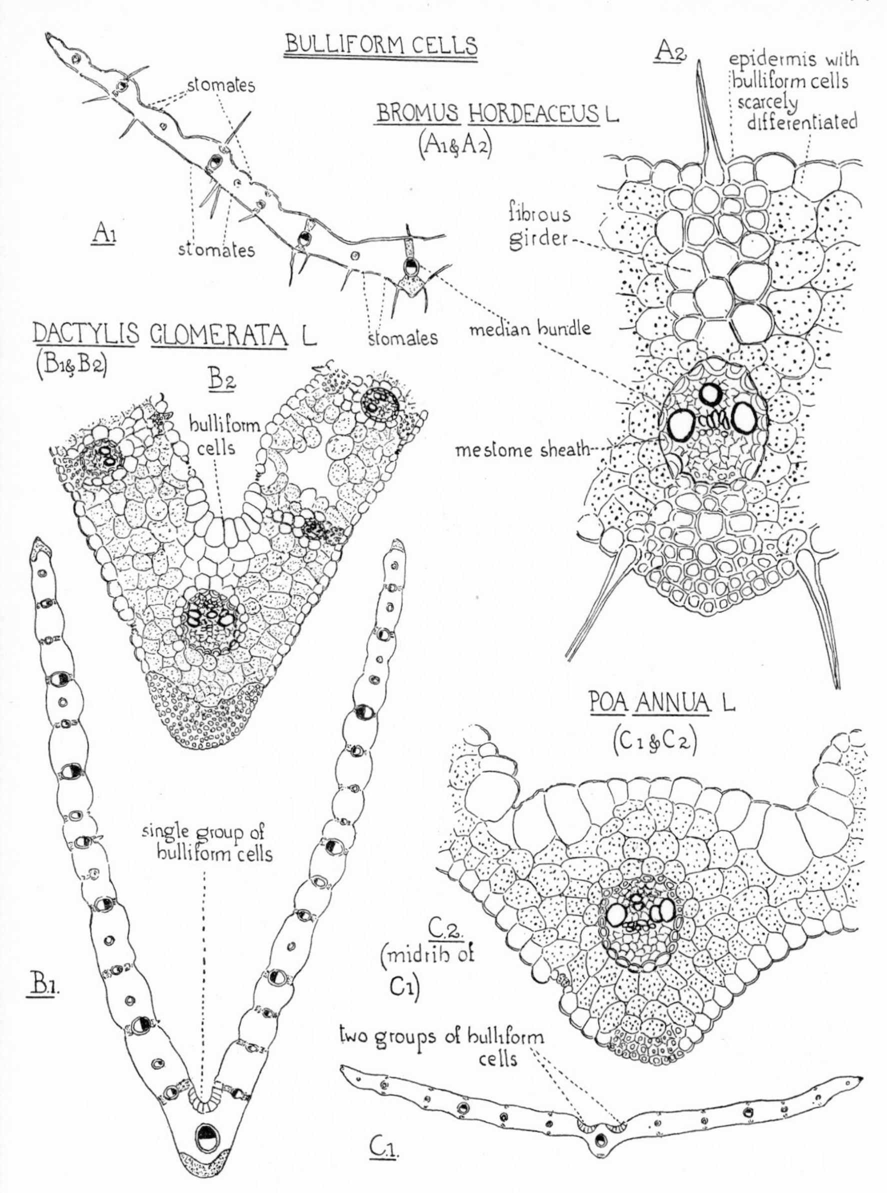

Fig. 157. Bulliform cells. A, *Bromus hordeaceus* L. A 1, transverse section of leaf-limb (× 23); A 2, midrib of A 1 (× 193 *circa*). B, *Dactylis glomerata* L. B 1, transverse section of limb (× 23); B 2, midrib of B 1 (× 77 *circa*). C, *Poa annua* L. C 1, transverse section of limb (× 23); C 2, midrib (× 193 *circa*). [A.A.]

forming each group are numerous, they are usually small; on the other hand, in such a grass as *Tragus racemosus* (L.) All., in which there are only three, they are correspondingly large (Fig. 160, C, p. 304).

Many grass blades open out flat in a damp atmosphere, but roll, or fold, if the air becomes dry. Duval-Jouve,[1] who was the first to

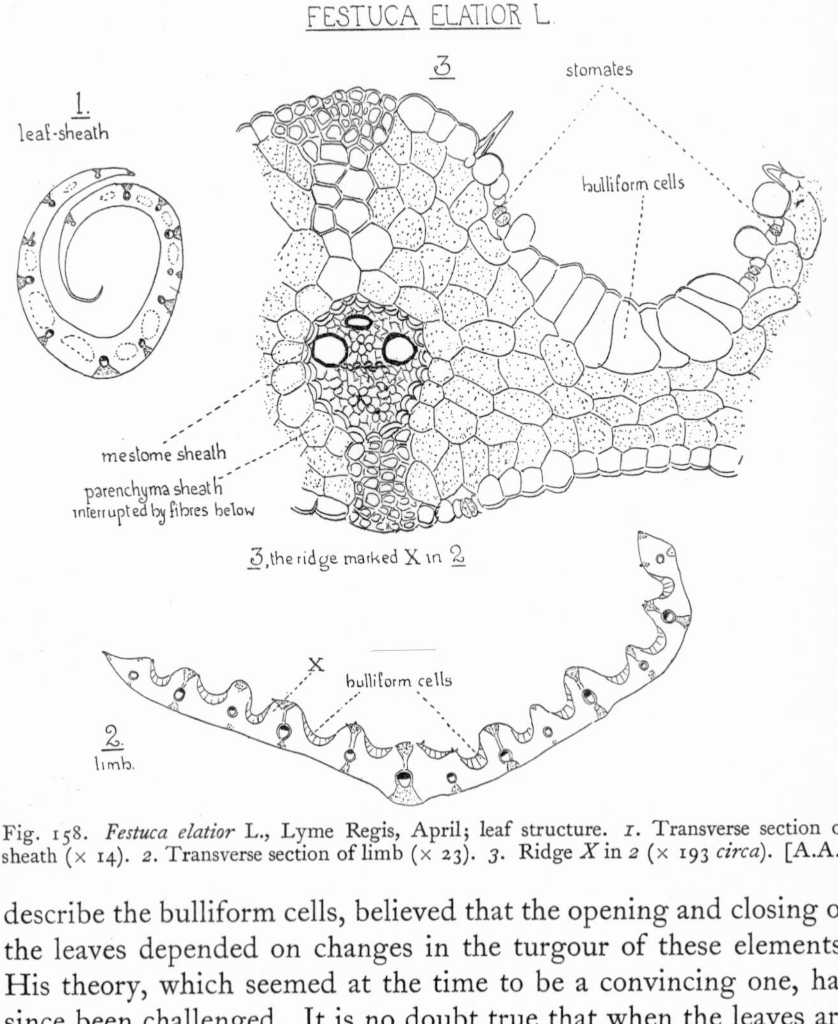

Fig. 158. *Festuca elatior* L., Lyme Regis, April; leaf structure. *1*. Transverse section of sheath (× 14). *2*. Transverse section of limb (× 23). *3*. Ridge *X* in *2* (× 193 *circa*). [A.A.]

describe the bulliform cells, believed that the opening and closing of the leaves depended on changes in the turgour of these elements. His theory, which seemed at the time to be a convincing one, has since been challenged. It is no doubt true that when the leaves are

[1] Duval-Jouve, J. (1875).

open, the bulliform cells are turgid, and that when they are closed, these cells are limp and flaccid; but whether it is the turgour of the bulliform cells which *causes* the movements is another matter. Later writers[1] agree that the rolling or unrolling of the leaves is influenced by the loss or uptake of water, but it appears that the movements are due, at least in some cases, to changes in the leaf fibres rather than in the epidermis; sometimes the movements take place perfectly well after the bulliform cells have been removed. It seems possible that the development of the bulliform cells may be merely an accidental outcome of the particular ontogenetic conditions prevailing at the bases of leaf-grooves. This speculation does not touch, however, on the part they play in the leaf, which will, I think, remain mysterious until botanists and physicists combine to interpret it.

The bulliform cell is only one example of the variety of cell types into which the grass epidermis may be differentiated.[2] As an example showing the complexity to which this layer sometimes attains, we may take *Oryza sativa* L. (Fig. 159, A, p. 302), in which there is a regular distribution of a series of cell forms in longitudinal bands. On either margin of each nerve, there is a file of short elements, in which there is an alternation of suberised cells with others which each include a saddle-shaped siliceous deposit. On either side, beyond these silica chains, is a band of long cells and stomates, while, midway between the nerves, there is a central strip of long cells whose ends are separated by pairs of short elements, each pair consisting of a siliceous cell towards the leaf-tip and a suberised cell towards the leaf-base. A pair of short cells may be replaced by a hair, which is either relatively long and slender, or short with a bulbous base. For these shorter excrescences, the French terms 'aiguillon' and 'denticule' are more appropriate than 'hair'. In grasses in which bulliform cells occur, they are found to be a specialised variety of the long cells. In *Oryza*, the long cells have 'ripple-walls', which also occur in the glumes[3] and leaves of various Gramineae, and in the petals of other

[1] Tschirch, A. (1882); Molliard, M. (1904); Leokadia, L. (1926).

[2] On the epidermis of grass leaves, see Grob, A. (1896), and Prat, H. (1932); the latter contains a fuller account of the examples here summarised.

[3] Winton, A. L. (1903).

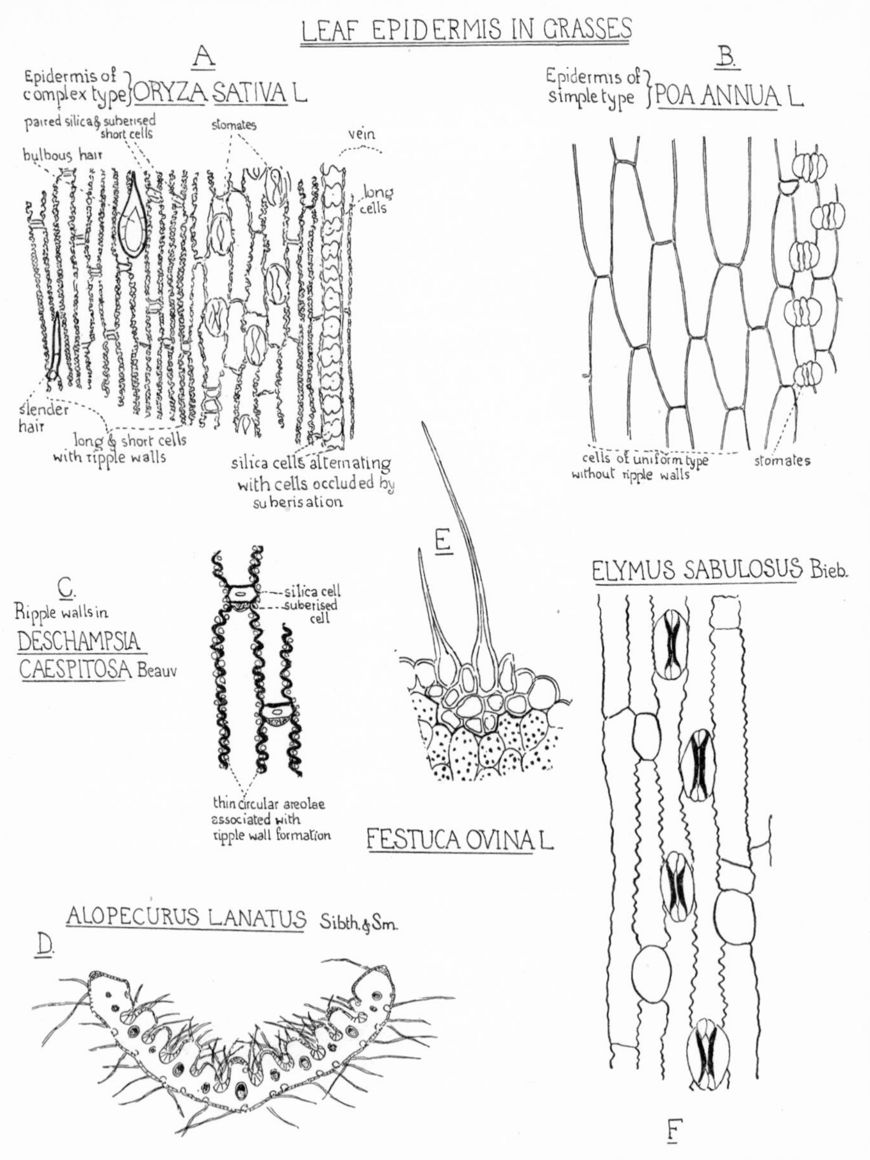

Fig. 159. Epidermis in grass leaves. A, *Oryza sativa* L. Epidermis of outer surface of upper part of leaf-sheath (× 193 *circa*). B, *Poa annua* L., epidermis of inner surface of upper part of leaf-sheath (× 193 *circa*). C, *Deschampsia caespitosa* Beauv., epidermal cells (× 318). D, *Alopecurus lanatus* Sibth. et Sm., transverse section of leaf-limb (× 23); for details see Fig. 154, p. 296. E, *Festuca ovina* L., hairs from upper surface of transverse section of leaf-limb (× 193 *circa*). F, *Elymus sabulosus* Bieb., leaf-epidermis with stomates (× 318). [A.A.]

families.[1] They are shown in surface view in Fig. 159, C, and Fig. 212, B3, p. 407, and in section in Fig. 212, B2. The undulation seems due to cutinisation taking an in-and-out course between thin patches in the cell-wall, which may remain roofed in as cavities. It is not always easy to reconcile the appearance of these ripple-walls in section and in surface view, and a fuller study of their development might be worth making.

Having considered the leaf epidermis of *Oryza* as an example of complexity, we may take that of *Poa annua* L. as a contrasting instance of simplicity; the homogeneous, smooth-walled character of its cells is seen in Fig. 159, B. This grass has been described[2] as being, as far as its epidermis is concerned, a permanently infantile form—a 'fixation' at a juvenile stage.

The epidermal hairs in many British grasses are sparse and relatively short, such as those which we have already noticed in Rice (e.g. *Festuca ovina* L., Fig. 159, E), but some of our native species are much more softly hairy. The American and English names of *Holcus lanatus* L.—Velvet-grass and Yorkshire-fog—both relate to this quality, for 'fog', in this connection, probably means 'moss', as in the proverb, "A ro'ing stane gathers nae fog".[3] An example of extreme hairiness is *Alopecurus lanatus* Sibth. et Sm. of Asia Minor, in which the leaves are clothed with a dense silky covering (Fig. 159, D). In a number of grasses the epidermis secretes wax; in Sorghum, Wheat, Maize and the Sugar-cane,[4] for instance, a whitish, waxy efflorescence may occur, especially between the leaf-sheath and the stem. Some Fescues, also, secrete a granular waxy layer, which frosts over the surface, and gives a peculiar bluish tint.[5]

The stomates of grasses are of a somewhat unusual form, which recurs in the Cyperaceae;[6] they are shown for *Elymus sabulosus* Bieb. in Fig. 159, F. The guard-cells have dumb-bell-shaped cavities, the median regions of which may be almost occluded by the thickening

[1] Hiller, G. H. (1884). [2] Prat, H. (1932).

[3] Pratt, A. (n.d. [1859]).

[4] Piédallu, A. (1923); Artschwager, E. (1925).

[5] Hackel, E. (1882).

[6] Schwendener, S. (1889); for an account of the mechanism with diagrams, see Haberlandt, G. (1914), pp. 451–2.

of the walls. It is the increase in width of the thin-walled cavities at the ends which draws the lips apart; this increase may amount to 3–5 μ.

The assimilating tissue of the leaf shows certain peculiarities. Sometimes the bundles are surrounded by palisade cells arranged in a radiating fashion, e.g. *Tragus racemosus* (L.) All. (Fig. 160, C),[1] while

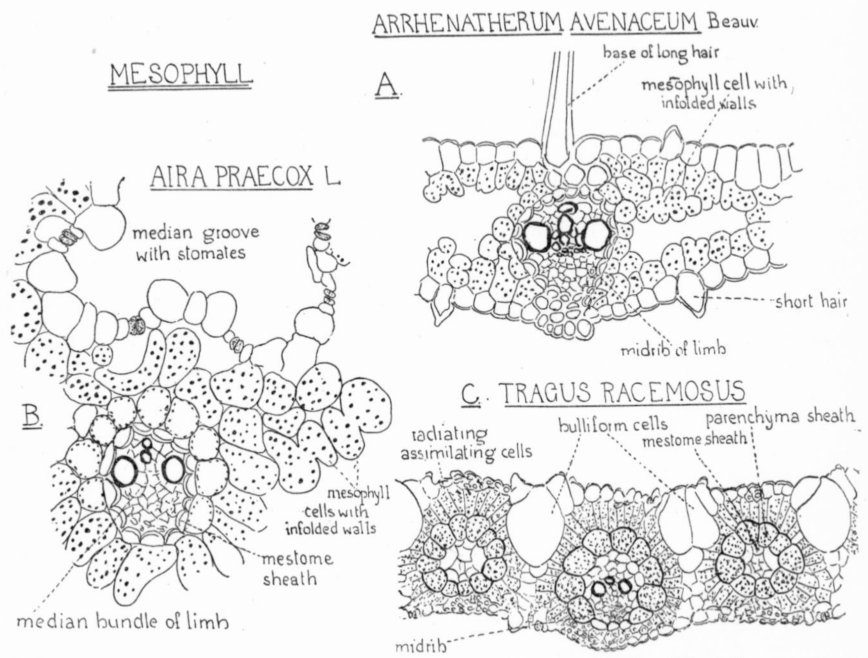

Fig. 160. Mesophyll. A, *Arrhenatherum avenaceum* Beauv., midrib of limb (× 193 *circa*). B, *Aira praecox* L., median bundle of leaf-limb and base of groove above it (× 318). C, *Tragus racemosus* (L.) All., transverse section of part of the leaf including the midrib (× 193 *circa*). [A.A.]

other Gramineae have mesophyll cells with infolded walls, recalling those of the bamboos (pp. 71–2) though they are less elaborated.[2] They are shown for *Arrhenatherum avenaceum* Beauv. and *Aira praecox* L. in Fig. 160, A and B.

Most of the grasses whose leaves are drawn in transverse section

1 The dotting used in this and other leaf sections to indicate green tissue is merely a conventional symbol; no attempt is made to represent the chloroplasts.

2 Karelstschicoff, S. (1868) and Haberlandt, G. (1882).

in the figures illustrating this and other chapters, have two cell layers enclosing the larger vascular bundles: an outer green or colourless *parenchyma sheath*, and an inner *mestome sheath* consisting of cells which often recall the endodermis in being strongly thickened on the inner face.[1] Examples are *Agrostis tenuis* Sibth. (Fig. 192, B, p. 383), *Deschampsia caespitosa* Beauv. (Fig. 156, A 2, p. 298), *Bromus hordeaceus* L. (Fig. 157, A 2, p. 299), etc. In some grasses, however, the mestome sheath is absent and the xylem and phloem abut directly upon the parenchyma sheath, e.g. *Setaria verticillata* Beauv. (Fig. 192, D). Where there is no mestome sheath, the parenchyma sheath may become to some degree thick-walled, as if assuming the characters of the absent layer, e.g. *Panicum Crus-galli* L. (Fig. 192, A), and *Zea Mays* L. (Fig. 188, p. 368). In the minor veins, the parenchyma sheath may consist of elements which are remarkably few and large; the left-hand bundles in Fig. 192, A and D, each have a sheath of four cells only, which are gigantic in proportion to those of the strand which they enclose.

The grasses share with other Monocotyledons[2] the capacity to develop a small amount of secondary tissue in the bundles of their shoots. Xylem elements, which must have had a cambial origin, can be recognised, for instance, in Figs. 115, A 3, p. 241, 143, C, p. 281, and 153, C 3, p. 295. The cross-connections between the veins appear to be established through this secondary tissue (Figs. 143, D, p. 281, and 188, B, p. 368).

One of the snares in the study of grass-leaf anatomy is the fact that quite different results may be obtained, according to the position on the plant of the leaf selected for sectioning. This has long been vaguely recognised, but the principle which underlies these differences was not defined until a Russian worker formulated the generalisation now known as Zalenski's Law,[3] according to which the *anatomical structure of the individual leaves of a shoot is a function of their distance from the root system*.[4] It is noteworthy that a somewhat

[1] Schwendener, S. (1890). [2] Arber, A. (1917), (1918), (1919[2]).

[3] Zalenski, W. von (1902) and Maximov, N. A. (1929).

[4] This is Maximov's formulation. It must be noted that he uses the expression "anatomical structure" in a narrow sense; he is concerned only with the system for the conduction of water, considered in its quantitative aspect.

similar relation has since been reported for the mechanical systems of successive internodes of the stem.[1] The measurable character on which Zalenski founded his results, was the total length of vascular bundles existing in a leaf segment of given area. He found, for instance, that in Cock's-foot-grass, *Dactylis glomerata* L., the total length of the strands corresponding to a unit of leaf surface, increased in the ratio of 1 : 1·68 in passing from the lowest leaf to the seventh. Zalenski also compared different species, and he found that, in general, shade plants had less bundle length per unit area than sun plants. His two conclusions, taken together, indicate that the upper leaves of a shoot—those further from the water supply—possess a more xeromorphic structure than the lower ones.

The physiological anatomy of grass leaves offers a wide field for research, as yet insufficiently explored. The subject has been liable to fall into the hands of teleologists, and thus to come to a dead end. Work such as Zalenski's, which follows statistical lines, is the most hopeful corrective for this tendency.

[1] Roelants, H. W. M. (1922); see also p. 258.

CHAPTER XIV

THE GRAMINEAE AND THE STUDY OF MORPHOLOGICAL CATEGORIES[1]

THE impulse to analyse the plant into component members seems, in the first instance, to have arisen out of the desire to establish a comparison between construction in the animal and the vegetable body; for the existence of a close analogy between the two was a fundamental postulate with the biologists of ancient Greece. The first extant attempt at such an analysis is, in some respects, strikingly alien to modern botanical thought. It is that of Theophrastus[2] who, in the fourth century B.C., stated that "the primary and most important parts...are these—root, stem, branch, twig; these are the parts into which we might divide the plant, regarding them as members, corresponding to the members of animals: for each of these is distinct in character from the rest, and together they make up the whole". Theophrastus then proceeds to distinguish, as subsidiary parts, the leaf, flower, fruit, etc. He was influenced in this discrimination by the fact that, in the tree, which he took as the standard of plant life, the trunk and its branches persist permanently, whereas the leaf, flower and fruit are ephemeral. The importance of the leaf was destined to remain for long unrecognised, and it was not until Goethe[3] turned his attention to botany, more than two thousand years later, that the equivalence of the foliage leaves and the parts of the flower came fully into the light. Yet, inadequate as the botanical analysis of Theophrastus may seem to us now, certain of its features have shown remarkable endurance: these long-lived notions are the conception of root and stem as primary organs, and the idea of the leaf as something wholly distinct from the stem.

[1] This chapter is based upon Arber, A. (1930[2]) where a fuller treatment with additional references will be found.

[2] Theophrastus (Hort, A.) (1916).

[3] Goethe, J. W. von (1790).

The discreteness of these units—*root*, *stem*, and *leaf*—was reiterated by A. P. de Candolle,[1] and it became the basis of the rigid morphology of the nineteenth century. The botanical history of the three categories has not, however, been uniform. The earlier botanists took for granted the importance of the root as an entity. They looked on plants primarily as the source of drugs, and the root was often the region to which they attached most value, and to which their attention was chiefly directed; the Greek herbalists were in consequence known as *ῥιζοτόμοι*, root-gatherers. In more recent times, however, when botany began to free itself from the shackles of medicine, the root lost this extraneous interest, and, being out of sight, tended to slip out of mind. Goethe, for instance, practically ignores the root in his botanical philosophy, while the majority of modern writers give little attention to it, though adopting it implicitly as a morphological unit. The position as regards stem and leaf is less simple; indeed, when we try to understand what has been thought of the status and relations of these two members, it is only too easy to get lost in a mist of controversy. One school maintains the view that neither stem nor leaf is a valid category, but that the plant is formed of a series of individuals (phytons), each consisting of an internode with its upper node and the leaf there attached; a root, generally secondary, may be associated with the base of each phyton. In the words of one of the advocates of this view,[2] "every stem is essentially built up of a sympodially developed succession of phyton units. The phyton is the true *Individual*". Another school of thought holds that the stem and leaf are independent categories, but that the true stem is a mere core, and that it is entirely enveloped by a cortex (pericaulome) derived from the leaf-bases. A third hypothesis, that of Lignier, will be discussed at a later point, since it is evolutionary, rather than purely morphological.

Considering that almost all botanists are concerned to some degree with the flowering plants, it is surprising that so few have expressed a clear idea of any kind about the relations of stem and leaf. In modern text-books one generally finds that the writers are content to

[1] Candolle, A. P. de (1827). [2] Worsdell, W. C. (1915–16).

drop the 'root', 'stem' and 'leaf' units, without substituting anything else for them.

In the preceding paragraphs, a few of the main standpoints of formal morphology have been indicated rapidly and without criticism. A closer analysis of the plant body is needed before we can evaluate the various morphological categories which have been proposed, or bring these concepts of formal morphology into relation with those of evolutionary morphology.

We have shown that the idea of a morphological distinction between stem and leaf was originally suggested by the deciduous character of foliage; but leaf detachment often occurs at a level which obviously cannot be taken to coincide with a morphological boundary. Many such erratic instances might be cited, but it will suffice here to recall one example from among the Gramineae; in the bamboo, *Arundinaria falcata* Nees, the disarticulation occurs *above* the base of the leaf-sheath, which is left as a collaret surrounding the axis.[1] If the simple plan of taking the absciss layer to mark the limit between stem and leaf thus proves to be inadmissible, where are we to set this limit? Alexander Braun,[2] more than eighty years ago, wrote of the "essential interconnection" of these two organs, and since his time no one has succeeded in delimiting them. Casimir de Candolle[3] suggested that the morphological unit was not the leaf, but *leaf+leaf-base+leaf-trace*. This formula may seem hopeful at first glance, but its application leads one inextricably into difficulties. We find, for instance, that the median bundle of a leaf in a bamboo[4] may descend for five or six internodes, in Maize[5] for six internodes, in the Sugar-cane[6] for about eight internodes, before it attaches itself to the shoot system, while in the minor bundles, on the other hand, such unions take place progressively earlier in the downward passage—the smaller bundles remaining free for one internode only, or even less. If the boundary between leaf and axis is to be determined by the traces, where are we to place it in such plants as these?

[1] Godron, D. A. (1880). [2] Braun, A. (1853). [3] Candolle, C. de (1868).
[4] Shibata, K. (1900); this author does not, however, describe the behaviour of the minor bundles.
[5] Strasburger, E. (1891). [6] Artschwager, E. (1925).

Nor is any settlement of the problem achieved by those who adopt the pericaulome or related theories. These writers are clear about the *outer* limit of the leaf tissue, which on their view coincides with the entire shoot surface, but they offer no clue which will help us to discover the *inner* limit which separates the foliar tissues from that central region to which they confine the term 'stem'. Others avoid the problem of boundaries by the simple expedient of dethroning the stem altogether. They take from it even its last stronghold, the axial core, and regard the region commonly called the stem as consisting merely of fused leaf-bases. On the opposite side, certain botanists, instead of treating the whole external surface as foliar, are prepared to allot to the stem various regions commonly attributed to the leaf. Hochstetter[1] and Bugnon,[2] for instance, treat the so-called leaf-sheath of grasses as a local transformation of the stem. The close relation of leaf-base and axis in this family is shown by the fact that the grass pulvinus may consist of leaf-sheath tissue, or haulm tissue, or a combination of both (pp. 256–7 and Fig. 124, p. 257).

We may conclude from this survey that no unanimity has ever been reached on the question of exactly what regions of the plant body are to be attributed respectively to stem and leaf; and even when we study members about whose stem or leaf nature there is no dispute, we frequently find that the characters which are supposed to differentiate these two entities prove, on close scrutiny, to be invalid. There are many such instances in botanical literature, but as the present book deals with the grasses alone, I will draw my illustrations mainly from this group; I have little doubt, however, that corresponding examples would be found in any family which was studied intensively.

One of the postulates of formal morphology is that a leaf cannot be terminal to a stem. We know that beside the base of the uppermost leaf borne by any axis, there is usually to be found an apical cone, representing the axis to which this uppermost leaf is lateral. In the Monocotyledons, this cone is sometimes reduced to very small dimensions; elsewhere I have figured sections of the shoot-tip of

[1] Hochstetter, C. F. (1847). [2] Bugnon, P. (1924[2]).

Uvularia, in which it is minimal.[1] Furthermore, this reduction may go so far that the axis vanishes entirely. In five species of bamboo, representing three genera, I have found, on examining the spikelet structure by means of serial sections, that the entire apex of the

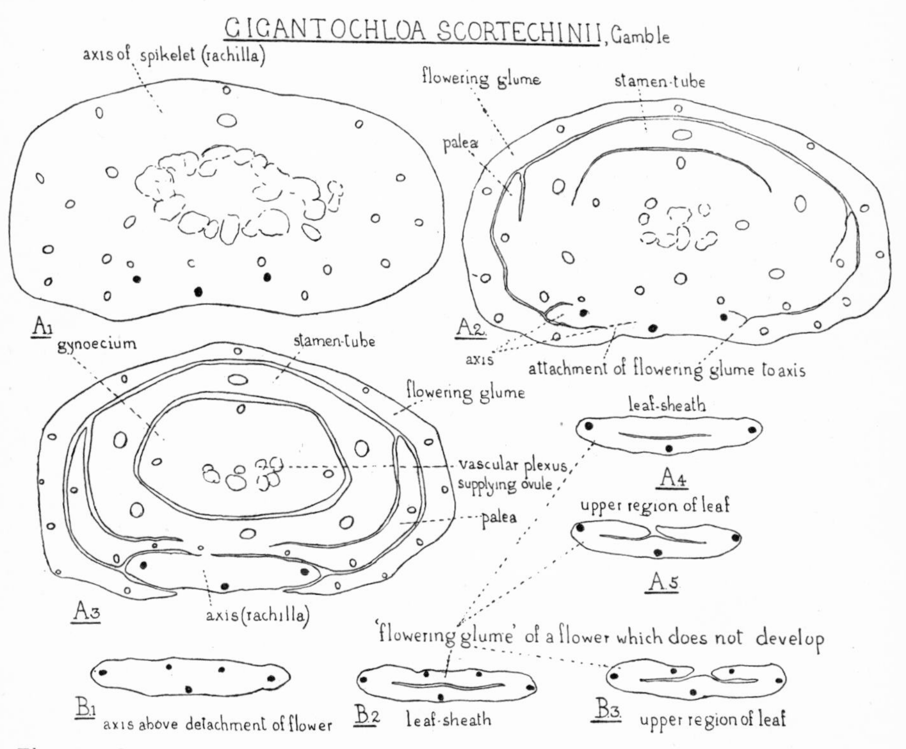

Fig. 161. *Gigantochloa Scortechinii* Gamble. A 1–A 5, sections from a transverse series through the apical region of a spikelet (× 42 *circa*). A 1, axis below uppermost flower; A 2 and A 3, basal region of flower; A 4 and A 5, the further history of the axis in A 3, showing its conversion into a leaf. B 1–B 3, sections (× 42 *circa*) from a transverse series through another spikelet to show development of leaf from apex of axis. (Throughout the figures the bundles which supply the continuation of the axis above the flower and the leaf into which it is transformed, are indicated in black.) [Arber, A. (1928[1]).]

rachilla (spikelet axis) is transformed into a leaf (lemma or flowering glume); no trace of apical tissue is left over, so there is no axis to which this uppermost leaf can be interpreted as lateral. This transformation is illustrated for a species of *Gigantochloa* in Fig. 161.

[1] Arber, A. (1925), Fig. xxxv, p. 58.

Another dictum of formal morphology is that the power of producing lateral shoots is confined to axes. When a leaf does, in fact, bear a shoot-bud, this shoot is described as 'adventitious', which means, literally, 'accidental'. This is a typical example of the tyranny exercised by words over thought; just because they have themselves labelled these buds 'accidental', botanists feel justified in dismissing them as of no morphological significance. There are certain instances, however, in which leaves produce shoots with such precision and regularity that we can scarcely shut our eyes to their behaviour, even though it is strikingly inconvenient from the point of view of the rigid demarcation of stem and leaf. 'Nepaul-barley' is a case in point. In this strange form (*Hordeum trifurcatum* Jacq.),[1] whose structure is illustrated in Figs. 162–5, the bract (lemma) in whose axil the flower arises, produces one or more accessory spikelets from near the apex of its upper (ventral) surface. This peculiarity is not casual or sporadic, but it is heritable, and definitely characterises the race. Those to whom the canons of morphology are sacred, explain these accessory spikelets as buds whose position is axillary to the flowering glume (lemma), but which have been carried up on its surface to a distance from their point of origin; but detailed study lends no support to this view. The orientation of the one-flowered accessory spikelet is always the same—its palea is addorsed to the ventral surface of the flowering glume, and its rachilla lies between the two; this is not what we should expect if the glume were the axillant leaf. There is not, moreover, the slightest anatomical indication that the lemma below the origin of an accessory spikelet consists of axillant leaf and bud fused together; on the contrary, the accessory spikelets receive their vascular supply direct from the simple collateral strand which forms the median bundle of the lemma. We can but conclude that the flowering glume, which is a *leaf* member, behaves to the accessory spikelet in all respects as if it were that spikelet's parent *axis*. And so we return—as we so often must—to the standpoint of Goethe,[2] whose morphological insight led him long ago to recognise the "fertility which lies hidden in a leaf".

1 Arber, A. (1929²).
2 Goethe, J. W. von (1790).

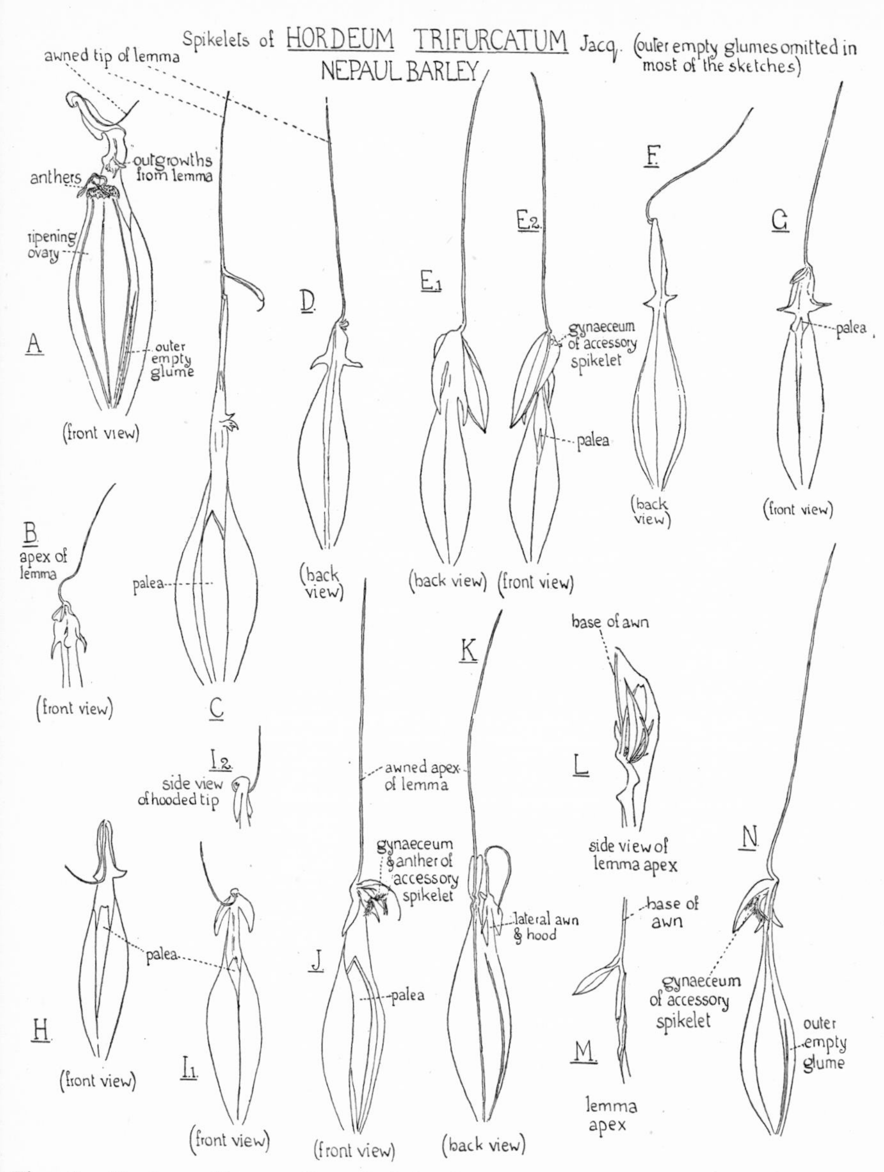

Fig. 162. *Hordeum trifurcatum* Jacq. Spikelets from plants grown in the Cambridge Botanic Garden from seed from Mr R. T. Pearl, South-Eastern Agricultural College, Wye (enlarged). [A.A.]

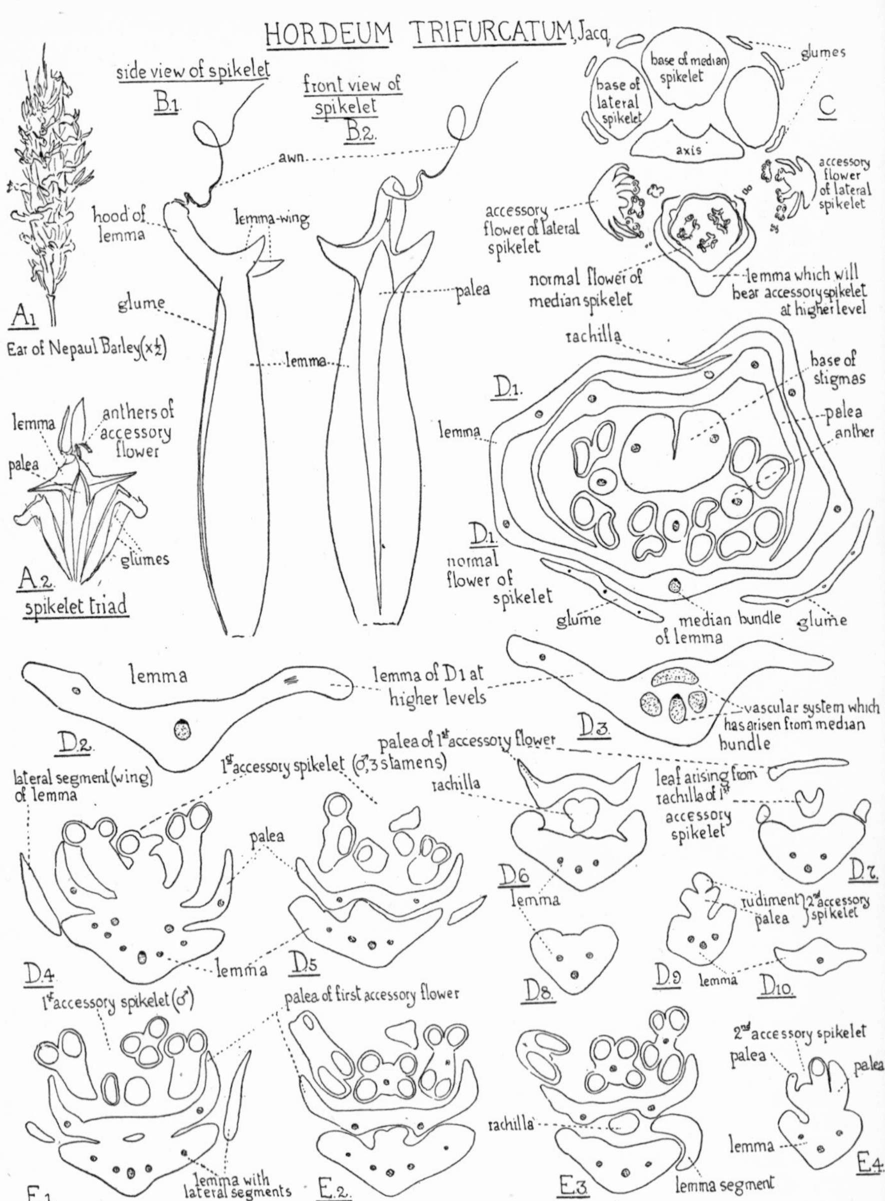

Fig. 163. *Hordeum trifurcatum* Jacq., Nepaul-barley. A 1, inflorescence (× ½ *circa*). A 2 and B, spikelets enlarged. C, transverse section (× 14) passing through two opposite spikelet triads. D 1–D 10, series of transverse sections from below upwards (× 48) through a lateral spikelet. E 1–E 4, series of transverse sections (× 48) through the accessory spikelets of a lateral spikelet. [For full legend, see Arber, A. (1929²).]

It is generally held that the stem conforms to radial symmetry, while the leaf is symmetrical about the antero-posterior plane only. Striking exceptions to the radial symmetry of the stem are found, however, among grass inflorescences. For instance, the rachis of *Tripsacum dactyloides* L. (Fig. 189, B 1, p. 370), the branches of the first order in the inflorescences of *Luziola Spruceana* Benth.[1] (Fig. 166, B 2, p. 317, Fig. 167, G, p. 318, and Fig. 168, A, p. 319) and the rachillae of *Arthrostylidium longiflorum* Mun. (Fig. 169, B, p. 320),

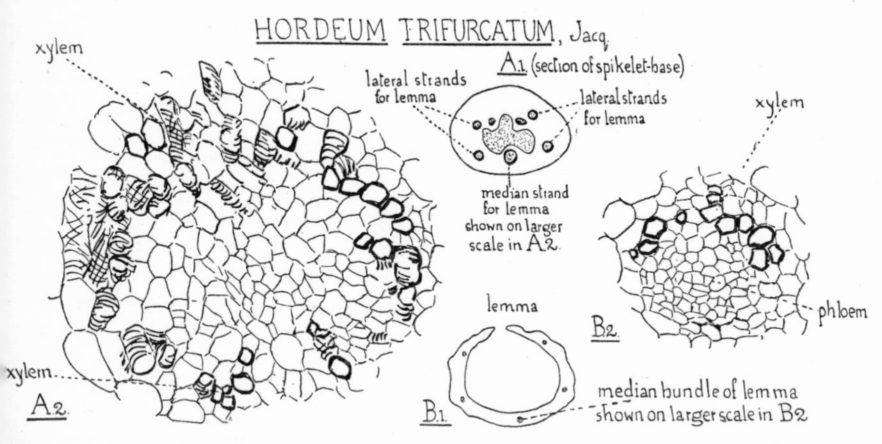

Fig. 164. *Hordeum trifurcatum* Jacq. A 1, transverse section of the base of the median spikelet of a well-developed triad below detachment of lemma (× 14) to show the five concentric bundles destined for the lemma. A 2, the median bundle of the lemma in A 1 on a larger scale (× 327) to show concentric structure. B 1, transverse section close to the base of the free region of the lemma (× 14). B 2, collateral median bundle of the lemma of B 1, on a larger scale (× 327). [Arber, A. (1929²).]

show an imperfect dorsiventral symmetry, both in form and anatomy. In the plan of their skeletal system these axes recall certain petioles. The same tendency to dorsiventrality may also be traced in other axes which are less richly vascular. In the Hordeums,[2] for instance, the inflorescence axis is liable to be strongly dorsiventral. The bundles may be arranged in a flattened ellipse with two main strands situated at the foci, e.g. *Hordeum jubatum* L. (Fig. 78, A 1, p. 173). Or, instead of an ellipse there may be a single row of bundles, as in *H. pratense* Huds. (Fig. 78, B 6); any anatomist who was shown an[3]

[1] Arber, A. (1928²). [2] Arber, A. (1929²). [3] Text continues on p. 321.

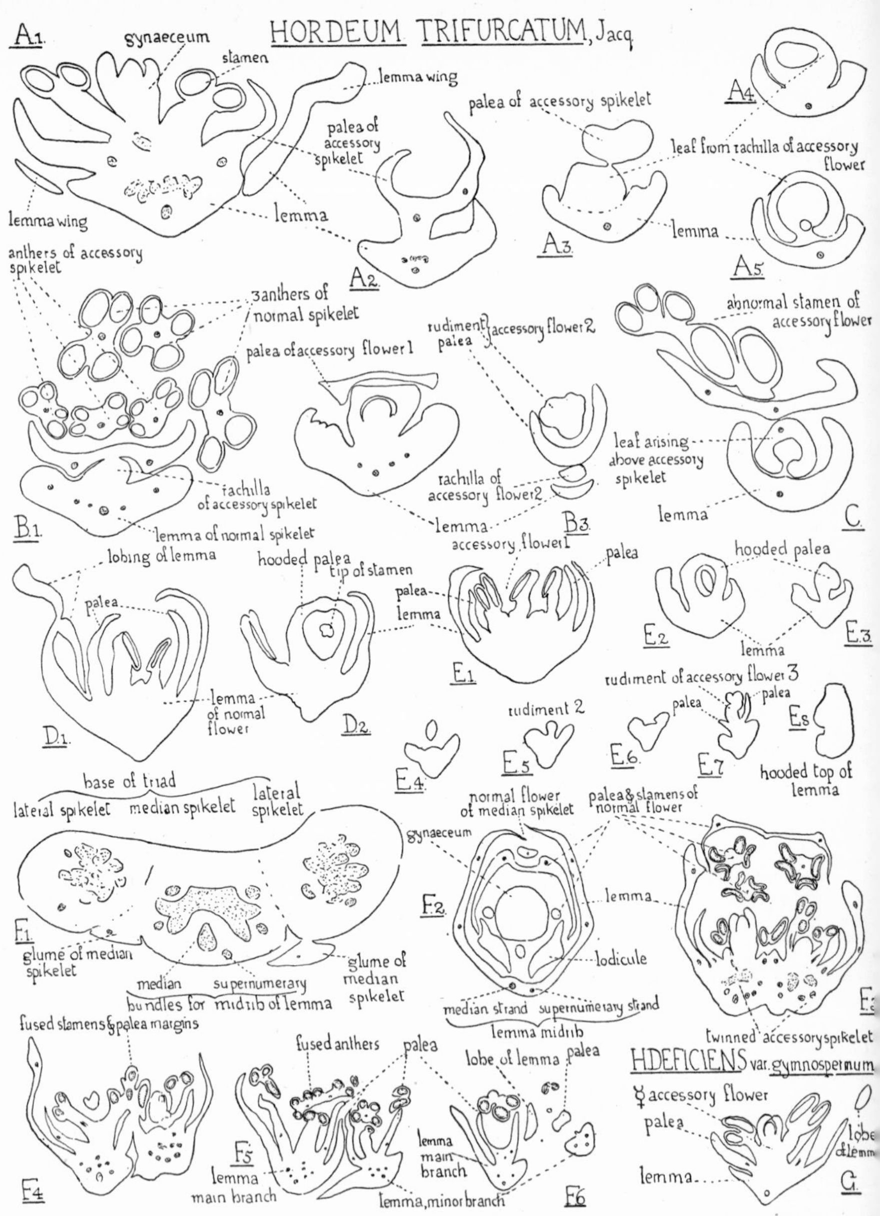

Fig. 165. A–F, *Hordeum trifurcatum* Jacq. A 1–A 5, series of transverse sections (× 48) through the young accessory spikelet of the central spikelet of a triad. B 1–B 3, transverse sections (× 48) from a series through the accessory flowers of a lateral spikelet. C, transverse section (× 48) of the accessory flower of a median spikelet to show a stamen with an extra pollen-sac. D 1 and D 2, transverse sections through an accessory spikelet (× 24) to show hooding of palea and lobing of side of lemma. E 1–E 8, series of transverse sections through the tip of a lemma bearing accessory flowers (× 24). F 1–F 6, series of transverse sections showing the origin of two hermaphrodite accessory spikelets from the apex of a single lemma (× 24). G, *H. sat. deficiens* Steud. var. *gymnospermum*. Transverse section of an accessory hermaphrodite flower borne by the lemma of a median spikelet (× 24). [For full legend, see Arber, A. (1929[2]).]

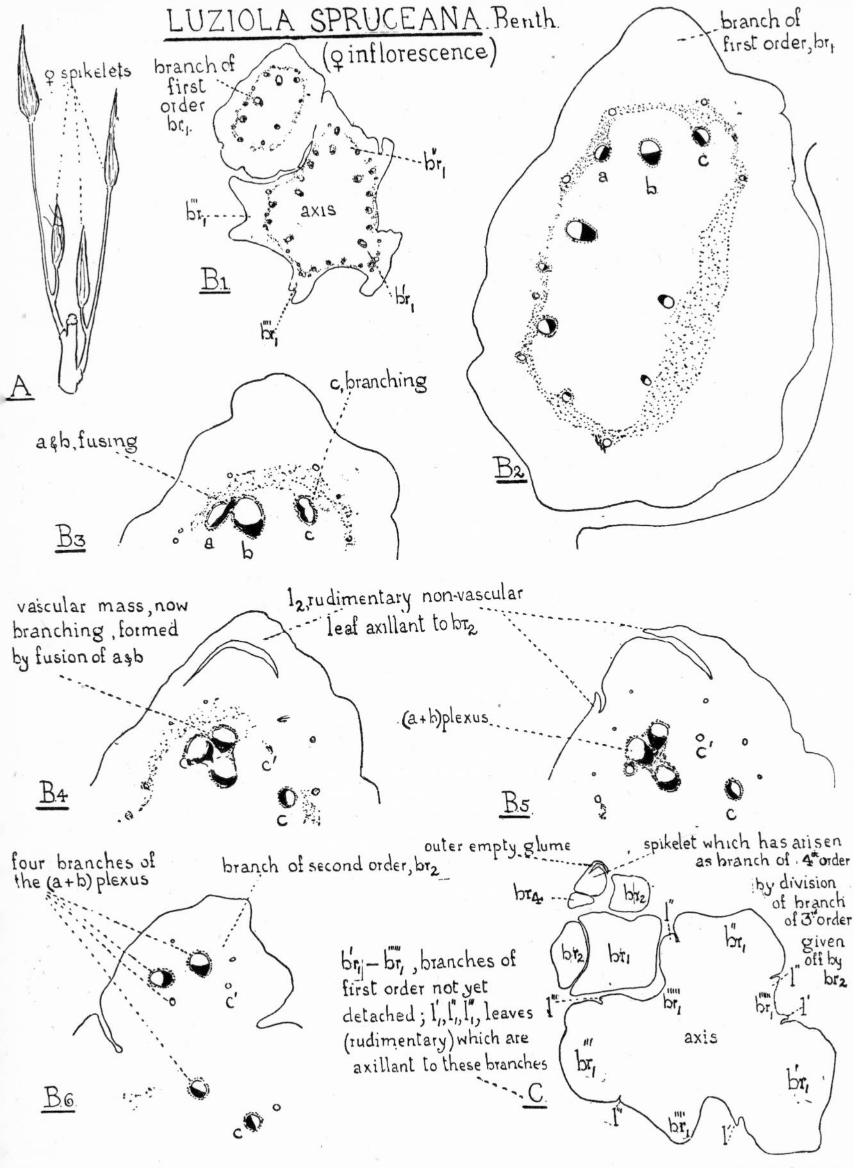

Fig. 166. *Luziola Spruceana* Benth., British Guiana. A, a very small part of a female inflorescence (enlarged). B and C, transverse sections of a small part of one female inflorescence. B1 (× 14); B2–B6 (× 47); C (× 14). [For full legend, see Arber, A. (1928[2]).]

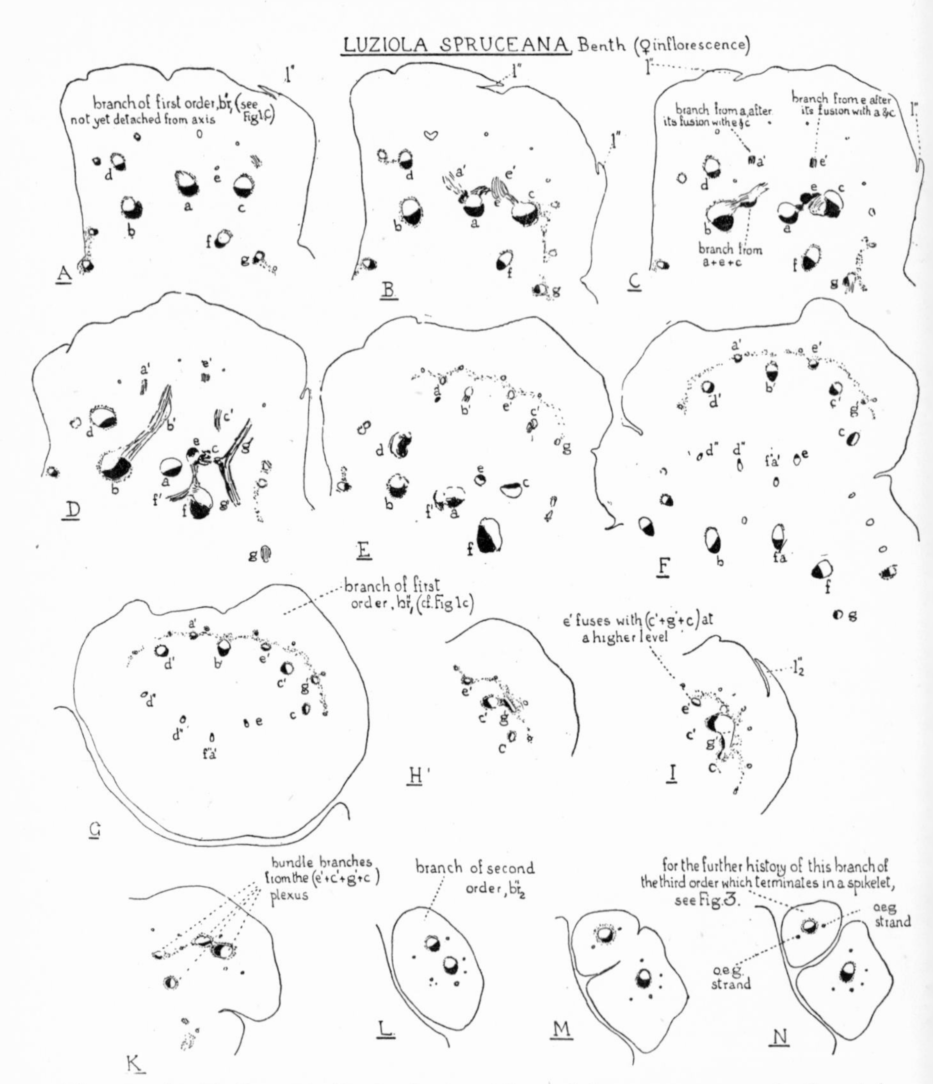

Fig. 167. *Luziola Spruceana* Benth. Sections (all × 31) from a transverse series from below upwards through the origin, and first and second branchings of the branch $br.''_1$, which is seen below its level of detachment in Fig. 166, B 1 and C, p. 317. (The Fig. 1 referred to in A and G is Fig. 166, p. 317, and the Fig. 3 referred to in N is Fig. 168, p. 319.) [For full legend, see Arber, A. (1928[2]).]

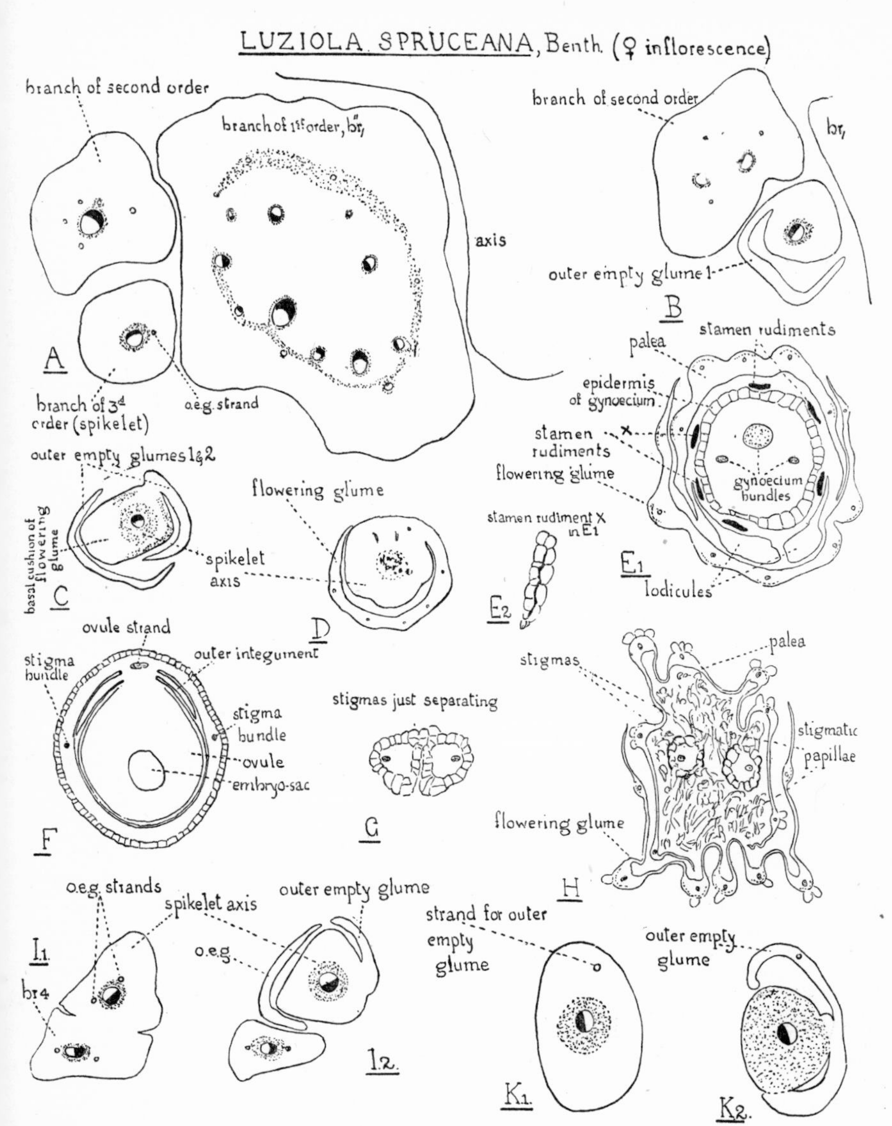

Fig. 168. *Luziola Spruceana* Benth. A–H, transverse sections from a series from below upwards to show a female spikelet and its origin (all except E 2 × 47; E 2 × 318). I 1 and I 2, sections (× 47); I 1, just below, and I 2, just above Fig. 166, C, p. 317, to show the two axes at the top left-hand corner of that figure, which are marked respectively *spikelet* and *br.* 4. K 1 and K 2, sections through the base of a spikelet (× 47) to show a single outer empty glume with a bundle —the only example of a vascular outer empty glume which was observed. [For full legend, see Arber, A. (1928²).]

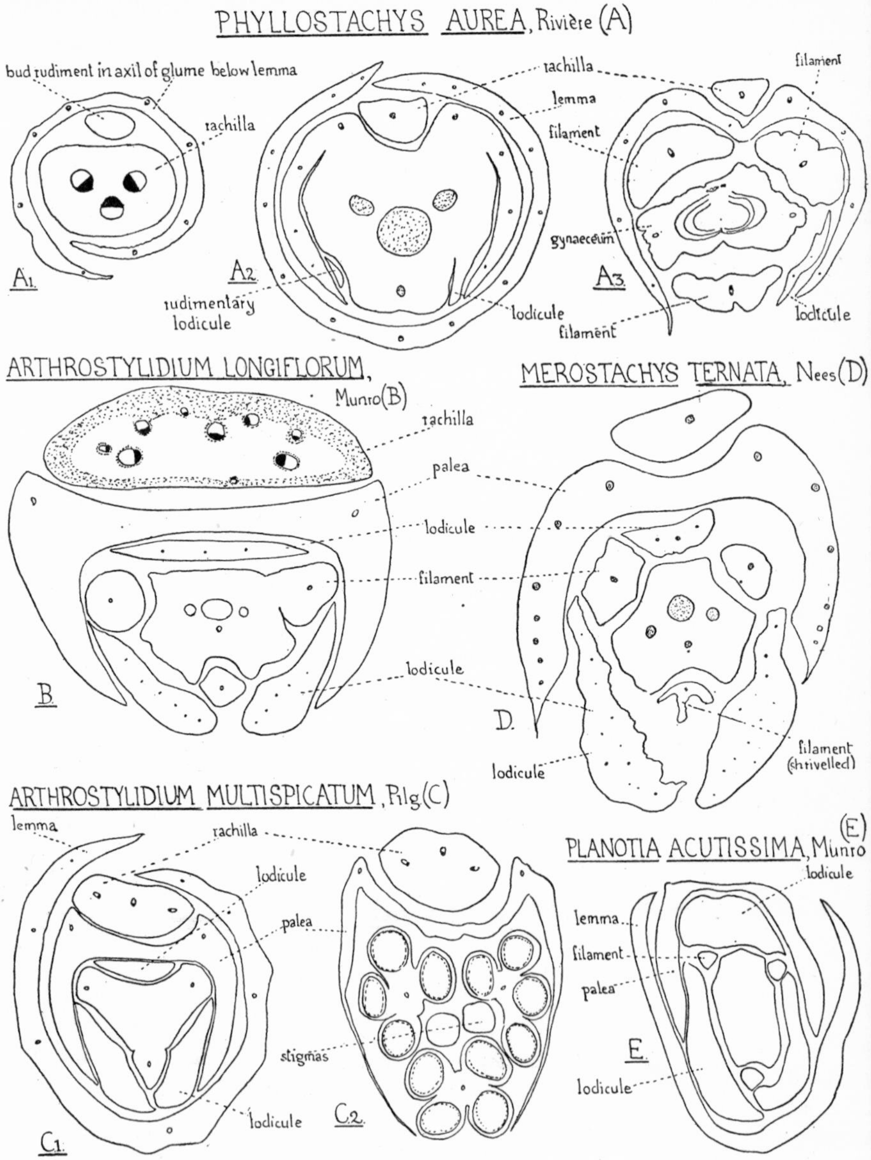

Fig. 169. Three-stamened bamboos. All diagrams from transverse series from below upwards through spikelets (× 47, except C 1 and C 2, which are × 77). A 1–A 3, *Phyllostachys aurea* Riv., Bot. Gard., Montpellier. B, *Arthrostylidium longiflorum* Mun., Colombia. C 1 and C 2, *Arthrostylidium multispicatum* Pilg. D, *Merostachys ternata* Nees, Brazil. E, *Planotia acutissima* Mun., Colombia. [Arber, A. (1929[3]).]

isolated section of an internode of the inflorescence axis in this species, might well suppose that he was looking at a foliar member.

Loss of radial symmetry is particularly noticeable in those grass shoots which show the greatest reduction in anatomy; the sterile spikelets of the Dog's-tail-grass, *Cynosurus cristatus* L.[1] (Fig. 170, B 1, p. 322), for instance, each consist of an axis bearing distichous leaves, but the vascular system is reduced to its lowest possible limit. The axial region is traversed by a single collateral bundle, which gives off leaf-traces on either flank (E and F). Among the grasses, such extreme vascular reduction is by no means confined to sterile spikelets. I have met with examples in which the stalk of a fertile one-flowered spikelet is supplied by means of a single collateral bundle alone, e.g. both male and female spikelets of *Luziola Spruceana* Benth. (Figs. 171, C8, p. 323, and 168, I2, p. 319), and hermaphrodite spikelets of *Alopecurus pratensis* L. (Fig. 83, A4, p. 179). A rachilla, moreover, may be one-bundled, even when it is destined to bear a series of flowers. In *Poa annua* L., for instance, there is a single collateral bundle in the spikelet axis, throughout the flowering region (Fig. 100, p. 202). A rachilla of this type, whose vascular system is reduced to a single vascular bundle, has become definitely dorsiventral in structure, and is thus foliar rather than axial in the character of its anatomy.

One-bundled rachillae, when seen in transverse section, may exactly simulate the awns of glumes, which are always treated as leaf members. This is shown for *Lamarkia aurea* Moench in Fig. 172, p. 324, cf. F 1, F 2. A one-bundled rachilla in association with the palea of the flower which it bears (e.g. *Hordeum distichon* L. var. *nigrum*, Fig. 67, B, p. 151) may closely resemble an awn with the bikeeled glume to which it belongs (e.g. *Anthoxanthum odoratum* L., Fig. 85, A7, p. 182, glume 4 with its awn). Such comparisons obviously cannot be pressed far, but they at least show that structures which rigid morphology regards as purely foliar, can, on occasion, develop on much the same lines as structures in which the axis is held to play a part.

Its radial symmetry is not the only distinguishing feature which

[1] Arber, A. (1928[1]).

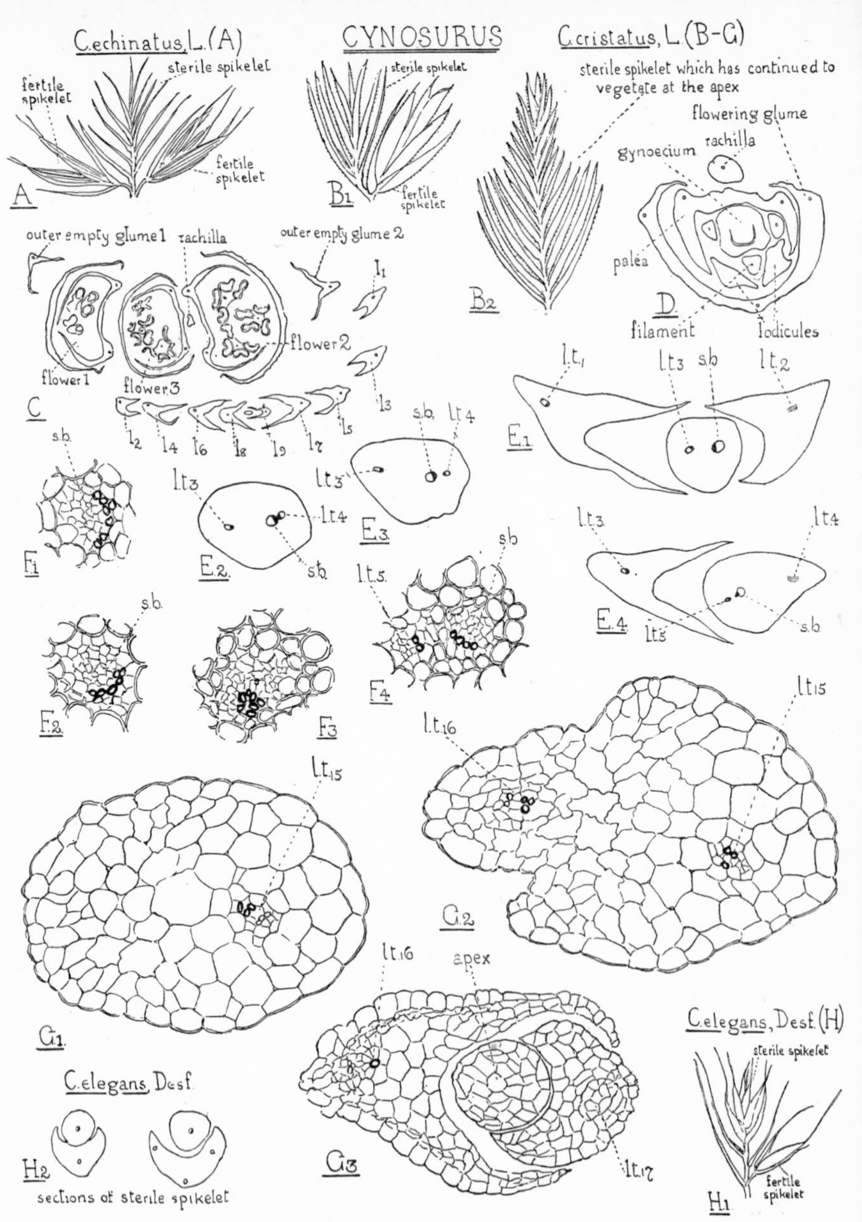

Fig. 170. *Cynosurus*. A, *C. echinatus* L., fascicle of two fertile and one sterile spikelet (enlarged). B–G, *C. cristatus* L. B 1, pair of young spikelets (enlarged). B 2, sterile spikelet in which vegetative growth has continued at the apex. C, transverse section of a fertile spikelet and its associated sterile spikelet bearing leaves l_1–l_9 (× 23 *circa*). D, transverse section of a flower (× 47). E 1–E 4, sections from a transverse series (× 47) through a sterile spikelet to show the vascular supply to the leaves. F 1–F 4, drawings on a larger scale to show the changes in the shoot-bundle, *s.b.*, between E 3 and E 4 (× 77). G 1–G 3, transverse sections from a series through the apex of a sterile spikelet (× 318). H, *C. elegans* Desf. H 1 (enlarged); H 2 (× 23). [For full legend, see Arber, A. (1928[1]).]

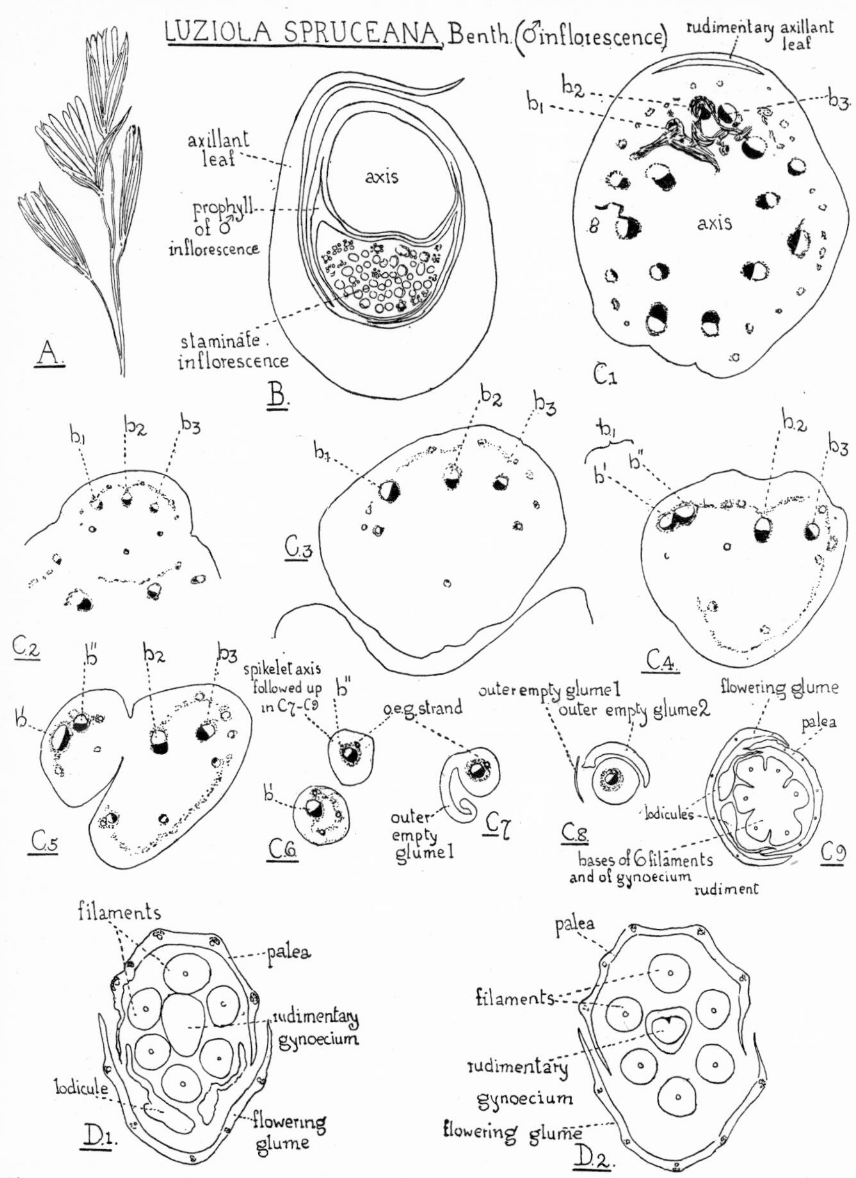

Fig. 171. *Luziola Spruceana* Benth. Male inflorescence. A, a few male spikelets (enlarged). B, transverse section through a young male inflorescence showing axillant leaf and prophyll (× 14). C 1–C 9, transverse sections from a series from below upwards through part of a male inflorescence; C 1 and C 2 (× 23); C 3–C 9 (× 47). C 1, main axis; C 9, base of flower. D 1 and D 2, two transverse sections (× 77) from a series from below upwards through a male spikelet to show the characters of the flower. [Arber, A. (1928[2]).]

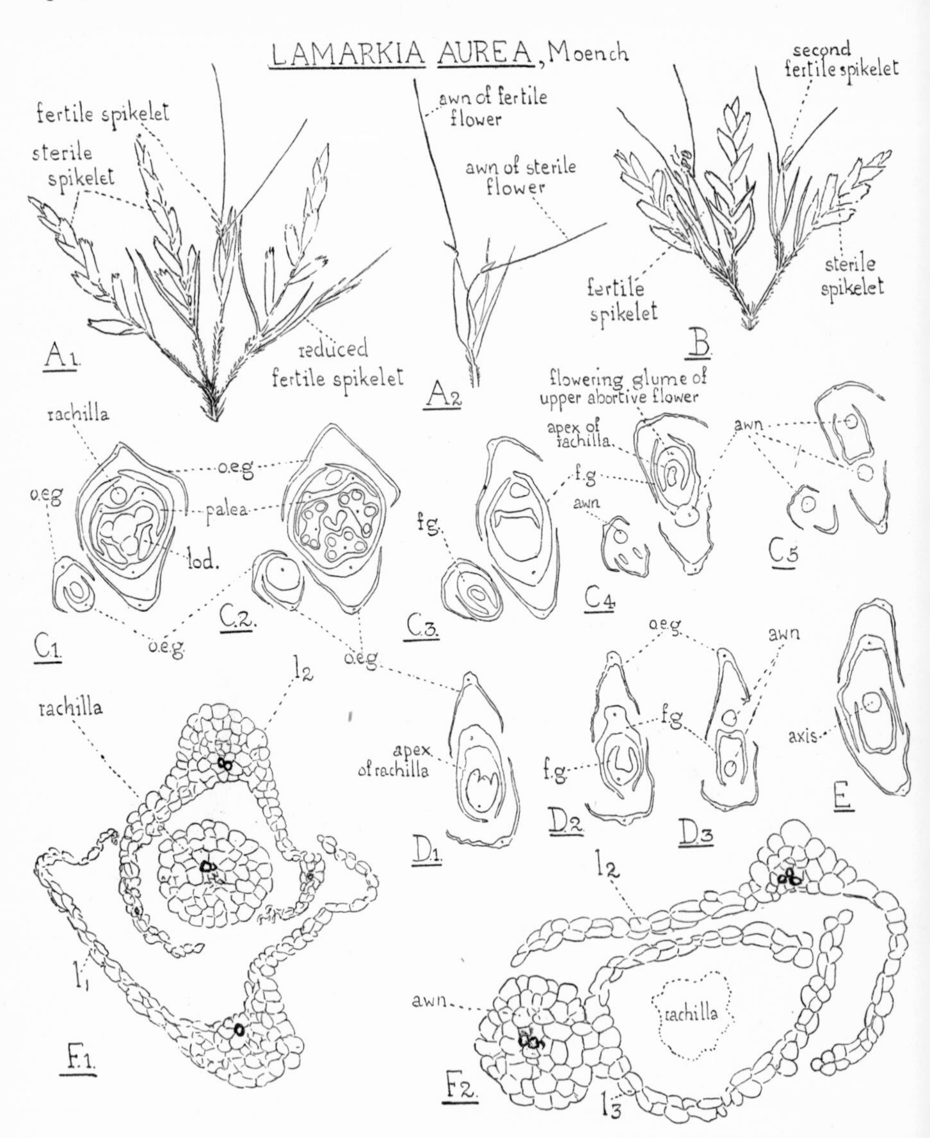

Fig. 172. *Lamarkia aurea* Moench. A 1, A 2, and B, partial inflorescences (enlarged). C 1–C 5, serial sections (× 95 *circa*) through two young spikelets, one (to the right) fertile, and the second, sterile. D 1–D 3, transverse sections of a young example of a reduced fertile spikelet like the one shown in A 1 (× 95 *circa*). E, transverse section (× 95 *circa*) of the top of a young, purely sterile spikelet with no awns. F 1 and F 2, transverse sections (× 390 *circa*) through an extremely small sterile spikelet, bearing three leaves only. [For full legend, see Arber, A. (1928[1]).]

the axis may lose. It may also forfeit its anatomical predominance in relation to the leaf. This happens, for instance, in the abortive lateral spikelets of *Hordeum distichon* L. (Fig. 173, B–D, p. 326). These spikelets consist of an axis bearing a few small leaves, equivalent to the glumes of the fertile spikelets. Two or three bundles enter the base of the minute shoot, but when they are followed up, they all prove to be leaf-traces for the first glume. A bundle for the continuation of the axis is given off from the median bundle of this glume, or even from one of its minor bundles. In the latter case, it is, indeed, strictly in accordance with the facts to speak of the vascular supply for the upper region of the axis as given off from the lateral bundle of a leaf; but such a description would be an infringement of the conventional canons of morphology. In this example, it is the leaf that usurps the status of the axis; but the converse may also occur. In grass inflorescences there is a strong tendency to leaflessness, and, in correlation with this, we find examples in which leaf-traces are replaced by branch-traces, which take a longitudinal course. In grasses, attention was first drawn to this feature by Bugnon,[1] who observed it in *Poa annua* L. It is illustrated for *Dactylis glomerata* L. in Fig. 174, p. 327, since in this species it happens to reveal itself diagrammatically.

Additional instances might be cited,[2] but those which I have brought forward are perhaps sufficient to show that 'leaf' and 'stem' are liable to adopt one another's characters in a disconcertingly protean way. Attempts to formulate their differentiating features lead us into the same confusion as those attempts to delimit their respective spheres which we have already discussed. There must surely be some explanation for the fact that, as soon as botanists try to arrive at any clear understanding of the relation of stem and leaf (considered as discrete morphological entities), they find themselves involved in a welter of contradictory opinions. Is not the explanation merely this—that *the problem itself is imaginary*? For the bitterest controversies always rage round problems which cannot be solved

[1] Bugnon, P. (1920[2]).

[2] E.g. the history of the rachilla in *Bromus mollis* L. (more correctly named *B. hordeaceus* L.) and *Cephalostachyum virgatum* Kurz, which is summarised from this point of view in Arber, A. (1927[2]), p. 486.

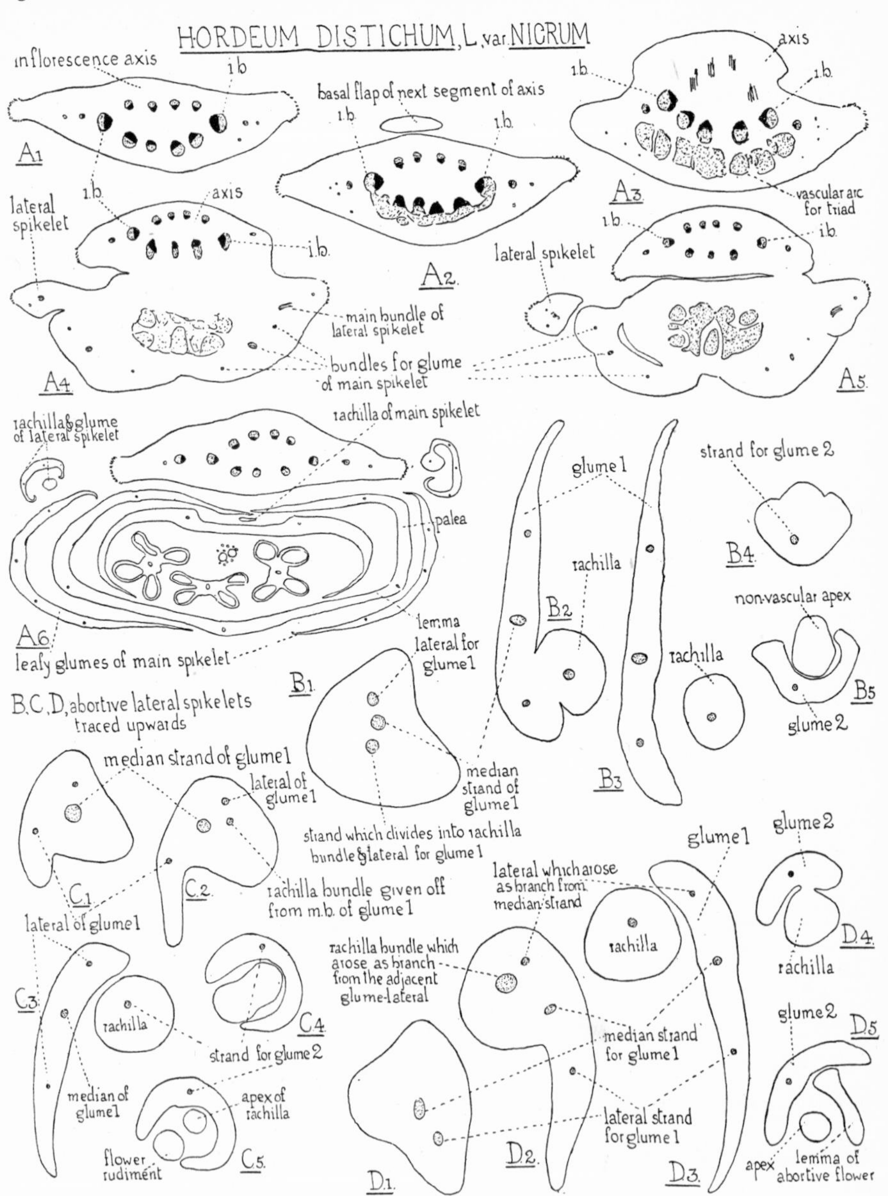

Fig. 173. *Hordeum* (*distichum*) *distichon* L. var. *nigrum*. A 1–A 6, series of transverse sections from below upwards through a segment of an inflorescence, to show the origin of a spikelet triad (× 24); *i.b.*, main bundles of inflorescence axis. B, C, D, series of transverse sections from below upwards through three lateral abortive spikelets (× 80) similar to those drawn on a smaller scale in A 5 and A 6, to show the origin of the glumes and of the abortive flower. [Arber, A. (1929[2]).]

for the simple reason that their very existence is an illusion. If we once accept the fact that 'stem' and 'leaf' are no more than convenient descriptive terms, which should not be placed in antithesis as if they corresponded to sharply opposed morphological categories, the problems of their delimitation and of their differentiating characters vanish into thin air.

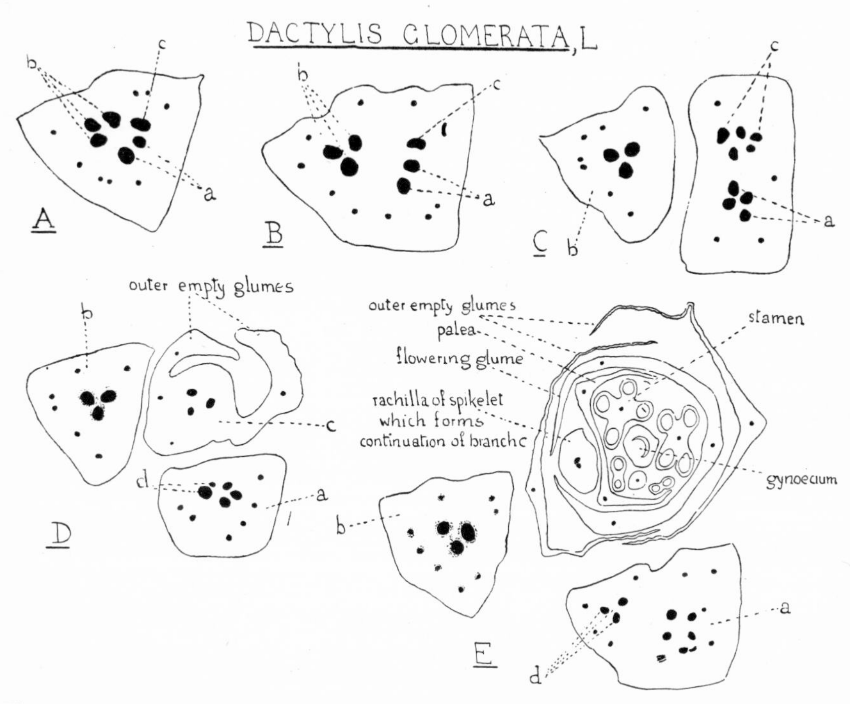

Fig. 174. *Dactylis glomerata* L. Transverse sections (× 47) from a series from below upwards through a branch of a very young inflorescence to show the behaviour of the vascular bundles in relation to further branching; *a*, *b*, and *c*, are not leaf-traces, but are traces for the branches so lettered. [Arber, A. (1928[2]).]

A study of the relation of the so-called axial and foliar members in the plant body thus leads us to the negative conclusion that stem and leaf cannot be accepted as valid morphological categories. If then we discard stem and leaf, what units are we to adopt? For the leaf-bearing region of the plant two alternative unit categories (not postulating a morphological distinction between stem and leaf) have been suggested at different times: *the phyton;* and *the individual shoot* (primary or lateral). The phyton theory commends itself at first glance, because it has a certain superficial simplicity, but it becomes excessively complicated as soon as it is applied to different types of flowering plants. In the shoot of a Monocotyledon with sheathing leaf-bases, the phytons are described as *superposed*, so that each internode belongs to one morphological unit. In a shoot, however, which bears whorled or spirally placed leaves whose attachments are relatively narrow, each stem segment must be regarded as being built up of as many *juxtaposed* units as there are orthostichies. This means that, in general, the internode of a Monocotyledon consists of a single morphological unit, while that of a Dicotyledon consists of a complex of such units. A theory which needs so much manipulation, before it will fit the facts of plant life, is an invitation to scepticism.

The conception of the shoot as a unit is much older than the phyton hypothesis. Goethe was so much absorbed by the idea of the leaf and its metamorphoses, that he paid scant attention to the rest of the plant body, but yet it is to him that we owe the first really lucid statement as to the importance of the individual shoot. He related lateral buds to seedlings, and regarded them as individual plantlets comparable with the parent shoot; but his notion remained incomplete because he seems to have forgotten the root.

If one accepts the shoot as a unit, the body of the flowering plant must be held to consist of parts belonging to two categories only—*root* and *shoot*. For a convincing presentation of this position, we must turn to the *Vorlesungen* of Julius von Sachs,[1] published half a century ago. This book marked an epoch in botanical thought, and it still lives through its masterly grasp of principles. I have been told by Dr D. H. Scott that, to the generation who were taught by Sachs, 'Wurzel und Spross' were household words—an antithetic couple, like 'Sea and Land'. It is regrettable that, since Sachs' day, the integral character of the shoot has often been forgotten.

A point which seems to me of special significance, though I have found no allusion to it in the literature, is that root and shoot are strictly comparable in the vital matter of reproduction. Each normally gives rise to other units *like itself*; the shoot produces lateral

[1] Sachs, J. von (1882).

shoots, which are essentially repetitions of the parent shoot, while the root produces lateral roots, similarly repeating the parent root. Moreover, both lateral shoots and lateral roots show an analogy in their position of origin, which seems in both cases to be determined by anatomical factors.

That shoots and roots, superfically dissimilar as they are, yet have a fundamental correspondence, is revealed by the fact that shoots can give rise to roots, and that roots, though more rarely, may produce shoot-buds. This suggests that in both shoot and root the potentiality of the other unit lies dormant beneath the character which predominates. Moreover, the cotyledons may be claimed as indicators of the essential unity of shoot and root, since they show that the root is not entirely devoid of the power to produce leaves. For the hypocotyl—the 'axis' on which the cotyledons are borne—is often internally more root-like than shoot-like, and the anatomy of the cotyledons themselves is apt to retain hints of root structure. The inner and outer differences between cotyledons and later leaves, may be due in part to the fact that cotyledons take their origin from a region which retains root characters, whereas the later leaves are derived from a region which is purely shoot.[1]

The striking differences in form between shoot and root originate out of the fact that—in addition to the uniform apical growth which both members share—the shoot has a second growth-rhythm, localised both in longitude and latitude, which leads to the production of outgrowths (leaves). We should be beginning to understand the differences between shoot and root if we could explain why the shoot shows this *localised and rhythmic cell-multiplication* in the superficial tissues, while in the root the tendency of the outer elements is not so much towards cell-division as towards that mode of *increase of surface area of individual cells* which produces root-hairs.

Up to this point I have considered morphological categories in the light of pure or formal morphology alone. I have done this deliberately, because I think that these categories ought to be evaluated quite apart from the doctrine of descent. Morphology existed long before it received an evolutionary explanation; more

[1] See the discussion on seedlings in Arber, A. (1925), p. 177, etc.

than two thousand years ago Aristotle had realised the homology between the forefeet of quadrupeds and the hands of man, though for him it had no evolutionary implication. The next stage in our argument must, then, be to discover how far the results at which we have arrived through the study of *formal morphology* are consistent with the findings of *evolutionary morphology*.

Our conclusion, that shoot and root are primary and equivalent categories, has already led us to discard one of the theories which claim to represent the historical course of evolution—the phyton theory. This theory was founded on the study of the living plant, and it cannot be said to have received any support from palaeobotany, or from general considerations about the course of evolution; there is thus no need to reconsider it at this point.

The pericaulome theory is in a somewhat different position, since it claims to be in harmony with the historical evidence drawn from fossil plants. It is difficult to attach any meaning to this theory unless one can believe that at some historical stage the leaf-bases possessed downward prolongations which were at first free from the axis, but which subsequently fused with it, and with each other, to form an enclosing rind. But in the Angiosperms, which are our special concern, there is no historical evidence that such a process has ever occurred; and nothing even remotely suggesting it takes place in the course of ontogeny.

There is yet another theory of the history of the plant body in the spermophytes, due primarily to Octave Lignier,[1] though earlier and later botanists have in part foreshadowed it, or applied it in special directions. It may be summarised in the briefest possible fashion as follows. The primaeval vascular plant is visualised as a cauloid, with no distinction of shoot and root. This cauloid—in the phylum which gave rise to the Angiosperms—became differentiated into a subterranean region (root) and an aerial region. The aerial cauloïd then suffered a reciprocal differentiation into what we now call stem and leaf, the leaf being merely a specialised cauloid-branch of limited growth. This idea of the essential equivalence of stem and leaf, which was maintained long ago by Henslow,[2] has recently received indirect

[1] Lignier, O. (1909). [2] Henslow, G. (1888).

confirmation from the work of Uittien,[1] who has shown that among the flowering plants there is a correlation between the relative lengths of the main and lateral axes of the branch system, and of the main and lateral nerves of the leaf. Moreover the palaeobotanical evidence, obtained since Lignier's day, is wholly favourable to this interpretation of the leaf. As a single example we may cite the Middle Devonian plant *Aneurophyton*, "which had the habit of a Tree-fern, with large, compound fronds; but the leaves had the same anatomical structure as the stem".[2] Indeed, Lignier's theory seems to have every advantage; it is simpler than the phyton theory or the pericaulome theory, and it can be applied without strain to a variety of cases. In accepting it, however, we must keep in mind the reservation that any such theory is in its very nature so speculative that it ought not to be treated as anything more than a provisional working hypothesis.

It is noticeable that two morphological conclusions are implicit in Lignier's theory: firstly, that, if one thinks back to origins, no fundamental distinction exists between stem and leaf; and secondly, that root and shoot are wholly equivalent entities. These results, however, are just those at which we had arrived independently from a consideration of formal morphology alone. That identical conclusions have thus been reached on two distinct lines of thought, adds to the probability that they represent at least an approximation to the truth.

[1] Uittien, H. (1928). [2] Scott, D. H. (1931).

CHAPTER XV

THE DISTRIBUTION AND DISPERSAL OF GRASSES

THE first point which strikes one about the mode of occurrence of the grasses on the face of the earth is their gregariousness. All meadows and many gardens in this country bear witness to the way in which the Gramineae, while living sociably among themselves, oust other plants. The same herd behaviour, coupled with exclusiveness, reappears among the bamboos. An American botanist,[1] who has made a special study of the trees and shrubs of Japan, writes that in that country, "the forest-floor is covered, even high on the mountains, and in the extreme north, with a continuous, almost impenetrable, mass of dwarf Bamboos of several species, which makes traveling in the woods, except over long-beaten paths and up the beds of streams, practically impossible. These Bamboos which vary in height from three to six feet in different parts of the country...prevent the growth of nearly all other under-shrubs, except the most vigorous species". The same writer attributes the climbing habit, which characterises so many Japanese plants, to this dense undergrowth of bamboos, which renders any other mode of life impracticable for relatively small species, if they are to keep a footing in the forests. The Gramineae are indeed formidable competitors in the struggle for space, not only because of their faculty for adopting a mass formation, but because they show an amazing tolerance about the external conditions of their lives. Only two flowering plants penetrate the Antarctic Circle, and one of these is a member of the family—*Deschampsia antarctica* (Hook.) Desv.[2] The grasses seem capable of existing at almost any altitude; *Festuca ovina* L.[3] grows at sea level in this country, but

[1] Sargent, C. S. (1894).

[2] Brown, R. N. Rudmose (1912); its companion is *Colobanthus crassifolius* Hook. f. var. *brevifolius* Engl., one of the Caryophyllaceae.

[3] This specific name is used here in the old, inclusive sense.

Hooker[1] noticed it at a height of nearly 18,000 ft. in the Himalayas. Apart from such extreme examples as this, we find that various Gramineae, which occupy low-lying habitats in temperate zones, are yet able to accommodate themselves to hotter countries by living at higher levels. Such European species, for instance, as *Avena fatua* L. are found in Afghanistan, but at a considerable elevation (2500–2700 m., i.e. between 8200 and 8900 ft.).[2] A more striking instance of adaptability is the way in which the grasses of hot regions are sometimes capable of flourishing in a chill mountain environment. When Hooker[1] was in Yoksun (Sikkim) at 5500 ft., the weather was very cold, the mean temperature being 39° F. and the lowest about 19° F. At eight o'clock in the morning, the thermometer, laid on the frosty grass, stood at 20° F. "I could not but regard with surprise", he wrote, "such half-tropical genera as...*Saccharum*,...large bamboos,...and cultivated millet, resisting such low temperatures." He thought, however, that their apparent hardiness might in part be accounted for by the relative warmth of the earth, beneath the surface, at these altitudes; at a depth of 3 ft., it might be 55° F.[1] Another plant capable of enduring a wide range of temperature at different altitudes is *Bouteloua gracilis* (H.B.K.) Lag., which occurs at the level of 1000 ft. in the prairies and at 7000 ft. in Arizona; it endures −40° F. in some places, and +100° F. in others.[3]

Even when varieties of altitude are not in question, we find that many grasses are able to face a surprising range of temperatures. *Cynodon dactylon* Pers., Bermuda-quick,[4] for instance, which is believed to have originated in Tropical Africa, is now to be found all over the world. It has travelled as far as Marazion Bay in Cornwall, and Studland Bay in Dorsetshire, and on the Continent it has penetrated even further north. It cannot survive strong frost, or imprisonment under a layer of snow—but these seem to be its only stipulations. Perhaps the extremest example that can be cited of tolerance, not only of high temperature, but also of noxious chemical

[1] Hooker, J. D. (1854).

[2] Vavilov, N. I. and Bukinich, D. D. (1929).

[3] Griffiths, D. (1912).

[4] Ridley, H. N. (1923), (1926), (1930); many original observations on the dispersal of grasses, and also references to the literature, will be found in the latter book.

conditions, is Ridley's discovery of Lalang-grass (*Imperata cylindrica* Beauv.) thriving in a volcanic fumarole in Java, in an overheated, steamy, sulphurous atmosphere, which no other plant could endure.[1] Plants of temperate regions do not furnish such sensational cases, but Sweet-vernal-grass, *Anthoxanthum odoratum* L., may be cited as a species which is remarkably catholic in its tastes. This grass is distributed over the whole of Europe from the furthest north to Southern Spain, Sicily and North Africa. It is found in the Caucasus, Siberia, Japan, the Canaries, Madeira and the Azores, Greenland and North America. It also inhabits Australia and Tasmania, but as an introduction. It flourishes in the most various climates, from the sea-shore to the snow regions of mountains. It can endure the long summer drought of the Mediterranean region, and it can winter as happily under the snow as in the mild atmosphere of the Atlantic Islands. It succeeds on sand, loam and clay, and is indifferent to the lime content of the soil.[2]

One result of their unexacting attitude towards their surroundings is that certain Gramineae are to be found in the small list of 'cosmopolitan' flowering plants—this word being used to indicate natural occurrence in the five continents (excluding Antarctica). The Reed, *Phragmites communis* Trin., and Manna-grass, *Glyceria fluitans* R.Br., may be reckoned among the cosmopolites;[3] they form an extreme example of the wide range which is often characteristic of aquatics.[4] Causeway-grass, *Poa annua* L., and Slender-foxtail-grass, *Alopecurus agrestis* L., may also be put into the same category, with the proviso that we cannot be certain that they have not been helped by man in attaining this position. Besides the interference of man, another difficulty met with in studying wide dispersals is to decide where specific boundaries are to be set. *Festuca ovina* L., Sheep's-fescue, is sometimes treated as a single, cosmopolitan species, which is found in such divergent climates as those of Greenland and New Guinea, but it is more reasonable to regard it as a species-complex.

[1] Ridley, H. N. (1930).
[2] Kirchner, O. von, Loew, E. and Schroeter, C. (1908, etc.).
[3] Hoeck, F. (1893).
[4] See Arber, A. (1920), p. 295, et seq.

DE STIRPIVM HISTORIA

Iuncus paluſtris maior. *Harundo.*

Groß weiher Bintzen. Rhor.

Fig. 175. "Harundo." The Reed (*Phragmites communis* Trin.) and other water plants. [From the *De Stirpium* of Jerome Bock (Hieronymus Tragus), 1552; initialled by David Kandel who worked for Bock.]

Ridley[1] believes the Reed (Fig. 175, p. 335) to be the most extensively distributed flowering plant in the world. It ranges all over Europe, Asia, Africa, America and Australia, but appears to be absent from New Zealand, Polynesia, and oceanic islands. It is found as far north as Finland, and it also lives in the hot, wet lowlands of the equator. In Tibet it reaches an altitude of 10,000 ft. It not only grows in swamps and water-courses, but it is capable of living in salt deserts as well as in salt marshes.[2] Clay and sand seem to satisfy its requirements equally well. When it is once established, this ubiquitous plant is most difficult to eradicate, owing to the persistence of its deep-set rhizomes. It was reported in 1836,[3] that, in the Carse of Gowrie in Scotland, there were several tracts of the best alluvial deposit, which had been under cultivation for upwards of a century, but yet the Reed grew as luxuriantly among the crops as at first.

The grasses do not all, however, possess the Reed's capacity for making themselves at home anywhere. Some, indeed, have very special tastes in the matter of locality, and sharply defined preferences as to soil.[4] The presence of Doddering-dillies, *Briza media* L., is indicative of lime in the substratum, while the Hair-grass, *Aira caryophyllea* L., likes sand, and the Melicks show a preference for localities with abundant humus. Special needs of this type sometimes result in peculiar distributions. *Nardus stricta* L.,[5] for instance, is characteristic, in Northern Britain, of a zone which follows the *margins* of the peat moors. This is due to its taste for the peat which has been redistributed from the exposed surface of peat-haggs by wind and water. Various grasses, again, are so constituted that they can endure strongly saline water, and hence are able to grow by the seaside, in salt marshes or on sandhills. Of these, *Psamma arenaria* Roem. et Schult. (*Ammophila arundinacea* Host), *Elymus arenarius* L. and *Glyceria* (*Sclerochloa*) *maritima* Mert. et Koch, are familiar on our own shores. Certain races of *Festuca rubra* Hack. are found in seaside stations; a variety of *F. rubra* sub-sp. *genuina* Hack. is con-

1 Ridley, H. N. (1923).
2 Salisbury, E. J. (1926).
3 Lawson, P. and Son [C.] (1836).
4 Ward, H. Marshall (1901).
5 Smith, W. G. (1918).

fined to the Severn Estuary, where it forms a compact turf on the salt marshes and older pebble ridges, periodically submerged by the highest tides.[1] Corresponding examples might be cited from all over the world; on the Natal coast, for instance, *Sporobolus pungens* Kunth grows where it is washed by the spray at all times and is covered by the sea at high tides.[2] We have to remember, however, that the presence of a grass at the seaside is not sufficient to prove that it is actually halophilous. In Belgium and other countries north of the Mediterranean area, *Phleum arenarium* L. occurs exclusively on dunes near the sea, and anyone, studying it in those countries alone, might assume that saline conditions were essential to it. In the Mediterranean region, and in Britain, on the other hand, it occurs in sandy stations inland, as well as on the coast. The explanation seems to be that the continental winters of Northern Europe are too severe for this species, and thus it can survive only in the coastal zone which is warmer. In the Mediterranean, and in the insular climate of Britain, on the other hand, the relative mildness of the winters permits it to flourish inland.[3] It is true also of other seaside grasses that they are by no means incapable of living in regions remote from salt water. Marram-grass (*Psamma arenaria* Roem. et Schult.), for instance, in this country a typical coastal grass, in America is not confined to salt-water areas, but grows abundantly along the shores of the Great Lakes.[4] Again, *Spartina versicolor* Fabre,[5] a Mediterranean grass, is uninjured by submergence in salt water, but yet it succeeds as well in places where the water is fresh as where it is salt.

One of the most remarkable of seaside grasses is *Poa flabellata* Hook.,[6] which grows in the Falkland Islands "wherever the waves beat with the greatest vehemence, and the saline spray is carried farthest". This—"the gold and the glory" of its island home, where it provides food for cattle and man[7]—is called Tussack-grass from

[1] Howarth, W. O. (1924).
[2] Bews, J. W. (1918).
[3] Massart, J. (1910), map, p. 52.
[4] Lamson-Scribner, F. (1895).
[5] Fabre, E. (1850).
[6] Originally described as *Dactylis caespitosa*; this account is taken from Hooker, W. J. (1843).
[7] See p. 272.

its unusual habit. With its densely crowded matted roots, it forms isolated hillocks or tumuli, 3–6 ft. in height and 3 or 4 ft. in diameter, from the top of which spring the stems and the great tuft of sweeping leaves. The tussocks generally grow apart, but within a few feet of one another, while the ground between tends to be bare of vegetation, so that in walking among the hillocks one is hidden from view, wandering in a complete labyrinth which also affords "a hiding-place for the sea-lions and sea-wolves". This tussock habit is no doubt well adapted to the rank, wet peat-bogs by the sea where *Poa flabellata* Hook. flourishes, but a somewhat different mode of growth is

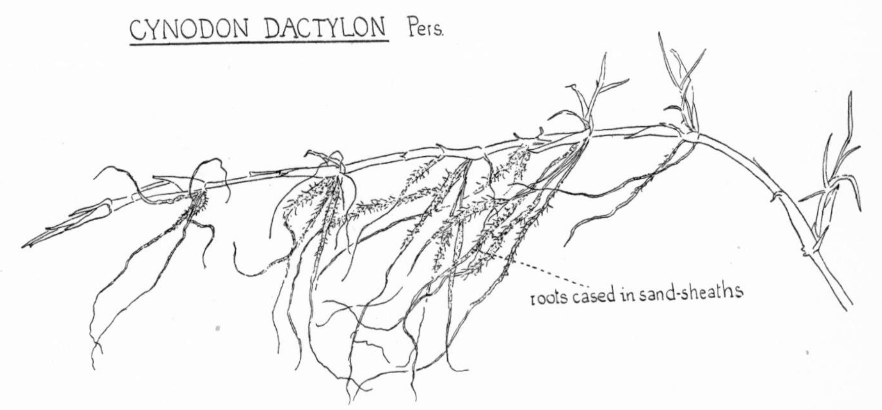

Fig. 176. *Cynodon dactylon* Pers. Mer de Sable, Carteret, Normandy, August 20. Shoot (× ½). [A.A.]

necessary for grasses growing on coastal dunes, where they have to keep a footing in moving sand. This particular sand-binding capacity is useful not only to the plant itself, but indirectly to man also, since it fixes the dunes and keeps the sand from overwhelming the hinterland. The sand form of Bermuda-quick, *Cynodon dactylon* Pers., from the 'Mer de Sable' near Carteret, Normandy, is shown in Fig. 176; it succeeds well on dunes, and agglutinates the sand-grains round its roots. Marram-grass is, however, more successful as a sand-binder, for as it gets buried in drifting sand, new branches arise from the higher nodes, so that the plant keeps up with the rising surface. The underground stems are said sometimes to reach a length of 20–30 ft., and they form a dense and resistant mass.[1] Another British grass, which is almost as valuable as Marram in binding sand, is Lyme-grass, *Elymus arenarius* L.

Sand-binding grasses, or 'oyats', have been of importance to man from very early days, but we need hardly go back so far as a certain writer,[1] who finds evidence for their primaeval use in the Story of Creation. The statement that 'grass' was called forth upon the face of the earth directly after the separation of land and water, he treats as indicating a providential arrangement "for the immediate purpose of binding the soil together and protecting it from the action of the winds and waves". Passing to the history of the sand-binding grasses in the New World, the same writer relates that, many years ago, it was customary every spring to warn the inhabitants of certain towns on Cape Cod, Massachusetts, to turn out to plant Marram-grass, just as, in inland districts, they were required to give their services in mending the roads. At one time there was a 'beach grass committee' at Provincetown, whose duty it was to enter any man's enclosure and set out Marram-grass, if the sand were uncovered or movable. In this way the sand-storms, which had been the terror of the inhabitants, were entirely prevented. In England, an act of the reign of Elizabeth prohibited the uprooting of Marram-grass.[2] In the reign of William III, the Scottish parliament passed an Act for the preservation of *Elymus arenarius* L. and *Psamma arenaria* Roem. et Schult. on the sea coasts of Scotland. These "provisions were, by the British Parliament, in the reign of George II., followed up by further enactments, extending the operation of the Scottish law to the coasts of England". Moreover, "it was rendered penal not only for any individual (without even excepting the lord of the manor) to cut the bent, but for any one to be in possession of any within eight miles of the coast".[3] Despite this legislation, in the latter part of the eighteenth century a large district on the east side of Scotland, near the Moray Firth, is said to have been turned into a desert by the advance of

[1] Lamson-Scribner, F. (1895).
[2] Knapp, J. L. (1804).
[3] Sinclair, G. (1824); on the use of grasses in connection with coast preservation, see Carey, A. E. and Oliver, F. W. (1918).

sand from the shore, owing to the destruction of the Marram-grass.[1]

Besides the grasses which flourish on dunes, there are a few which can endure the even less sympathetic conditions provided by shingle. A British example is *Festuca arundinacea* Schreb., whose strong growth is shown in Fig. 177. The horizontal axis is clothed with an inextricably tangled coat of old roots and tough leaf-fibres. On dissection one finds that the roots run down among the wefts of leaf-fibres, taking a course parallel with the axis.

Shingle grasses form a transition to those less familiar members of the family which colonise and fix moving screes. *Stipa Calamagrostis* Wahl., whose behaviour has been studied in Switzerland,[2] is a good example. This grass has tough rhizomes which creep at the surface and are often half-covered by descending pebbles. The terminal part of the rhizome forms the annual shoot; it is sheltered by strong, fibrous scales, and it is highly resistant to cold and the shock of falling pebbles. The dense weft of rhizomes is held in place by long wiry roots, with extremely thick, sclerised walls in the endodermis and inner cortex. Earth collects on the slope above and below the rhizome reticulum. The sudden drop in level—which may be as much as 20 cm.—below a colony of *Stipa Calamagrostis* Wahl., is an indication of the degree of protection which the network of rhizomes affords.

Dune grasses are not the only Gramineae whose stabilising power is useful to man; there are various grasses growing in tidal muds which are of essential service in land reclamation. The most renowned of these species, Townsend's-grass, will be considered in a later chapter (pp. 372–8); but we will now take an American plant belonging to the same genus, *Spartina glabra* Muhl., as an example. *S. glabra* has been studied in detail in Cold Spring Harbour, New York.[3] It has a range of 1 ft. 6 in. to 6 ft. 6 in. above low water, and is succeeded by a narrower belt of a second species, *S. patens* Muhl.—the two grasses together occupying most of the zone between tide

[1] Syme, J. T. B. (afterwards Boswell) (1873).
[2] Porta, N. H. (1928).
[3] Johnson, D. S. and York, H. H. (1915).

Fig. 177. *Festuca arundinacea* Schreb. Part of a huge tuft growing on the shingle, Lyme Regis, Dorset, April 21, 1925, collected by Miss G. Lister (× ½). [A.A.]

marks. By its particular mode of growth, *S. glabra* brings about the matting and stabilising of the silt. The terminal bud of the main axis runs horizontally at between 1 and 2 dm. below the surface of the mud, while the lateral offshoots turn upwards and become the aerial shoots. These shoots, which rise to heights varying from 1 to 2 m., may be so numerous that there are 300–600 stalks to the square metre. Since a hidden rhizome branch corresponds to each of these visible shoots, a most complex subterranean network is formed, which may be from four to five layers deep. From this rhizome reticulum, roots penetrate downwards into the mud, so that it is firmly matted for a depth of 2 or 3 dm. below the surface.

Although *Spartina glabra* Muhl., and other grasses to which we have referred, will tolerate submergence in salt water, they do not take to life in the sea with the same completeness as some of the other Monocotyledons.[1] In fresh water they seem to be more at home, and many are to be found living an amphibious existence, but none of them have their pollen-grains carried to the stigmas by water, so that they have not become aquatic in the extremest sense. *Glyceria fluitans* R.Br., Manna-grass (Fig. 1, p. 1) is as completely aquatic as any British grass. The anatomy of its long upper leaves, which float on the water-surface, is shown in Fig. 178. *G. aquatica* Wahl. also lives in wet places, but its leaves do not float. *Alopecurus geniculatus* L., Elbowit-grass (Fig. 179) and *Agrostis palustris* Huds., Fiorin-grass, are other species which are often semi-aquatic. When *A. palustris* grows at the edge of a ditch or stream, it may be found late in the summer putting out long, trailing vegetative shoots into the water. The leaf anatomy of an aquatic of hot countries, the Rice plant, *Oryza sativa* L., is indicated in Fig. 180. The large air-spaces in the tissues, seen here and in Fig. 178, are highly characteristic of water plants.[2]

Hitherto we have only been considering individual grasses in relation to their habitats. When we turn from this limited subject to the broader questions arising out of the geographical distribution of the members of the family, we at once find ourselves in the midst of

[1] On marine Monocotyledons, see Arber, A. (1920).
[2] On the anatomy of aquatics, see Arber, A. (1920).

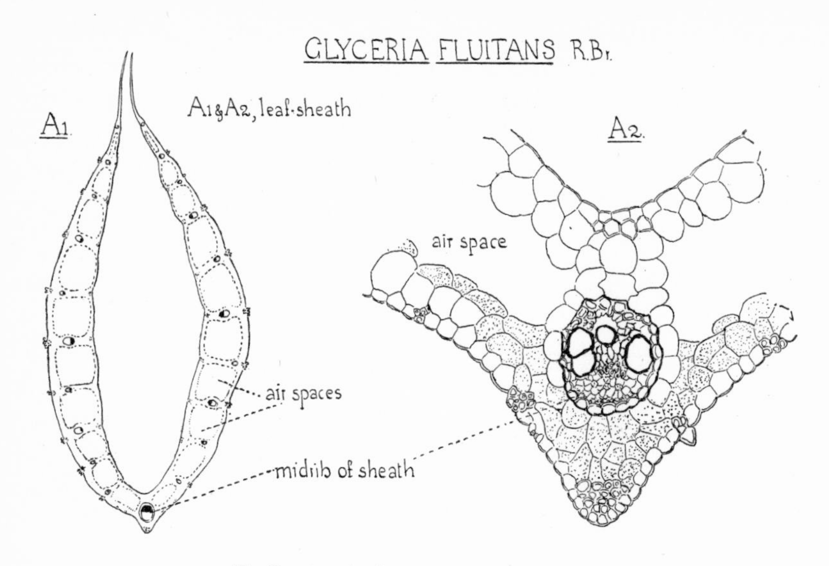

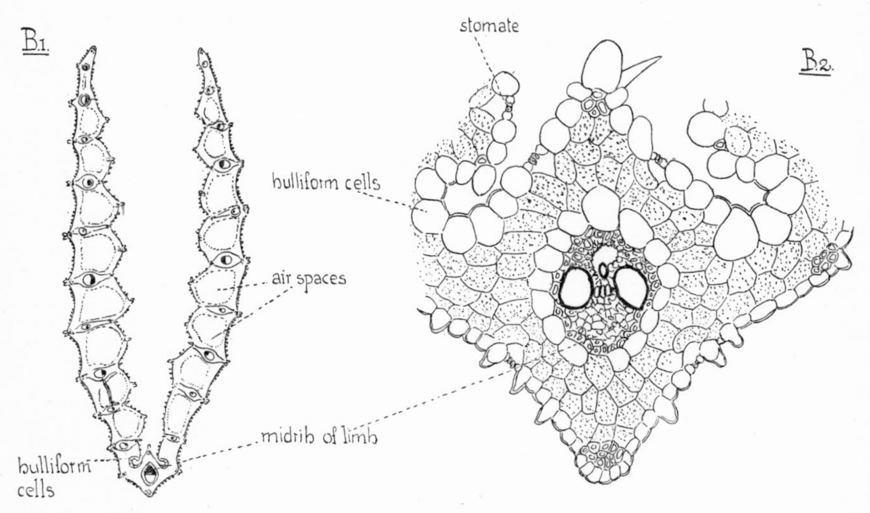

Fig. 178. *Glyceria fluitans* R.Br. A 1, transverse section of leaf-sheath (× 23). A 2, midrib of sheath in A 1 (× 193 *circa*). B 1, transverse section of limb (× 23). B 2, midrib of limb (× 193 *circa*). [A.A.]

difficulties. An idea of the complexities involved can be gained, for instance, by studying Stapf's account of the grasses of Africa;[1] while, as examples of problematic distribution, we may recall that four genera—*Amphibromus*, *Distichlis*, *Triodia* and *Leptoloma*—are found in America and Australasia, but not elsewhere,[2] while *Arundinaria* occurs discontinuously over a considerable part of the tropical zone.[3] It seems as though we have not as yet reached a view-point from which the general principles of the subject can be discerned clearly; and this is scarcely surprising, for interpretative distribution studies,

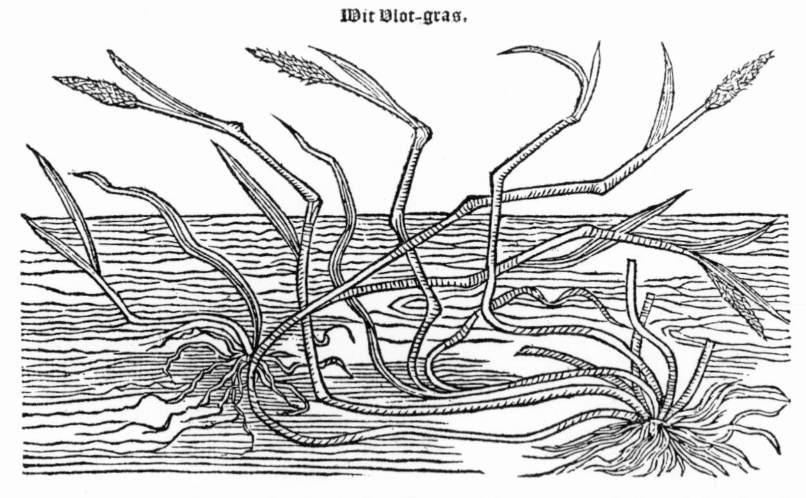

Fig. 179. "Wit Vlot-gras", Marsh-foxtail-grass (*Alopecurus geniculatus* L.) reduced from the *Kruydtboeck* of Mathias Lobel (de l'Obel), 1581.

worthy of the name, can only come into existence as the final outcome of the most exact and exhaustive taxonomic work.

When geographical analysis is limited to a single genus, the problem is simplified into something less unmanageable. A good example of such a study is the gallant attempt made by Hackel,[4] fifty years ago, to grapple with the distribution of the genus *Festuca*. In the half-century that has elapsed since his monograph of the European Fescues was published, our ideas about plant distribution

[1] Stapf, O. (1904[1]).
[2] Swallen, J. R. (1931).
[3] Good, R. D'O. (1927).
[4] Hackel, E. (1882).

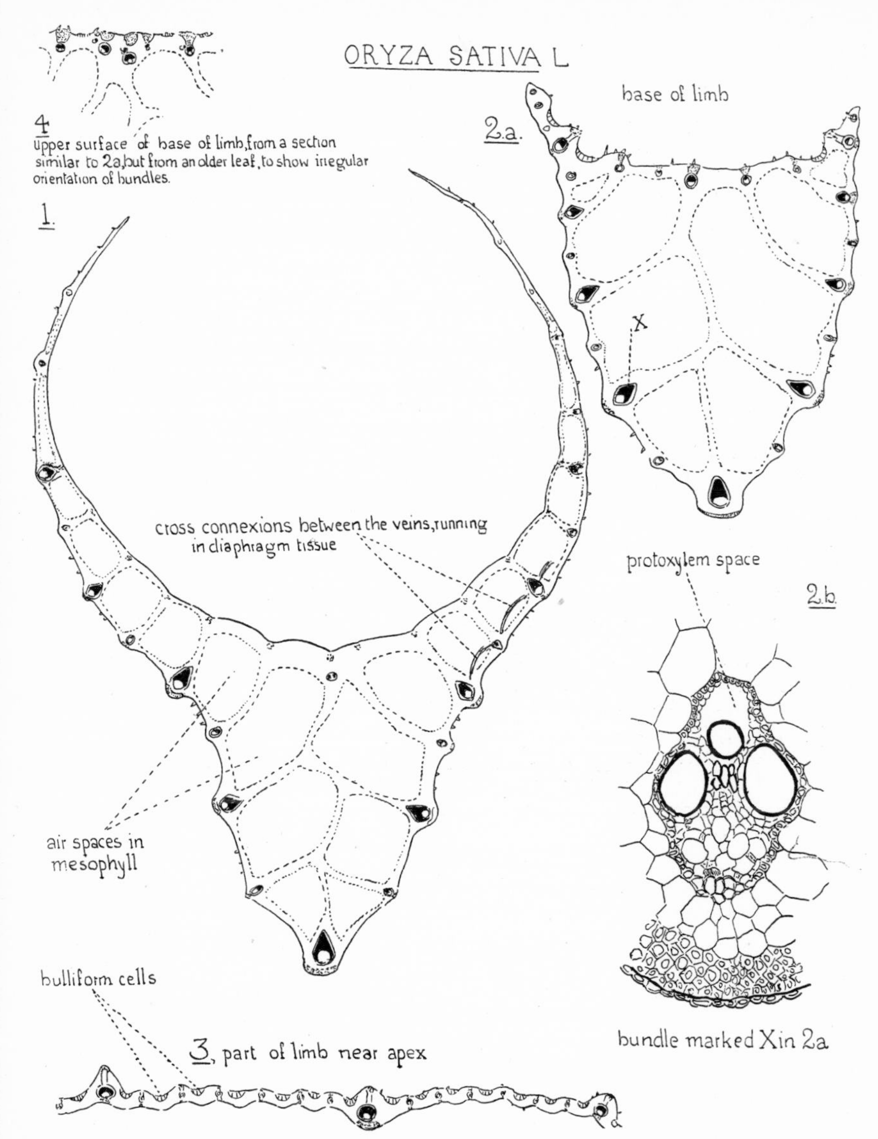

Fig. 180. *Oryza sativa* L., leaf structure. *1* and *2 a*, transverse sections of sheath and base of limb above ligule (× 23). *2 b*, bundle *X* in *2 a* (× 193 *circa*). *3*, part of transverse section of limb near apex (× 23). *4*, part of upper surface of a transverse section of the base of a limb similar to *2 a* but from an older leaf (× 14) to show that some of the upper bundles are irregularly orientated. For sketches of the ligule, see Fig. 148, p. 287. [A.A.]

and its evolutionary significance have suffered a gradual change, and his argument, though well worth study, does not now carry the same conviction as it did to the botanists of the 'eighties. The genus *Festuca* contains species of all grades, from those that are widely disseminated and consist of numerous individuals, to rare and strictly localised endemics. *Festuca ovina* L. in the broad sense, whether we regard it as a species or a species complex, may be taken as a type of the polymorphic wide-ranging Fescues, for it occurs in Europe, Asia, Africa, and America, and is found in many sub-species, varieties, and sub-varieties. On the other hand, there are a number of species, some of which are of great rarity, occupying narrow areas in certain of the mountain ranges of the South European peninsulas. These relatively small species Hackel regards as of great age, while he supposes such species as *ovina* to be the most 'recent' of the genus. It seems, however, more logical to suppose that, before the mountains were colonised by the numerous small species which now inhabit them, one or more species—comparable in their wide range and polymorphic character with *ovina*—must have been in existence to serve as the source from which the varied Fescue flora of the mountains could be derived. The distribution of *Festuca* indeed opens up a whole series of difficult problems, which seem less easily soluble than they did in Hackel's day. The genus is essentially European; of the twenty-eight species recognised by Hackel, sixteen are peculiar to Europe, while the Iberian peninsula is the richest region, containing seventeen species, i.e. 58 per cent. of the European forms. There is, however, no real necessity to assume that this abundance of species implies that Spain has been the centre of distribution of the genus, since the numerical richness may be, at least in part, accounted for on two grounds; firstly that Spain comes in for a share in the African element in the genus in addition to the European element; and, secondly, that the Iberian peninsula has been to some extent isolated from the rest of Europe. This is illustrated, among the Fescues, by the fact that the widely distributed *F. silvatica* Vill. and *F. varia* Haen. have apparently failed to pass over the Pyrenees into Spain. Behind the barrier of the Pyrenees, no less than eight endemic Fescues have been developed, besides peculiar sub-species and varieties, just as endemic species of the genus have arisen under insular conditions in the Azores and Canary Islands.

When we turn from the existing distribution of grasses to the mechanism by which this distribution has been brought about, we again find that our knowledge is in many ways defective. The evidence is mainly circumstantial, and it happens to be closely related to conclusions drawn from the recolonisation of Krakatau—conclusions which many botanists now regard as 'non-proven'. As is well known, the island of Krakatau, between Java and Sumatra, was devastated in 1883 by a volcanic eruption of extraordinary violence. No lava flows occurred, but the island was covered with pumice and ash. It was assumed by the earlier investigators that the "last remnants of plant-life which had withstood the first outbursts were everywhere destroyed and buried under a thick covering of glowing stones".[1] Twenty-five years later, the island was largely re-clothed with vegetation, and the impression arose that botanists were being presented with the results of a gigantic experiment performed by Nature for their benefit, from which it could be learned "how a volcanic island, which has lost the whole of its flora as the result of an eruption, acquires a new vegetation".[1] This view of the significance of Krakatau was generally accepted until recent times; in 1921, for example, the Director of the Buitenzorg Garden wrote of the eruption as having resulted in the "absolute annihilation of all that lived".[2] In the last decade, however, two writers,[3] approaching the subject independently, have thrown grave doubt on the "absolute annihilation" of plant life, which has been assumed as a postulate in the argument from Krakatau. It is not denied that the lower parts of the island were covered by so thick a layer of hot pumice that it probably destroyed every vestige of vegetation, but it is by no means certain that the much thinner layer of ashes and fine grit, that clothed the upper regions, would necessarily annihilate all plant life; and if there were any survival of seeds, or organs from which buds could arise, it at once disposes of the argument that the new vegetation of Krakatau came entirely from

[1] Ernst, A. (1908). [2] Docters van Leeuwen, W. M. (1921).
[3] Scharff, R. F. (1925); Backer, C. A. (n.d. [1930]).

over the sea. Moreover, it has now been shown that the earlier examinations of the 'new' flora were imperfect in character. This was not due to any remissness on the part of those who undertook them, but to the impossibility of spending an adequate length of time on the island, and to other practical hindrances. Results obtained under such difficulties cannot be accepted as on a par with those of field work in accessible regions. Later expeditions to the island have been made, but at long intervals, and the plants have not always been collected from comparable localities, so that we possess no exact taxonomic record of the history of the vegetation subsequent to the eruption. It seems, then, that the information about Krakatau, which was long treated as of basic importance for our knowledge of wind-and-sea carriage in general, and for that of the grasses in particular,[1] cannot be accepted at its face value. It is clear that observations and experiments of a much more critical type are essential before our knowledge of plant dispersal reaches a scientific plane. About the dissemination of grasses, with which we are here specially concerned, we know many individual facts, but they form at present a welter of unrelated detail, and the best we can do is to pick out something relevant, here and there, from this chaos of information.

Wind is no doubt one of the principal agents by which grasses are distributed; the caryopses with their enclosing glumes are often light and easily carried. The fact that many of our British grasses may be found growing on high walls and roofs affords a presumption that they are wind borne, but this presumption does not amount to proof. Though *Poa annua* L. may grow in house gutters, ten feet or more from the ground, it is not safe to claim this as evidence for wind dispersal, for it must be remembered that a sparrow, carrying some treasure in its beak, may often be seen to perch on the edge of a gutter, and then, in capriciousness or alarm, drop its burden, and fly away.

Sometimes the buoyancy of the 'seed' is increased by tufts of hairs, but we need experimental proof of their exact degree of effectiveness in wind carriage. Such evidence as we have is mainly indirect. Ridley has shown that there are certain genera in which a

[1] E.g. in Ridley, H. N. (1930).

species with a plumed 'seed' has a wider distribution than other members of the genus, which are not so provided. For example, in the genus *Calamagrostis*, the spikelets are often not plumed at all, or have hairs too short to be of much use in flight; but *C. epigeios* Roth has a whorl of silky hairs, twice as long as the glumes, at the base of the spikelets. It is a native of Europe, Africa and India, and is thus the most widely distributed species, as well as the one that is best furnished for flight. The exact mode of dispersal of grasses with plumed fruits needs, however, more rigorous testing than it has hitherto received. There are various difficulties connected with the subject. Ridley points out, for instance, that plumed grasses are relatively rare on oceanic islands, though other grasses, with apparently much less facility for travelling, are often present.

In individual cases, gales and whirlwinds may do great feats in conveying plants from one place to another, and such transferences may occasionally lead to an extension of the area of distribution of a species. In Lincolnshire a whirlwind has been known to carry a tuft of *Agropyron repens* Beauv., torn up by the roots, for a distance of 20 or 25 miles. A naturalist,[1] to whom we owe this and other records of storm dispersal, has concluded that 'storm-column' carriage, coming as it does between mid-July and mid-September, when grasses are in full seed, must be a far more potent means of dispersal than ordinary wind-drift.

The fate of those grass fruits which are blown along the ground is easier to observe than are the travels of those that are purely wind-borne. *Spinifex squarrosus* L.,[2] the so-called Water-pink, which grows on the sandy shores of India, Burma, China and Java, is cultivated on the Madras coast as a sand-binder. The fruiting head, which is 10 to 12 in. in diameter, forms a very light, spiky ball. These heads, "becoming detached, are propelled by the wind, assisted by the elasticity of the peduncles, with great velocity along the sandy shores, dropping the seeds in transitu. One may be followed by the eye for miles on its journey". The balls run on the tips of the spines, and frequently bound into the air. A fox-terrier,

[1] Woodruffe-Peacock, E. A. (1917).
[2] Trimen, H. (1900); Ridley, H. N. (1930).

who pursued them, is described as having had great difficulty in overtaking them. They are also so buoyant as to float lightly on water, while the upper peduncles, acting as sails, assist them across estuaries. Another grass fruit, whose transportation by sea has actually been observed, is *Thuarea sarmentosa* Pers.,[1] which is distributed by currents on the coasts of the Malay Archipelago. The inflorescence consists generally of five to seven spikelets, which are each two-flowered. The lowest spikelet contains one female and one male flower, while the higher spikelets are exclusively male. After flowering, the upper male spikelets fall off, and the part of the inflorescence axis which has borne them, curves round and encloses the basal fertile spikelet. During the maturation of the seed, an air-tissue, which serves eventually as a float for the fruit, develops in the axis.

It is not only the 'seeds' of grasses which may be transported by the sea; there are instances, believed to be well-attested, in which whole plants have been conveyed in this way. Ridley[2] states that all the Sugar-cane which grew on Cocos Keeling Island was derived from a clump of the plant which was washed up on the shore. He considers that it must have come from Java, 700 miles away. Living clumps of Lalang-grass (*Imperata cylindrica* Beauv.) and living bamboos, have also been washed up there. In Europe, certain grasses, such as *Glyceria* (*Sclerochloa*) *maritima* Mert. et Koch, *Elymus arenarius* L. and *Psamma arenaria* Roem. et Schult., are said to be dispersed by the drifting of rhizome branches in the sea.

Sea ice may occasionally act as a dispersal agent. It has been observed in Cold Spring Harbour, Long Island, that the rhizome mats and stubble of *Spartina glabra* Muhl. may be frozen in blocks of ice at low tide. As the tide rises, these blocks float up, and the *Spartina* may thus be carried bodily to other parts of the harbour.[3]

Rivers play an important part in the dispersal of grasses. Ridley[2] records that he has seen, in the East, clumps of bamboos washed down by the stream and growing successfully when stranded. Grasses form a large proportion of the Sudd—the mass of vegetation which,

[1] Nieuwenhuis-Uexkuell, M. (1902). [2] Ridley, H. N. (1930).
[3] Johnson, D. S. and York, H. H. (1915).

in African rivers, extends inwards from the shallow margins. This dense tangle may entirely block the channels, until the rise of the water tears off fragments and carries them down stream. Floating masses of plants corresponding to the African Sudd are also found in rivers in India and South America. A recent traveller[1] has described the great tangles of coarse grasses, chiefly species of *Paspalum* and *Panicum*, which occur along the margins of the Amazon. These Paniceae have rhizomes, $\frac{1}{2}$ in. thick, which root in the mud and grow out into the river. Masses of these grasses are torn loose from their moorings by the floods and float down stream as green carpets of vegetation. River carriage may also be observed, on a smaller scale, in temperate countries. Tufts of *Poa annua* L.[2] have been seen to drift down the Thames and establish themselves eventually at its margin. This grass travels the country in a variety of ways. Its spikelets may be blown about, the grains remaining enclosed in the glumes which act as wings. They are also carried by rain-wash along paths and roadsides and even streets; indeed, the way in which this species may spring up between flags or cobble-stones, justifies its name of Causeway-grass. *Poa annua* L. is one of a number of grasses which have been found in birds' nests in Willows near Cambridge.[3] In the Himalayas, Hooker[4] found it on a path leading from Walloong to Wallanchoon at 12,000 ft. He concluded that man or the yak must have imported this little wanderer from the north.

Unintentional conveyance by animals is one of the principal means by which grasses are carried about the globe. A large number of Gramineae have been observed to come up from cattle manure, while birds often carry the seeds internally and drop some of them unharmed. Grass fruits are also liable to adhere to fur, feathers or clothing, and may thus be carried for long distances. In *Tragus racemosus* (L.) All. (*Lappago racemosa* Willd.),[5] the Small-burr-grass of Australia, the stalked spikelet-group falls as a little burr. The structure of the hooks borne on the second glume is indicated in Fig. 181, p. 352. The range of this grass has been extended by the agency of

[1] Gates, R. Ruggles (1925). [2] Ridley, H. N. (1930).
[3] Willis, J. C. and Burkill, I. H. (1893).
[4] Hooker, J. D. (1854). [5] Maiden, J. H. (1898).

sheep, in whose wool it becomes entangled. Ridley[1] describes another striking example—*Centotheca lappacea* Desv., a grass which occurs in India, China, Java, Australia, Africa, etc. Its wide distribution is attributable to the fact that the glumes bear stiff reflexed spines, which catch very readily in textiles or fur, and adhere strongly when they have once taken hold. The Malays call it 'Cloth-spoiling-grass'. It is an example of the Gramineae of dense tropical forests, where wind

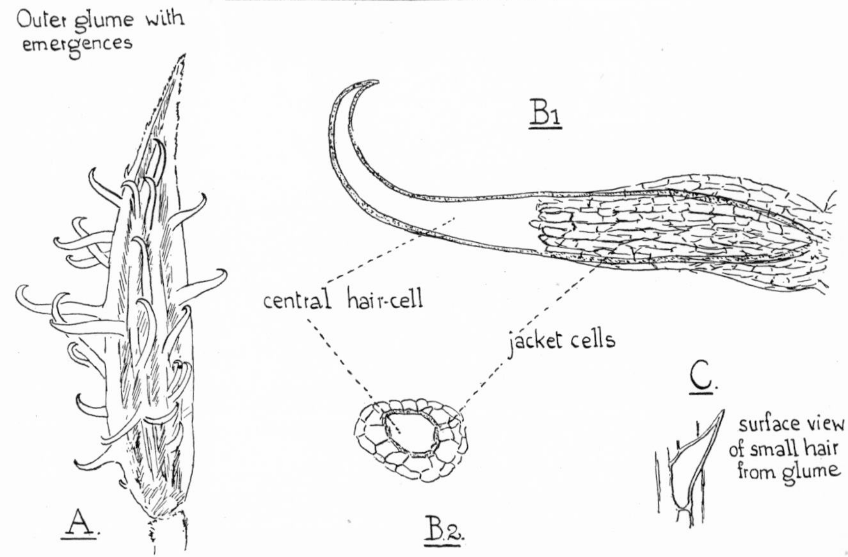

Fig. 181. *Tragus racemosus* (L.) All. A, second outer glume (× 19 *circa*). B 1, emergence in surface view, showing the jacket of cells enclosing the lower part of the large central hair-cell (× 103 *circa*). B 2, transverse section of another emergence (× 103 *circa*). C, surface view of a small hair (× 258). [A.A.]

carriage is impossible, and adhesion to passing animals is the only means whereby migration of the species can come about. In our own country, *Sieglingia decumbens* Bernh. (*Triodia decumbens* Beauv.) is distributed in Lincolnshire by wild ducks, which are specially fond of the seed, and frequent the peaty heaths for it in July and August.[2]

An example of dispersal in which both birds and wind play a

[1] Ridley, H. N. (1930). [2] Woodruffe-Peacock, E. A. (1916).

part, has been recorded from Ascension.[1] In 1917 the Farm Superintendent of the Island wrote that "a new grass has suddenly appeared in great abundance on the lower parts of the island. It first appeared to windward of the plain which the wide-a-wakes (sooty terns) frequent during their periodical visits, and has spread from there by the prevailing south-east trade wind to the Garrison three or four miles on. It has quite altered the appearance of Garrison and the intervening country....It is climbing up the craters and turning them from red to green hills". This grass proved to be *Enneapogon mollis* Lehm. (*Pappophorum molle* Kunth). Its light fruits, with their feathery awns, are capable of clinging to plumage, fur and clothing, and are also suited to wind carriage.

Not only mammals and birds, but even insects may play a part in disseminating the Gramineae, for it has been observed[2] in the Belgian Congo that termites harvest grass seeds and thus localise them. There is a termite that stores large amounts of a species of *Cynodon* in its nest, and, on the open primitive veld, it is found that this particular grass occurs only on the circular patches where the remains of old nests are found, or where they formerly stood.

Like the termites, man transports the grasses which he uses for food, but he also conveys others from place to place in all sorts of unintentional ways. Sometimes this process can be traced in its curious detail. It is known, for instance, that *Agrostis tenuis* Vasey,[3] the Red-top of North America, was introduced into New Zealand by certain Nova Scotians who emigrated to the Cape of Good Hope, and thence to Australia, and thence to New Zealand. Before starting they filled their mattresses with hay, which included Red-top. When they finally discarded the mattresses in New Zealand, the seeds germinated and this American grass became abundant. Here the part played by man was direct, though unwitting. In other cases, human action, involving no direct interference, has yet accidentally opened up, for some particular species, increased facilities for travelling. In the Malay Archipelago, for example, the Lalang-grass (*Imperata cylindrica* Beauv.)[3] is quite unable to cross the narrowest

[1] Stapf, O. (1917). [2] Davy, J. Burtt (1928).
[3] Ridley, H. N. (1930).

belt of forest. This failure seems due to the fact that its plumed spikelets are not, as a rule, wafted for distances of more than about 16 yards from the parent plant, even in a fairly strong breeze. If, however, a wide open path happens to be cut through the forest, the plant makes its way along it, and can thus reach a point which would otherwise have been inaccessible.

Another way in which man may assist dispersal, is that, in making use of a grass for his own purposes, he may accidentally afford it opportunities for spreading. At Singapore there was, for many years, a patch of the American grass, *Axonopus compressus* Beauv.,[1] growing along the roadside near the Botanic Gardens; this was the only known occurrence of this grass in the Malay Peninsula. As it was a broad-leaved plant, suitable for edging flower borders, some of it was dug up and planted in the Gardens. In a few years it had spread, not only all over the Gardens, and along the roadsides all over Singapore, but into the Malay Peninsula so far as Selangor. The original patch had been hemmed in by more strongly growing grasses of a creeping habit, so that it had been quite unable to extend its area; but as soon as it was transferred to a spot where there was little or no competition, it made a start, and was then able to dominate even other Gramineae. Ecological analysis of the progress of such a migrant might furnish clues towards the solution of some of the many outstanding problems connected with the distribution and dispersal of grasses.

[1] Ridley, H. N. (1930).

CHAPTER XVI

MAIZE AND TOWNSEND'S CORD-GRASS: TWO PUTATIVE HYBRIDS

IN an earlier chapter[1] something has been said about the agricultural history of Maize (*Zea Mays* L.). Its peculiarities call, however, for further study, since—as was already recognised in Lyte's *Nievve Herball* of 1578—"This corne is a marveilous strange plante, nothing resembling any other kinde of grayne". Maize is a vigorous annual, whose tall shoot bears a succession of broad distichous leaves; the lowest of these leaves arise at or below the surface of the ground, the buds in their axils producing suckers or tillers.[2] In its general vegetative growth, Maize does not differ essentially from the Gramineae of the Old World, but it was its mode of reproduction which perplexed the botanists of Europe when it was first introduced from America: "for it bringeth foorth his seede cleane contrarie from the place where as the flowers growe, which is agaynst the nature and kindes of all other plantes, which bring foorth their fruite there, where as they have borne their flower". The main shoot terminates in the male inflorescence, or, in Lyte's words, "at the highest of the stalkes growe idle and barren eares, which bring foorth nothing but the flowers or blossome". Since in the sixteenth century the function of the stamens was still undiscovered, these blossoms, which were not succeeded by fruits, naturally seemed objectless and "idle". The female reproductive shoots (Fig. 182, B, p. 356) are borne laterally in the leaf axils at some distance below the male inflorescence. As compared with the main axis, each fertile lateral is much abbreviated. The bikeeled prophyll is succeeded by a number of leaves called 'husks', whose sheaths are very large in proportion to the diminished blades. The ear itself remains hidden, but the long styles or 'silks' emerge at the top. When the husks are dissected off (Fig. 182, C), it is seen that the ear has a thickened axis, the 'cob', symmetrically armoured in longitudinal rows of paired spikelets.

[1] See pp. 28–33. [2] Weatherwax, P. (1918).

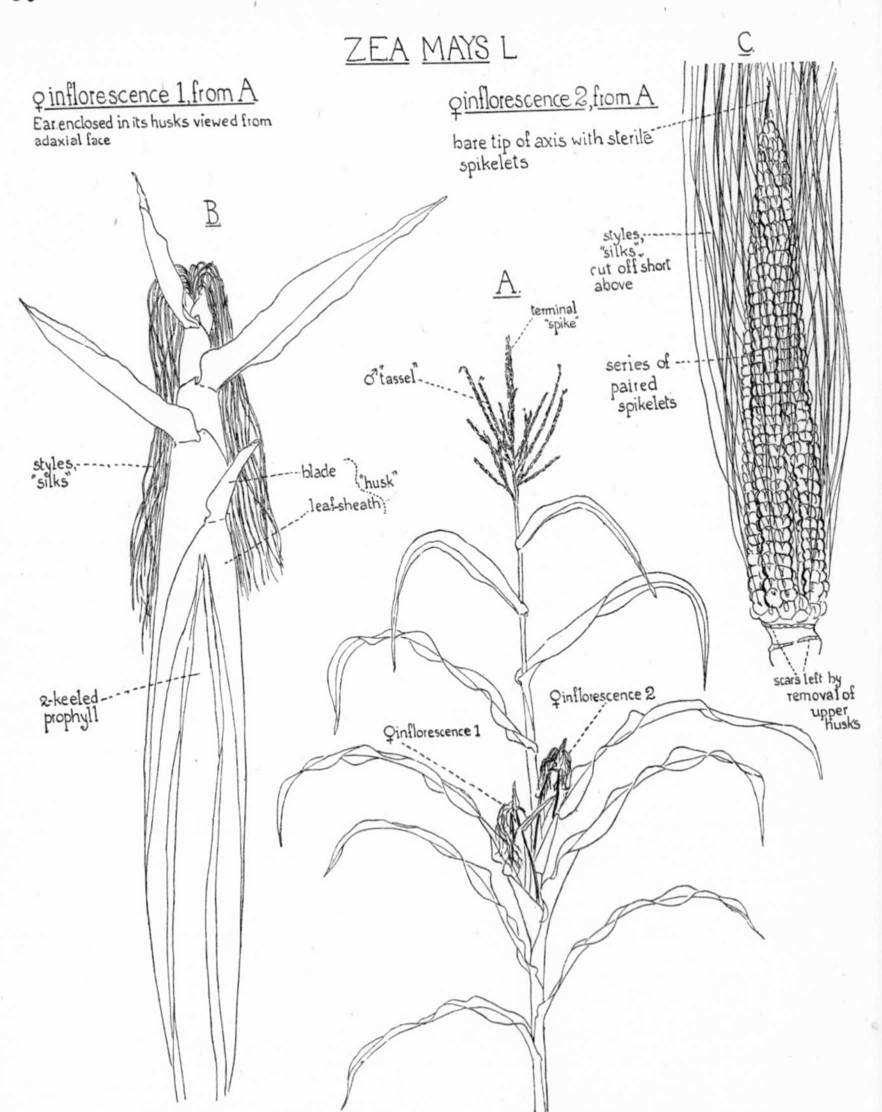

Fig. 182. *Zea Mays* L. A, habit drawing of a plant grown in the Cambridge Botanic Garden, showing almost the whole of the above-ground region (× about $\frac{1}{15}$). B, ♀ *inflorescence 1*, from A (× about $\frac{2}{5}$); axillant leaf removed. In the axils of the husks of this ear two axillary buds were found containing young female inflorescences, whose structure is illustrated in Fig. 183, p. 358. C, *inflorescence 2*, from A (× about $\frac{2}{5}$) dissected out of the husks; this ear had five double rows of spikelets. [A.A.]

The fact that the "grayne or seede" grew "orderly about the ears" was noted by Lyte.

Each female spikelet includes, as a rule, only one functional flower. Very young spikelets are seen in transverse and longitudinal section in Fig. 183, p. 358. The spikelet begins, as usual, with two outer empty glumes. Then follows an abortive flower, which, in the example figured, had a lemma, palea, and rudimentary stamens and gynaeceum (flower 1 in Fig. 183, A, B).[1] In the race of Maize called 'Country Gentleman', this first flower is often fertile, so that there are two normal flowers in the spikelet;[2] in most varieties, however, it is only the second flower which is fertile. This flower includes no lodicules, but it has three rudimentary stamens, and a fully formed gynaeceum. It will be seen in the series B 1–B 4 that the free part of the palea of the fertile flower is reduced to two lobes (B 3), the median region remaining fused with the flattened rachilla. This rachilla, at a slightly higher level (B 4), dies out in forming a lemma, as if for a third flower, which, however, does not come into existence. The fact that the rachilla continues in this way above the fertile flower, seems to have been hitherto overlooked. The gynaeceum, unlike that of most grasses, terminates in a single style, the 'silk'. Since this style is traversed by two bundles, and is bifurcated at the tip, it seems that it is equivalent to two lateral styles in a state of union.[3] In transverse section it is seen that each stylar bundle of Maize is accompanied by a strand of conducting tissue for the pollen-tubes. These strands have been interpreted as representing the continuation of the loculi of the two sterile carpels.[4] The styles bear hairs, each of which arises from a single epidermal cell, which divides anticlinally (Fig. 183, D), giving rise to four, or occasionally five, elements. Each of these cells, by repeated transverse division, produces a long filament. Since there is only a slight connection between the four filaments, a canal is left in the middle of the hair.[5] The silks appear to be in some degree receptive throughout their

[1] Lodicules may be developed in this flower (Weatherwax, P. (1916), p. 135), but they did not happen to be present in the examples I examined.

[2] Weatherwax, P. (1916).

[3] Guéguen, F. (1901).

[4] Randolph, F. R. (1926).

[5] Weatherwax, P. (1917).

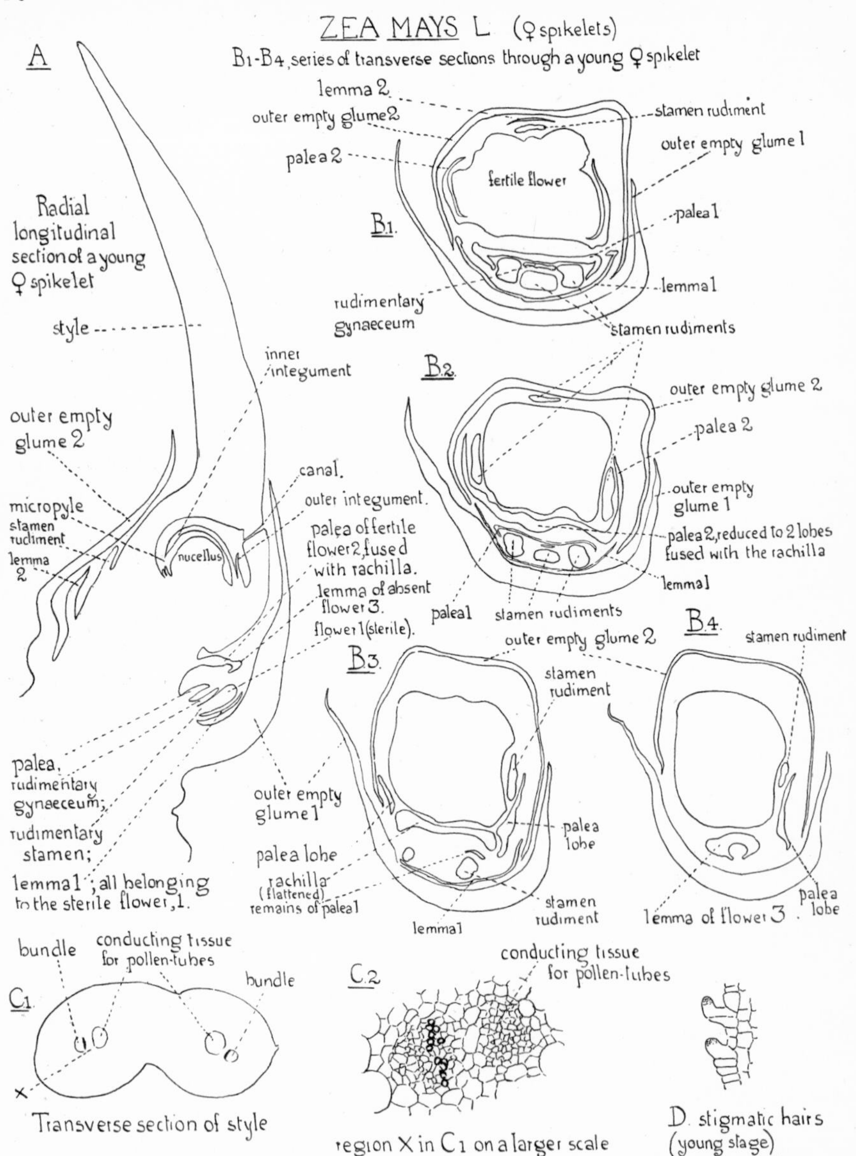

Fig. 183. *Zea Mays* L. A and B, structure of fertile spikelets from very young inflorescences found in the axils of the husks of *inflorescence 1*, Fig. 182, A, B, p. 356. A, radial longitudinal section of a very young fertile spikelet (× 47). B 1–B 4, transverse sections from a series passing upwards from below through another very young fertile spikelet (× 47). C, structure of style. C 1, transverse section (× 47). C 2, one bundle and its associated strand of pollen-tube conducting tissue (× 193 *circa*). D, young stigmatic hairs from a longitudinal section of a style from the same ear as A (× 193 *circa*). [A.A.]

entire length, and hence it is a matter of indifference whether they are called stigmas or styles.

The male inflorescence, or 'tassel' (Fig. 184, p. 360), consists of a relatively thick central 'spike', below which there are a number of slenderer lateral branches, some of which branch again. On both the central axis and the branches, the spikelets are, as a rule, arranged in pairs. A spikelet pair is drawn in Fig. 185, A, p. 361; one member of the pair is stalked and the other is sessile. Fig. 186, p. 362, is from the engraving of the male spikelet of Maize in Malpighi's *Anatome Plantarum*.[1] This seventeenth-century writer gives a description of the structure, which, for its date, is remarkably exact. He was clear about the absence of the style, and he observed the lodicules: "In medio, loco styli nil observatur; circa staminum exortum foliola subalba, molliaq;". The spikelet which he figures is, however, unusual in containing one flower alone; the normal structure is shown in Fig. 185, B and C. The thick lodicules have numerous well-differentiated bundles (C 1 and C 2). Owing to the limited development of the fibrous layer in the wall of the anther, it opens by a small pore only.[2] It has been calculated[3] that there are 2500 pollen-grains to an anther, and that the tassel may have 7200 stamens, giving a total of 18 million pollen-grains to the plant; according to other calculations, an especially vigorous plant may produce 30 million to 60 million.[4] At the lowest of these estimates there will be 9000 pollen-grains to each ovule. The pollen is apparently air-borne, and it is said to be able to travel a quarter of a mile in a high wind,[2] but the noticeably sweet scent of the tassel makes one wonder if insects ever assist in pollination.

The part played by plant pollen is now so well understood, that it requires an effort of the imagination to realise that its function is not self-evident; it gives one, indeed, a slight shock to learn that the Zuñi Indians used to go to great trouble to distribute the spores of Corn-smut in their fields—thus spreading a serious disease—in the belief that it was these fungal spores which fertilised the corn.[5]

[1] Malpighi, M. (1675).
[2] Weatherwax, P. (1923).
[3] Sturtevant, E. L. (1881).
[4] Davy, J. Burtt (1914).
[5] Collins, G. N. (1923).

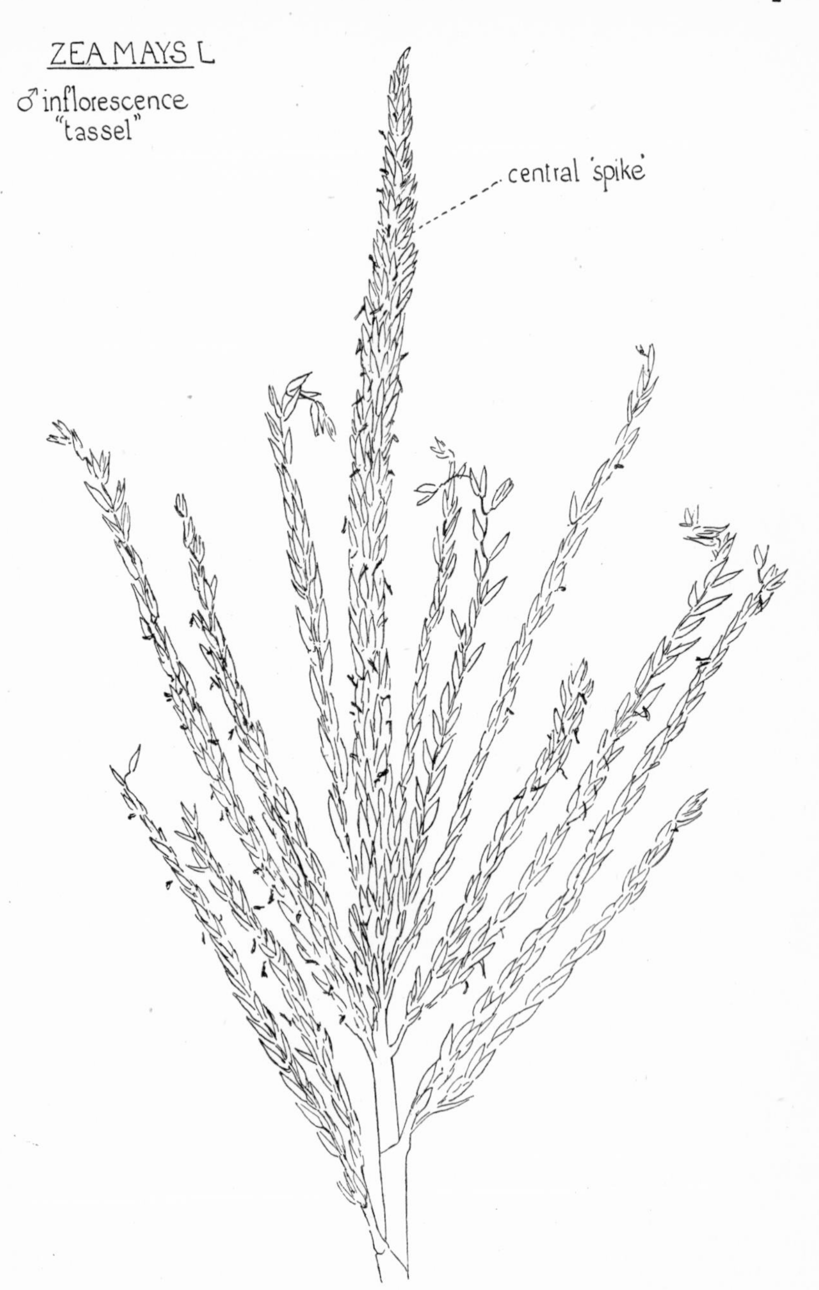

Fig. 184. *Zea Mays* L. Terminal male inflorescence from the plant drawn on a smaller scale in Fig. 182, A, p. 356 (× ½). [A.A.]

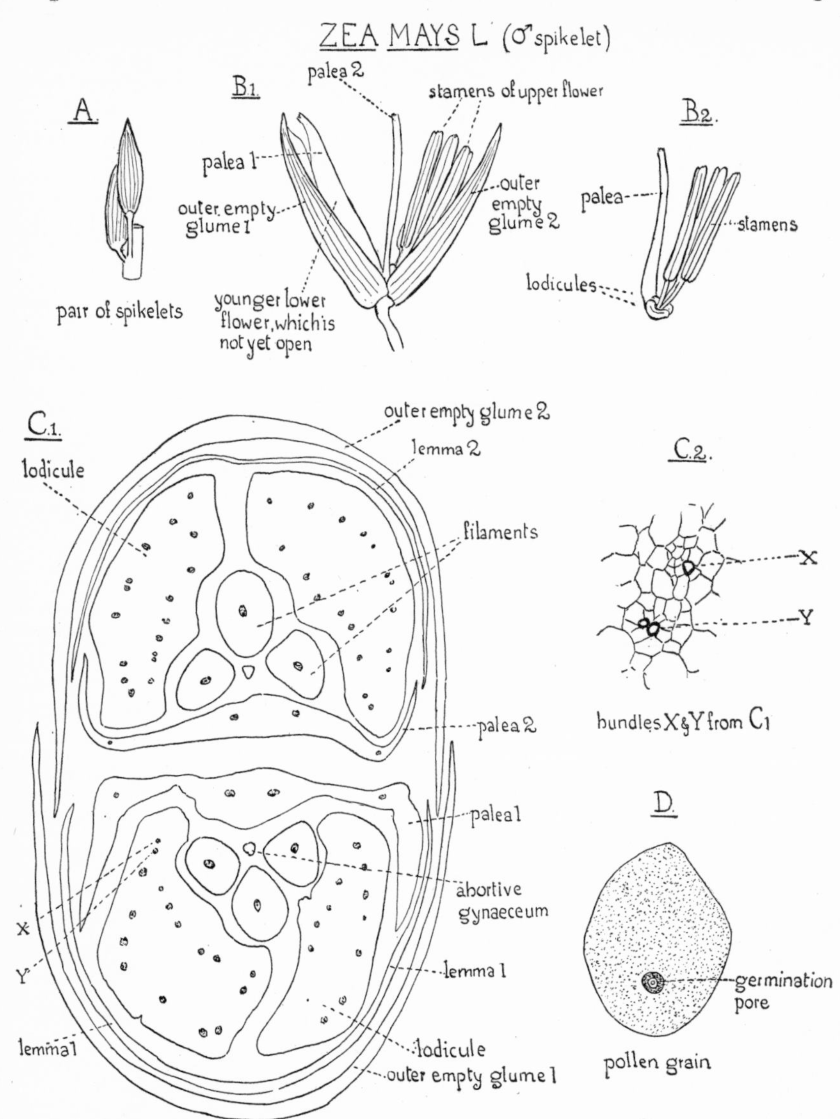

Fig. 185. *Zea Mays* L. Male spikelet. A, a short segment of one of the lateral axes of the 'tassel' (Fig. 184), to show a pair of spikelets. Of the pair, one spikelet is sessile and the other shortly stalked (about nat. size). B 1, ♂ spikelet enlarged. B 2, right-hand flower dissected out, to show the two kidney-shaped lodicules, grooved along their upper surfaces. C 1, transverse section of a male spikelet (drawn from a number of sections) older than the ♀ spikelets drawn in Fig. 183, p. 358 (× 47). Bundles omitted in the glumes, except the principal bundles of the palea. The rachis does not continue between the two flowers. C 2, the two lodicule bundles marked *X* and *Y* in C 1, to show lignified xylem (× 318). D, pollen-grain with germination pore (× 318). [A.A.]

A few hours after the pollen-grains lodge in the hairs of the silk, the pollen-tubes emerge. The behaviour of the tubes is variable. They may at once enter the cavity of the hair; or they may pass down the outside of the hair to its base and then enter the silk; or they may even penetrate the silk direct, without the intervention of a hair.[1] The styles may attain a length of 50 cm. or more. Fertilisation has been observed to occur as early as 25 hours after pollination in a gynaeceum in which the silks were 25 cm. long,[2] so both the rate of migration of the pollen-tube, and the actual distance which it can cover, greatly exceed the average for the flowering plants. It is

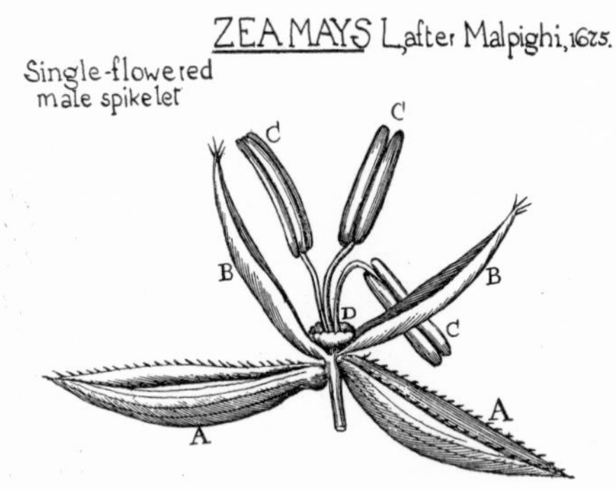

Fig. 186. "Triticum Indicum" (*Zea Mays* L.). One-flowered male spikelet. [Reduced from the *Anatome Plantarum* of Marcello Malpighi, 1675.]

probable that the pollen-tube dies away behind as it proceeds, and that not more than a centimetre or two is alive at any given moment.[2]

When the pollen-tube has entered the embryo-sac,[3] it extends and expands until the tip is near the polar nuclei. The wall of the tube then dissolves and fusion occurs between one sperm nucleus and the egg, and between the second sperm nucleus and one of the polar nuclei. The two polar nuclei unite at the same time, or a little later, thus completing the triple fusion. Traces of the pollen-tube remain until they are crowded out by the developing endosperm and

[1] Miller, E. C. (1919); see also Crozier, A. A. (1888), on the receptivity of the lower part of the silk.

[2] Weatherwax, P. (1919[1]).

[3] This account of double fertilisation in Maize is taken from Miller, E. C. (1919); see also Guignard, L. (1901) and Weatherwax, P. (1919[1]).

embryo. Almost immediately after fertilisation, the triple nucleus begins to divide; in ten to twelve hours it may have given rise to twenty to thirty nuclei, which line the embryo-sac. At this stage the egg nucleus, which is slower in development, undergoes its first division. The cells of the endosperm increase very rapidly, and within thirty-six hours after fertilisation, when the embryo does not consist of more than sixteen cells, the sac is completely filled. The antipodals at first remain intact and increase in number, but are soon suppressed by the encroaching endosperm cells. When the grain[1] arrives at maturity, no nucellar or integumental tissue remains, except in some scattered patches; over the greater part of its surface, the aleurone layer of the endosperm lies in close contact with the inner epidermis of the pericarp.[2] It happens that there are a number of races of Maize which differ in their endosperm characters, and, owing to the translucency of the pericarp, these characters can be distinguished in the ripe grain, even with the naked eye. The aleurone layer, for instance, may give a blue, purple, or black colour,[3] and the sugary or starchy character of the endosperm also affects the external appearance of the grain. This is of some use in genetic research, since, as a result of double fertilisation, hybrid characters are visible in the endosperm, within the grain itself, as soon as it is ripe, whereas corresponding characters in the embryo are not revealed until after germination in the next season.[4] Particoloured ears, in which the individual grains belong to divergent types, can be obtained by pollinating various regions of the ear from the tassels of different races.

Though in Maize the terminal inflorescences are normally male and the lateral female, countless examples have been recorded in which the sex segregation does not follow this simple ruling.[5] Hermaphrodite flowers may occur; the tassels may include female flowers; or the ear may be partly male. A mixed inflorescence is sketched in Fig. 187, p. 364. The reproductive shoots produced by the

[1] See Fig. 107, p. 225.
[2] Randolph, F. R. (1926).
[3] See Davy, J. Burtt (1914) for references on this and other points of Maize structure.
[4] This was first pointed out by Vries, H. de (1900).
[5] Penzig, O. (1921–2), and later papers such as Weatherwax, P. (1925²), etc.

basal suckers of the plant are often mixed—so commonly, indeed, that an androgynous inflorescence in this position can scarcely be called abnormal.[1] A botanist,[2] who examined 1651 plants of Maize

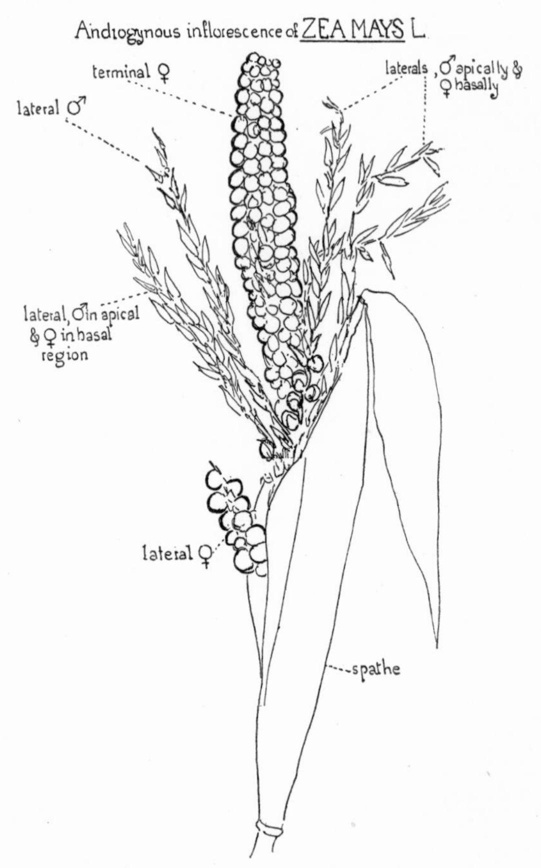

Fig. 187. Sketch (× ½) of an androgynous inflorescence of *Zea Mays* L., from herbarium material from Mr W. C. Worsdell. In this inflorescence the main axis is exclusively female, and has two basal lateral branches which are also female. At least three of the higher laterals are female at the base and male above. [A.A.]

of mixed races growing out-of-doors under ordinary conditions, found that in more than 98 per cent. the main axis terminated in a purely male tassel, while less than 2 per cent. bore terminal inflorescences of a mixed type. On the other hand, in the 721 basal suckers

[1] Weatherwax, P. (1918). [2] Werth, E. (1922).

produced by the same plants, only 377 ended, like the main axis, in a male panicle, while 344 had a mixed inflorescence. Nearly 48 per cent. of the suckers thus bore mixed inflorescences, as against less than 2 per cent. of the main axes. For comparison with the production of androgynous inflorescences by these basal suckers, we may recall the behaviour of lateral shoots in water plants which bear more than one type of leaf; in these side branches, we sometimes meet with a regression to a less differentiated leaf-form, while the main shoot is bearing the mature type of foliage.[1] The androgynous inflorescence of Maize is less specialised than the unisexual reproductive shoot, and may perhaps be regarded as showing 'juvenile' characters, like the lateral shoots of these heterophyllous aquatics.

It is tempting to associate the characters of the male and female inflorescences with the conditions under which they are formed. A French author, who has made a special study of the development of Maize, states that, a month after sowing, the young plants show only four or five leaves, and the stem above the first node is only 1 cm. long; but careful dissection reveals that, even at this early stage, all the leaves and also the male panicle are already in existence. By the time the stem reaches 7 cm., the panicle is completed. In the first fortnight of July, male inflorescences 15 cm. in length may be found, terminating stems which have themselves only reached 10 cm. The male panicle thus develops at a time when the roots are few, and the leaves assimilate little more than is needed for their own use; its elongated and branched character has been attributed to the consequent lack of nourishment.[2] It is uncertain whether this view is tenable, but there is at least no doubt that in Maize elongation is associated with maleness and abbreviation with femaleness. This is shown by statistics[3] as to the length of the main axis and of the different types of basal shoot in a mixed population of Maize plants. The average result was: main axis with terminal *male* inflorescence,

[1] Arber, A. (1920), pp. 160–1.

[2] Blaringhem, L. (1908); but this view is difficult to reconcile with the results of certain starvation experiments in which female inflorescences predominated; see Werth, E. (1922).

[3] Werth, E. (1922).

171·5 cm.; basal shoot with terminal *male* inflorescence, 157·1 cm.; basal shoot with terminal *androgynous* inflorescence, 112·7 cm.; basal shoot with terminal *female* inflorescence, 42·5 cm.

The normal female inflorescences come into existence at a time when the main stem is already almost completely developed. They are abundantly provided with water and salts from the thick tuft of roots, and with carbohydrates from the large leaves. Their shortened and thickened construction has been attributed to two influences[1] acting in combination—the compression under which they develop within the firm leaf-sheaths, and the excessive stream of nourishment which is poured into them; but whether these factors are actually *causal* is still an open question.

Speculation has been rife as to the homologies of the ear in Maize, for it diverges widely from any other reproductive shoot met with in the Gramineae. Some writers hold that it is equivalent to the central spike alone of the male inflorescence.[2] Others regard it as the result of fasciation, or as formed by the 'congenital' union of the central spike with its own lateral branches, and thus as equivalent to the male tassel as a whole.[3] Those who take this 'fusion' view, point to a type of abnormal female inflorescence, in which 'disruption' has occurred, the ear being represented by a copiously branched panicle. This form has long been known; it was figured by Boccone in the seventeenth century.[4] Azara,[5] who came upon it during his travels in Paraguay (1781–1801), described the transformed ear as consisting of "une espèce de discipline à plusieurs cordes", each covered with grains. On boiling in oil, all the grains burst, after the manner of pop-corn, though remaining in place, and "il en résulte un superbe bouquet, capable d'orner la nuit la tête d'une dame, sans que l'on puisse reconnaître ce que c'est". Such branched ears—interesting as they are—do not, however, throw much light upon the origin of the ear of Maize, for they recur in other cereals,[6] in which there is

[1] Blaringhem, L. (1908).
[2] Montgomery, E. G. (1906); Weatherwax, P. (1918).
[3] Harshberger, J. W. (1893); Schumann, K. (1904); Worsdell, W. C. (1915–16).
[4] Boccone, P. (1674), Pl. 16. [5] Azara, F. de (1809).
[6] See p. 385.

nothing peculiar in the normal inflorescence; moreover there is no reason to believe that teratological forms can be taken, in general, to represent ancestral stages.[1] We are driven to conclude that the problem of the nature of the ear of Maize, like so many other questions connected with this cereal, still awaits solution.[2]

Zea Mays L. is generally treated as a single 'species', consisting of a complex of races, which have sometimes, however, been elevated to specific rank. In an earlier chapter we have spoken of the obscurity which surrounds the question of its origin.[3] If there were evidence for the existence of any truly wild form of Maize at the present day, it might go far towards dispelling the mystery. There is, indeed, one peculiar and long-recognised race, which has sometimes been claimed as a wild form from which the cultivated Maizes are derived. This is Pod-corn, *Zea tunicata* Sturt., in which the mature grain is enclosed in enlarged glumes, instead of being exposed as in other races.[4] Saint-Hilaire,[5] who described it more than a century ago, believed it to be wild in Paraguay, but the evidence which he cites is inconclusive. Modern writers regard it, not as a wild plant, but as a mutation arising under culture and dominant to normal Maize, or else as a heterozygous condition, comparable with the Blue Andalusian fowl.[6] The idea that it is the primaeval type of the genus is thus excluded.

If no wild species of *Zea* now exists, it is clear that the only hope of light upon the origin of this cereal must come from an exhaustive study of its systematic relations and of the genera which appear to have the closest affinity with it; in this study American botanists have specialised. Fortunately there is no difficulty in deciding whereabouts in the Gramineae *Zea* should be placed. That it belongs to the sub-family known as the Panicoideae is shown by the fact that, in the female spikelet, an abortive flower occurs *below* the female flower (Fig. 183, A, p. 358), and also by such a detail as the absence

[1] For a discussion of teratology and atavism, see Arber, A. (1931[2]), pp. 197–200.
[2] Collins, G. N. (1919). [3] See pp. 31–2.
[4] Weatherwax, P. (1918).
[5] Saint-Hilaire, A. de (1829).
[6] See discussion in Collins, G. N. (1917), and Jones, D. F. and Gallastegui, C. A. (1919).

of a mestome sheath in the leaf-bundles (Fig. 188) .[1]Within the Panicoideae, *Zea* is placed in the Andropogoneae,[2] or in a somewhat heterogeneous tribe, the Maydeae, consisting of two groups, one of which is American, while the other is primarily Asiatic. The three New World genera—*Zea*, *Tripsacum* and *Euchlaena*[3]—form a related trio, distinct in character from the rest of the tribe. It is this

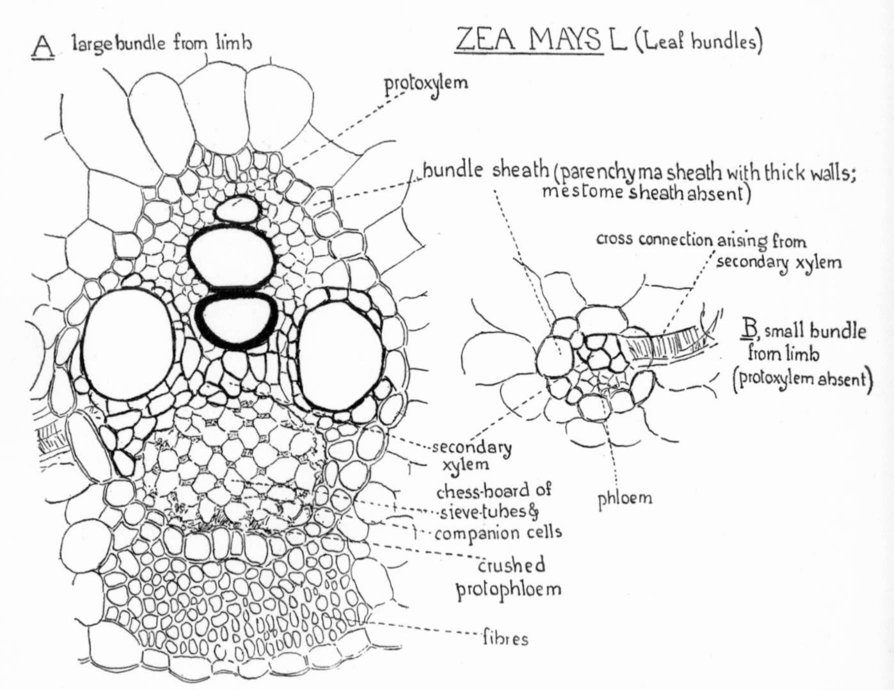

Fig. 188. *Zea Mays* L. Bundles from leaf-limb in transverse section (×193 *circa*). A, a larger, and B, a smaller bundle. The smaller bundle has no protoxylem and no large vessels. It consists of the 'secondary' type of xylem and it is from this xylem that branching is taking place. [A.A.]

group with which alone we are concerned, since there is no reason to doubt that the evolution of Maize took place in the American continent.

[1] On these points, see pp. 382–3.

[2] For a general account of the Andropogoneae, see Hackel, E. (1889).

[3] For a comparative study of these three genera, which has been largely drawn upon in the account that follows, see Weatherwax, P. (1918), and, on Teosinte, Collins, G. N. (1923).

Tripsacum includes two or three species, of which *T. dactyloides* L., Gama-grass, may be taken as an example. It is a perennial, bearing short, sterile, leafy shoots, and long, fertile shoots, which branch and produce numerous panicles. Two inflorescences are seen in Fig. 189, A 1, A 2, p. 370. The upper region of each branch bears paired male spikelets, while, lower down, we come to the female spikelets. C 1 and C 2, Fig. 189, are from sections through a pair of two-flowered male spikelets. The female spikelet (B 1), like that of *Zea*, first produces an abortive flower, and then a female flower with rudimentary stamens. The gynaecea are unlike those of *Zea* in possessing two distinct styles (B 2). Each female spikelet is sunk in a deep excavation in the axis, which, at maturity, breaks into segments enclosing the grains. Though *Tripsacum* shows a certain resemblance to *Zea*, this resemblance is less close than that between *Zea* and *Euchlaena mexicana* Schrad.,[1] the only representative of the third genus of New World Maydeae, which, under the name of Teosinte, is cultivated for forage in the warmest parts of North America.[2] In habit this species shows a general similarity to Maize, but many more suckers arise from the base of the stem, and produce tall shoots. The main axis and some of the taller laterals are terminated by male tassels, closely recalling those of *Zea*, except in showing no differentiated central spike (Fig. 190, A 1, B, p. 371). The female inflorescence (A 2) is, however, very different from that of Maize, for it is usually a simple spike, with alternating spikelets deeply embedded in segments of the rachis, which—as in *Tripsacum*—fall apart on ripening. The single styles with their paired bundles, however, recall those of Maize, and, like Maize, Teosinte is able to produce mixed inflorescences, as well as those that are exclusively male or female (A 4). Furthermore, Maize and Teosinte are hosts to the same parasitic diseases, and their chromosomes show certain similarities.[3] They hybridise freely; Maize, indeed, responds as easily and completely to the pollen of Teosinte as to that of its own race. The capacity of the

[1] Also called *Euchlaena luxurians* Dur. and *Reana luxurians* Dur.

[2] It is doubtful whether Teosinte is strictly a 'wild' plant, for "evidence is lacking of its existence in nature outside of cultivated regions"; see Collins, G. N. (1925).

[3] Collins, G. N. (1931).

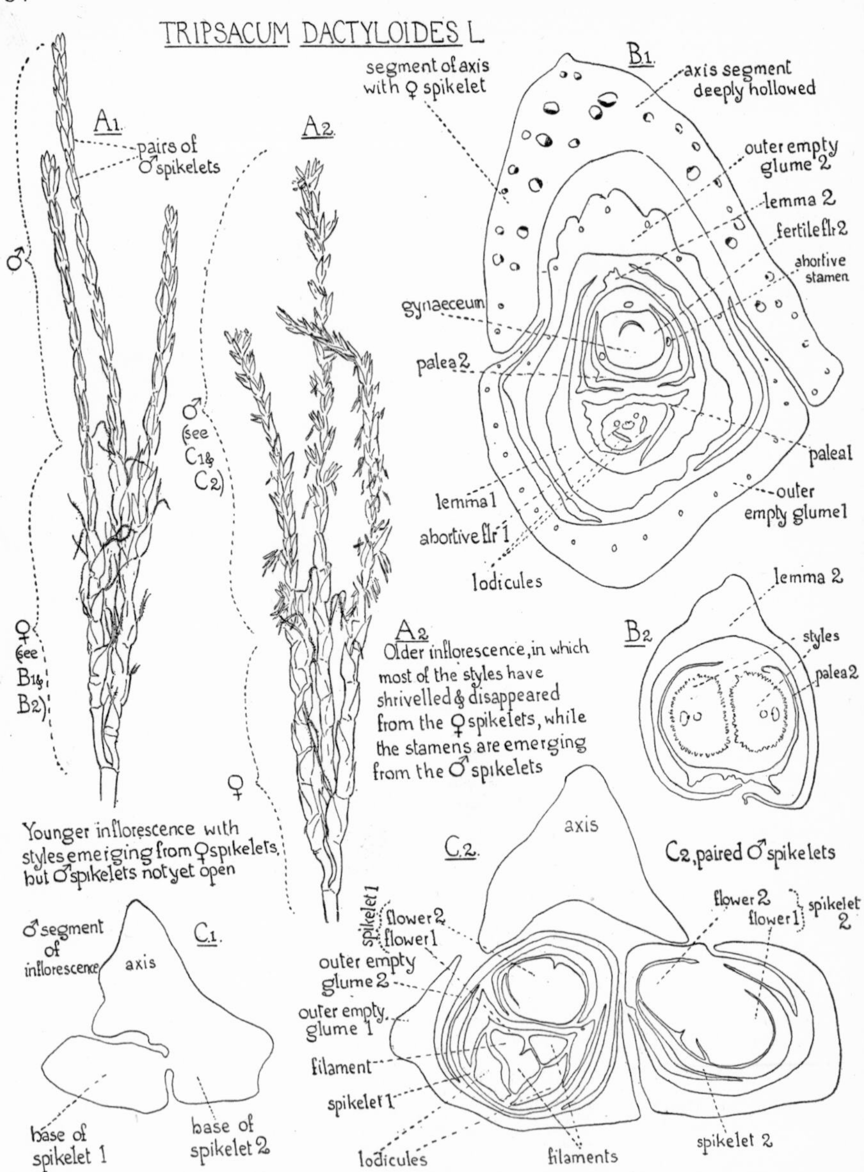

Fig. 189. *Tripsacum dactyloides* L. A 1 and A 2, younger and older inflorescences (× ½), Cambridge Botanic Garden, September. B 1 and B 2, sections from a transverse series through a segment of the axis with a fertile spikelet. B 1 (× 14), bundles not indicated except in axis and outer empty glumes. (First outer glume broken in section and somewhat reconstructed.) B 2 (× 23), fertile flower only, vascular tissue omitted, except in styles. C 1 and C 2, sections from a transverse series through a segment of the male part of the inflorescence. C 1, through the base of a spikelet pair (× 14); C 2, higher in the same segment (× 23). [A.A.]

two genera for interbreeding, their geographical relationship, and their resemblances in certain characters, indicate a close genetic

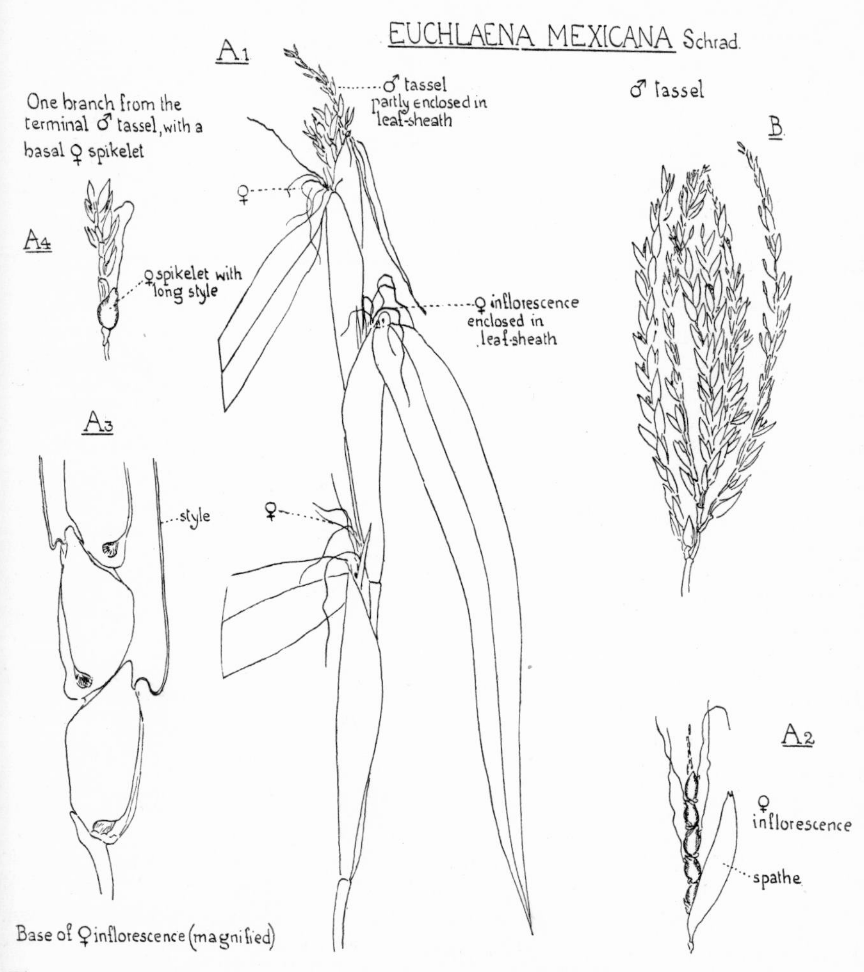

Fig. 190. *Euchlaena mexicana* Schrad. Herbarium material of cultivated plants from Professor Weatherwax. A 1, A 2, A 4 and B (× ½), A 3 (enlarged). A 1, reproductive shoot with male and female inflorescences partially enclosed in leaf-sheaths; A 2, female inflorescence; A 3, base of female inflorescence; A 4, androgynous branch from the male tassel. B, a male tassel. [A.A.]

connection. It has even been suggested that Maize is a mutation from Teosinte, or a teratological variety[1] of this grass, "née et

[1] Schumann, K. (1904).

propagée par les soins de l'homme".[1] Another view is that Maize arose by selection from Teosinte; it has indeed been claimed that Maize has been re-created by this method in recent times; but there is reason to suppose that the Teosinte employed in the experiment in question, was contaminated with Maize, so that the 'selection' was, in reality, segregation from a hybrid.[2] Moreover Teosinte is not now eaten by man, and it seems improbable that it can ever have been eaten in the past, so there is no apparent reason why selection should have been brought to bear upon it. A more plausible idea is that the qualities adapting Teosinte to the uses of man may have resulted from accidental hybridisation with another species.[3] It has been suggested that this hypothetical parent may have belonged to the Andropogoneae, and that it may have borne naked, or nearly naked grains on a rigid rachis. The difficulties offered by this theory are great—though they are perhaps less formidable than those presented by other views.[4] The chromosome numbers do not appear to give much help, though they do not preclude the hybrid theory; 10 is the haploid number in most strains of Maize and in annual Teosinte, and it has also been found in *Andropogon*.[5]

Another grass useful to man, which is believed, on stronger evidence than Maize, to be a hybrid, is *Spartina Townsendi* Groves,[6] the reclaimer of mud-flats. Before considering the question of its origin, however, we must review its history and behaviour. The members of the genus *Spartina*, Cord-grass,[7] occur chiefly in salt

[1] Blaringhem, L. (1908).

[2] Blaringhem, L. (1924); Collins, G. N. (1925); Weatherwax, P. (1925[1]).

[3] Collins, G. N. (1923); for arguments against the hybridisation view, see Weatherwax, P. (1919[2]).

[4] For an interesting modern summary of the "confusing array of conflicting hypotheses" concerning the evolutionary history of Maize, see Collins, G. N. (1931).

[5] For references, see Fisk, E. L. (1927).

[6] The following account of *Spartina Townsendi* Groves is derived from Bugnon, P. (1920[1]); Carey, A. E. and Oliver, F. W. (1918); Corbière, L. (1927); Groves, H. and J. (1881); Groves, J. (n.d.); Oliver, F. W. (1920), (1924), (1925), (1926), (1927–8); Stapf, O. (1914), (1926). According to the modern International Rules of Botanical Nomenclature, the specific name should have been written "*Townsendii*", but as the name dates from 1881, it appears that it should retain its original form.

[7] On the inappropriateness of the names Cord-grass and Rice-grass, see Stapf, O. (1914).

Fig. 191. *Spartina Townsendi* H. and J. Groves. Vire Estuary, Normandy, August 27, 1927. Part of a clump (× ½). This specimen does not show the long anchoring roots. [A.A.]

marshes on the Atlantic coast of America. In Europe one species, *S. stricta* Roth, is indigenous; in Britain it is found here and there on muddy shores on our southern and eastern coasts. In 1836 a colony of an American Cord-grass, *S. alterniflora* Lois., presumably introduced by shipping, was noticed at the mouth of the river Itchen, whence it gradually spread. Then, in 1870, a third Cord-grass, hitherto unknown, was discovered at Hythe, on Southampton Water. At first it was supposed that it was merely a form of our native *S. stricta* Roth, but more careful examination showed that it had a good claim to be regarded as a distinct species, and it received the name of *S. Townsendi* Groves, in honour of Frederick Townsend, who had been the first to suggest that it deserved specific rank. In the 'nineties it reached various localities in the Isle of Wight, and it has since spread along the south coast of England to east and west. In 1899 it reached Poole Harbour, where it alarmed the authorities by its rapid spread and the silting which it induced. On the other side of the Channel, it was first observed about thirty-six years after its début in England. On September 22, 1906, Louis Corbière noticed fifteen to twenty tufts of it in the neighbourhood of Carentan (Normandy) near the junction of the Douve and the Taute, and, a little later in that autumn, another small station was discovered elsewhere in the same area. In fifteen years, from this unobtrusive beginning, Townsend's-grass had spread so as to occupy an area of more than 2000 acres in the Baie des Veys, the joint estuary of the Douve and the Vire. On August 27, 1927, I had the good fortune to visit the Baie des Veys under the guidance of Louis Corbière and Pierre Bugnon. When we came to the estuary, we looked over an area entirely clothed, as far as the eye could reach, by *Spartina Townsendi* in full flower, lighted by the golden tint which it assumes at this season. The beauty of the infinite ranks of "spear-like spikes wrapped in a veil of long chenille-like whitish stigmas" was so great, that it was its charm, rather than its economic importance, which made the first impression. When it was realised, however, that twenty years earlier this vast *Spartina* tract had been represented merely by a monotonous sheet of barren tidal mud, admiration for the beauty of the Cord-grass paled before surprise at its power to work such a miracle with no help from man; it is not without reason that it has been called a "superplant". To grasp the qualities which enable it to create dry land, it is necessary to consider its structure and life-history. It is apparently the rich system of air spaces throughout the plant which permits it to flourish in soft "bottomless" muds, which—partly on account of their poverty in oxygen—"have resisted for ages the entry and establishment of other halophytes". The soft viscous texture of these muds makes anchorage a great difficulty, but Townsend's-grass overcomes this by the rapid establishment of a deep root system. There is dimorphism among the roots.[1] Some are elongated and descend vertically; the greatest length recorded for these anchoring roots is 4 ft., but 2 or 3 ft. is more common. The remaining roots are short and much branched, spreading all round the base of the stem and the nodes of the stolons, close to the surface of the mud; these branched roots are shown in Fig. 191, p. 373.

The region which *Spartina Townsendi* colonises is that from a little below high water at spring tides to a line about a metre below this level; six hours' consecutive immersion per tide seems to be the most that it can endure. According to whether the slope is rapid or gradual, the belt occupied may vary from some metres to some kilometres in breadth.

The process of mud colonisation has been followed in the seedling, but a fuller account of the history from embryo to maturity is still needed. Anthesis takes place from July to November—an unusual flowering season, vying with that of the Mediterranean *S. versicolor* Fabre, which blooms from October to April.[2] There are difficulties to be overcome in analysing the life-history of Townsend's-grass, since, in the words of F. W. Oliver, who has added so much to our knowledge of this plant,—"From the special nature of the habitat Spartina studies are an acquired taste. Every visit to a clump requires the use of mud boards on the feet". November seems to be a suitable time to look for germinating seeds, which are sometimes seen sprouting viviparously from fruits which are still attached to the

[1] For other cases of root dimorphism, see p. 251.

[2] Fabre, E. (1850).

spike; the majority are to be found, however, in the tidal drift. One might suppose that the detached fruits would sink too deeply in the soft mud and so lose their chance of germination, but this catastrophe is averted by the delicate weft of filamentous algae which clothes the slime, and entangles and holds the spikelets. The young plants, by their first summer, are 6 or 8 in. high, with ten or twelve basal lateral shoots and a dense tuft of roots. The lateral branches, growing out from the base of the plant in all directions, produce a characteristic circular tussock, which, in the course of years, may reach a diameter of 15 ft. or more. In an area of 100 sq. cm. (15½ sq. in.) of *Spartina* meadow, there are, on an average, 200 anchor roots, 24 erect shoots, and 30 underground rhizomes.[1] It is, indeed, reasonable to describe the plant as "an astonishingly vigorous and pushful halophyte"; its vigour is proved by counts such as these, and its "pushfulness" by various observations upon its relations with the few other plants which are capable of living under the same searching conditions. In the area which *S. Townsendi* has invaded on the south coast of Britain, the tidal muds, before its appearance, were generally clothed with a close felt of *Zostera nana* Roth, but the Cord-grass, in the process of establishing a footing, has ruthlessly effaced the Grass-wrack. *Spartina Townsendi*, when in close formation, will also overwhelm *Scirpus maritimus* L., the Sea-club-rush, and *Glyceria maritima* Mert. et Koch, the Sea-meadow-grass. The mode in which it ousts the latter species has been followed on the French coast.[2] At the periphery of the circular tufts of *Spartina*, the turf of *Glyceria* is, little by little, raised up and forced outwards by the centrifugal growth of the vigorous vertical axes arising in serried ranks from the rhizomes of the Cord-grass. Occasionally, however, the position is reversed, and it is the *Spartina* that goes under. An instance has been placed on record in which *Glyceria maritima* was apparently able to invade and overrun a strip of salt marsh on the Seine, formerly occupied by *Spartina Townsendi*.[3] Moreover, in New Zealand, where it has been introduced, this species cannot make its way in competition with the native Sea-rush, *Juncus maritimus*

[1] On the growth of *S. glabra* Muhl., see p. 342.

[2] Bugnon, P. (1920[1]).

[3] Oliver, F. W. (1926).

Lam. var. *australiensis*, or with *Leptocarpus simplex* A. Rich.[1] (Restionaceae); but this failure in New Zealand seems to have occurred in positions which are above the level best suited to the needs of *Spartina*.

As Townsend's Cord-grass becomes established, its "stems and lower leaves and leaf-bases act as a very effective strainer on the water, with the result of an accelerated and increased deposition of mud over the area tenanted by the grass". The rise in level will sometimes reach as much as 4 or 5 in. a year. The stabilising power of its roots and rhizomes, joined to its mud-filtering capacity, make *Spartina Townsendi* an invaluable ally in land reclamation. It has been used in Holland for the last ten years in the reclaiming of land from the sea or from tidal rivers, and there is no doubt that there is a great future for it in such work.

There has been much controversy as to the origin of Townsend's Cord-grass. It has been suggested that it is a foreign species, accidentally introduced; but no member of the genus, with which it can be exactly matched, is known from any part of the world. Another view is that it is a sport or mutation from *S. stricta* Roth. There seems, however, to be no evidence for this idea, for *S. stricta* is a species of a conservative type, which has not been known to give rise to new forms.

A third alternative is that *S. Townsendi* is a hybrid between our native *S. stricta* and the American introduction, *S. alterniflora*, both of which inhabited Southampton Water before the appearance of *S. Townsendi*. As their flowering seasons do not usually synchronise, crossing can only have come about through some change in the periodicity of one or more individuals. The only other locality where these two species are known to overlap in their distribution is the innermost corner of the Bay of Biscay, and here another form, *S. Neyrautii* Fouc., different from either, also occurs; it is similar to *S. Townsendi*, though not identical with it. Like *S. Townsendi*, it seems to be a hybrid between *S. alterniflora* and *S. stricta*; the differences between these two putative hybrids have been accounted for on the (perhaps unnecessary) theory, that they have resulted from

[1] Allan, H. H. (1930).

reciprocal crosses. Until 1930, the evidence for the hybrid origin of *S. Townsendi* was circumstantial only, but recent cytological work[1] seems to have settled the question beyond any reasonable doubt. It has been shown that this grass possesses a very high somatic chromosome number—126. Now *S. alterniflora* has, as haploid number, 35, while *S. stricta* has 28. Interspecific hybridisation would thus give 63 for the somatic number, and if this were followed by chromosome doubling,[2] the number, 126, found in *S. Townsendi*, would be obtained. A hybrid origin would account for the fact that Townsend's-grass is taller and more vigorous than either of its supposed parents; while the doubling of its chromosomes would explain its perfect fertility, and the fact that, on the whole, it breeds true—points which have sometimes been regarded as obstacles in the way of the hybrid theory. It is pleasant to find modern cytological work confirming the conclusion as to hybrid origin, at which systematists had arrived as the result of their own studies. We still await the chromosome count for *S. Neyrautii*, and the experimental production of the two reciprocal hybrids between *S. stricta* and *S. alterniflora*; when these have been achieved, the story will be, in essentials, complete.

Among the plants useful to man, there are many others which are believed to be natural hybrids; examples are Pearl-millet,[3] the large American form of Timothy-grass,[4] Toowoomba-canary-grass,[5] Germiston- or Florida-grass,[6] and various Wheats.[7] To emphasise these examples, however, may give a misleading impression; for—now that we are becoming aware of the part which hybridisation has apparently played in evolution—we have to disembarrass ourselves of "the notion that a form is *either* a species *or* a hybrid".[8] The new

[1] Huskins, C. L. (1930).
[2] On the process of chromosome-doubling in hybrids, called allo-polyploidy, see Darlington, C. D. (1928).
[3] See p. 25.
[4] Gregor, J. W. and Sansome, F. W. (1930).
[5] Jenkin, T. J. and Sethi, B. L. (1932).
[6] Bews, J. W. (1929).
[7] See p. 8.
[8] Darlington, C. D. (1928); on hybridisation in general, see also Hurst, C. C. (1933), and on cytological evidence for hybridity in grasses, see Church, G. L. (1929^1), (1929^2).

conceptions of race constitution, which have arisen out of genetics and cytology,[1] are not easy to fit into the orthodox systematic framework; but the old disharmony between the standpoint of the field and herbarium worker, and of the genetical cytologist, is rapidly being resolved by co-operation in research.[2]

[1] On the cytological side, see Darlington, C. D. (1932[1]).

[2] On the nuclei of the Gramineae, considered in relation to systematics, see Avdulow, N. P., "Karyo-systematische Untersuchung der Familie Gramineen", *Bull. Applied Bot., Gen. and Plant Breeding*, Supp. 44, 1931, 428 pp. (Russian, with German abstract, 72 pp.) This memoir came to my notice too late for inclusion in the bibliography.

CHAPTER XVII

PATTERN AND RHYTHM IN THE GRAMINEAE

THE grass family is remarkable both for its naturalness and for its distinctness from neighbouring groups. Though other families have sometimes been included with it in the order[1] Glumales or Glumiflorae, the most recent tendency is to treat it as a group apart,[2] and to regard the resemblances which it shows, for instance, to the Sedge family (Cyperaceae), as indicating parallel development rather than affinity. If this view be accepted, I should like to suggest that the established word *Gramineae* might be retained for the *order*; the name *Graminaceae*[3] could then perhaps be used for the only *family* included in this order.

Within the group itself, classification presents special difficulties; the subdivisions have even been described as "complètement artificielles".[4] The species are very numerous—they are estimated as 8000, or even more, grouped in about 550 genera[5]—and their structural characteristics do not lend themselves to any obvious classificatory scheme. Indeed, as Kunth[6] wrote, more than a hundred years ago, "dans les familles éminemment naturelles, comme celles des Graminées,...on ne trouve que très-peu de caractères...qui

[1] The word 'order' is used here in the sense in which 'cohort' is used in Arber, A. (1925).

[2] Fritsch, K. (1932).

[3] Dr J. Burtt Davy has suggested to me that, though "Gramineae" has been retained as the name for the family by the International Rules, it might be an advantage to return to Lindley's "Graminaceae", since the name would then fall into line with those of other families formed from an old generic name with the ending "aceae". Tournefort's "Gramen" (1719), on which the family name would then be based, appears to have been a duly described and valid genus, though it included a number of grasses allotted subsequently to different Linnean genera. "Graminaceae" would be equivalent to "Poaceae" used by Hitchcock, A. S. (1920), etc.

[4] Camus, E. G. (1913).

[5] The late Dr O. Stapf gave me these figures, as provisional for the family, in a letter dated May 5, 1933; he pointed out, at the same time, that the estimation of the number of species presents great difficulties, owing to the changes which have taken place in the species concept.

[6] Kunth, C. (1815).

puissent servir à distinguer les genres, et le plus souvent ces caractères sont aussi vagues que minutieux". In classifying the grasses, botanists have thus been obliged to rely, even more than usual, upon that intuitive faculty for detecting affinities, which can be cultivated to a high pitch by "la grande habitude de voir".[1] There is nothing magical about this faculty; it is the subconscious result of long labour acting upon an instinctive interest. It leads to the 'placing' of the plant as a whole, for reasons which are no more definable than those which enable one to recognise one's friends at the most cursory glance. As contrasted with the method from 'characters', the process is holistic rather than analytic. It must have been this intuitive faculty which led Robert Brown,[2] early in the nineteenth century, to distinguish the Pooideae and Panicoideae as sub-families of the Graminaceae. One of the main distinctions on which these subdivisions are based is that, in the Pooideae, the upper flowers of the spikelet are liable to be rudimentary (e.g. *Melica nutans* L., Fig. 81, p. 177), while, in the Panicoideae, there is a tendency to imperfection in the lower flower (e.g. *Zea*, Fig. 183, A and B, p. 358; *Coix*, Fig. 97, p. 198; *Ischaemum*, Fig. 98, p. 200). Another distinguishing feature is that, in the Pooideae, the mature spikelet is usually detached above the outer empty glumes, so that it leaves them behind when it falls; but in the Panicoideae, on the other hand, the outer glumes remain in connection with the spikelet, even when it is shed. These differences may seem, at first glance, to be somewhat artificial points on which to base a classification, but experience shows that—though exceptions can be found to both of them—on the whole they separate the grasses into two groups, which have a fair claim to be natural divisions.[3]

The Pooideae consist of the bamboos and a series of other tribes, which include most of the British grasses, and of the cereals cultivated in Europe. The Panicoideae include the Andropogoneae (e.g.

[1] Turpin, P. J. F. (1820).

[2] Brown distinguished the two groups in 1810, adding, as a doubtful third group, a set of genera including *Anthoxanthum*, *Phalaris* and *Ehrharta*. In 1814 he named the two sub-families Poaceae and Paniceae; see Brown, R. (1810) and (1866).

[3] A table showing the classificatory position of the various genera mentioned in this book will be found on p. 410.

Sorghum), the Maydeae (e.g. *Zea*), and the Paniceae (e.g. *Setaria*). The two tribes Phalarideae and Oryzeae are generally placed in the Pooideae, but it is possible that they have more affinity with the Panicoideae. The naturalness of the division of the Graminaceae into the two sub-families is attested by a curious little fact which was discovered in the laboratory long after these divisions had been established by systematists; in the leaves of the Pooideae, the bundles are enclosed in a thick-walled mestome sheath, surrounded by a parenchyma sheath, while in the majority of the Panicoideae there is a parenchyma sheath alone.[1] This distinction is illustrated in Fig. 192, in which B and C are taken from the Pooideae, and A and D from the Panicoideae.

The differences between the sub-families, which we have been considering, are of a *qualitative* character; but in the spikelets belonging to different tribes there are also certain differences which may be described as of a *geometric* type. In some tribes there is a tendency towards lateral compression of the spikelets, while in others they look as if they had been pinched from back to front. These recurrent trends are illustrated by the contrasting pairs of spikelets sketched in section in Figs. 193, p. 384, and 82, p. 178. Early in the nineteenth century, Turpin[2] was intrigued by variations of this kind on a given type, and his account of their significance is worthy of attention, even today. In his own words, "il suffit, si je puis m'exprimer ainsi, de tirailler ce type plutôt dans un sens que dans un autre, pour en obtenir toutes les modifications possibles, et sans que pour cela ce type cesse d'être un instant le même". As an example of the 'pulling about' of a type in order to obtain a series of derivatives, he instances a head drawn on a piece of parchment of some elasticity, which, if stretched to different degrees in different directions, shows absurd grimaces and caricature effects, but without losing its individuality. A century later, this conception was given an exact mathematical guise by D'Arcy Thompson.[3] He showed that related forms may often be described, on the method of co-ordinates, as "deformations" of one fundamental type.

[1] Schwendener, S. (1890); see also p. 305.
[2] Turpin, P. J. F. (1820).
[3] Thompson, D'Arcy W. (1917).

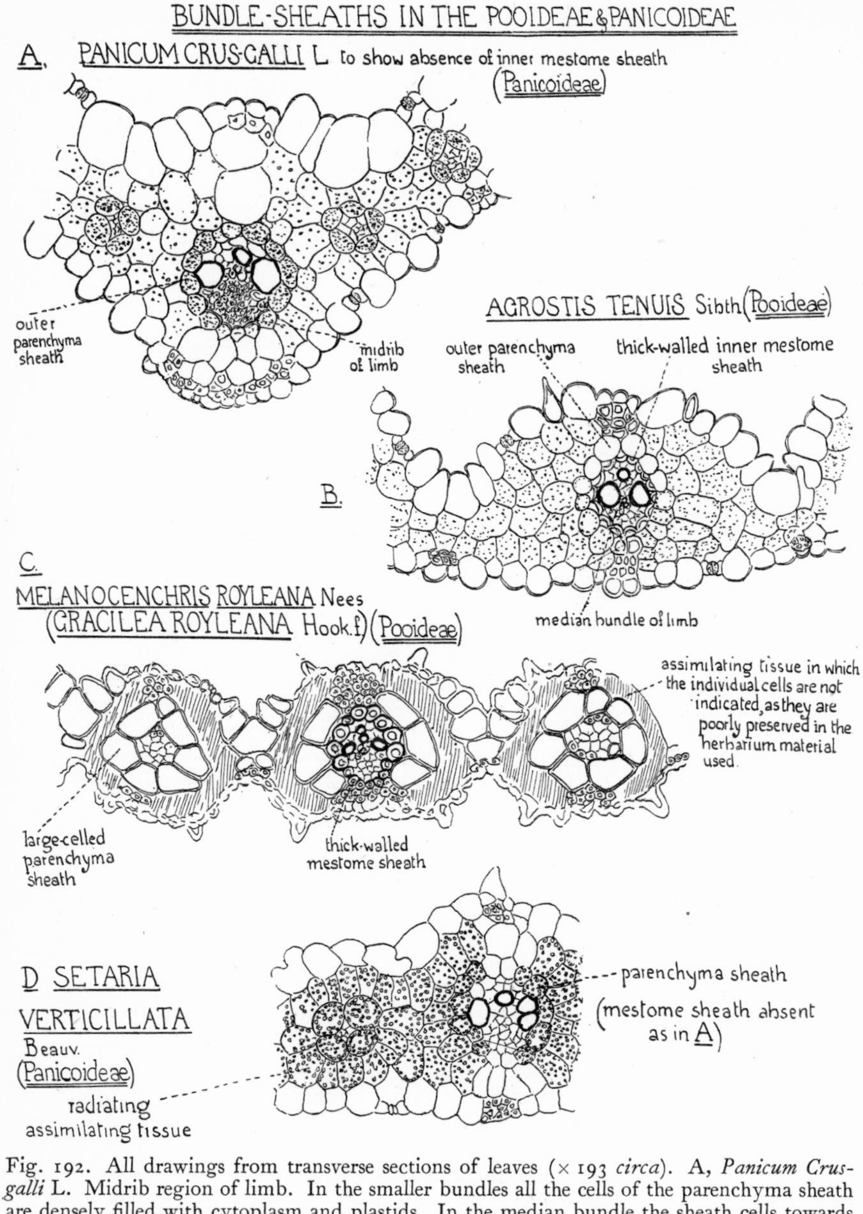

Fig. 192. All drawings from transverse sections of leaves (× 193 *circa*). A, *Panicum Crus-galli* L. Midrib region of limb. In the smaller bundles all the cells of the parenchyma sheath are densely filled with cytoplasm and plastids. In the median bundle the sheath cells towards the phloem have thickened inner walls, and both they, and the sheath cells at the opposite pole (towards the xylem), have less dense contents than the cells on the flanks. B, *Agrostis tenuis* Sibth., midrib of limb. C, *Melanocenchris Royleana* Nees (*Gracilea Royleana* Hook. f.). From Bombay Island (Professor Blatter). Part of limb showing three ribs. D, *Setaria verticillata* Beauv. Part of limb, fresh material. [A.A.]

Comparative analysis of species and races shows that whole series of variations repeat themselves, time after time, in the different branches of the family. By realising these repetitions we arrive at a sense of the rhythmical pattern underlying the group of phenomena which we call 'the Gramineae'. Such parallelisms have been studied especially in the cereals, but they are also readily discernible in wild grasses. Numerous examples might be cited, but it will suffice

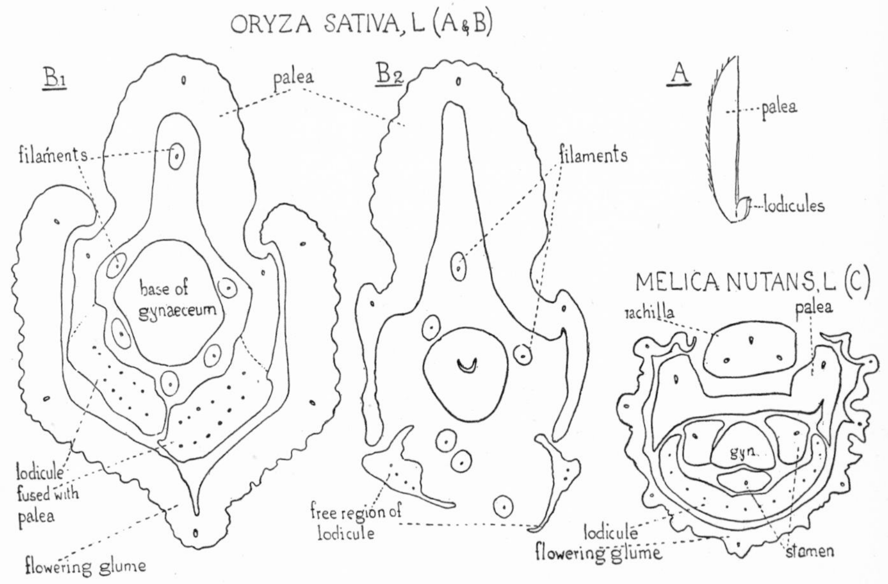

Fig. 193. A and B, *Oryza sativa* L. A, external view of the folded palea to which the two lodicules are attached at the base (enlarged). B 1 and B 2, two sections from a transverse series from below upwards through a flower (× 41) passing through the filaments of the six stamens. The lodicules are fused with the palea in B 1 and free in B 2. C, *Melica nutans* L., transverse section through a flower showing the single anterior lodicule (× 41). [Arber, A. (1927^2).]

to pick out one here and there. A French botanist[1] long ago noticed an example of parallelism between three species of *Agropyron* and three species of *Brachypodium*. He considered that *Agropyron caninum* Beauv. corresponded to *Brachypodium sylvaticum* Beauv.; *Agropyron repens* Beauv. to *Brachypodium pinnatum* Beauv.; and *Agropyron campestre* Godr. to *Brachypodium ramosum* Roem. et Schult. "Ce parallélisme", he wrote, "est si complet et si parfait, que, pour le

[1] Duval-Jouve, J. (1875).

décrire, il n'y aurait qu'à répéter sur une espèce ce qu'on aurait dit de la correspondante."

In reproductive behaviour, striking parallelisms may be found in different tribes of the Graminaceae. Both cleistogamy and separation of the sexes must have arisen independently in unrelated genera. Such a detail, also, as the pairing of the spikelets—one sessile (or nearly so) and one stalked—recurs repeatedly, e.g. *Luziola* (Fig. 166, A, p. 317) and *Zea* (Fig. 185, A, p. 361).

It is not only in normal characters that repetition and parallelism can be traced; they are expressed at least as clearly in the field of teratology. 'Wonder-wheats' with branched ears have long been known. Lobel gave a woodcut of one of these forms in 1581 (Fig. 194, p. 386). This change from a 'spike' to a branched inflorescence can be paralleled in other members of the Gramineae, for instance, *Zea Mays* L. (p. 366) and Rye-grass (*Lolium perenne* L., Fig. 195, C, E, p. 387).[1] In the latter species I have seen an example with as many as twelve lateral branches bearing spikelets. Another peculiarity of *L. perenne* L. is the elongation of the spikelet axis, without production of lateral spikelets (Fig. 195, F); this is repeated in *Trisetum flavescens* Beauv. (Fig. 197, D 1, p. 389). A more familiar abnormality is the conversion of the spikelet, above the first two glumes, into a leafy shoot (Fig. 195, G, H). Proliferated spikelets of this type can be paralleled throughout the Gramineae; *Poa alpina* L. f. *vivipara* L. a familiar example. Inflorescences of Timothy-grass[2] in this state are shown in Fig. 197. The specimens came from a roadside near Cambridge, where these leafy 'spikes' are developed season after season; I have found them there in eight different years. The best time to look for them is between September and December. They illustrate the conservatism of the outer empty glumes, which retain their normal character without sharing in the proliferation; this is shown in Fig. 197, A 2 and A 3, and it can be paralleled for *Festuca ovina* L. in Fig. 197, C 2, and for *Arrhenatherum avenaceum* Beauv. in Fig. 196, A 3.

One of the extremest instances of proliferation, which have been

[1] See also Figs. 73, p. 167, and 74, p. 168.
[2] Toumey, J. W. (1891); Evans, M. W. (1927).

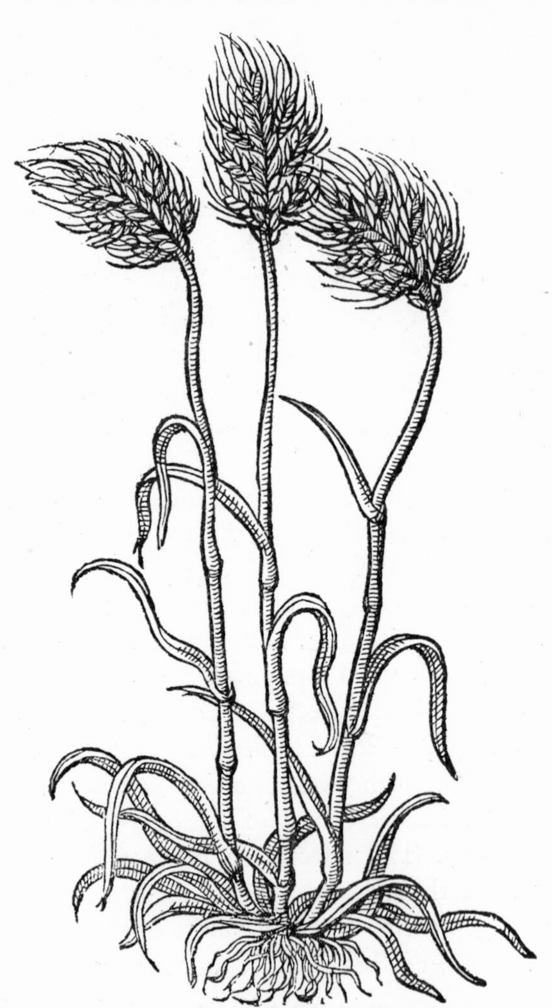

Andere Ghebaerde Terwe/Blé loca.

Fig. 194. Wheat with branched ears, 'Wonder-wheat'.
[From the *Kruydtboeck* of Mathias Lobel (de l'Obel), 1581.]

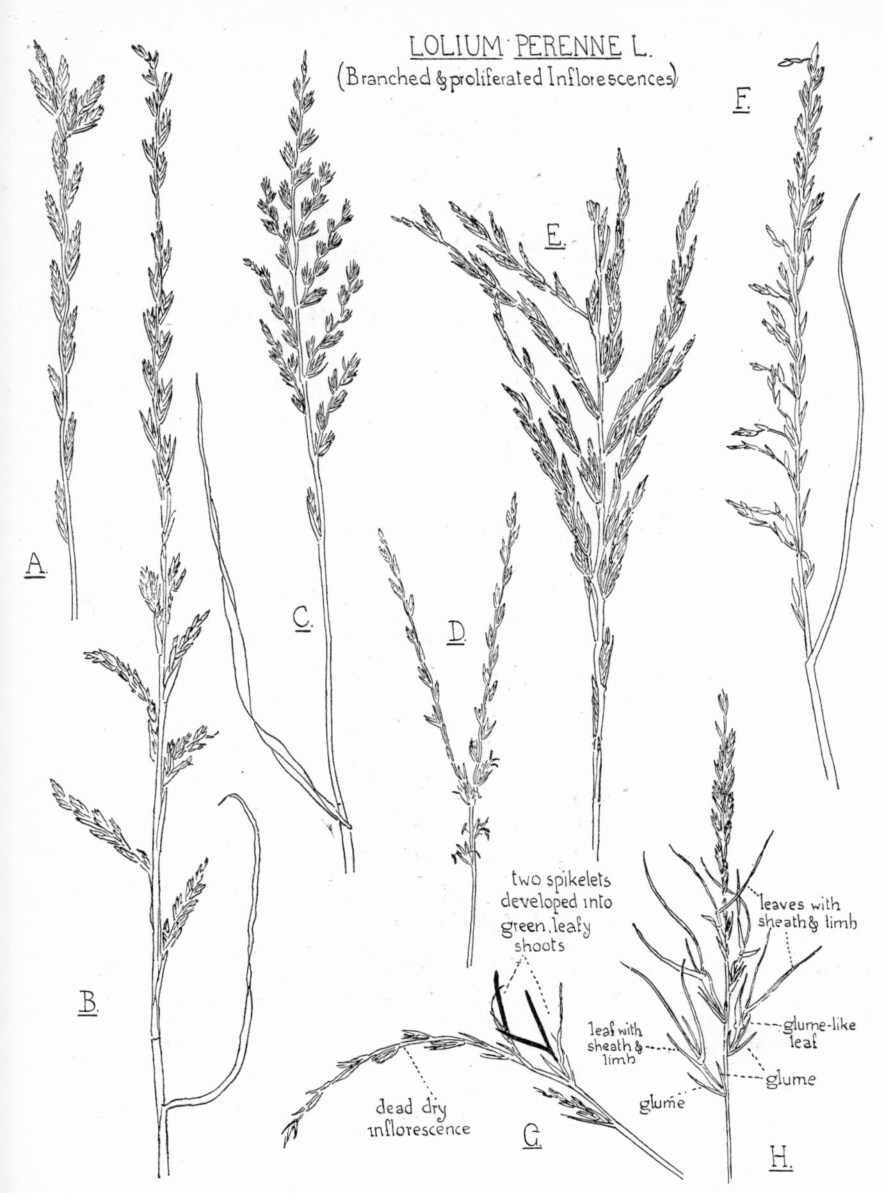

Fig. 195. Abnormal forms of *Lolium perenne* L. (all × ½). A, B, D, F, Whitwell, Isle of Wight, August, 1924; E and H, Whitwell, Isle of Wight, August, 1925; C, near Cambridge, October, 1924; G, near Cambridge, September 28, 1924. [A.A.]

described, is a sterile head of Sorghum,[1] in which the lower spikelet of each pair bore, in succession to the outer glumes, a series of as

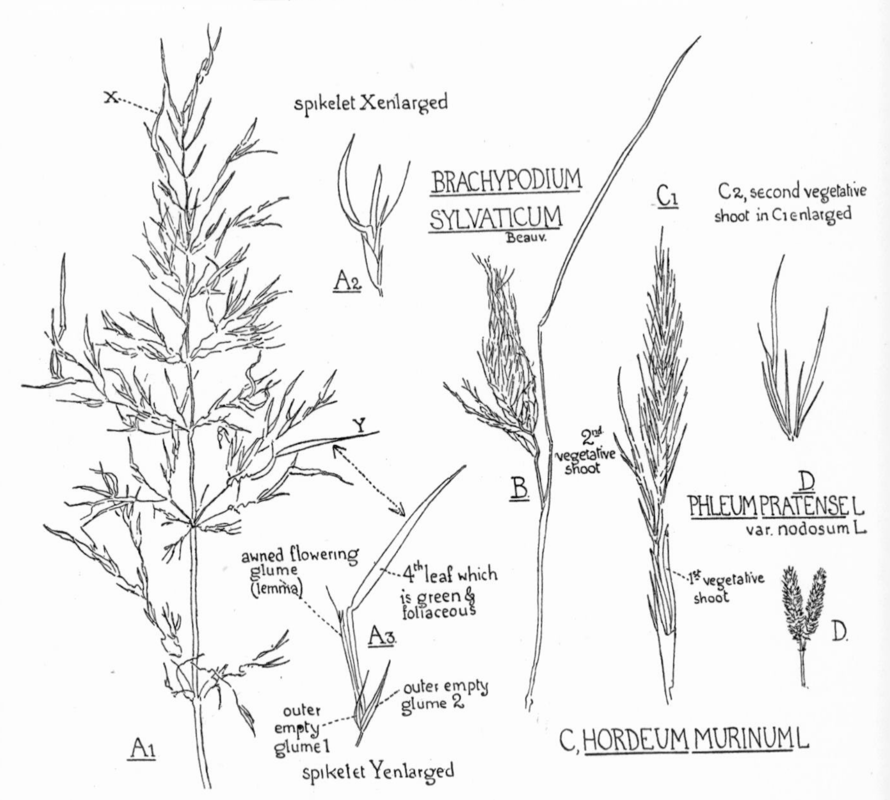

Fig. 196. Abnormal inflorescences. A, *Arrhenatherum avenaceum* Beauv., Girton, Cambridge, December 5, 1926. A 1, faded inflorescence with proliferating shoots ($\times \frac{1}{2}$). B, *Brachypodium sylvaticum* Beauv., inflorescence ($\times \frac{1}{2}$), Whitwell, Isle of Wight, August 5, 1926. The spikelets are crowded on a short, bent and curved main axis. C, *Hordeum murinum* L., Cambridge, December 28, 1926. C 1, proliferating inflorescence ($\times \frac{1}{2}$); C 2, second vegetative shoot (enlarged). D, *Phleum pratense* L. var. *nodosum* L. inflorescence with twin 'spikes' ($\times \frac{1}{2}$). Kettlewell, Yorks, August 24, 1928. [A.A.]

many as 28 to 41 scale-leaves closely inserted on the spikelet axis. In *Cynosurus cristatus* L., I have seen a comparable example. In this

[1] Laude, H. H. and Gates, F. C. (1929).

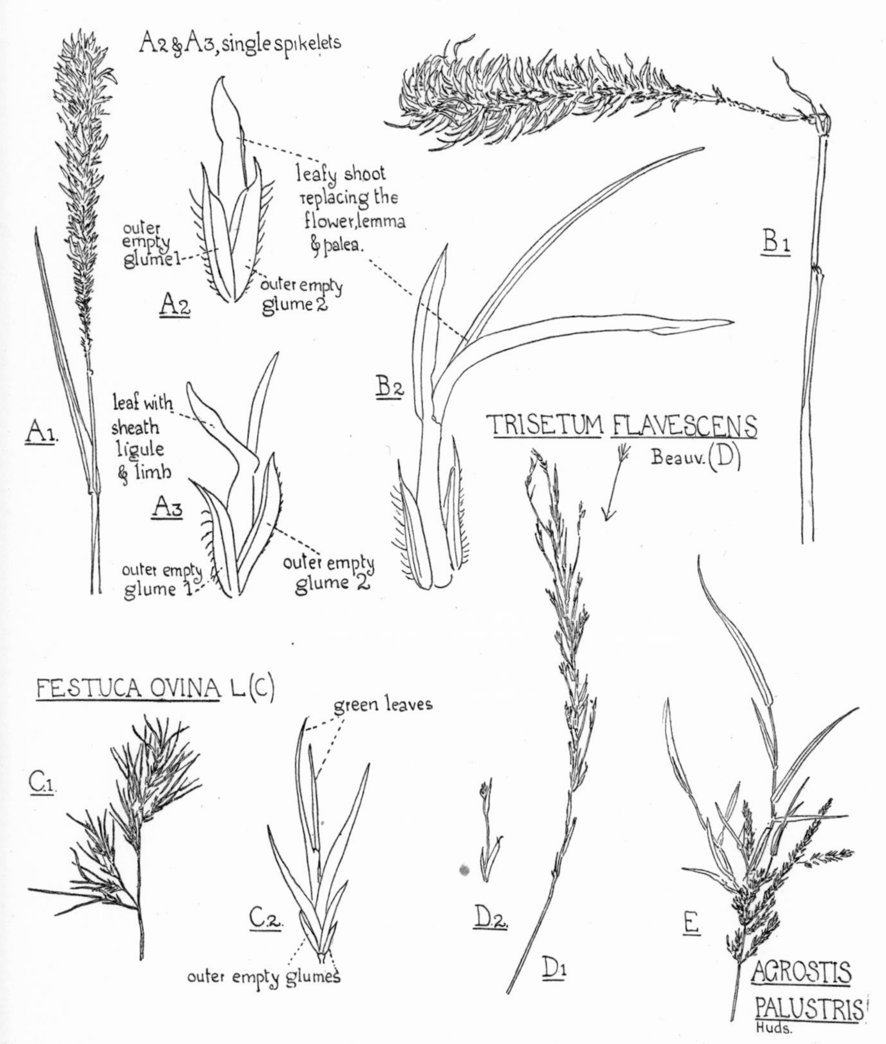

Fig. 197. Proliferating inflorescences. A and B, *Phleum pratense* L. A 1 and B 1, inflorescences near Madingley, Cambridge, November, 1924 (× ½); A 2, A 3, B 2, individual proliferating spikelets on a larger scale. A 2 is the least modified. C, *Festuca ovina* L., proliferating inflorescence from the Cambridge Botanic Garden (× ½). C 2, one spikelet enlarged. D, *Trisetum flavescens* Beauv. D 1, inflorescence from Carteret, Normandy, September 3, 1927 (× ½). D 2, single spikelet with an elongated axis and 'bouquet' tip, on a slightly larger scale. E, *Agrostis palustris* Huds., Kettlewell, Yorks, late in August, 1929. A branched vegetative shoot, rooting at the nodes, is borne laterally on the completely dry and shrivelled inflorescence axis (× ½). [A.A.]

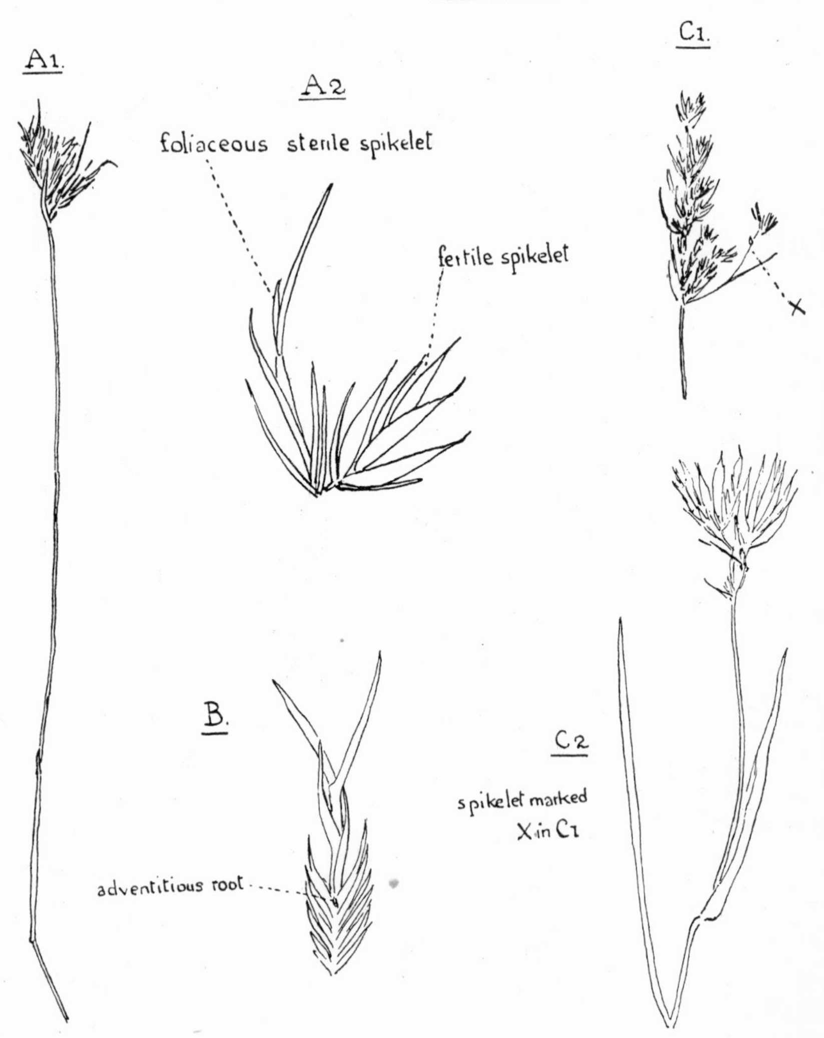

Fig. 198. Proliferation in *Cynosurus cristatus* L. A 1, depauperated example, with some of the sterile spikelets foliaceous, downs above St Lawrence, Isle of Wight, August 29, 1925 (nat. size). A 2, one pair of spikelets enlarged; the fertile member pushed aside to show the foliaceous sterile member. B, proliferated sterile spikelet, Carteret, Normandy, August 22, 1927 (enlarged). C 1, proliferating inflorescence, Eynsham Park, Oxon, September, 1930. C 2, spikelet marked *X* in C 1, on a larger scale, to show that it is of the 'bouquet' type. [A.A.]

species the sterile spikelets, which constantly accompany those that are fertile, may be prolonged so as to bear numerous leaves; in the spikelet drawn in Fig. 170, B2, p. 322, there were twenty-six. This example from the Dog's-tail-grass, since it affects spikelets which are sterile even in the normal spike, indicates that spikelet-proliferation falls into place as a further stage in that sterilisation process which we have already noticed in the normal reproductive shoot in the Gramineae.[1] A special vegetative development in the inflorescence, which I have called the 'bouquet' abnormality, can also be illustrated from the Dog's-tail-grass; in this form, lateral axes, which are more or less naked, terminate in tufts of spikelets (Fig. 198, C2). The 'bouquet' can be studied more easily in another species in which it recurs—the Cock's-foot-grass, *Dactylis glomerata* L.; a characteristic form is seen in Fig. 199, D2, p. 392. The axis, from the remains of the glumes at the base to the top of the terminal spikelet cluster, may measure 3 cm.

It is unfortunate that, even in recent literature, the word 'vivipary' is used to describe the vegetative proliferation of the spikelets of grasses, whereas this term should be confined to the germination of seeds *in situ*, as in *Melocalamus*.[2] Under natural conditions, indeed, proliferated spikelets seem seldom to grow into new plants.

At first glance, one might be inclined to attribute proliferation entirely to external conditions, for it is certainly easiest to find examples in the autumn, more especially after a wet summer, and there is no doubt that excess of water may sometimes be a predisposing cause. In *Deschampsia caespitosa* Beauv. var. *rhenana* Grem., for instance, a submerged plant has been noticed, in which flower-production was entirely inhibited in favour of proliferation, while neighbouring examples, which had not been swamped, bore flowers and fruit. It is not safe, however, to generalise from such examples, for, in the same variety, a race is known in which this peculiarity is heritable and does not depend upon the environment.[3] These two classes of proliferation—that controlled by the environment, and

[1] See Chapter IX.
[2] See p. 65.
[3] Schroeter, C. and Kirchner, O. von (1902).

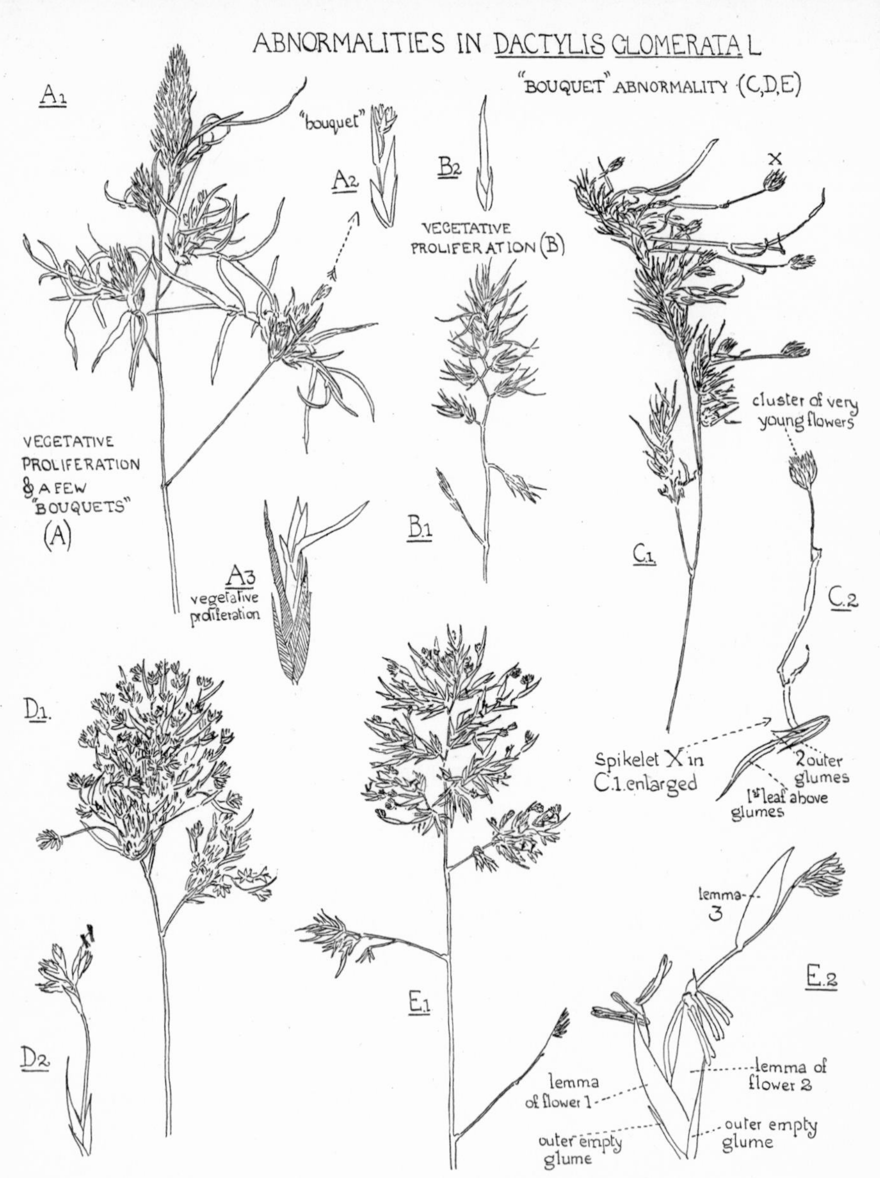

Fig. 199. Abnormalities in *Dactylis glomerata* L. A 1, near Cambridge, November 11, 1924 (× ½); A 2 and A 3, single spikelets on a larger scale; in A 3, the four basal leaves, which are shaded, are dry and shrivelled, but the succeeding leaves are bright green and have a distinct sheath and limb. B 1, near Cambridge, January, 1927 (× ½); B 2, a single spikelet on a larger scale. C 1, one of two inflorescences showing 'bouquet' abnormality, from a shady lane near St Catherine's Down, Whitwell, Isle of Wight, August 9, 1925 (× ½); the shoot was drooping right over, and to give the natural position, the main axis should have been placed horizontally in the figure, with shoots such as *X* rising vertically. The normal inflorescences near by were dead and faded. D 1, from near Cambridge, September 25, 1924 (× ½). D 2, a single modified spikelet (enlarged). E 1, from Girton, Cambridge, August 31, 1924 (× ½); E 2, a single modified spikelet (enlarged). [A.A.]

that which is innate and hereditary—recur in other genera.[1] Recently some of these grasses have been subjected to cytological analysis, and it has been shown for *Festuca ovina* L.[2] that, in a normal sexual form the number of chromosomes was 7, but in a type which had never been known to produce flowers, a high number, 42 (hexaploid), was reached. Forms in which proliferation was not altogether complete, were intermediate as regards chromosomes (21, triploid, or 28, tetraploid). A chromosomal basis for proliferation is also indicated by certain experiments in which ears with proliferated spikelets occurred as a result of crossing Wheats with 28 chromosomes with others having 42.[3] We are indeed far from a complete understanding of the meaning of proliferation in the grass spikelet. It is a subject which demands further research; but, from the work that has been done already, it is at least perfectly clear that in different genera of grasses this abnormality follows parallel courses.

Another grass in which teratology reveals parallelism, is 'Nepaul-barley' (Fig. 200, p. 394). This strange form of *Hordeum*, with its hooded lemma, whose tip encloses supernumerary sexual organs, has already been considered in another connection.[4] The point to which I wish to draw attention here, is that this cereal shows convergence towards *Triticum* in the nakedness of the grain; when it was first imported in 1817 from near the line of perpetual snow in the Himalayas, it went under the name of 'Nepaul-wheat'.[5] Another peculiar cereal form, *Zea tunicata* Sturt., Pod-corn,[6] finds its parallel in Polish-wheat, *Triticum polonicum* L., which has exaggerated glumes.[7] This race of Wheat illustrates parallelism in another direction, for its grain recalls that of Rye in form, and it has sometimes been grown under the name of Rye.[8]

Up to this point we have said nothing about abnormality within

[1] Hopkirk, T. (1817); Schuster, J. (1910); Jenkin, T. J. (1922); Howarth, W. O. (1925); Turesson, G. (1926).
[2] Turesson, G. (1930), (1931).
[3] Biffen, R. H. and Engledow, F. L. (1926).
[4] See pp. 312–16.
[5] Lawson, P. and Son [C.] (1836).
[6] See p. 367.
[7] Biffen, R. H. and Engledow, F. L. (1926), Fig. 1.
[8] Percival, J. (1921).

the flower itself. I happen to have made a special study of the teratology of the flowers of certain bamboos, and the conclusion to which I have come is that the abnormalities which they show are by no means erratic, but belong to certain recurrent types, which run

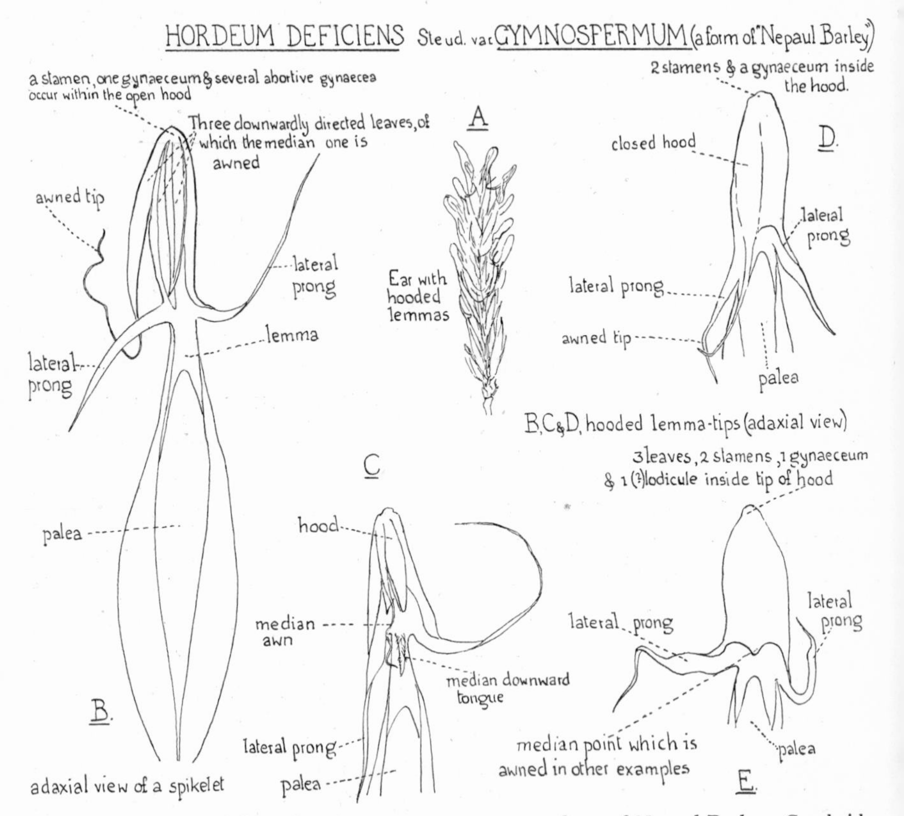

Fig. 200. *Hordeum deficiens* Steud. var. *gymnospermum*; a form of Nepaul Barley. Cambridge Botanic Garden, 1926. Professor Engledow's seed. A, inflorescence (× ½); B, single spikelet enlarged; C, D and E, the tips of three other lemmas (enlarged). [A.A.]

closely parallel, even when they arise in members of the tribe which are not intimately related to one another.[1] I have found, for instance, that a rare Burmese bamboo, *Cephalostachyum virgatum* Kurz, under cultivation in Calcutta, showed a number of peculiarities which are illustrated in Figs. 202, 204, 205, 206, 209, 210, and in Figs. 41,

[1] See Arber, A. (1927[1]), (1928[1]), (1929[3]).

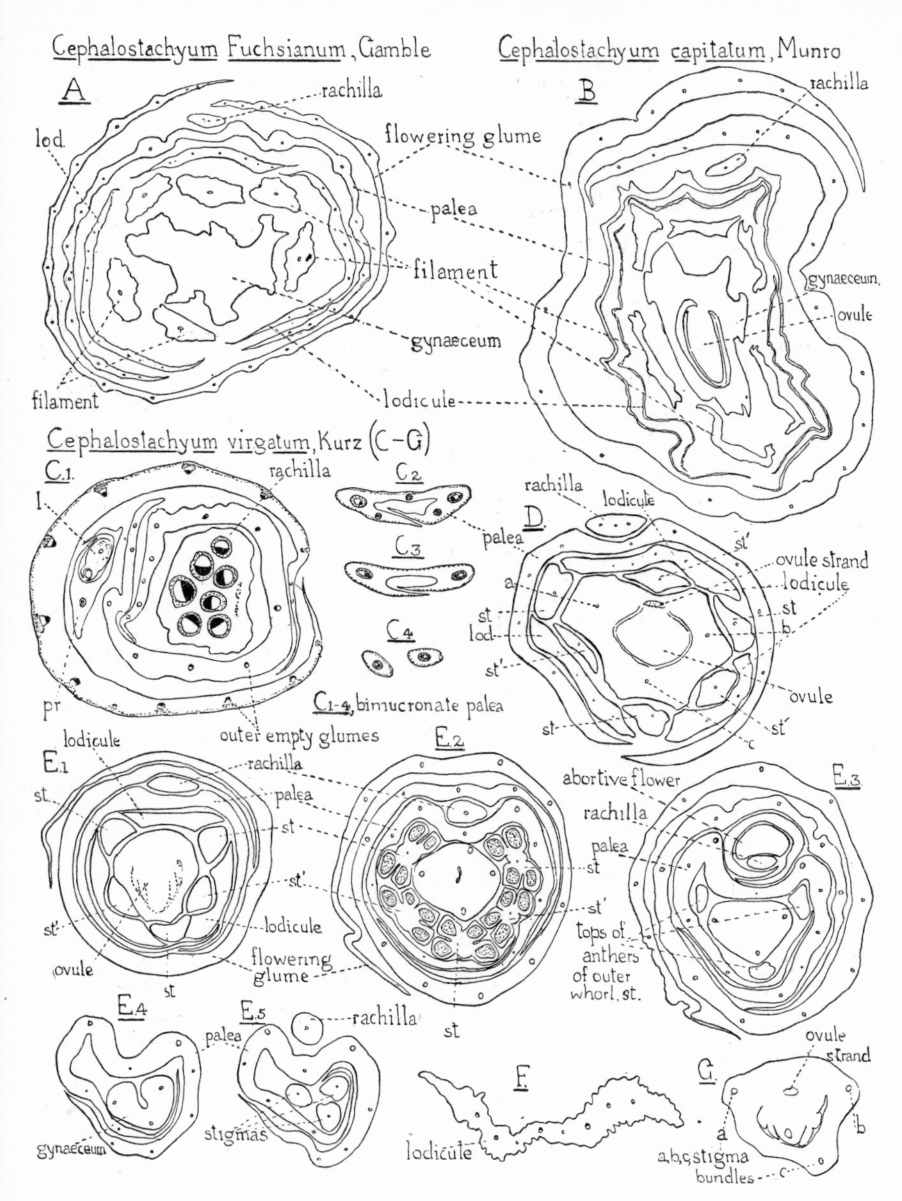

Fig. 201. Normal structure of the spikelet in the genus *Cephalostachyum*. [For full legend, see Arber, A. (1927[1]).]

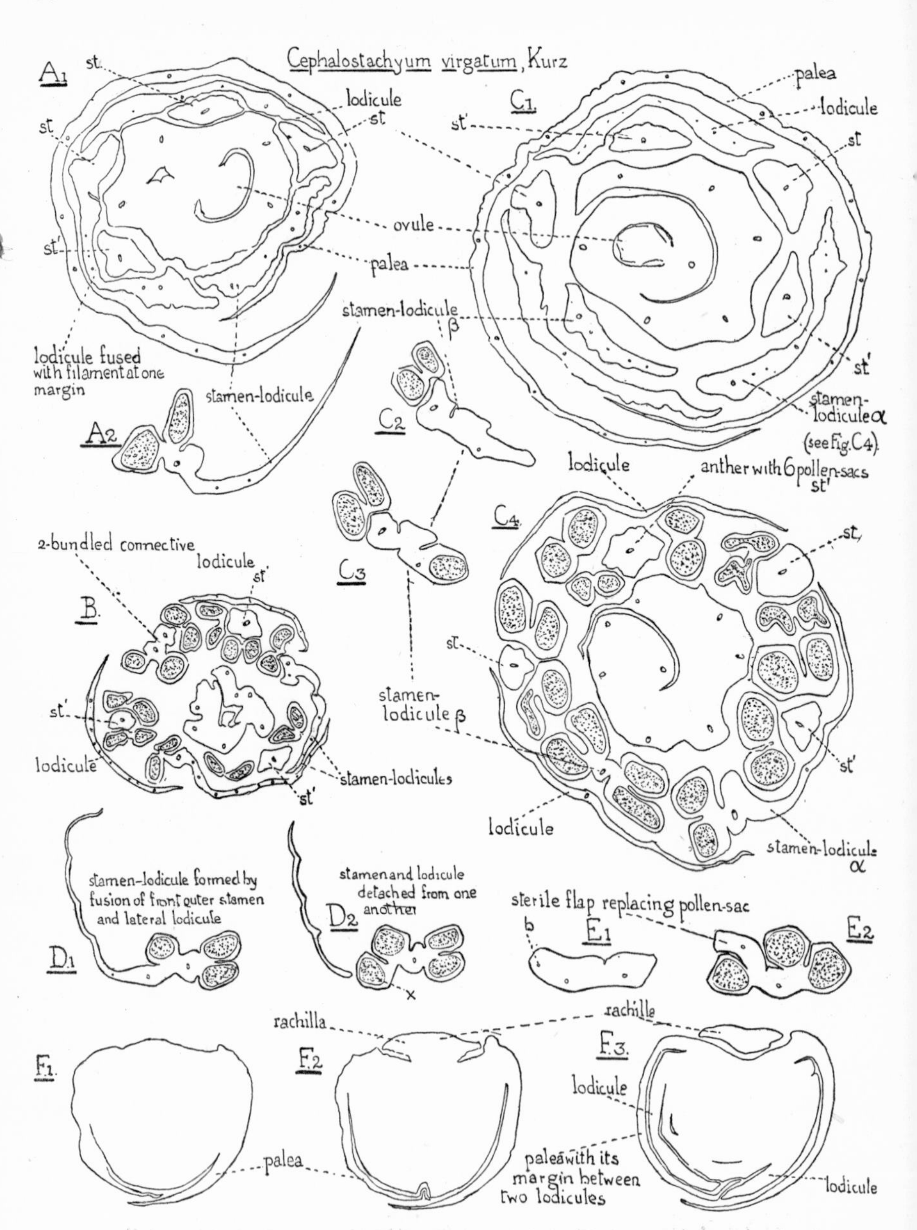

Fig. 202. *Cephalostachyum virgatum* Kurz. Sections of abnormal flowers showing stamen-lodicules. [For full legend, see Arber, A. (1927[1]).]

p. 118, and 43, p. 119. The normal structure is shown in Fig. 201, D, which is drawn from a wild specimen. There are three lodicules and six stamens; two other species sketched in the same figure are similar in structure. In the abnormal flowers, however, we find that the lodicules sometimes bear pollen-sacs. Habit drawings of these

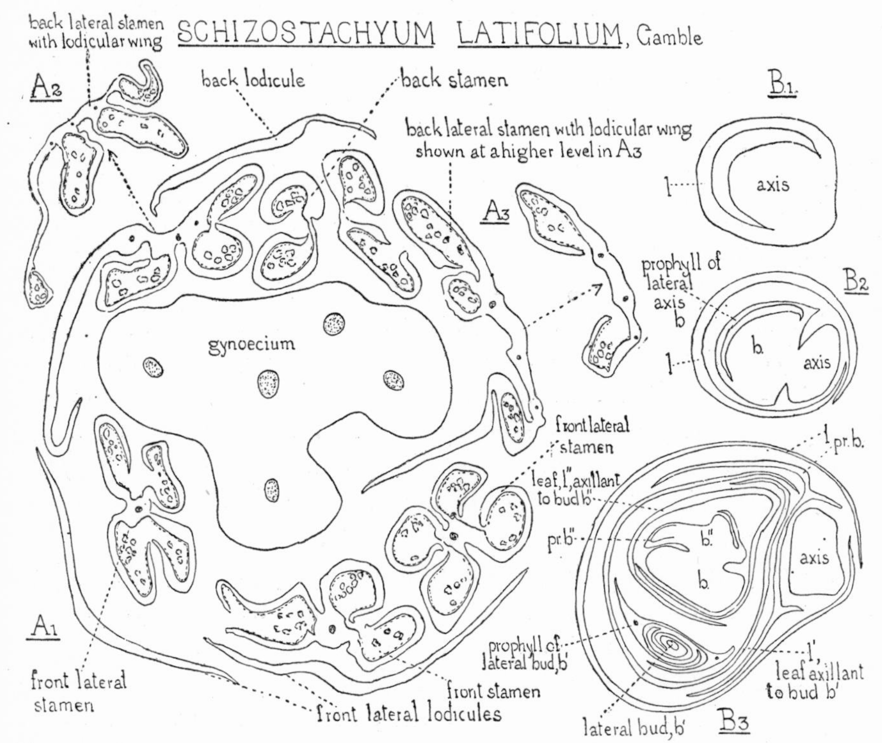

Fig. 203. *Schizostachyum latifolium* Gamble. Sections of an abnormal flower showing stamen-lodicules, for comparison with those of *Cephalostachyum virgatum* Kurz. [For full legend, see Arber, A. (1928[1]).]

stamen-lodicules will be found in Fig. 41, C2–C4, p. 118, while they are seen in section in Fig. 202, A2, B, C4, D1 and D2. These strange structures are exactly repeated in species belonging to a different bamboo genus—*Schizostachyum latifolium* Gamble (Fig. 203, A) and *S. chilianthum* Kurz (Fig. 207, p. 402).

In addition to the stamen-lodicules, the abnormal flowers of *Cephalostachyum virgatum* show another surprising structure—a flange developed from the gynaeceum wall and bearing pollen-sacs.[1] This structure, which is illustrated in Figs. 204, 205 and 206, is repeated in *Schizostachyum chilianthum* Kurz (Fig. 207, 2) and in *Bambusa nana* Roxb. (Fig. 208, p. 403). In the latter figure (C 1) another peculiarity is shown—a biovulate gynaeceum; and this, again, recurs in *Cephalostachyum virgatum* (Fig. 209, p. 404). As a final point we may mention the increase in the number of stigmas; an example with five is shown in Fig. 210, B 5, p. 404. A comparable and even greater increase may be found in *Bambusa nana* Roxb. (Fig. 208, C 3 and C 4).

I think that the instances which have now been cited suffice to show that, both in normal and in teratological structure, *parallelism* and *repetition* play a decisive part within the grass family. As a corresponding example, from the animal kingdom, of parallel but independent developments recurring within a large group, we may quote Wheeler's[2] instance of the origin of insect societies. He states that "social habits have arisen no less than 24 different times in as many different groups of solitary insects". He adds that "while each of the 24 different societies has its own peculiar features, we nevertheless observe that all of them have arisen in the same manner and have the same fundamental structure".

So far, in discussing the Gramineae and parallelism, we have confined the argument to relations *within* the group. When we enter a larger field, and institute a comparison between this natural and clearly delimited group and the other orders of Monocotyledons—Palms, Aroids, Orchids, Liliiflorae, etc.—we again meet with parallelisms and recurrences, which, though naturally less frequent than those within the unit itself, are yet sufficiently striking. A few of these parallelisms have been mentioned incidentally earlier in this book, but it may be worth while to bring them together at this point.[3] When we consider mode of life in

[1] I have found this structure also in the flower of an abnormal Crucifer; Arber, A. (1931²), pp. 192–5.

[2] Wheeler, W. M. (n.d. [1923]).

[3] Details of the examples here cited for comparison from Monocotyledons outside the Gramineae will be found in Arber, A. (1925).

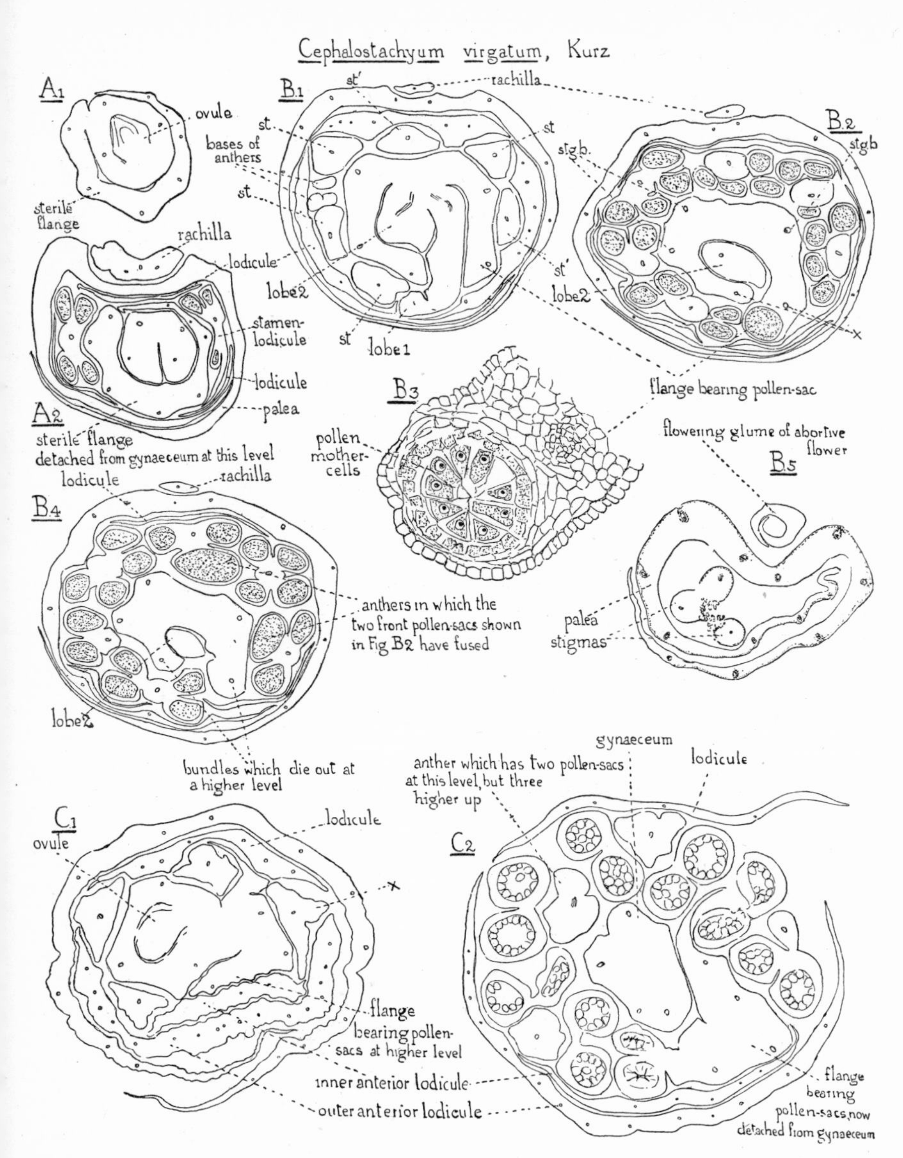

Fig. 204. *Cephalostachyum virgatum* Kurz. Transverse sections of abnormal spikelets to show examples of the flange bearing pollen-sacs which is developed from the gynaeceum wall. [For full legend, see Arber, A. (1927[1]).]

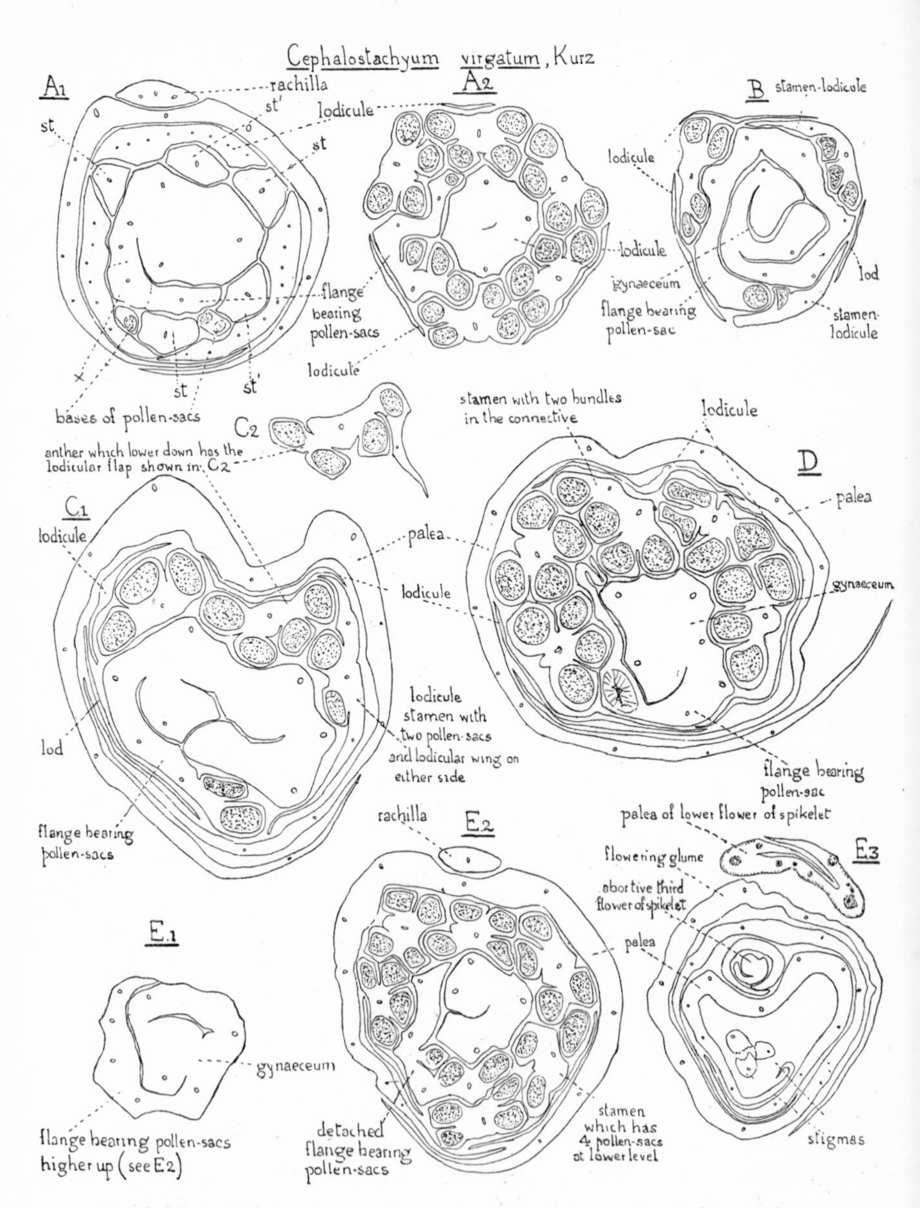

Fig. 205. *Cephalostachyum virgatum* Kurz. Sections of abnormal spikelets, to show stamen-lodicules, and flanges bearing pollen-sacs developed from the gynaeceum. [For full legend, see Arber, A. (1927[1]).]

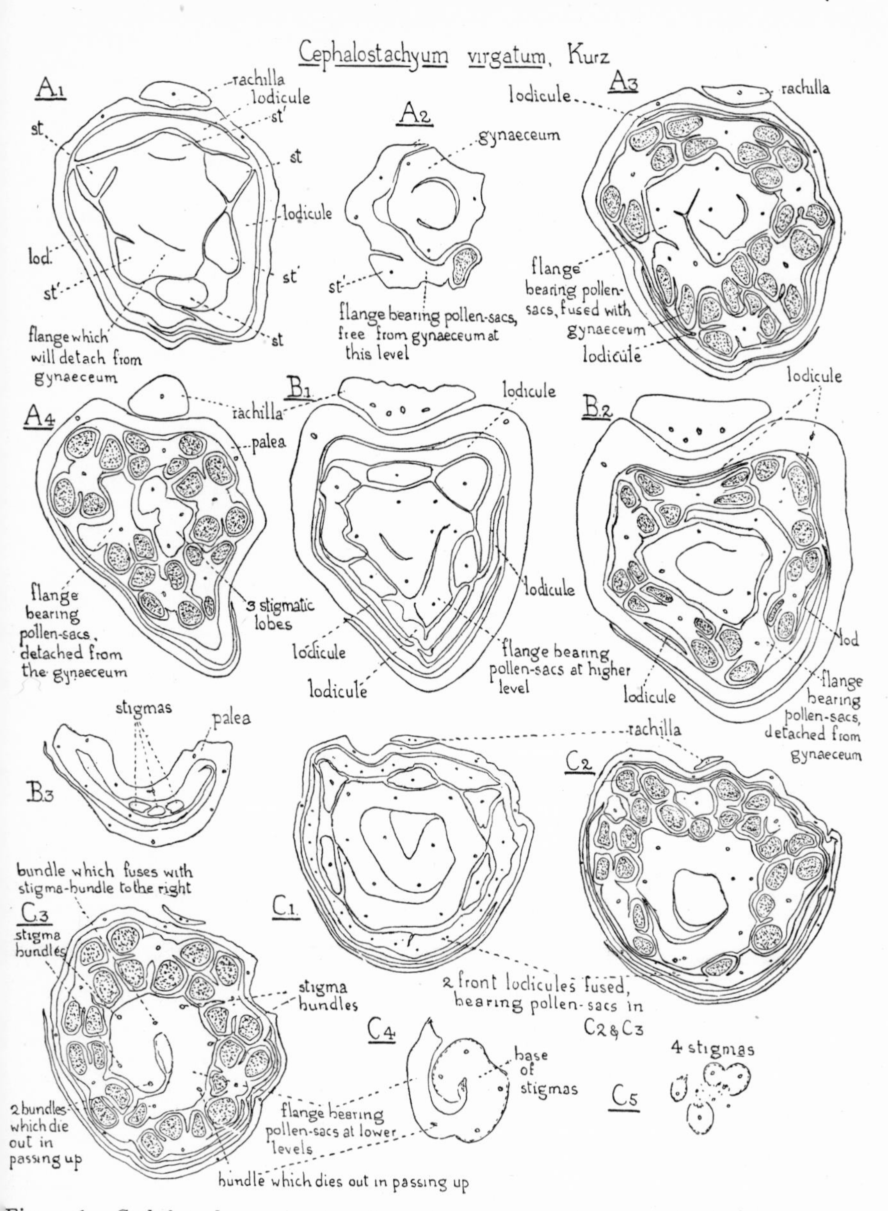

Fig. 206. *Cephalostachyum virgatum* Kurz. Sections through abnormal spikelets to show flanges bearing pollen-sacs developed from the gynaeceum. [For full legend, see Arber, A. (1927[1]).]

general, we find that the occurrence of tree forms (bamboos) in the Gramineae, which is prevailingly an herbaceous group, may be compared with the appearance of arboreal types in the Liliiflorae; while the aquatic, seaside and climbing Gramineae can also be paralleled in other groups of Monocotyledons. Even certain specialised structural types, which occur as rarities among the grasses, reappear in other orders. For instance, the curious root-thorns of bamboos (Fig. 28, B, p. 78) can be paralleled in the Liliiflorae and Palms, while the positively geotropic rhizome of *Festuca caerulescens* Desf.

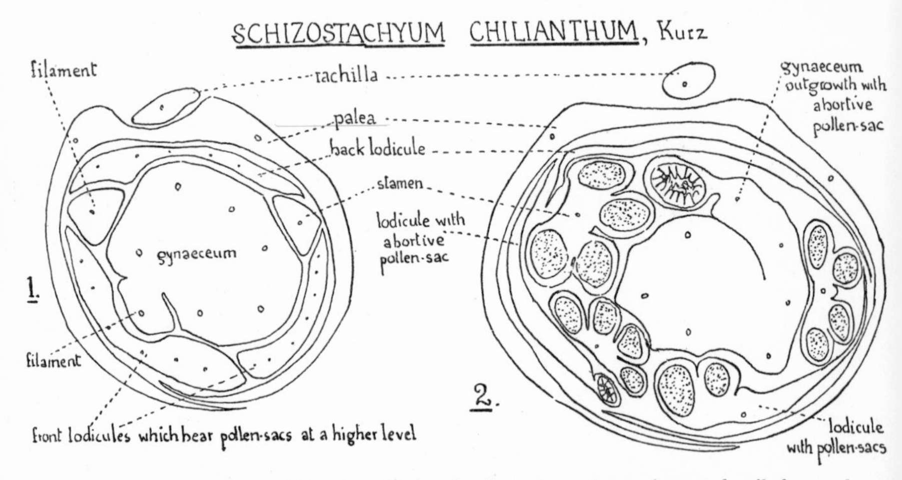

Fig. 207. *Schizostachyum chilianthum* Kurz. Sections through an abnormal spikelet to show lodicules with pollen-sacs, and a gynaeceum flange with a pollen-sac, for comparison with *Cephalostachyum virgatum* Kurz. [For full legend, see Arber, A. (1929[3]).]

(Fig. 137, p. 272) recalls that of certain Palms of the genus *Sabal*. Though the grass leaf is distinctive in character, it, also, may repeat features met with in other groups. The leaf-base, for example, may enwrap the axis more than once (e.g. Figs. 143, A 3, p. 281 and 155, A and B 1, p. 297), thus recalling the 'overlap' in the spathe of *Arum maculatum* L. (Spathiflorae) and in the leaf of *Thamnochortus* (Farinosae), while the pseudo-axis formed by the concentric leaf-sheaths of certain grasses is repeated in *Veratrum album* L. (Liliiflorae), and in the Musaceae and Marantaceae (Scitamineae). The development of the leaf-bases into a bulb, seen in *Poa bulbosa* L.

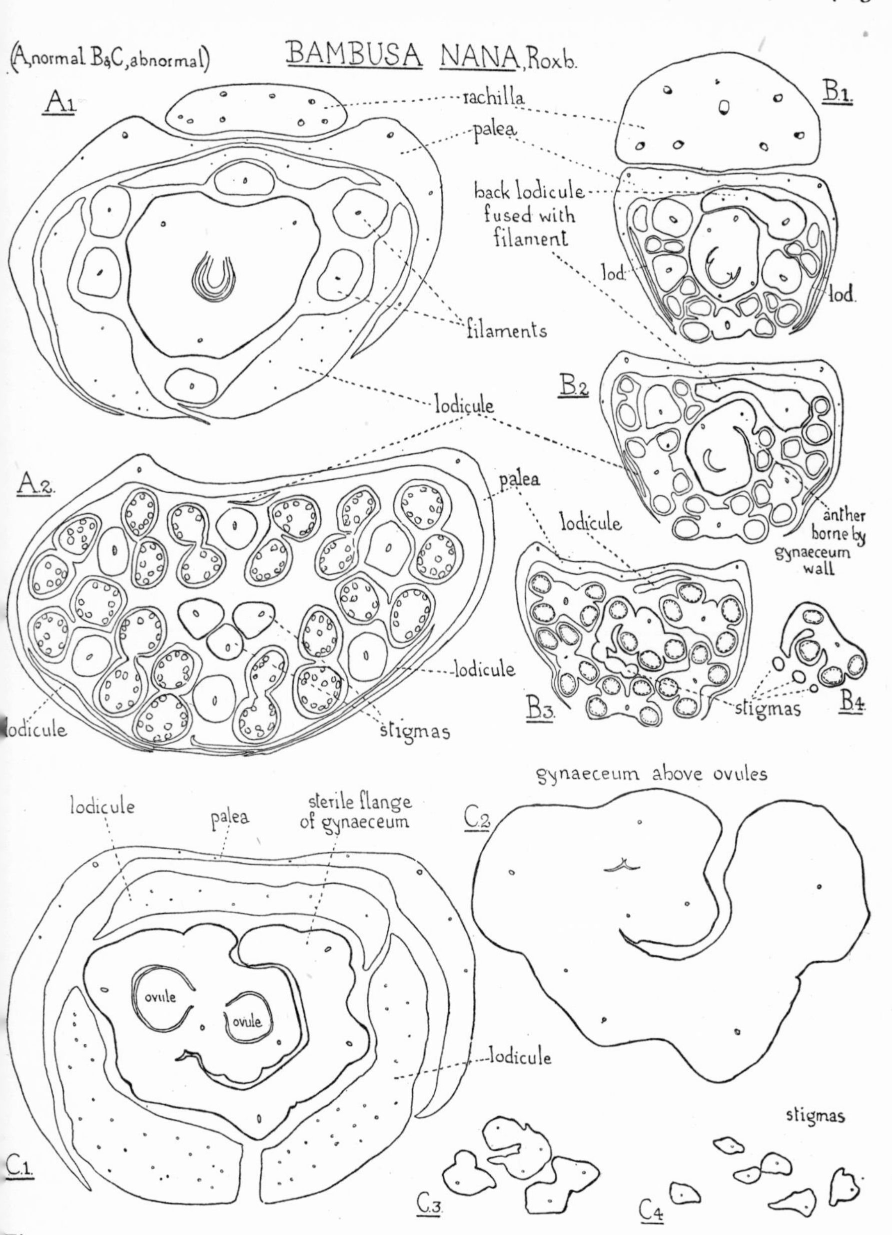

Fig. 208. *Bambusa nana* Roxb. Normal and abnormal spikelet structure for comparison with *Cephalostachyum virgatum* Kurz. [For full legend, see Arber, A. (1929³).]

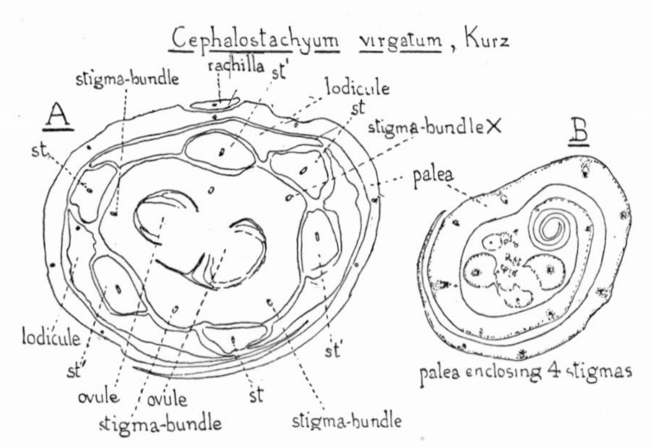

Fig. 209. *Cephalostachyum virgatum* Kurz. Sections through an abnormal flower with a biovulate gynaeceum and four stigmas. [For full legend, see Arber, A. (1927[1]).]

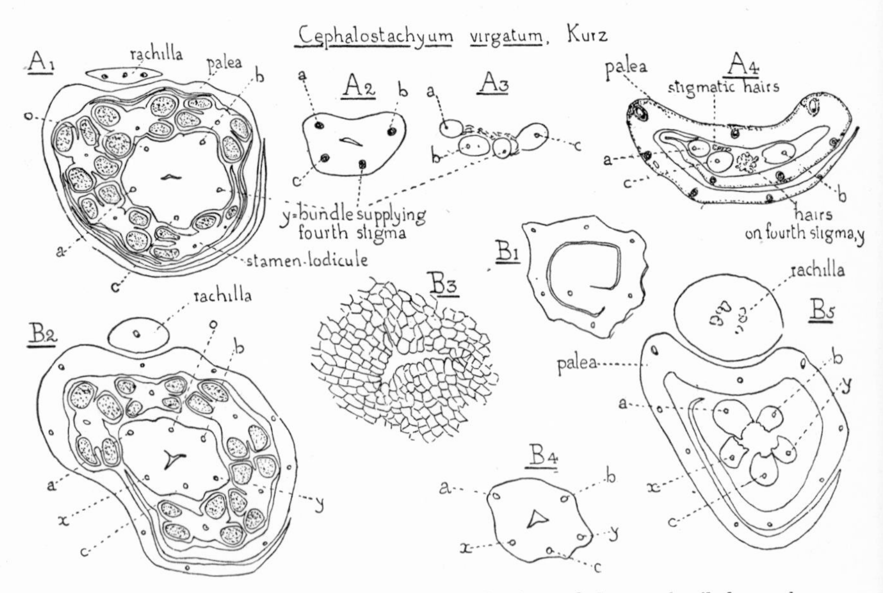

Fig. 210. *Cephalostachyum virgatum* Kurz. Sections of abnormal spikelet to show stigma variation. [For full legend, see Arber, A. (1927[1]).]

(Fig. 138, p. 273), finds a parallel in *Allium ursinum* L. (Liliiflorae). The narrow leaves of our meadow grasses are not universal in the Gramineae; in tropical forests the 'blade' may be replaced by a wide limb. A Swedish botanist,[1] who has made a special study of these broad-leaved grasses in the primaeval forests of Brazil, finds that the leaf-limb is generally lanceolate to ovate or almost heart-shaped. He points out that in the leaf-form, and the whole construction of the shoot, there is remarkable parallelism with certain Commelinaceae (Farinosae). Some broad-leaved forest grasses, belonging to the genera *Oplismenus*, *Panicum*, *Streptochaeta*, *Olyra*, *Pharus*, *Orthoclada* and *Pseudechinolaena*, are illustrated in Figs. 150, p. 290, and 151, p. 291, and 211, p. 406. In Fig. 211, F, a member of the Commelinaceous genus *Dichorisandra* is shown among them; it will be seen that it is scarcely distinguishable from them in type.

Blades of a definitely sagittate character occur, though rarely, among the grasses, e.g. *Phyllorachis sagittata* Trim. (Fig. 211, E) and this type of form and venation is repeated in *Monochoria* (Farinosae). The recognition that the broad-leaved grasses carry many reminders of other orders, is, indeed, no new thing. A hundred years ago, Link[2] pointed out that the nervation of the limb of *Panicum plicatum* Lam. recalled that of leaves belonging to the Scitamineae. We are also reminded of this order, which includes the Zingiberaceae, by the aromatic oils, sometimes with a ginger-like scent, which occur in certain grasses; *Cymbopogon flexuosus* Stapf has a Tamal name which means Ginger-grass.[3]

When we turn to the detailed structure of the grass blade, we find that one of its salient features—the presence of 'bulliform' cells (pp. 298–301)—is repeated in other orders.[4] Examples from the Farinosae (*Joinvillea*), the Liliiflorae (*Curculigo*) and the (possibly) less remote genus *Cyperus*,[5] are shown in Fig. 212, p. 407, for comparison with those of the Gramineae.

In the reproductive shoots, parallelism with other orders is less obvious, owing to the special elaboration and complexity of the

[1] Lindman, C. A. M. (1899).
[2] Link, H. F. (1825).
[3] Stapf, O. (1906).
[4] Leokadia, L. (1926).
[5] Duval-Jouve, J. (1875).

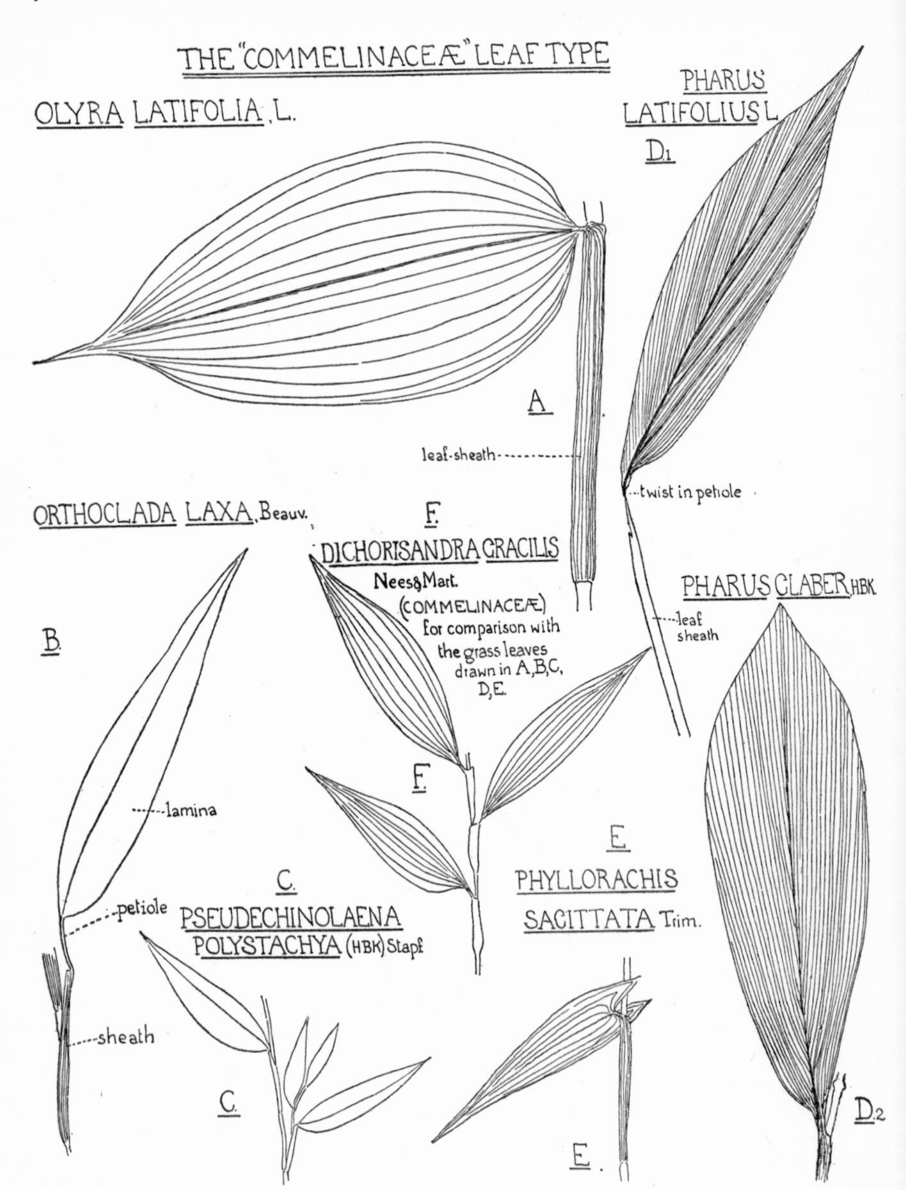

Fig. 211. Leaves of the Commelinaceous type, drawn from herbarium material. A–E, Gramineae; F, Commelinaceae (all × ½). A, *Olyra latifolia* L. (Uganda). B, *Orthoclada laxa* Beauv. (Mexico). C, *Pseudechinolaena polystachya* (H.B.K.) Stapf (Sumatra). D 1, *Pharus latifolius* L. (Cuba), showing leaf torsion; D 2, *P. glaber* H.B.K. (Paraguay) (this dried leaf does not happen to show the torsion described on p. 292). E, *Phyllorachis sagittata* Trim. (Welwitsch, Iter Angolense). F, *Dichorisandra gracilis* Nees et Mart. [A.A.]

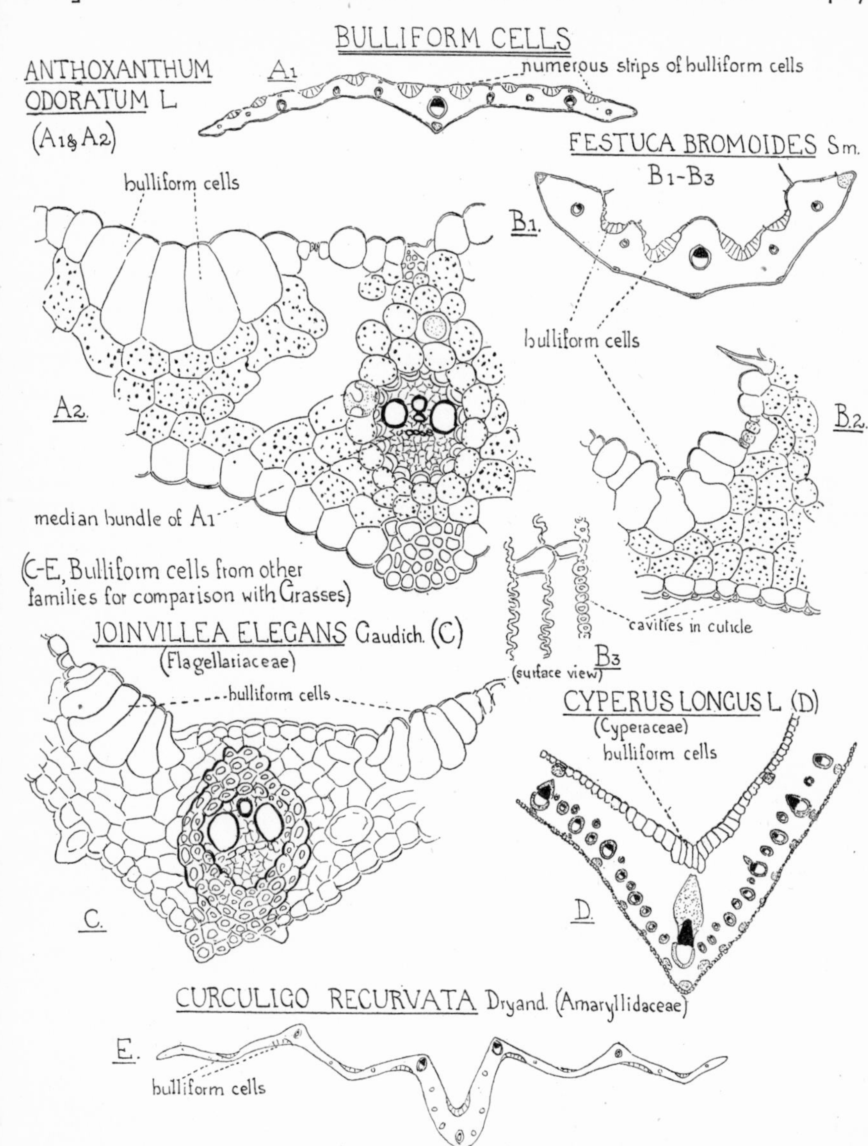

Fig. 212. Bulliform cells seen in transverse sections of leaves belonging to various Monocotyledonous families. A, *Anthoxanthum odoratum* L. (Gramineae). A 1, limb (× 23); A 2, midrib (×193 *circa*). B, *Festuca bromoides* Sm. B 1, limb (×77 *circa*); B 2, bulliform cells and cuticle cavities (× 318). B 3, surface view of epidermis near leaf margin to show that the sinuosities in the cuticle may be converted into a series of rings (× 318). C, *Joinvillea elegans* Gaudich. (Flagellariaceae); rib of limb (×193 *circa*). D, *Cyperus longus* L. (Cyperaceae), midrib of lamina (×23). E, *Curculigo recurvata* Dryand. (Amaryllidaceae) near base of limb (×14). [A.A.]

inflorescence, and the simplification of the flower. The character of the grasses, which has been mainly instrumental in conditioning these divergences from the other great groups, is their particular mode of branching.[1] The tendency to abbreviation of the basal internodes of each axis, associated with the precocious production of laterals of successive orders, affects both vegetative and reproductive shoots.[2] In the vegetative shoots it gives the characteristic 'tillering', while in the inflorescence it produces a branch system of a complicated type, in which each ultimate member terminates in a small partial inflorescence—the spikelet. The complex inflorescence is enclosed, in its young stages, within a tight bandage of leaf-sheaths. The bracts and bracteoles (glumes, lemmas and paleas) harden early—a feature which may be associated with the general tendency of the family to silica-deposition. The result of these factors is that the crowded flowers suffer from confinement in youth. With the consequent compression, we may, in some degree, associate the reduction within the inflorescence, which is so conspicuous in the grasses; but there is also evidence for the existence of an inherent trend towards sterilisation,[3] which is independent of external conditions. In a somewhat different form, the tendency to sterilisation of the reproductive shoot repeats itself in the Aroids,[4] and there is little doubt that it would be detected in other Monocotyledonous groups, if they were examined from this point of view.[5]

This slight sketch of the trends[6] revealed in the grasses, opens up the idea that this group is a microcosm, in which are displayed, in little, all those tendencies of the corresponding macrocosm—the Monocotyledons—which can be developed within the definite, though elastic, framework of the ordinal type. If this be true of the Gramineae, it would probably prove to be true of any other order which was scrutinised from this standpoint. Within each of the

[1] See pp. 75 and 261. [2] See p. 134.

[3] See Chapter IX.

[4] For an account of Engler's analysis of this order, see Arber, A. (1925), p. 212, et seq.

[5] Among the Dicotyledons, a corresponding trend has been recognised in the Leguminosae; see p. 180.

[6] For a further study of parallelism, with references, see Arber, A. (1925), p. 223, et seq.

great groups, Nature seems to have allowed the development of every variant of mode of life, and of every scheme of structure, which could be fitted into the limits imposed by the fundamental pattern. Using an analogy from music, we may suppose that a certain theme with variations is played in a succession of keys, and that, as played in each key, it is equivalent to one of the orders. The differences of key may be taken to symbolise the basic discreteness of the great groups. To some extent we are able, within each of these groups, to analyse the series of forms which represent the variations on the theme. In this process the findings of the geneticist give powerful help; but the mystery that still remains untouched is that of the distinctive character of the 'key'—or, in other words, of the underlying 'configuration' which gives each group its individual rhythm. What is the *meaning* of the differences that separate the Gramineae so delicately, yet so definitely, from any other order, and that so prevail that a grass remains a grass, however freely the type may vary? To attribute these differences to genic constitution is an 'explanation' of a merely descriptive kind; it enables us, indeed, to assign a place to them in the mental framework which we impose upon reality, but in so doing we have shelved, not solved, the problem. The mystery abides.

TAXONOMIC TABLE

[The following scheme shows the position of those genera of Gramineae which are mentioned in this book. This table does not pretend to be critical, and is included merely for the convenience of the reader, who must refer to the works of systematists for authoritative discussions on taxonomy. The classification of the bamboos is adapted from Camus, E. G. (1913). The members of the other tribes are, with a few exceptions, placed in the order followed by Bews, J. W. (1929), where an account will be found of the distribution, ecology, and economics of these and other genera. It is impossible in the present book to give more than incidental references to the extensive systematic literature on grasses. Hackel, E. (1887) (of which a second edition is announced), forms a useful outline; Hackel, E. (1890[2]) is an English version of the first edition. See the Note at foot of p. 411.]

Order GRAMINEAE

Family GRAMINACEAE

Sub-family I. Pooideae

Tribe	Genus
Bambuseae-Arundinariae	*Sasa*
	Arundinaria
	Phyllostachys
	Glaziophyton
	Arthrostylidium
	Aulonemia
	Merostachys
	Chusquea
	Planotia
	Semiarundinaria
	Shibataea
	Sinobambusa
	Yadakeya
Bambuseae-Eubambuseae	*Guadua*
	Bambusa
	Gigantochloa
	Oxytenanthera
Bambuseae-Bacciferae	*Dendrocalamus*
	Melocalamus
	Pseudostachyum
	Teinostachyum
	Cephalostachyum
	Schizostachyum
	Melocanna
	Ochlandra
	Pseudocoix
	Streptochaeta *
	Anomochloa *
Phareae	*Pharus*
	Olyra

Tribe	Genus
Festuceae	*Centotheca*
	Orthoclada
	Arundo
	Phragmites
	Gynerium
	Cortaderia
	Glyceria
	Poa
	Festuca
	Amphibromus
	Sclerochloa
	Bromus
	Brachypodium
	Briza
	Dactylis
	Distichlis
	Uniola
	Lamarkia
	Cynosurus
	Monantochloë
	Anthochloa
	Melica
	Triodia
	Triplasis
	Eragrostis
	Molinia
	Enneapogon (*Pappophorum*)
Aveneae	*Danthonia*
	Sieglingia
	Avena
	Deschampsia
	Corynephorus

* These genera are of uncertain position, and should possibly be placed with the Oryzeae rather than with the Bambuseae.

Tribe	Genus	Tribe	Genus
Aveneae (*continued*)	*Trisetum* *Koeleria* *Aira* *Holcus* *Arrhenatherum*	Agrostideae	*Calamagrostis* *Agrostis* *Psamma* (*Ammophila*) *Lagurus* *Epicampes* *Sporobolus* *Phippsia* *Coleanthus* *Mibora* (*Chamagrostis*) *Crypsis* *Cornucopiae* *Heleochloa* *Phleum* *Alopecurus* *Stipa* *Aristida*
Chlorideae	*Eleusine* *Bouteloua* *Melanocenchris* *Munroa* *Cynodon* *Spartina* *Buchloë*		
Hordeae	*Pariana* *Elymus* *Hordeum* *Agropyron* *Secale* *Triticum* *Aegilops* *Lolium* *Jouvea* *Lepturus* *Pholiurus* *Nardus*	Zoysieae	*Anthephora* *Tragus*
		Phalarideae	*Ehrharta* *Anthoxanthum* *Hierochloë* *Phalaris*
		Oryzeae	*Oryza* *Leersia* *Lygeum* *Zizania* *Luziola*

Sub-family II. Panicoideae

Tribe	Genus	Tribe	Genus
Paniceae	*Leptoloma* *Ichnanthus* *Panicum* *Paspalum* *Oplismenus* *Pseudechinolaena* *Axonopus* *Amphicarpon* *Stenotaphrum* *Setaria* *Pennisetum* *Cenchrus* *Phyllorachis* *Thuarea* *Spinifex*	Andropogoneae	*Imperata* *Saccharum* *Pollinia* *Sorghum* *Vetiveria* *Andropogon* *Cymbopogon* *Ischaemum* *Urelytrum*
		Maydeae	*Tripsacum* *Coix* *Euchlaena* *Zea*

Note. A revised classification of the Gramineae by C. E. Hubbard will be found in a volume which has been announced since the present book was in print: J. Hutchinson, "The Families of Flowering Plants. II. Monocotyledons."

BIBLIOGRAPHY

[This bibliography is limited to the titles of memoirs actually cited in the text; it thus includes only a small proportion of the existing literature relating to the Gramineae. References are given to the pages in this book where each memoir is mentioned, and to any figures derived from it.]

Aldrich-Blake, R. N. (1927) Some Factors Influencing the Increment of Forests. Forestry (The Journal of the Soc. of For. of Great Britain), vol. I, 1927, pp. 82–96.
[p. 91.]

Allan, H. H. (1930) *Spartina Townsendii*. 2. Experience in New Zealand. New Zealand Journ. Agric., vol. XL, 1930, pp. 189–96.
[p. 377.]

Allard, H. A. See **Garner, W. W. and Allard, H. A.** (1920).

Alsberg, C. L. and Black, O. F. (1915) Concerning the Distribution of Cyanogen in Grasses, especially in the Genera *Panicularia* or *Glyceria* and *Tridens* or *Sieglingia*. Journ. Biol. Chem., vol. XXI, 1915, pp. 601–9.
[p. 24.]

Ammann, P. (1910) Sur l'existence d'un riz vivace au Sénégal. Comptes rendus de l'acad. d. sci., Paris, vol. CLI, 1910, pp. 1388–9.
[p. 37.]

Andrews, C. W. (1903) Some Suggestions on Extinction. Geol. Mag., dec. IV, vol. X, 1903, pp. 1, 2.
[p. 90.]

Anon. (1766) Histoire des Plantes de l'Europe, et des plus usitées qui viennent d'Asie, d'Afrique et d'Amérique. 2 vols. Lyons, 1766.
[p. 53 (Fig. 21).]

Anthony, S. and Harlan, H. V. (1920) Germination of Barley Pollen. Journ. Agric. Res., vol. XVIII, 1919–20, No. 10, Feb. 16, 1920, pp. 525–36.
[p. 202.]

Aquinas, St Thomas (1854) Opera omnia (1852–73), vol. IV, Summa Theologica, Part III, Parma, 1854.
[pp. 15, 16.]

Arber, A. (1917) On the Occurrence of Intrafascicular Cambium in Monocotyledons. Ann. Bot., vol. XXXI, 1917, pp. 41–5.
[p. 305.]

Arber, A. (1918) Further Notes on Intrafascicular Cambium in Monocotyledons. Ann. Bot., vol. XXXII, 1918, pp. 87–9.
[p. 305.]

Arber, A. (1918–19) The 'Law of Loss' in Evolution. Proc. Linn. Soc. Lond., Session 131, 1918–19, pp. 70–8.
[p. 85.]

Arber, A. (1919[1]) On Atavism and the Law of Irreversibility. Amer. Journ. Sci., vol. XLVIII, 1919, pp. 27–32.
[p. 85.]

Arber, A. (1919[2]) Studies on Intrafascicular Cambium in Monocotyledons (III and IV). Ann. Bot., vol. XXXIII, 1919, pp. 459–65.
[p. 305.]

Arber, A. (1920) Water Plants: a Study of Aquatic Angiosperms. Cambridge, 1920.
[pp. 334, 342, 365.]

Arber, A. (1923) Leaves of the Gramineae. Bot. Gaz., vol. LXXVI, 1923, pp. 374–88.
[pp. 280 (Fig. 142), 283, 297 (Fig. 155).]

Arber, A. (1925) Monocotyledons: a morphological study. Cambridge Botanical Handbooks. Cambridge, 1925.
[pp. 18, 78, 86, 102, 179, 212, 225 (Fig. 107), 247, 251, 284 (Legend of Fig. 145), 311, 329, 380, 398, 408.]

Arber, A. (1926) Studies in the Gramineae. I. The Flowers of certain Bambuseae. Ann. Bot., vol. XL, 1926, pp. 447–69.
[pp. 96 (Fig. 32), 108, 109 (Fig. 33), 111 (Fig. 35), 112 (Fig. 36), 113 (Fig. 37), 115 (Fig. 39), 116 (Fig. 40), 122 (Fig. 45), 123 (Fig. 46), 154, 221 (Fig. 106).]

Arber, A. (1927[1]) Studies in the Gramineae. II. Abnormalities in *Cephalostachyum virgatum* Kurz, and their Bearing on the Interpretation of the Bamboo Flower. Ann. Bot., vol. XLI, 1927, pp. 47–74.
[pp. 118 (Fig. 41), 119 (Fig. 43), 394, 395 (Fig. 201), 396 (Fig. 202), 399 (Fig. 204), 400 (Fig. 205), 401 (Fig. 206), 404 (Figs. 209, 210).]

Arber, A. (1927[2]) Studies in the Gramineae. III. Outgrowths of the Reproductive Shoot, and their Bearing on the Significance of Lodicule and Epiblast. Ann. Bot., vol. XLI, 1927, pp. 473–88.
[pp. 142 (Fig. 59), 143 (Fig. 60), 145 (Fig. 62), 182 (Fig. 85), 183 (Fig. 87), 184, 239 (Fig. 113), 241 (Fig. 115), 325, 384 (Fig. 193).]

Arber, A. (1928[1]) Studies in the Gramineae. IV. 1. The Sterile Spikelets of *Cynosurus* and *Lamarkia*. 2. Stamen-lodicules in *Schizostachyum*. 3. The Terminal Leaf of *Gigantochloa*. Ann. Bot., vol. XLII, 1928, pp. 173–87.
[pp. 186, 311 (Fig. 161), 321, 322 (Fig. 170), 324 (Fig. 172), 394, 397 (Fig. 203).]

Arber, A. (1928[2]) Studies in the Gramineae. V. 1. On *Luziola* and *Dactylis*. 2. On *Lygeum* and *Nardus*. Ann. Bot., vol. XLII, 1928, pp. 391–407.
[pp. 146 (Fig. 63), 169 (Fig. 75), 184, 315, 317 (Fig. 166), 318 (Fig. 167), 319 (Fig. 168), 323 (Fig. 171), 327 (Fig. 174).]

Arber, A. (1928[3]) The Tree Habit in Angiosperms: its Origin and Meaning. New Phyt., vol. XXVII, 1928, pp. 69–84.
[p. 81.]

Arber, A. (1929[1]) Studies in the Gramineae. VI. 1. *Streptochaeta*. 2. *Anomochloa*. 3. *Ichnanthus*. Ann. Bot., vol. XLIII, 1929, pp. 35–53.
[pp. 127, 128 (Fig. 49), 129 (Fig. 50), 130 (Fig. 51), 131 (Fig. 52), 132 (Fig. 53), 144 (Fig. 61), 291 (Fig. 151).]

Arber, A. (1929[2]) Studies in the Gramineae. VII. On *Hordeum* and *Pariana*, with Notes on 'Nepaul Barley'. Ann. Bot., vol. XLIII, 1929, pp. 508–33.
[pp. 151 (Fig. 67), 170, 172 (Fig. 77), 173 (Fig. 78), 174 (Fig. 79), 175 (Fig. 80), 312, 314 (Fig. 163), 315 (Fig. 164), 316 (Fig. 165), 326 (Fig. 173).]

Arber, A. (1929[3]) Studies in the Gramineae. VIII. On the Organisation of the Flower in the Bamboo. Ann. Bot., vol. XLIII, 1929, pp. 765–81.
[pp. 108, 114 (Fig. 38), 119 (Fig. 42), 121 (Fig. 44), 124, 125 (Fig. 47), 126 (Fig. 48), 320 (Fig. 169), 394, 402 (Fig. 207), 403 (Fig. 208).]

Arber, A. (1930[1]) Studies in the Gramineae. IX. 1. The Nodal Plexus. 2. Amphivasal Bundles. Ann. Bot., vol. XLIV, 1930, pp. 593–620.
[pp. 79 (Fig. 29), 231, 232 (Fig. 109), 254 (Fig. 122), 259 (Fig. 125), 260 (Fig. 126), 261 (Fig. 127), 262 (Fig. 128), 263 (Fig. 129), 267, 269 (Fig. 134), 270 (Fig. 135), 271 (Fig. 136).]

Arber, A. (1930[2]) Root and Shoot in the Angiosperms: a study of morphological categories. New Phyt., vol. XXIX, 1930, pp. 297–315.
[p. 307.]

Arber, A. (1931[1]) Studies in the Gramineae. X. 1. *Pennisetum*, *Setaria*, and *Cenchrus*. 2. *Alopecurus*. 3. *Lepturus*. Ann. Bot., vol. XLV, 1931, pp. 401–20.
[pp. 170, 171 (Fig. 76), 179 (Fig. 83), 186, 188 (Fig. 90), 189 (Fig. 91), 190 (Fig. 92), 192 (Fig. 93), 193 (Fig. 94), 194 (Fig. 95).]

Arber, A. (1931[2]) Studies in Floral Morphology. II. On some normal and abnormal Crucifers: with a discussion on teratology and atavism. New Phyt., vol. XXX, 1931, pp. 172–203.
[pp. 367, 398.]

Arber, A. See also **Sargant, E. and Robertson (Arber), A. (1905)**, and **Sargant, E. and Arber, A. (1915)**.

Arber, E. See **Smith, J. (1612)** in **Arber, E. (1884)**.

Archer, T. C. (1856) The Useful Plants of the Natural Order Graminaceae. Pharm. Journ., vol. XV, 1856, pp. 548–9.
[pp. 2, 33.]

Armstrong, S. F. (1917) British Grasses and their Employment in Agriculture. Cambridge, 1917.
[p. 163.]

Artschwager, E. (1925) Anatomy of the Vegetative Organs of Sugar Cane. Journ. Agric. Res., vol. XXX, 1925, pp. 197–221.
[pp. 303, 309.]

Ashley, W. (1928) The Bread of our Forefathers. An Enquiry in Economic History. Oxford, 1928.
[pp. 4, 13, 15.]

Askenasy, E. (1879) Ueber das Aufblühen der Gräser. Verhandl. d. naturhist.-med. Vereins zu Heidelberg, N.F., vol. II, 1880, pt. 4, 1879, pp. 261–73.
[p. 158.]

Aubrey, J. (1847) The Natural History of Wiltshire (written between 1656 and 1691), edited by John Britton. Wiltshire Topographical Society, London, 1847.
[p. 208.]

Avery, G. S. (1928) Coleoptile of *Zea Mays* and other grasses. Bot. Gaz., vol. LXXXVI, 1928, pp. 107–10.
[p. 236.]

Avery, G. S. (1930) Comparative Anatomy and Morphology of Embryos and Seedlings of Maize, Oats, and Wheat. Bot. Gaz., vol. LXXXIX, 1930, pp. 1–39.
[pp. 233, 235, 246.]

Azara, F. de (1809) Voyages dans l'Amérique Méridionale (1781–1801). Publiés d'après les manuscrits de l'auteur par C. A. Walckenaer, vol. I, Paris, 1809.
[p. 366.]

"B., A. H." (1882) Notices regarding the flowering of Bamboos. Journ. Agric. and Hort. Soc. India, vol. VI, 1882 for 1878–81, pp. 230–7. (See also Indian Forester, vol. VII, 1881–2, pp. 376–9.)
[pp. 99, 102, 104.]

"B., F. G. R." (1902) An Interesting Bamboo. Indian Forester, vol. XXVIII, 1902, pp. 432–3.
[pp. 65, 124.]

"B., T. F." (1887) Seeding of Bamboos. Indian Forester, vol. XIII, 1887, pp. 409 and 579.
[p. 97.]

"B., T. F." (1893) The flowering of Bamboos in Travancore. Indian Forester, vol. XIX, 1893, p. 20.
[p. 97.]

Backer, C. A. (n.d. [1930]) The Problem of Krakatao as seen by a Botanist. 299 pp. Published by the author, Java and The Hague, n.d. [1930].
[pp. 217, 347.]

Bailey, I. W. See **Sinnott, E. W. and Bailey, I. W.** (1914).

Baker, O. E. See **Finch, V. C. and Baker, O. E.** (1917).

Baldensperger, L. See **Crowfoot, G. M. and Baldensperger, L.** (1932).

Ball, C. R. (1910) The History and Distribution of Sorghum. U.S. Dept. of Agric. Bureau of Plant Industry, Bull. No. 175, 1910, 63 pp.
[p. 23.]

Bancroft, H. (1930) The Arborescent Habit in Angiosperms: a Review. New Phyt., vol. XXIX, 1930, pp. 153–69, 227–75.
[p. 81.]

Barth, H. (1857–8) Travels and Discoveries in North and Central Africa, 1849–1855. 5 vols., London, 1857–8.
[pp. 2, 34.]

Beal, W. J. (1900) Some Monstrosities in Spikelets of *Eragrostis* and *Setaria*. Bull. Torr. Bot. Club, vol. XXVII, 1900, pp. 85–6.
[p. 140.]

Beddows, A. R. (1931) *Triodia decumbens*, Beauv. (*Sieglingia decumbens*, Bernh.). Ann. Bot., vol. XLV, 1931, pp. 443–51.
[p. 204.]

Bentham, G. and Hooker, J. D. (1883) Genera Plantarum, vol. III, pt. II, London, 1883.
[p. 164.]

Bergevin, E. de (1891) Remarques sur les Variations de *Lolium perenne* L. dans ses Sous-Variétés *cristatum* Coss. et Germ. Fl. Et *ramosum* P.Fl. Bull. de la soc. d. Amis d. sci. nat. de Rouen, 1891 for 1890, pp. 161–85.
[p. 166.]

Besler, B. (1613) Hortus Eystettensis. [s.l.] 1613.
[p. 35 (Fig. 16).]

Bews, J. W. (1918) The Grasses and Grasslands of South Africa. Pietermaritzburg, 1918.
[pp. 249, 337.]

Bews, J. W. (1925) Plant Forms and their Evolution in South Africa. London, 1925.
[pp. 207, 208.]

Bews, J. W. (1929) The World's Grasses. Their Differentiation, Distribution, Economics and Ecology. London, 1929.
[pp. 24, 209, 378, 410.]

Bhalerao, S. G. (1926) The Morphology of the Rice Plant and of the Rice Inflorescence. Journ. Indian Bot. Soc., vol. V, 1926, pp. 13–15.
[p. 262.]

Bicknell, E. P. (1898) Two New Grasses from Van Cortlandt Park, New York City. Bull. Torr. Bot. Club, vol. XXV, 1898, pp. 104–7.
[p. 212.]

Bidder, G. P. (1927) The Ancient History of Sponges and Animals. Pres. Address to Section D, Zoology. Rep. Brit. Ass. Adv. Sci., Leeds, 1927, pp. 58–74.
[p. 87.]

Biffen, R. H. and Engledow, F. L. (1926) Wheat Breeding Investigations at the Plant Breeding Institute, Cambridge. Ministry of Agriculture and Fisheries. Research Monograph 4, 1926.
[pp. 227, 393.]

Black, O. F. See **Alsberg, C. L. and Black, O. F.** (1915).

Blane, G. (1790) An Account of the *Nardus Indica*, or Spikenard. Phil. Trans. Roy. Soc. Lond., vol. LXXX, pt. II, 1790, pp. 284–92. [p. 57.]

Blaringhem, L. (1908) Mutation et Traumatismes. 248 pp. Paris, 1908. [pp. 365, 366, 372.]

Blaringhem, L. (1924) Note sur l'Origine du Maïs. Métamorphose de l'*Euchlaena* en *Zea* obtenue au Brésil par Bento de Toledo. Ann. d. sci. nat., sér. 10, Bot., vol. VI, 1924, pp. 245–63. [p. 372.]

Blatter, E. and Parker, R. N. (1929) The Indian Bamboos brought up-to-date. Indian Forester, vol. LV, 1929, pp. 541–62, 586–613. [p. 95.]

Boccone, P. (1674) Icones et Descriptiones rariorum plantarum. E Theatro Sheldoniano. 1674. [p. 366.]

Bock, J. See **Tragus, H.** (**Bock, J.**) (1552).

Bonafous, M. (1836) Histoire naturelle, agricole et économique du Maïs. Paris, 1836. [p. 32.]

Borissow, G. (1924) Ueber die eigenartigen Kieselkörper der Wurzelendodermis bei *Andropogon*-Arten. Ber. d. deutsch. Bot. Gesellsch., vol. XLII, 1924, pp. 366–80. [p. 253.]

Boswell, J. T. See **Syme, J. T. B.** (afterwards **Boswell**) (1873).

Bower, F. O (1918) Hooker Lecture: On the Natural Classification of Plants, as exemplified in the Filicales. Journ. Linn. Soc. Lond., Bot. vol. XLIV, 1917–20, No. 296, 1918. [p. 86.]

Boyd, L. (1931) Evolution in the Monocotyledonous Seedling: A New Interpretation of the Morphology of the Grass Embryo. Trans. and Proc. Bot. Soc. Edinburgh, vol. XXX, pt. IV, 1931, pp. 286–303. [p. 235.]

Boyd, L. (1932) Monocotylous Seedlings. Trans. and Proc. Bot. Soc. Edinburgh, vol. XXXI, pt. I, 1932, pp. 1–224. [p. 235.]

Bradley, J. W. (1914) Flowering of Kyathaung Bamboo (*Bambusa polymorpha*) in the Prome Division, Burma. Indian Forester, vol. XL, 1914, pp. 526–9. [pp. 98, 99, 102.]

Brandis, D. (1887) Tabasheer. Indian Forester, vol. XIII, 1887, pp. 107–15. [p. 60.]

Brandis, D. (1899) Biological Notes on Indian Bamboos. Indian Forester, vol. XXV, 1899, pp. 1–25.
[pp. 65, 95, 98, 99, 101, 102, 106, 219.]

Brandis, D. (1906[1]) On some Bamboos in Martaban south of Toungoo between the Salwin and Sitang Rivers. Indian Forester, vol. XXXII, 1906, pp. 179–86, 236–45, 288–95.
[p. 97.]

Brandis, D. (1906[2]) Indian Trees. London, 1906.
[p. 108.]

Branthwaite, F. J. (1902) The Flowering of stool shoots of *Dendrocalamus strictus*. Indian Forester, vol. XXVIII, 1902, p. 233.
[p. 105.]

Braun, A. (1841) Bemerkungen über die Flora von Abyssinien. I. Die abyssinischen Culturpflanzen. Flora, vol. XXIV, pt. I, 1841, pp. 260–72.
[p. 34.]

Braun, A. (1853) The Phenomenon of Rejuvenescence in Nature. Bot. and Phys. Mem., No. 1. Edited by A. Henfrey. Ray Soc., London, 1853.
[p. 309.]

Bremekamp, C. E. B. (1915) Der dorsiventrale Bau des Grashalmes nebst Bemerkungen über die morphologische Natur seines Vorblattes. Recueil des travaux botaniques néerlandais, vol. XII, 1915, pp. 31–43.
[p. 75.]

Brenchley, W. E. (1912) The Development of the Grain of Barley. Ann. Bot., vol. XXVI, 1912, pp. 903–28.
[p. 226.]

Bretschneider, E. (1870) On the Study and Value of Chinese Botanical Works. Foochow, 1870.
[pp. 18, 26, 29, 32, 54.]

Brewster, D. (1828) On the Natural History and Properties of Tabasheer, the Siliceous concretion in the Bamboo. Edinburgh Journ. of Science, vol. VIII, 1828, pp. 285–94.
[p. 60.]

Britten, J. and Holland, R. (1886) A Dictionary of English Plant-names. English Dialect Society. London, 1886.
[pp. 49 (Legend of Fig. 18), 134, 294.]

Britton, J. See **Aubrey, J.** (1847).

Brockmann-Jerosch, H. (1913) Die Trichome der Blattscheiden bei Gräsern. Ber. d. deutsch. Bot. Gesellsch., vol. XXXI, 1913, pp. 590–4.
[p. 286.]

Bromfield, W. A. (1844) Curious form of the common Reed. Phytologist, vol. I, 1844, p. 146.
[p. 208.]

Brongniart, A. (1851) Description d'un nouveau genre de Graminées du Brésil. Ann. d. sci. nat., sér. III, Bot., vol. XVI, 1851, pp. 368–72. [p. 127.]

Brongniart, A. (1860) Note sur le sommeil des feuilles dans une plante de la famille des Graminées, le *Strephium guianense*. Bull. de la soc. bot. de France, vol. VII, 1860, pp. 470–2. [p. 289.]

Broun, A. F. (1886) Seeding of Bamboos. Indian Forester, vol. XII, 1886, pp. 413–14. [p 103.]

Brown, C. A. (1929) Notes on *Arundinaria*. Bull. Torr. Bot. Club, vol. LVI, 1929, pp. 315–18. [pp. 59, 103, 105.]

Brown, R. (1810) Prodromus Florae Novae Hollandiae, vol. I (all published), London, 1810. [p. 381.]

Brown, R. (1866) General Remarks...on the Botany of Terra Australis. (Reprinted from A Voyage to Terra Australis by Matthew Flinders, London, 1814.) Miscellaneous Botanical Works, vol. I, Ray Soc., London, 1866. [p. 381.]

Brown, R. N. Rudmose (1912) Report on the Scientific Results of the Voyage of S.Y. "Scotia" during the years 1902–4. Vol. III, Bot., 1912. I. The Problems of Antarctic Plant Life, pp. 1–20. [p. 332.]

Bruce, C. W. A. (1904) The flowering of *Dendrocalamus strictus*. Indian Forester, vol. XXX, 1904, pp. 269–70. [p. 103.]

Brunfels, O. (1530) Herbarum vivae eicones. Strasburg, 1530. [pp. 14 (Fig. 7), 15.]

Buchanan, F. (afterwards Hamilton) (1807) A Journey from Madras through...Mysore, Canara, and Malabar. 3 vols. London, 1807. [p. 97.]

Buchet, S. (1912) Le cas du *Lolium temulentum*, L. Bull. de la soc. bot. de France, vol. LIX (sér. 4, vol. XII), 1912, pp. 188–91. [p. 50.]

Bugnon, P. (1920[1]) Contributions à la connaissance de la flore de Normandie: Observations faites en 1920. Bull. de la soc. linn. de Normandie (sér. VII, vol. III), 1920, pp. 315–24. [pp. 372, 376.]

Bugnon, P. (1920[2]) Dans la tige des Graminées, certains faisceaux libéroligneux longitudinaux peuvent être des faisceaux gemmaires. Comptes rendus de l'acad. d. sci., Paris, vol. CLXX, 1920, pp. 1201–2 [p. 325.]

Bugnon, P. (1921) La feuille chez les Graminées. Thèses...docteur ès sciences. ...Université de Paris. Sér. A, No. 877; No. d'ordre, 1691, Caen, 1921, 109 pp.
[pp. 235, 279, 283, 288.]

Bugnon, P. (1924[1]) Contribution à la connaissance de l'appareil conducteur chez les Graminées. Mém. de la soc. linn. de Normandie, vol. XXVI (sér. 2, vol. X), Caen, 1924, pp. 21–40.
[pp. 279, 283.]

Bugnon, P. (1924[2]) Sur les homologies de la feuille chez les Graminées. Bull. de la soc. bot. de France, vol. LXXI (sér. 4, vol. XXIV), 1924, pp. 246–51.
[pp. 283, 310.]

Bukinich, D. D. See **Vavilov, N. I. and Bukinich, D. D.** (1929).

Buller, A. H. R. (1928) The Plants of Canada Past and Present. Trans. Roy. Soc. Canada, vol. XXII, 1928, pp. xxxiii–lviii.
[p. 222.]

Bureau, E. (1903) Étude sur les Bambusées. Végétation et floraison de l'*Arundinaria Simoni* Riv. Bull. du muséum d'hist. nat., 1903, pp. 403–9.
[pp. 75, 99.]

Burkill, I. H. (1923) A Spiny Yam from Sumatra. Gardens' Bull., Straits Settlements, Singapore, vol. III, Nos. 1–3, August 1923, pp. 3, 4.
[p. 78.]

Burkill, I. H. See also **Willis, J. C. and Burkill, I. H.** (1893).

Burnell, A. C. and Hopkins, E. W. (1884) The Ordinances of Manu. Translated from the Sanskrit. London, 1884.
[p. 53.]

Burnell, A. C. See also **Yule, H. and Burnell, A. C.** (1903).

Burr, S. and Turner, D. M. (1933) British Economic Grasses: Their Identification by Leaf Anatomy. London, 1933.
[p. 294.]

"C., E. G." (1881) The Flowering of the Bamboo. Indian Forester, vol. VII, 1881–2, p. 162.
[p. 101.]

Caius Plinius Secundus See **Pliny (Holland, P.)** (1601).

Camus, A. (1922) Note sur les genres "*Lepturus*" R.Br. et "*Pholiurus*" Trinius. Ann. de la soc. linn. de Lyon, vol. LXIX, 1922, pp. 86–90.
[p. 170.]

Camus, A. (1924) *Hickelia* et *Pseudocoix* genres nouveaux de Bambusées malgaches. Bull. de la soc. bot. de France, vol. LXXI, 1924, pp. 899–906.
[p. 127.]

Camus, A. (1925) *Hitchcockella*, genre nouveau de Bambusées malgaches. Comptes rendus de l'acad. d. sci., Paris, vol. CLXXXI, 1925, pp. 253–5.
[p. 127.]

Camus, E. G. (1913) Les Bambusées. Text and Atlas. Paris, 1913.
[pp. 73, 110, 380, 410.]

Candolle, A. de (1884) Origin of Cultivated Plants. International Scientific Series, vol. XLIX, London, 1884.
[pp. 3, 18, 29, 31.]

Candolle, A. P. de (1827) Organographie végétale. Vol. I, Paris, 1827.
[p. 308.]

Candolle, C. de (1868) Théorie de la Feuille. Arch. d. sci. de la bibl. universelle, vol. XXXII, 1868, pp. 32–64.
[p. 309.]

Carey, A. E. and Oliver, F. W. (1918) Tidal Lands: A Study of Shore Problems. London, 1918.
[pp. 339, 372.]

Carruthers, W. (1911) On the Vitality of Farm Seeds. Journ. Roy. Agric. Soc. of England, vol. LXXII, 1911, pp. 168–83.
[p. 224.]

Cavendish, F. H. (1905) A Flowering of *Dendrocalamus Hamiltonii* in Assam. Indian Forester, vol. XXXI, 1905, p. 479.
[pp. 97, 102.]

Čelakovský, L. J. (1897) Ueber die Homologien des Grasembryos. Bot. Zeit., vol. LV, 1897, pp. 141–74.
[p. 233.]

Chase, A. (1908) Notes on Cleistogamy of Grasses. Bot. Gaz., vol. XLV, 1908, pp. 135–6.
[pp. 203, 205.]

Chase, A. (1914) Field Notes on the Climbing Bamboos of Porto Rico. Bot. Gaz., vol. LVIII, 1914, pp. 277–9.
[pp. 61, 76.]

Chase, A. (1918) Axillary Cleistogenes in some American Grasses. Amer. Journ. Bot., vol. V, 1918, pp. 254–8.
[pp. 203, 204.]

Chase, A. (1921[1]) The Linnaean Concept of Pearl Millet. Amer. Journ. Bot., vol. VIII, 1921, pp. 41–9.
[p. 25.]

Chase, A. (1921[2]) The North American Species of *Pennisetum*. Contributions from the United States National Herbarium, vol. XXII, pt. 4, 1921, pp. i–x, 209–34.
[p. 186.]

Chase, A. (1924) *Aciachne*, a cleistogamous grass of the high Andes. Journ. Wash. Acad. Sci., vol. XIV, 1924, pp. 364–6.
[p. 203.]

Chase, A. See also **Hitchcock, A. S. and Chase, A.** (1910), (1931).

Christ, H. (1879) Das Pflanzenleben der Schweiz. Zurich, 1879.
[p. 15.]

Church, A. H. (1919) Thalassiophyta and the Subaerial Transmigration. Bot. Memoirs, No. 3, Oxford, 1919.
[p. 92.]

Church, G. L. (1929[1]) Meiotic Phenomena in certain Gramineae. Bot. Gaz., vol. LXXXVII, No. 5, 1929, pp. 608–29.
[p. 378.]

Church, G. L. (1929[2]) Meiotic Phenomena in certain Gramineae. II. Paniceae and Andropogoneae. Bot. Gaz., vol. LXXXVIII, No. 1, 1929, pp. 63–84.
[p. 378.]

Churchill, A. and J. (1732) A Collection of Voyages. Vol. v. Appendix containing...an Account of the first Discoveries of *America* by the *Europeans*; with a brief Relation of Admiral *Christopher Columbus's* Voyages. London, 1732.
[p. 31.]

Clavigero, F. S. (1787) The History of Mexico. Translated by C. Cullen. Vol. I, London, 1787.
[p. 31.]

C[ogan], H. See **Diodorus Siculus (H. C.)** (1653).

Coldstream, W. (1889) Illustrations of some of the Grasses of the Southern Punjab. London, Calcutta, and Bombay, 1889.
[pp. 2, 191.]

Coleridge, S. T. (E. L. Griggs) (1932) Unpublished Letters. Edited by E. L. Griggs. 2 vols. London, 1932.
[p. 82.]

Collins, G. N. (1909) A New Type of Indian Corn from China. U.S. Dept. of Agric. Bureau of Plant Industry, Bull. No. 161, 1909, 30 pp.
[p. 32.]

Collins, G. N. (1912) The Origin of Maize. Journ. Wash. Acad. Sci., vol. II, 1912, pp. 520–30.
[p. 32.]

Collins, G. N. (1914) A Drought-resisting Adaptation in Seedlings of Hopi Maize. Journ. Agric. Res., vol. I, 1914, pp. 293–302.
[p. 233.]

Collins, G. N. (1917) Hybrids of *Zea ramosa* and *Zea tunicata*. Journ. Agric. Res., vol. IX, 1917, pp. 383–95.
[p. 367.]

Collins, G. N. (1919) Structure of the Maize Ear as indicated in *Zea-Euchlaena* Hybrids. Journ. Agric. Res., vol. XVII, No. 3, 1919, pp. 127–35.
[p. 367.]

Collins, G. N. (1923) Notes on the Agricultural History of Maize. Annual Rep. Amer. Hist. Assoc., 1923 (for 1919), vol. I, pp. 409–29.
[pp. 31, 359, 368, 372.]

Collins, G. N. (1924) The Prophyllum of Grasses. Bot. Gaz., vol. LXXVIII, 1924, pp. 353–4.
[p. 279.]

Collins, G. N. (1925) The "metamorphosis" of *Euchlaena* into Maize. Journ. Heredity, vol. XVI, 1925, pp. 378–80.
[pp. 369, 372.]

Collins, G. N. (1931) The Phylogeny of Maize. Bull. Torr. Bot. Club, vol. LVII, 1931, pp. 199–210.
[pp. 369, 372.]

Columella (1541) L. Junii Moderati Columellae de re rustica Libri XII. Lugduni apud Seb. Gryphium, 1541.
[pp. 15, 43.]

Copeland, E. B. (1924) Rice. London, 1924.
[pp. 26, 27, 28.]

Corbière, L. (1927) Le *Spartina Townsendi* en Normandie. Bull. de la soc. linn. de Normandie, sér. VII, vol. IX, 1927, pp. 92–117.
[p. 372.]

Crowfoot, G. M. and Baldensperger, L. (1932) From Cedar to Hyssop: A Study in the Folklore of Plants in Palestine. London, 1932.
[p. 23.]

Crozier, A. A. (1888) Silk seeking pollen. Bot. Gaz., vol. XIII, 1888, p. 242.
[p. 362.]

"D., J. C." (1883) Note on the *Dendrocalamus strictus* in the Central Provinces. Indian Forester, vol. IX, 1883, pp. 529–39.
[pp. 65, 95, 98, 103.]

Darlington, C. D. (1928) Studies in *Prunus*, I and II. Journ. Genetics, vol. XIX, 1927–8, pp. 213–56.
[p. 378.]

Darlington, C. D. (1932[1]) Recent Advances in Cytology. London, 1932.
[p. 379.]

Darlington, C. D. (1932[2]) Chromosomes and Plant-breeding. London, 1932.
[p. 7.]

Daveau, J. (1922) *Phyllostachys aurea* Rivière. Sa floraison à Montpellier. Bull. de la soc. bot. de France, vol. LXIX (sér. 4, vol. XXII), 1922, pp. 232–6.
[p. 110.]

Davis, T. A. W. and Richards, P. W. (1933) The Vegetation of Moraballi Creek, British Guiana: an Ecological Study of a Limited Area of Tropical Rain Forest. Part I. Journ. Ecology, vol. xxi, 1933, pp. 350–84.
[p. 161.]

Davy, J. Burtt (1898) *Stapfia*, a new genus of Meliceae, and other noteworthy grasses. Erythea, vol. vi, 1898, pp. 109–13.
[pp. 186, 288.]

Davy, J. Burtt (1914) Maize: its History, Cultivation, Handling, and Uses. London, 1914.
[pp. 33, 359, 363.]

Davy, J. Burtt (1922) The Suffrutescent Habit as an Adaptation to Environment. Journ. Ecology, vol. x, 1922, pp. 211–19.
[p. 84.]

Davy, J. Burtt (1928) A Visit to the Belgian Congo. Journ. Oxford Univ. Forestry Soc., ser. 1, No. 8, pp. 28–34, 1928 (for Michaelmas Term, 1927).
[p. 353.]

de l'Obel See **Lobel (de l'Obel), M. (1581)**.

Derby, O. A. (1879) Rats in Brazil and their connection with the flowering of the Bamboo. Indian Forester, vol. v, 1879–80, pp. 177–8.
[p. 97.]

Digby, K. (1661) A Discourse Concerning the Vegetation of Plants. Spoken by Sir K.D., at Gresham College, on the 23. of January, 1660. At a Meeting of the Society for promoting Philosophical Knowledge by Experiments. London, 1661.
[p. 222.]

Diodorus Siculus (H. C.) (1653) The History of Diodorus Siculus. Done into English by H[enry] C[ogan] Gent. London, 1653.
[p. 9.]

Docters van Leeuwen, W. M. (1921) The Flora and Fauna of the Islands of the Krakatau-Group in 1919. Ann. du jardin bot. de Buitenzorg, vol. xxxi, 1921, pp. 103–39.
[p. 347.]

Dodoens, R. (Lyte, H.) (1578) A Nievve Herball...nowe first translated out of French into English, by Henry Lyte Esquyer. London, 1578.
[pp. 6 (Fig. 3), 7, 18, 49 (Fig. 18), 50, 52 (Fig. 20), 218 (Fig. 105), 355.]

Doell, J. C. (1868) Beiträge zur Pflanzenkunde. I. Untersuchungen über den Bau der Grasblüthe, insbesondere über die Stellung derselben innerhalb des Aehrchens. Jahresber. d. Mannheimer Vereins f. Naturkunde, vol. xxxiv, 1868, pp. 30–59.
[p. 181.]

Drabble, E. (1906) Notes on Cereals. New Phyt., vol. v, 1906, pp. 17–21.
[p. 228.]

Duchartre, P. (1852) Note sur la germination des Céréales récoltées avant leur maturité. Comptes rendus de l'acad. d. sci., Paris, vol. XXXV, 1852, pp. 940–2.
[p. 223.]

Duval-Jouve, J. (1866) Étude sur le genre *Crypsis* et sur ses espèces françaises. Bull. de la soc. bot. de France, vol. XIII, 1866, pp. 317–26.
[p. 162.]

Duval-Jouve, J. (1869) Sur les parois cellulaires du *Panicum vaginatum* Godr. et Gren. Bull. de la soc. bot. de France, vol. XVI, 1869, pp. 110–14.
[p. 272.]

Duval-Jouve, J. (1871) Étude anatomique de l'arête des Graminées. Mém. de l'acad. des sci. et lettres de Montpellier, Sect. des Sci., vol. VIII, 1876 for 1872–5, pp. 33–78. [This paper appears to have been published in 1871.]
[pp. 149, 256.]

Duval-Jouve, J. (1875) Histotaxie des feuilles de Graminées. Ann. d. sci. nat., sér. 6, Bot., vol. I, 1875, pp. 294–371.
[pp. 251, 289, 294, 300, 384, 405.]

Dyer, W. T. Thiselton (1885) The Square Bamboo. Nature, vol. XXXII, 1885, pp. 391–2.
[p. 71.]

Edgworth, M. P. (1862) Florula Mallica. Journ. Proc. Linn. Soc. Lond., Bot., vol. VI, 1862, pp. 179–210.
[p. 2.]

Ellis, E. V. (1907) *Cephalostachyum pergracile* in Flower (Tinwa). Indian Forester, vol. XXXIII, 1907, pp. 323–4.
[p. 102.]

Engelmann, G. (1859) Two new dioecious Grasses of the United States. Trans. Acad. Science St Louis, vol. I, 1856–60, No. 3, 1859, pp. 431–42.
[p. 195.]

Engledow, F. L. (1927) The Winter Hardiness of Crops. Agricultural Research in 1926. Roy. Agric. Soc. London, 1927, pp. 20–35.
[p. 38.]

Engledow, F. L. See also **Biffen, R. H. and Engledow, F. L.** (1926).

Engler, A. (1892) Die systematische Anordnung der monokotyledoneen Angiospermen. Abhandl. d. k. Preuss. Akad. d. Wiss. zu Berlin, pt. II, 1892, 55 pp.
[p. 149.]

Ernst, A. (1908) The New Flora of the Volcanic Island of Krakatau. Trans. by A. C. Seward. Cambridge, 1908.
[p. 347.]

Ettingshausen, C. Ritter von (1866) Beitrag zur Kenntniss der Nervation der Gramineen. Sitzungsber. d. Math.-Nat. Cl., k. Akad. d. Wiss., Wien, vol. LII, pt. I, 1866 (for 1865), pp. 405–32.
[p. 294.]

Evans, M. W. (1927) The Life History of Timothy. U.S. Dept. of Agric. Dep. Bull. No. 1450, Washington, March, 1927.
[pp. 159, 233, 274, 385.]

Evans, M. W. (1931) Relation of Latitude to Time of Blooming of Timothy. Ecology, vol. XII, 1931, pp. 182–7.
[p. 214.]

Evans, M. W. See also **Oakley, R. A. and Evans, M. W. (1921).**

Everard, T. (1692 and 1693/4) A Collection for Improvement of Husbandry and Trade. Letters from Thomas Everard communicated by John Houghton, No. 14, May 25, 1692, and No. 85, March 16, 1693/4.
[p. 222.]

Fabre, E. (1850) D'une nouvelle espèce de Graminée, *Spartina versicolor*, propre à fixer les sables maritimes et à fertiliser les terrains salés du littoral méditerranéen. Bull. de la Soc. d'Agric. de l'Hérault, Montpellier, Jan. Feb. and March, 1850, 9 pp.
[pp. 337, 375.]

Finch, V. C. and Baker, O. E. (1917) Geography of the World's Agriculture. U.S. Dept. of Agric., Washington, 1917.
[pp. 28, 41.]

Fingerhuth, K. A. (1839) Beiträge zur Synonymie der Pflanzen des Alterthums. II. Ueber Zucker und Zuckerrohr der Alten. Flora, Jahrg. XXII, Bd. II, Regensburg, 1839, pp. 529–40.
[p. 60.]

Fisk, E. L. (1927) The Chromosomes of *Zea Mays*. Amer. Journ. Bot., vol. XIV, 1927, pp. 53–75.
[p. 372.]

Fortune, R. (1857) A Residence among the Chinese. London, 1857.
[p. 62.]

Fournier, E. (1876) Sur les Graminées mexicaines à sexes séparés. Bull. de la soc. roy. de bot. de Belgique, vol. XV, 1876, pp. 459–76.
[p. 195.]

Franchet, A. (1889) Note sur deux Nouveaux Genres de Bambusées. Journ. de Bot., vol. III, 1889, pp. 277–88.
[pp. 109, 219.]

Freeman, E. M. (1904) The Seed-Fungus of *Lolium temulentum*, L., the Darnel. Phil. Trans. Roy. Soc. Lond., Ser. B, vol. CXCVI, 1904, pp. 1–27.
[p. 50.]

Freeman-Mitford, A. B. (Lord Redesdale) (1896) The Bamboo Garden. London, 1896.
[p. 69.]

Fritsch, K. (1932) Die systematische Gruppierung der Monokotylen. Ber. d. deutsch. Bot. Gesellsch., vol. L*a* (Festschrift), 1932, pp. 162–84.
[p. 380.]

Frohmenger, M. (1914) Die Entstehung und Ausbildung der Kieselzellen bei den Gramineen. Bibl. Bot., vol. XXI, pt. LXXXVI, 1914, 41 pp.
[pp. 256, 258.]

Fuchs, J. (1912) Beitrag zur Kenntnis des Loliumpilzes. Hedwigia, vol. LI, 1912, pp. 221–39.
[p. 50.]

Fuchs, L. (1542) De Historia Stirpium. Basle, 1542.
[pp. 5 (Fig. 2), 19 (Fig. 9), 20 (Fig. 10), 22 (Fig. 12), 29, 30 (Fig. 15), 147 (Fig. 64).]

Gallastegui, C. A. See **Jones, D. F. and Gallastegui, C. A.** (1919).

Gamble, J. S. (1890) The Treatment of Bamboo Forests. Indian Forester, vol. XVI, 1890, pp. 418–19.
[p. 105.]

Gamble, J. S. (1896) The Bambuseae of British India. Ann. Roy. Bot. Gard. Calcutta, vol. VII, 1896, xvii + 133 pp.
[pp. 61, 95, 98.]

Gamble, J. S. (1900) Flowering of *Arundinaria Falconeri*. Indian Forester, vol. XXVI, 1900, p. 386.
[p. 97.]

Gamble, J. S. (1904) The Flowering of the Bamboo. Nature, vol. LXX, Sept. 1, 1904, p. 423.
[p. 95.]

Garman, H. (1900) Kentucky Forage Plants. The Grasses. Kentucky Agric. Exp. Station of the State College of Kentucky. Bull. No. 87, No. 1, Lexington, May, 1900, pp. 55–122.
[pp. 16, 46.]

Garner, W. W. and Allard, H. A. (1920) Effect of the relative length of day and night and other factors of the environment on growth and reproduction in plants. Journ. Agric. Res., vol. XVIII, 1919–20, No. 11, March 1, 1920, pp. 553–606.
[p. 212.]

Gates, F. C. See **Laude, H. H. and Gates, F. C.** (1929).

Gates, R. Ruggles (1925) The Vegetation of the Amazon Basin. Proc. Linn. Soc. Lond., Session CXXXVIII, 1925–6, Dec. 17, 1925, pp. 13–15.
[p. 351.]

Gathorne-Hardy, G. M. (1921) The Norse Discoverers of America: The Wineland Sagas translated and discussed. Oxford, 1921.
[p. 29.]

Gerard, J. (1597) The Herball, or Generall Historie of Plantes. London, 1597.
[pp. 36, 196, 258.]

Godron, D. A. (1873) De la Floraison des Graminées. Mém. de la soc. des sci. nat. de Cherbourg, vol. XVII, 1873, pp. 105–97.
[pp. 134, 155.]

Godron, D. A. (1879) Études Morphologiques sur la Famille des Graminées. Rev. d. sci. nat., Montpellier, vol. VII, 1878–9, pp. 393–411; vol. VIII, 1879, pp. 14–30.
[p. 180.]

Godron, D. A. (1880) Les Bourgeons axillaires et les Rameaux des Graminées. Rev. d. sci. nat., vol. VIII (N.S. vol. I), 1880, pp. 429–42.
[pp. 264, 309.]

Goebel, K. von (1884) Beiträge zur Entwickelungsgeschichte einiger Inflorescenzen. Prings. Jahrb. f. wiss. Bot., vol. XIV, 1884, pp. 1–42.
[p. 164.]

Goebel, K. von (1922) Organographie der Pflanzen, vol. III, pt. I, Jena, 1922.
[p. 238.]

Goebel, K. von (1926) *Leersia hexandra*, ein Gras mit nyktinastischen Bewegungen. Ann. du jardin bot. de Buitenzorg, vol. XXXVI, 1926, pp. 186–201.
[p. 292.]

Goethe, J. W. von (1790) Versuch die Metamorphose der Pflanzen zu erklären. Gotha, 1790.
[pp. 307, 308, 312, 328.]

Goliński, S. J. (1893) Ein Beitrag zur Entwicklungsgeschichte des Androeceums und des Gynaeceums der Gräser. Bot. Centralbl., vol. LV, 1893, pp. 1–17, 65–72, 129–35.
[p. 154.]

Good, R. D'O. (1927) A Summary of Discontinuous Generic Distribution in the Angiosperms. New Phyt., vol. XXVI, 1927, pp. 249–59.
[p. 344.]

Gordon, M. (1922) The Development of Endosperm in Cereals. Proc. Roy. Soc. Victoria, vol. XXXIV (N.S.), pt. II, 1922, pp. 105–16.
[p. 226.]

Goudot, J. (1846) *Aulonemia*, nouveau genre de la tribu des Bambusées. Ann. d. sci. nat., sér. III, Bot., vol. V, 1846, pp. 75–7.
[p. 76.]

Graham, R. J. D. (1927) Studies in Wild Rice—*Oryza sativa*, Linn. Trans. and Proc. Bot. Soc. Edinburgh, vol. XXIX, pt. 4, 1927, pp. 349–51.
[pp. 223, 224.]

Gray, A. (1887) The Botanical Textbook (Sixth edition), Part I, Structural Botany. London, 1887.
[p. 92.]

Gray, L. C. (1933) History of Agriculture in the Southern United States To 1860. 2 vols., 1933, Carnegie Institution of Washington.
[p. 28.]

Green, T. (n.d. [1816–20]) The Universal Herbal. Liverpool, n.d. [1816–20].
[pp. 28, 54.]

Gregor, J. W. and Sansome, F. W. (1930) Experiments on the Genetics of Wild Populations. II. *Phleum pratense* L. and the Hybrid *P. pratense* L. × *P. alpinum* L. Journ. Genetics, vol. XXII, 1930, pp. 373–87.
[pp. 45, 378.]

Griffiths, D. (1912) The Grama Grasses: *Bouteloua* and Related Genera. Contributions from the U.S. Nat. Herb., vol. XIV, pt. 3, 1912, pp. i–xi, 343–428.
[p. 333.]

Griggs, E. L. See **Coleridge, S. T. (E. L. Griggs)** (1932).

Grob, A. (1896) Beiträge zur Anatomie der Epidermis der Gramineenblätter. Bibliotheca Botanica, Stuttgart, vol. VII, pt. XXXVI, 1896, 123 pp.
[p. 301.]

Groves, H. and J. (1881) *Spartina Townsendi nobis*. Rep. Bot. Exchange Club for 1880, Manchester, 1881, p. 37.
[p. 372.]

Groves, J. (n.d.) Grasses. School Nature Study Union, Publication No. 42, 5 pp.
[p. 372.]

Guéguen, F. (1901) Anatomie comparée du style et du stigmate des Phanérogames. I. Monocotylédones, apétales et gamopétales. Journ. de bot., vol. XV, 1901, pp. 265–300.
[p. 357.]

Guérin, P. (1898[1]) Sur la Présence d'un Champignon dans l'Ivraie (*Lolium temulentum* L.). Journ. de bot., vol. XII, 1898, pp. 230–8.
[p. 50.]

Guérin, P. (1898[2]) Sur le développement des téguments séminaux et du péricarpe des Graminées. Bull. de la soc. bot. de France, vol. XLV (sér. III, vol. V), 1898, pp. 405–11.
[p. 162.]

Guérin, P. (1898[3]) Structure particulière du fruit de quelques Graminées. Journ. de bot., vol. XII, 1898, pp. 365–74.
[p. 162.]

Guérin, P. (1899) Recherches sur le développement du tégument séminal et du péricarpe des Graminées. Thèses...Faculté des sciences de Paris...docteur ès sciences naturelles, sér. A, No. 330, No. d'ordre 990, 1899, 59 pp.
[p. 162.]

Guignard, L. (1901) La double fécondation dans le Maïs. Journ. de bot., vol. XV, 1901, pp. 37–50.
[p. 362.]

Guillaud, M. (1924) Sur la préfeuille des Graminées. Bull. de la soc. linn. de Normandie, sér. 7, vol. VII, 1924, pp. 41–99.
[p. 279.]

"H., H. H." (1892) The Seeding of Bamboos. Indian Forester, vol. XVIII, 1892, p. 304.
[p. 97.]

Haberlandt, G. (1882) Vergleichende Anatomie des assimilatorischen Gewebesystems der Pflanzen. Pringsheim's Jahrb., vol. XIII, 1882, pp. 74–188.
[pp. 72, 304.]

Haberlandt, G. (1890) Die Kleberschicht des Gras-Endosperms als Diastase ausscheidendes Drüsengewebe. Ber. d. deutsch. Bot. Gesellsch., vol. VIII, 1890, pp. 40–8.
[p. 227.]

Haberlandt, G. (1914) Physiological Plant Anatomy, translated from the fourth German edition by Montagu Drummond. London, 1914.
[p. 303.]

Hackel, E. (1880) Ueber das Aufblühen der Gräser. Bot. Zeit., Jahrg. XXXVIII, 1880, pp. 431–7.
[p. 157.]

Hackel, E. (1882) Monographia Festucarum europaearum. Kassel and Berlin, 1882.
[pp. 134, 220, 263, 283, 303, 344.]

Hackel, E. (1887) Gramineae, in Die Natürlichen Pflanzenfamilien, Teil II, Abt. 2, 1887, pp. 1–97. [A new edition is announced.]
[pp. 3, 12, 410.]

Hackel, E. (1889) Andropogoneae in Candolle, A. and C. de, Monographiae Phanerogamarum Prodromi..., vol. VI, Paris, 1889.
[p. 368.]

Hackel, E. (1890[1]) Ueber...Eigenthümlichkeiten der Gräser trockener Klimate. Verhandl. d. k.-k. zool.-bot. Gesellsch. Wien, vol. XL, 1890, pp. 125–38.
[pp. 270, 286.]

Hackel, E. (1890[2]) The True Grasses. Translated from Die Natürlichen Pflanzenfamilien by F. Lamson-Scribner and E. A. Southworth. New York, 1890.
[p. 410.]

Hackel, E. (1906) Ueber Kleistogamie bei den Gräsern. Oesterr. Bot. Zeit., vol. LVI, 1906, pp. 81–8, 143–54, 180–6.
[pp. 203, 204.]

Hamada, H. (1933) Wachstumsverhältnisse der Keimorgane von verschiedenen Gramineen im Dunkel und bei Belichtung mit besonderer Berücksichtigung ihrer systematischen Stellung. Mem. Coll. Sci., Kyoto Imp. Univ., ser. B, vol. IX, No. 2, 1933, pp. 71–128.
[p. 233.]

Hamilton, F. See **Buchanan, F.** (afterwards **Hamilton**) (1807).

Hanausek, T. F. (1898) Vorläufige Mittheilung über den von A. Vogl in der Frucht von *Lolium temulentum* entdeckten Pilz. Ber. d. deutsch. Bot. Gesellsch., vol. XIV, 1898, pp. 203–7.
[p. 50.]

Hannig, E. (1907) Ueber pilzfreies *Lolium temulentum*. Bot. Zeit., vol. LVI, pt. I, 1907, pp. 25–38.
[p. 50.]

Harlan, H. V. See **Anthony, S. and Harlan, H. V. (1920).**

Harshberger, J. W. (1893) Maize: A Botanical and Economic Study. Contributions from the Bot. Lab. of the University of Pennsylvania, vol. I, 1893, No. 2, pp. 75–202.
[pp. 29, 31, 366.]

Hartig, R. (1888) Ueber den Einfluss der Samenproduktion auf Zuwachsgrösse und Reservestoffvorrat der Bäume. Bot. Centralbl., Jahrg. IX, vol. XXXVI, 1888, pp. 388–91.
[p. 106.]

Hartmann, F. (1923) L'Agriculture dans l'ancienne Égypte. Thèse pour le doctorat d'Université, Paris, 1923.
[p. 9.]

Head, B. V. (1881) British Museum. Department of Coins and Medals. Guide to the Principal Gold and Silver Coins of the Ancients from circ. B.C. 700 to A.D. I. 2nd ed., 1881.
[p. 11 (Fig. 5).]

Henslow, G. (1888) The Origin of Floral Structures. London, 1888.
[p. 330.]

Hildebrand, F. (1872) Ueber die Verbreitungsmittel der Gramineen-Früchte. Bot. Zeit., Jahrg. XXX, 1872, pp. 853–63, 869–75.
[pp. 143, 163.]

Hildebrand, F. (1873) Beobachtungen über die Bestäubungsverhältnisse bei den Gramineen. Monatsber. d. k. preuss. Akad. d. Wiss. zu Berlin, 1873 (for 1872), pp. 737–64.
[pp. 160, 194.]

Hiller, G. H. (1884) Untersuchungen über die Epidermis der Blüthenblätter. Pringsheim's Jahrb. f. wiss. Bot., vol. XV, 1884, pp. 411–51.
[p. 303.]

Hitchcock, A. S. (1892) Some depauperate grasses. Bot. Gaz., vol. XVII, 1892, p. 194.
[p. 211.]

Hitchcock, A. S. (1895) Note on buffalo grass. Bot. Gaz., vol. XX, 1895, p. 464.
[p. 195.]

Hitchcock, A. S. (1920) The Genera of Grasses of the United States. U.S. Dept. of Agric., Bull. No. 772, March 1920, 307 pp.
[pp. 2, 45, 46, 55, 59, 207, 249, 380.]

Hitchcock, A. S. (1925) The North American Species of *Stipa*. Synopsis of the South American Species of *Stipa*. Contributions from the U.S. Nat. Herbarium, vol. XXIV, pt. 7, 1925, pp. xi + 215–89.
[p. 203.]

Hitchcock, A. S. (1926) *Eragrostis hypnoides* and *E. reptans*. Rhodora, vol. XXVIII, 1926, pp. 113–15.
[p. 195.]

Hitchcock, A. S. (1929) The Relation of Grasses to Man. South African Journ. Sci., vol. XXVI, 1929, pp. 133–8.
[pp. 53, 258.]

Hitchcock, A. S. and Chase, A. (1910) The North American Species of *Panicum*. Contributions from the U.S. Nat. Herbarium, vol. XV, 1910, xiv + 396 pp.
[p. 184.]

Hitchcock, A. S. and Chase, A. (1931) Grass. Old and New Plant Lore, Smithsonian Scientific Series, vol. XI, 1931, pp. 201–50.
[pp. 2, 12, 23, 32, 53, 217.]

Hitchcock, A. S. See also **Lyon, T. L. and Hitchcock, A. S.** (1904).

Hochstetter, C. F. (1847) Aufbau der Graspflanze. Jahreshefte d. Vereins f. vaterl. Naturkunde in Württemberg, vol. III, 1847, pp. 1–84.
[pp. 289, 310.]

Hoeck, F. (1893) Kosmopolitische Pflanzen. Naturwiss. Wochenschrift, vol. VIII, 1893, pp. 135–8.
[p. 334.]

Hoehnel, F. von (1884) Ueber die Art des Auftretens einiger vegetabilischer Rohstoffe in den Stammpflanzen. Sitzungsber. d. Math.-Nat. Klasse d. k. Akad. d. Wiss., Wien, vol. LXXXIX, pt. I, 1884, pp. 6–16.
[p. 55.]

Hole, R. S. (1911) On some Indian Forest Grasses and their Oecology. The Indian Forest Memoirs, Forest Bot. Ser., vol. I, pt. I, ii + 126 pp., Calcutta, 1911.
[pp. 160, 256, 283.]

Holland, P. See **Pliny (Holland, P.)** (1601).

Holland, R. See **Britten, J. and Holland, R.** (1886).

Hooke, R. (1665) Micrographia. London, 1665.
[p. 143.]

Hooker, J. D. (1854) Himalayan Journals. 2 vols. London, 1854.
[pp. 34, 333, 351.]

Hooker, J. D. See also **Bentham, G. and Hooker, J. D.** (1883).

Hooker, W. J. (1843) Notes on the Botany of H.M. Discovery Ships, Erebus and Terror in the Antarctic Voyage; with some account of the Tussac Grass of the Falkland Islands. Hooker's London Journ. of Bot., vol. II, 1843, pp. 247–329.
[pp. 272, 347.]

Hoops, J. (1905) Waldbäume und Kulturpflanzen im germanischen Altertum. Strassburg, 1905.
[pp. 3, 5, 11, 12.]

Hopkirk, T. (1817) Flora Anomala. A General View of the Anomalies in the Vegetable Kingdom. Glasgow, 1817.
[p. 393.]

Hort, A. See **Theophrastus (Hort, A.)** (1916).

Houard, C. (1904) Recherches anatomiques sur les galles de tiges: Acrocécidies. Ann. d. sci. nat., sér. 8, Bot., vol. XX, 1904, pp. 289–384.
[p. 283.]

Howarth, W. O. (1924) On the Occurrence and Distribution of *Festuca rubra*, Hack. in Great Britain. Journ. Linn. Soc. Lond., Bot., vol. XLVI, 1924, pp. 313–31.
[p. 337.]

Howarth, W. O. (1925) On the Occurrence and Distribution of *Festuca ovina* L., sensu ampliss., in Britain. Journ. Linn. Soc. Lond., Bot., vol. XLVII, 1925, pp. 29–39.
[p. 393.]

Howarth, W. O. (1927) The Seedling Development of *Festuca rubra*, L., var. *tenuifolia* mihi, and its Bearing on the Morphology of the Grass Embryo. New Phyt., vol. XXVI, 1927, pp. 46–57.
[pp. 224, 235.]

Hudson, W. (1762) Flora Anglica. London, 1762.
[p. 50.]

Hunter, H. (1924) Oats: their Varieties and Characteristics. London, 1924.
[p. 12.]

Hunter, H. (1926) The Barley Crop. London, 1926.
[p. 9.]

Hunter, H. (1931) Baillière's Encyclopaedia of Scientific Agriculture, edited by H. Hunter. 2 vols. London, 1931.
[pp. 26, 227.]

Hunter, H. and Leake, H. Martin (1933) Recent Advances in Agricultural Plant Breeding. London, 1933.
[pp. 12, 13, 53, 227.]

Hurst, C. C. (1933) The Mechanism of Creative Evolution. Cambridge, 1932 (reprinted 1933).
[pp. 7, 222, 378.]

Huskins, C. L. (1927) On the Genetics and Cytology of Fatuoid or False Wild Oats. Journ. Genetics, vol. XVIII, 1927, pp. 315–64.
[p. 17.]

Huskins, C. L. (1928) On the Cytology of Speltoid Wheats in relation to their origin and genetic behaviour. Journ. Genetics, vol. XX, 1928–9, pp. 103–22.
[p. 18.]

Huskins, C. L. (1930) The Origin of *Spartina Townsendii*. Genetica, vol. XII, 1930, pp. 531–8.
[p. 378.]

Huskins, C. L. and Smith, S. G. (1932) A Cytological Study of the Genus *Sorghum* Pers. I. The Somatic Chromosomes. Journn. Genetics, vol. XXV, 1931–2, pp. 241–9.
[p. 21.]

Jackson, V. G. (1922) Anatomical Structure of the Roots of Barley. Ann. Bot., vol. XXXVI, 1922, pp. 21–39.
[p. 251.]

Jagor, F. (1866) Reiseskizzen. Singapore-Malacca Java. Berlin, 1866.
[p. 59.]

Jefferies, T. A. (1915) Ecology of the Purple Heath Grass (*Molinia caerulea*). Journ. Ecology, vol. III, 1915, pp. 93–109.
[p. 247.]

Jefferies, T. A. (1916) The Vegetative Anatomy of *Molinia caerulea*, the Purple Heath Grass. New Phyt., vol. XV, 1916, pp. 49–71.
[pp. 249, 275.]

Jenkin, T. J. (1922) Notes on Vivipary in *Festuca ovina*. Rep. of the Bot. Soc. and Exchange Club of the Brit. Isles, vol. VI, pt. 3, 1922 (for 1921), pp. 418–32.
[p. 393.]

Jenkin, T. J. (1924) The Artificial Hybridisation of Grasses. University College of Wales, Aberystwyth. Welsh Plant Breeding Station, Series H, No. 2, 1924, 18 pp.
[p. 160.]

Jenkin, T. J. and Sethi, B. L. (1932) *Phalaris arundinacea*, *Ph. tuberosa*, their *F* 1 Hybrids and Hybrid Derivatives. Journ. Genetics, vol. XXVI, 1932, pp. 1–36.
[p. 378.]

Johannsen, W. (1903) Ueber Erblichkeit in Populationen und in reinen Linien. Jena, 1903.
[p. 39.]

Johnson, C. (n.d. [1857–61]) The Grasses of Great Britain, illustrated by John E. Sowerby. London, n.d. Plates dated 1857–61.
[p. 16.]

Johnson, D. S. and York, H. H. (1915) The Relation of Plants to Tide-levels. Carnegie Inst. of Washington, Publ. No. 206, 1915, 162 pp.
[pp. 340, 350.]

Johnston, H. H. (1913) A History of the Colonization of Africa. Cambridge, 1913.
[p. 33.]

Jones, D. F. and Gallastegui, C. A. (1919) Some Factor Relations in Maize with reference to linkage. Amer. Nat., vol. LIII, 1919, pp. 239–46.
[p. 367.]

Jones, E. T. (1930) Morphological and Genetical Studies of Fatuoid and other Aberrant Grain-types in *Avena*. Journ. Genetics, vol. XXIII, 1930, pp. 1–68.
[p. 18.]

Jones, L. R. (1902) Vermont Grasses and Clovers. Vermont Agric. Exp. Station, Burlington, Vt. Bull. No. 94, 1902, pp. 139–84.
[p. 45.]

Kanjilal, U. (1891) The Bhalkua Bans (*Bambusa balcooa*) of Bengal. Indian Forester, vol. XVII, 1891, pp. 53–5.
[p. 62.]

Karelstschicoff, S. (1868) Ueber die faltenfoermigen Verdickungen in den Zellen einiger Gramineen. Bull. de la Soc. Imp. des Naturalistes de Moscou, vol. XLI, 1868, pp. 180–90.
[pp. 72, 304.]

King, F. H. (1927) Farmers of Forty Centuries or Permanent Agriculture in China, Korea and Japan. London, 1927.
[p. 59.]

Kirchner, O. von, Loew, E. and Schroeter, C. (1908, etc.) Lebensgeschichte der Blütenpflanzen Mitteleuropas. Gramineae (unfinished). Stuttgart, 1908, etc.
[pp. 69, 154, 202, 211, 220, 223, 229, 275, 334.]

Kirchner, O. von See also **Schroeter, C. and Kirchner, O. von** (1902).

Knapp, J. L. (1804) Gramina Britannica. London, 1804.
[pp. 211, 339.]

Koernicke, F. (1885) Die Arten und Varietäten des Getreides. Vol. I of Koernicke, F. and Werner, H., Handbuch des Getreidebaues. Berlin, 1885.
[pp. 2, 3, 12, 18, 28, 33, 34, 37, 157, 158, 204, 207, 226.]

Koernicke, F. (1908) Die Entstehung und das Verhalten neuer Getreidevarietäten. Archiv f. Biontologie. Herausgegeben von d. Gesellsch. naturf. Freunde zu Berlin, vol. II, Part 2, 1908–9, No. 5, 1908, pp. 393–437.
[pp. 3, 5, 11.]

Kokkonen, P. (1931) Untersuchungen über die Wurzeln der Getreidepflanzen. I. Die Wurzelformen, ihr Bau, ihre Aufgabe und Lage im Wurzelsystem. Acta Forestalia Fennica, vol. XXXVII, 1931, 123 pp.
[p. 251.]

Kouznetsov, E. S. See **Vavilov, N. I. and Kouznetsov, E. S.** (1922).

Kraus, G. (1895) Physiologisches aus den Tropen. Das Längenwachstum der Bambusrohre. Ann. du jardin bot. de Buitenzorg, vol. XII, 1895, pp. 196–210.
[pp. 62, 63.]

Kraus, G. (1908) Gynaeceum oder Gynoecium? Verhandl. der Physik-Med. Gesellsch. zu Würzburg, N.F., vol. XXXIX, 1908, pp. 9–14.
[pp. 120, 152.]

Ktêsias (McCrindle, J. W.) (1882) Ancient India as described by Ktêsias the Knidian. Calcutta, 1882.
[pp. 58, 71.]

Kugler, H. (1928) Ueber invers-dorsiventrale Blätter. Planta, vol. v, 1928, pp. 89–134.
[p. 292.]

Kunth, C. (1815) Considérations générales sur les Graminées. Mém. du mus. d'hist. nat., vol. II, 1815, pp. 62–75.
[p. 380.]

Kurz, S. (1876) Bamboo and its use. Indian Forester, vol. I, 1876, pp. 219–69, 335–62.
[pp. 59, 60, 61, 62, 64, 73, 77, 78, 97, 98, 100, 103, 133.]

Kurz, S. (1877) Forest Flora of British Burma. 2 vols. Calcutta, 1877.
[p. 101.]

Lamson-Scribner, F. (1895) Grasses as Sand and Soil Binders. Yearbook of the U.S. Dept. of Agric., 1895 (for 1894), pp. 421–36.
[pp. 337, 339.]

Lamson-Scribner, F. (1898) Economic Grasses. U.S. Dept. of Agric., Division of Agrostology, Bull. No. 14 (Agros. 34), Washington, 1898.
[pp. 16, 56.]

Lang, W. D. (1920) The Pelmatoporinae, an Essay on the Evolution of a Group of Cretaceous Polyzoa. Phil. Trans. Roy. Soc. Lond., Ser. B, vol. CCIX, 1920, pp. 191–228.
[p. 93.]

Langdon, S. (1927) Wheat in 3500 B.C. The Times, Jan. 29 and Feb. 3, 1927.
[p. 3.]

Larger, R. (1917) Théorie de la Contre-Évolution ou Dégénérescence par l'Hérédité Pathologique. Paris, 1917.
[p. 91.]

Laude, H. H. and Gates, F. C. (1929) A head of Sorghum with greatly proliferated spikelets. Bot. Gaz., vol. LXXXVIII, 1929, pp. 447–50.
[p. 388.]

Laufer, B. (1907) The Introduction of Maize into Eastern Asia. Congrès Internat. des Américanistes, Session XV, 1907 (for 1906), vol. I, pp. 223–57.
[p. 32.]

Lawson, P. and Son [C.] (1836) The Agriculturalist's Manual. Edinburgh, 1836.
[pp. 143, 336, 393.]

Lawson, P. and Son [C.] (1852) Synopsis of the Vegetable Products of Scotland. Edinburgh. Private Press of Peter Lawson and Son, 1852.
[p. 40.]

Lea, A. M. (1915) An Insect-Catching Grass (*Cenchrus australis*, R.Br.). Trans. and Proc. Roy. Soc. S. Australia, vol. XXXIX, 1915, pp. 92–93.
[p. 191.]

Leake, H. Martin See **Hunter, H. and Leake, H. Martin** (1933).

Leeke, P. (1907) Untersuchungen über Abstammung und Heimat der Negerhirse [*Pennisetum americanum* (L.) K. Schum.]. Zeitschrift f. Naturwissenschaften. Organ des naturwiss. Vereins f. Sachsen und Thüringen zu Halle, vol. LXXIX, 1907, pp. 1–108.
[pp. 24, 25, 37.]

Lehmann, E. (1906) Zur Kenntnis der Grasgelenke. Ber. d. deutsch. Bot. Gesellsch., vol. XXIV, 1906, pp. 185–9.
[pp. 71, 256.]

Leokadia, L. (1926) Zur Kenntnis der Entfaltungszellen monokotyler Blätter. Flora, G.R. vol. CXX, N.F. vol. XX, 1926, pp. 283–342.
[pp. 301, 405.]

Lestiboudois, T. (1819–22) Notice sur la plus interne des enveloppes florales des Graminées. Lille. Travaux, 1819–22, pp. 174–80.
[p. 180.]

Lewton-Brain, L. (1904) On the Anatomy of the Leaves of British Grasses. Trans. Linn. Soc. Lond., ser. II, Bot., vol. VI, 1901–5, pt. VII, 1904, pp. 315–59.
[p. 294.]

Lignier, O. (1909) Essai sur l'Évolution morphologique du Règne végétale. Assoc. Franç. pour l'avance. des sciences, Compte rendu, 37e session, Clermont-Ferrand, 1909 (for 1908), pp. 530–5.
[p. 330.]

Lindau, G. (1904) Ueber das Vorkommen des Pilzes des Taumellolchs in altägyptischen Samen. Sitzungsber. d. k. preuss. Akad. d. Wiss., Berlin, Jahrg. 1904, pp. 1031–5.
[p. 50.]

Lindman, C. A. M. (1899) Zur Morphologie und Biologie einiger Blätter und belaubter Sprosse. Bihang till k. Svenska Vet.-Akad. Handlingar, vol. XXV, pt. 3, no. 4, 1899, 63 pp.
[pp. 289, 292, 405.]

Link, H. F. (1825 Ueber die natürliche Ordnung der Gräser. Abhandlung d. k. Akad. d. Wiss. Berlin, 1828 (for 1825), pp. 17–44.
[p. 405.]

Lisle, E. (1757) Observations in Husbandry. Edition II, 2 vols. London, 1757.
[pp. 48, 235.]

Lobel (de l'Obel), M. (1581) Kruydtboeck. Christophe Plantin, Antwerp, 1581.
[pp. 1 (Fig. 1), 10 (Fig. 4), 386 (Fig. 194).]

Loew, E. See **Kirchner, O. von, Loew, E. and Schroeter, C.** (1908, etc.).

Lyon, T. L. and Hitchcock, A. S. (1904) Pasture, Meadow, and Forage Crops in Nebraska. U.S. Dept. of Agric. Bureau of Plant Industry, Bull. No. 59, 64 pp., Washington, 1904.
[p. 45.]

Lyte, H. See **Dodoens, R. (Lyte, H.)** (1578).

Macloskie, G. (1895[1]) Antidromy of Plants. Bull. Torr. Bot. Club, vol. XXII, 1895, pp. 379–87.
[p. 292.]

Macloskie, G. (1895[2]) Vegetable Spiralism. Bull. Torr. Bot. Club, vol. XXII, 1895, pp. 466–70.
[p. 143.]

Macmillan, H. F. (1908) Flowering of *Dendrocalamus giganteus*, the "Giant Bamboo". Ann. Roy. Bot. Gard. Perideniya, vol. IV, 1908, pp. 123–9.
[pp. 61, 105.]

Maiden, J. H. (1898) A Manual of the Grasses of New South Wales. Sydney, 1898.
[pp. 148, 351.]

Malpighi, M. (1675) Anatome Plantarum. Published by the Royal Society, London, 1675.
[pp. 139 (Fig. 57), 359, 362 (Fig. 186).]

Martius, C. F. P. de (1848) Ueber das Längenwachsthum von Schossen des Bambusrohrs. Gelehrte Anzeigen k. bayer. Akad. der Wiss., Munich, vol. XXVI, 1848, pp. 763–6.
[p. 62.]

Massart, J. (1898) Un voyage botanique au Sahara. Bull. soc. roy. de bot. de Belgique, vol. XXXVII, 1898, pp. 202–339.
[p. 249.]

Massart, J. (1910) Esquisse de la Géographie Botanique de la Belgique. Recueil de l'Institut botanique Léo Errera, Brussels, vol. VII bis, 1910, xi + 332 pp.
[p. 337.]

Mattioli, P. A. (1560) Commentarii in libros sex Pedacii Dioscoridis. Venice, 1560.
[pp. 17 (Fig. 8), 21 (Fig. 11), 27 (Fig. 14), 50 (Footnote).]

Maximov, N. A. (1929) The Plant in Relation to Water. English trans., edited by R. H. Yapp. London, 1929.
[p. 305.]

McLennan, E. I. (1920) The Endophytic Fungus of *Lolium*. I. Proc. Roy. Soc. Victoria, vol. XXXII (N.S.), pt. II, 1920, pp. 252–301.
[pp. 50, 226.]

McLennan, E. I. (1926) The Endophytic Fungus of *Lolium*. II. The Mycorrhiza on the Roots of *Lolium temulentum*. Ann. Bot., vol. XL, 1926, pp. 43–68.
[p. 50.]

Menon, K. G. (1918[1]) *Bambusa arundinacea* growing epiphytically. Indian Forester, vol. XLIV, 1918, p. 130.
[p. 77.]

Menon, K. G. (1918[2]) Flowering and After of *Bambusa arundinacea*. Indian Forester, vol. XLIV, 1918, pp. 519–20.
[p. 100.]

Messer, M. (1932) An Agricultural Atlas of England and Wales. Ed. 2. Ordnance Survey, 1932.
[p. 41.]

Miller, E. C. (1919) Development of the Pistillate Spikelet and Fertilization in *Zea Mays* L. Journ. Agric. Res., vol. XVIII, 1919–20, No. 5, Dec. 1, 1919, pp. 255–66.
[p. 362.]

Molliard, M. (1904) Sur une des conditions de développement du tissu bulliforme chez les Graminées. Bull. de la soc. bot. de France, vol. LI (sér. 4, vol. IV), 1904, pp. 76–80.
[p. 301.]

Montgomery, E. G. (1906) What is an Ear of Corn? Pop. Sci. Monthly, vol. LXVIII, 1906, pp. 55–62.
[p. 366.]

Montgomery, E. G. (1920) The Corn Crops. A Discussion of Maize, Kafirs, and Sorghums as grown in the United States and Canada. Revised Edition. New York, 1920.
[pp. 23, 24.]

Mortier, B. Du (1868) Étude agrostographique sur le genre *Michelaria*, et la classification des Graminées. Bull. de la soc. roy. de bot. de Belgique, vol. VII, 1868, pp. 42–70.
[p. 110.]

Müller, C. (1878) Fragmenta Historicorum Graecorum, vol. II, Paris, 1878.
[p. 8.]

Murbach, L. (1900) Note on the Mechanics of the Seed-burying Awns of *Stipa avenacea*. Bot. Gaz., vol. XXX, 1900, pp. 113–17.
[p. 145.]

Nelmes, E. See **Sprague, T. A. and Nelmes, E.** (1931).

Nestler, A. (1898) Ueber einem in der Frucht von *Lolium temulentum* L. vorkommenden Pilz. Ber. d. deutsch. Bot. Gesellsch., vol. XVI, 1898, pp. 207–14.
[p. 50.]

Nestler, A. (1904) Zur Kenntnis der Symbiose eines Pilzes mit dem Taumellolch. Sitzungsber. d. Math.-Naturwiss. Klasse d. k. Akad. d. Wissensch., vol. CXIII, pt. 1, 1904, pp. 529–46.
[p. 50.]

Nicholls, J. (1895) The Flowering of the Thorny Bamboo. Indian Forester, vol. XXI, 1895, pp. 90–5.
[p. 98.]

Nicholson, J. W. (1922) Note on the Distribution and Habit of *Dendrocalamus strictus* and *Bambusa arundinacea* in Orissa. Indian Forester, vol. XLVIII, 1922, pp. 425–8.
[p. 103.]

Nieuwenhuis-Uexkuell, M. (1902) Die Schwimmvorrichtung der Früchte von *Thuarea sarmentosa* Pers. Ann. du jardin bot. de Buitenzorg, vol. XVIII (N.S. vol. III), 1902, pp. 114–23.
[p. 350.]

Nisbet, J. (1895) Bamboo-Seeding and Fever. Indian Forester, vol. XXI, 1895, p. 151.
[p. 95.]

Nishimura, M. (1922) Comparative Morphology and Development of *Poa pratensis*, *Phleum pratense* and *Setaria italica*. Jap. Journ. Bot., vol. I, No. 2, 1922, pp. 55–85.
[pp. 224, 275.]

Noerner, C. (1881) Beitrag zur Embryoentwicklung der Gramineen. Flora, vol. LXIV, 1881, pp. 241–51, 257–66, 273–84.
[p. 228.]

Oakley, R. A. and Evans, M. W. (1921) Rooting Stems in Timothy. Journ. Agric. Res., vol. XXI, 1921, pp. 173–8.
[p. 254.]

Obel, M. de l' See **Lobel (de l'Obel), M.** (1581).

Oliver, F. W. (1920) *Spartina* Problems. Annals of Applied Biology, vol. VII, 1920, pp. 25–39.
[p. 372.]

Oliver, F. W. (1924) *Spartina Townsendi*. Gard. Chron., ser. III, vol. LXXV, 1924, pp. 148, 162–3.
[p. 372.]

Oliver, F. W. (1925) *Spartina Townsendii*; its mode of establishment, economic uses and taxonomic status. Journ. Ecology, vol. XIII, 1925, pp. 74–91.
[pp. 229, 251, 372.]

Oliver, F. W. (1926) *Spartina* in France. Gard. Chron., ser. III, vol. LXXIX, 1926, pp. 212–13.
[pp. 372, 376.]

Oliver, F. W. (1927–8) Vignettes from Holland. Trans. Norfolk and Norwich Nat. Soc., vol. XII, 1927–8, pp. 502–19.
[p. 372.]

Oliver, F. W. See also **Carey, A. E. and Oliver, F. W.** (1918).

Osmaston, B. B. (1918) Rate of Growth of Bamboos. Indian Forester, vol. XLIV, 1918, pp. 52–7.
[p. 62.]

Pallis, M. (1916) The Structure and History of Plav: the Floating Fen of the Delta of the Danube. Journ. Linn. Soc. Lond., Bot., vol. XLIII, 1916, pp. 233–90.
[pp. 217, 218, 251.]

Pantin, C. F. A. (1932) The Origin of Body Fluids. Rep. Brit. Ass. Adv. Sci., York, 1932, p. 336.
[p. 93.]

Parker, R. N. See **Blatter, E. and Parker, R. N.** (1929).

Parnell, F. R. (1921) Notes on the Detection of Segregation by the examination of the Pollen of Rice. Journ. Genetics, vol. XI, 1921, pp. 209–12.
[p. 28.]

Parodi, L. R. (1924) Notas sobre Flores Cleistógamas Axilares en las Aveneas Platenses. Revista de la Fac. de Agronomía y Vet., Univers. de Buenos Aires, vol. IV, 1924, pp. 508–17.
[p. 203.]

Parodi, L. R. (1926) Dos nuevas especies de Gramíneas de la Flora argentina. Physis (Revista de la Sociedad Argentina de Ciencias Naturales), vol. VIII, 1926, pp. 372–9.
[p. 136.]

Parodi, L. R. (1928) Revisión de las Gramíneas argentinas del género *Sporobolus*. Rev. de la Fac. de Agronomía y Vet., Univers. de Buenos Aires, vol. VI, 1928, pp. 115–68.
[p. 294.]

Payne, E. J. (1892) History of the New World called America, vol. I. Oxford, 1892.
[p. 31.]

Pée-Laby, E. (1898) Étude anatomique de la feuille des Graminées de la France. Ann. d. sci. nat., sér. 8, Bot., vol. VIII, 1898, pp. 227–346.
[p. 294.]

Penzig, O. (1921–2) Pflanzen-Teratologie. 3 vols., Edition 2. Berlin, 1921–2.
[p. 363.]

Percival, J. (1921) The Wheat Plant. A Monograph. London, 1921.
[pp. 4, 5, 6 (Legend of Fig. 3), 7, 154, 159, 160, 223, 224, 231, 249, 292, 393.]

Percival, J. (1926[1]) Agricultural Botany: Theoretical and Practical. Seventh edition. London, 1926.
[p. 48.]

Percival, J. (1926[2]) The Morphology and Cytology of some Hybrids of *Aegilops ovata* L. ♀ × Wheats ♂. Journ. Genetics, vol. XVII, 1926–7, pp. 49–68.
[pp. 8, 39.]

Percival, J. (1927) The Coleoptile Bundles of Indo-Abyssinian Emmer Wheat (*Triticum dicoccum*, Schübl.). Ann. Bot., vol. XLI, 1927, pp. 101–5.
[p. 236.]

Percival, J. (1930) Cytological Studies of some Hybrids of *Aegilops* sp. × Wheats, and of some Hybrids between different Species of *Aegilops*. Journ. Genetics, vol. XXII, 1930, pp. 201–78.
[pp. 8, 12.]

Percival, J. (1934) Wheat in Great Britain. Published by the author: Leighton, Shinfield, Reading, 1934.
[pp. 4, 38.]

Peter, A. (1900) Ueber hochzusammengesetzte Stärkekörner im Endosperm von Weizen, Roggen und Gerste. Oest. Bot. Zeitschr., vol. I, 1900, pp. 315–18.
[p. 226.]

Petiver, J. (1702) Mr. Sam. Brown his seventh Book of East Indian Plants. Phil. Trans. Roy. Soc. Lond., vol. XXII, 1702, pp. 1251–62.
[p. 56.]

Piédallu, A. (1923) Le Sorgho. Thèses prés. Faculté des Sci. de Paris, Sér. A, No. 937, No. d'ordre 1760, 1923, 383 pp.
[pp. 23, 303.]

Pilger, R. (1905) Beiträge zur Kenntnis der monöcischen und diöcischen Gramineen-Gattungen. Bot. Jahrbücher (Engler), vol. XXXIV, 1905, pp. 377–416.
[p. 194.]

Pillai, M. V. (1905) The Dimensions of *Bambusa arundinacea*. Indian Forester, vol. XXXI, 1905, p. 153.
[p. 61.]

Plank, E. N. (1892) *Buchloe dactyloides*, Englm., not a Dioecious Grass. Bull. Torr. Bot. Club, vol. XIX, 1892, pp. 303–6.
[p. 195.]

Pliny (Holland, P.) (1601) The Historie of the World. Commonly called, the Naturall Historie of C. Plinius Secundus, Translated into English by Philemon Holland. London, 1601.
[pp. 15, 16, 36, 39, 43.]

Plot, R. (1677) The Natural History of Oxford-shire. Oxford, 1677.
[p. 45.]

Porta, N. H. (1928) Esquisse de Géographie botanique et d'Écologie des "Rochers du Coin". Thèse presentée à la Faculté des Sciences de l'Université de Genève, No. 843, 1928, 146 pp.; also Bull. de la Soc. Bot. de Genève, vol. XX, 1928, pt. I, pp. 10–148.
[pp. 292, 340.]

Porterfield, W. M. (1923) A New Feature in Vascular Anatomy as displayed by Bamboo, particularly by the young sheath leaf. China Journ. of Sci. and Arts, vol. I, 1923, pp. 273–9.
[p. 80.]

Porterfield, W. M. (1925) The Square Bamboo. A Preliminary Study of *Phyllostachys quadrangularis*. China Journal, vol. III, 1925, pp. 333–5.
[p. 71.]

Prat, H. (1932) L'épiderme des Graminées. Étude anatomique et systématique Ann. d. sci. nat., sér. 10, vol. XIV, 1932, pp. 118–325.
[pp. 233, 235, 301, 303.]

Pratt, A. (n.d. [1859]) The British Grasses and Sedges. London [1859].
[p. 303.]

Prescott, W. H. (1847) History of the Conquest of Peru. 2 vols. London, 1847.
[p. 31.]

Preston, T. A. (1888) The Flowering Plants of Wilts. (Wiltshire Arch. and Nat. Hist. Soc.), 1888.
[p. 209.]

Price, S. R. (1911) The Roots of some North African Desert-Grasses. New Phyt., vol. x, 1911, pp. 328–40.
[pp. 249, 251.]

Priestley, J. H. and Radcliffe, F. M. (1924) A Study of the Endodermis in the Filicineae. New Phyt., vol. XXIII, 1924, pp. 161–93.
[p. 251.]

Radcliffe, F. M. See **Priestley, J. H. and Radcliffe, F. M.** (1924).

Randolph, F. R. (1926) A Cytological Study of two Types of Variegated Pericarp in Maize. Cornell University Agric. Experiment Station, Memoir 102, 1926, 14 pp.
[pp. 357, 363.]

Raspail, F. V. (1824) Sur la formation de l'Embryon dans les Graminées. Ann. d. sci. nat., vol. IV, 1824, pp. 271–319.
[p. 208.]

Rayner, M. C. (1927) Mycorrhiza. New Phytologist Reprint, No. 15, vi + 246 pp. London, 1927.
[pp. 50, 69.]

Rebsch, B. A. (1910) The Bamboo (*Dendrocalamus strictus*) forests of the Ganges Division, U.P. Indian Forester, vol. XXXVI, 1910, pp. 202–21.
[p. 103.]

Redesdale, Lord See **Freeman-Mitford, A. B.** (**Lord Redesdale**) (1896).

Reid, C. (1899) The Origin of the British Flora. London, 1899.
[p. 163.]

Richard, A. (1811) Analyse botanique des embryons Endorhizes..., et particulièrement de celui des Graminées. Annales du Muséum, vol. XVII, 1811, pp. 223–51, 442–87.
[p. 162.]

Richards, P. W. See **Davis, T. A. W. and Richards, P. W.** (1933).

Ridgeway, W. (1892) The Origin of Metallic Currency and Weight Standards. Cambridge, 1892.
[pp. 9, 59, 157.]

Ridley, H. N. (1923) The Distribution of Plants. Ann. Bot., vol. XXXVII, 1923, pp. 1–29.
[pp. 333, 336.]

Ridley, H. N. (1924) Poisoning by Bamboo Hairs. Journ. Trop. Med. and Hygiene, vol. XXVII, 1924, p. 278.
[p. 73.]

Ridley, H. N. (1926) Discussion on E. J. Salisbury's paper on The Geographical Distribution of Plants in Relation to Climatic Factors. Geog. Journ., vol. LXVII, 1926, pp. 336–8.
[p. 333.]

Ridley, H. N. (1930) The Dispersal of Plants throughout the World. Ashford, 1930. [pp. 145, 333, 334, 348–354.]

Rimbach, A. (1922) Die Wurzelverkürzung bei den grossen Monokotylenformen. Ber. d. deutsch. Bot. Gesellsch., vol. XL, 1922, pp. 196–202. [p. 251.]

Rimpau, W. (1877) Die Selbst-Sterilität des Roggens. Landwirthsch. Jahrb., Berlin, vol. VI, 1877, pp. 1073–6. [p. 12.]

Rimpau, W. (1882) Das Blühen des Getreides. Landwirthsch. Jahrb., Berlin, vol. XI, 1882, pp. 875–919. [pp. 62, 157, 158.]

Rivière, A. and C. (1878) Les Bambous. Paris, 1878, 364 pp. [pp. 63, 66, 69, 70, 73, 95, 99, 100, 110, 219.]

Roelants, H. W. M. (1922) Ueber das mechanische System in den Stengeln der Gramineen. Rec. des travaux bot. néerlandais, vol. XVIII, pt. III, 1922, pp. 322–32. [pp. 258, 306.]

Roesler, P. (1928) Histologische Studien am Vegetationspunkt von *Triticum vulgare*. Planta, vol. v, 1928, pp. 28–69. [p. 279.]

Rogers, C. G. (1900) Flowering of bamboos in the Darjeeling District. Indian Forester, vol. XXVI, 1900, pp. 331–2. [p. 102.]

Rogers, C. G. (1901) Flowering of *Arundinaria Falconeri* in the Darjeeling District in 1900. Indian Forester, vol. XXVII, 1901, pp. 185–7. [p. 97.]

Ross, H. (1883) Beiträge zur Anatomie abnormer Monocotylenwurzeln (Musaceen, Bambusaceen). Ber. d. deutsch. Bot. Gesellsch., vol. I, 1883, pp. 331–7. [p. 68.]

Rouville, P. de (1853) Monographie du genre *Lolium*. Montpellier, 1853, 57 pp. [p. 166.]

Royer, C. (1883) Flore de la Côte d'Or, vol. II. Paris, 1883. [pp. 54, 143, 220.]

Ruellius, J. (1536) De Natura Stirpium Libri tres. Paris, 1536. [p. 29.]

Rumphius, G. E. (1750) Het Amboinsch Kruid-boek (Herbarium Amboinense). 6 vols. Amsterdam, 1750. [pp. 34, 58.]

Russell, P. (1790) An Account of the Tabasheer. Phil. Trans. Roy. Soc. Lond., vol. LXXX, pt. II, 1790, pp. 273–83. [p. 60.]

Ryan, G. (1901) Flowering and Seeding of Manwell Bamboos (*Dendrocalamus strictus*) in the Central Thana Division, Bombay Presidency. Indian Forester, vol. XXVII, 1901, pp. 428–9. [p. 95.]

"S." (1882) Flowering of the Ringal Bamboo in Jaunsar. Indian Forester, vol. VII, 1881–2, p. 258.
[p. 97.]

Sablon, M. Leclerc du (1900) Recherches sur les fleurs cléistogames. Rev. gén. de bot., vol. XII, 1900, pp. 305–18.
[pp. 180, 203.]

Sachs, J. von (1882) Vorlesungen über Pflanzen-physiologie. Leipzig, 1882.
[pp. 225 (Fig. 107), 328.]

Saint-Hilaire, A. de (1829) Lettre sur une variété remarquable de Maïs du Brésil. Ann. d. sci. nat., vol. XVI, 1829, pp. 143–5.
[p. 367.]

Saint-Hilaire, A. de (1847) Leçons de Botanique. Paris, 1847.
[p. 100.]

Sakamura, T. (1918) Kurze Mitteilung über die Chromosomenzahlen und die Verwandtschaftverhältnisse der *Triticum*-Arten. Bot. Mag. Tokyo, vol. XXXII, 1918, pp. 150–3.
[p. 7.]

Salisbury, E. J. (1926) The Geographical Distribution of Plants in Relation to Climatic Factors. Geog. Journ., vol. LXVII, 1926, pp. 312–35.
[p. 336.]

Sansome, F. W. See **Gregor, J. W. and Sansome, F. W. (1930)**.

Sargant, E. (1900) Recent Work on the Results of Fertilization in Angiosperms. Ann. Bot., vol. XIV, 1900, pp. 689–712.
[p. 224.]

Sargant, E. and Robertson (Arber), A. (1905) The Anatomy of the Scutellum in *Zea Maïs*. Ann. Bot., vol. XIX, 1905, pp. 115–23.
[pp. 225 (Fig. 107), 231, 235, 245 (Fig. 118).]

Sargant, E. and Arber, A. (1915) The Comparative Morphology of the Embryo and Seedling in the Gramineae. Ann. Bot., vol. XXIX, 1915, pp. 161–222.
[p. 235.]

Sargent, C. S. (1894) Forest Flora of Japan. Boston and New York, 1894.
[p. 332.]

Schaffner, J. H. (1920) The diecious nature of buffalo-grass. Bull. Torr. Bot. Club, vol. XLVII, 1920, pp. 119–24.
[p. 195.]

Schaffner, J. H. (1927) Control of sex reversal in the tassel of Indian corn. Bot. Gaz., vol. LXXXIV, 1927, pp. 440–9.
[p. 214.]

Schaffner, J. H. (1930) Sex reversal and the experimental production of neutral tassels in *Zea Mays*. Bot. Gaz., vol. XC, 1930, pp. 279–98.
[p. 214.]

Scharff, R. F. (1925) Sur le problème de l'île de Krakatau. Assoc. Franç. pour l'avance. des sciences. Compte rendu, 49e Session, Grenoble, 1925, pp. 746–50.
[p. 347.]

Schellenberg, H. C. (1897) Ueber die Bestockungsverhältnisse von *Molinia coerulea* Mönch. Bull. de la soc. bot. Suisse (Ber. d. Schweiz. Bot. Gesellsch.), vol. VII, 1897, pp. 69–82.
[pp. 275, 278.]

Schnarf, K. (1926) Kleine Beiträge zur Entwicklungsgeschichte der Angiospermen. VI. Ueber die Samenentwicklung einiger Gramineen. Oesterr. Bot. Zeitschrift, vol. LXXV, 1926, pp. 105–13.
[p. 162.]

Schnarf, K. (1929) Embryologie der Angiospermen, in K. Linsbauer, Handbuch der Pflanzenanatomie, sect. II, pt. 2, vol. X. 2, Berlin, 1929.
[pp. 162, 228.]

Schoute, J. C. (1910) Die Bestockung des Getreides. Verhandel. d. Kon. Akad. v. Wetenschappen te Amsterdam, sect. II, vol. XV, No. 2, 1910, xix + 492 pp.
[p. 75.]

Schroeter, C. (1885) Der Bambus. Zurich, 1885, 55 pp.
[pp. 61, 62, 71, 78, 85, 95, 97.]

Schroeter, C. and Kirchner, O. von (1902) Die Vegetation des Bodensees. Vol. II, Bodensee-Forschungen, Abschn. 9, viii + 86 pp., 1902.
[p. 391.]

Schroeter, C. See also **Kirchner, O. von, Loew, E. and Schroeter, C. (1908, etc.).**

Schulz, A. (1911) Die Geschichte des Roggens. Jahresber. Westfäl. Prov.-Vereins f. Wiss. und Kunst, vol. XXXIX, Münster, 1911 (for 1910–11), pp. 153–63.
[pp. 12, 37.]

Schulz, A. (1913) Die Geschichte der kultivierten Getreide. I. Halle a.d.S., 1913.
[pp. 3, 7, 11.]

Schumann, K. (1895) Die Gräser Ostafrikas (Engler, Die Pflanzenwelt Ost-Afrikas). Deutsch-Ost-Afrika, vol. V, B. II, 1895, pp. 31–87.
[p. 209.]

Schumann, K. (1904) Mais und Teosinte. Festschrift zur Feier des siebzigsten Geburtstages...P. Ascherson. Herausgegeben Urban, I. und Graebner, P., 1904, pp. 136–57.
[pp. 366, 371.]

Schuster, J. (1910) Ueber die Morphologie der Grasblüte. Flora, vol. C, 1910, pp. 213–66.
[p. 393.]

Schweinfurth, G. (1883) The Flora of Ancient Egypt. Nature, vol. XXVIII, 1883, pp. 109–14.
[p. 56.]

Schweinfurth, G. (1884) Ueber Pflanzenreste aus altaegyptischen Gräbern. Ber. d. deutsch. Bot. Gesellsch., vol. II, 1884, pp. 351–71.
[pp. 3, 56.]

Schwendener, S. (1874) Das mechanische Princip im anatomischen Bau der Monocotylen. Leipzig, 1874.
[pp. 80, 258.]

Schwendener, S. (1889) Die Spaltöffnungen der Gramineen und Cyperaceen. Sitzungsber. d. k. preuss. Akad. d. Wiss., Berlin, Jahrg. 1889, pp. 65–79.
[p. 303.]

Schwendener, S. (1890) Die Mestomscheiden der Gramineenblätter. Sitzungsber. d. k. preuss. Akad. d. Wiss., Berlin, Jahrg. 1890, pp. 405–26.
[pp. 305, 382.]

Scott, D. H. (1897) On two new instances of Spinous Roots. Ann. Bot., vol. XI, 1897, pp. 327–32.
[p. 78.]

Scott, D. H. (1931) Fossil Plants and Evolution. Proc. Bournemouth Nat. Sci. Soc., vol. XXIII, 1931, pp. 46–52.
[p. 331.]

Scott, J. A. (1920) Herodotus and the Fertility of Babylonia. Class. Journ., vol. XV, 1919–20, No. 6, March 1920, pp. 370–2.
[p. 8.]

Seifriz, W. (1923) Observations on the Causes of Gregarious Flowering in Plants. Amer. Journ. Bot., vol. X, 1923, pp. 93–112.
[pp. 98, 104.]

Sethi, B. L. See **Jenkin, T. J. and Sethi, B. L.** (1932).

Seward, A. C. (1933) Plant Life through the Ages. Edition II. Cambridge, 1933.
[pp. 83, 91.]

Seward, A. C. See also **Ernst, A.** (1908).

Seybold, A. (1925) Ueber die Drehung bei der Entfaltungsbewegung der Blätter. Bot. Abhandlungen herausgegeben von K. Goebel, vol. VI, 1925, 80 pp.
[p. 75.]

Shadowsky, A. E. (1926) Der antipodale Apparat bei Gramineen. Flora, N.F. vol. XX (G.R. vol. CXX), 1926, pp. 344–70.
[p. 162.]

Shibata, K. (1900) Beiträge zur Wachstumsgeschichte der Bambusgewächse. Journ. Coll. Sci. Imp. Univ. Tokyo, vol. XIII, 1900–1, pp. 427–96.
[pp. 62, 67, 69, 78, 79, 80, 309.]

Shirreff, P. (1873) Improvement of the Cereals. Privately printed, Edinburgh and London, 1873.
[p. 40.]

Sifton, H. B. (1920) Longevity of the Seeds of Cereals, Clovers, and Timothy. Amer. Journ. Bot., vol. VII, 1920, pp. 243–51.
[p. 223.]

Sinclair, G. (1824) Hortus Gramineus Woburnensis. London, 1824.
[p. 339.]

Sinnott, E. W. (1916) Comparative Rapidity of Evolution in Various Plant Types. Amer. Nat., vol. L, 1916, pp. 466–78.
[p. 83.]

Sinnott, E. W. and Bailey, I. W. (1914) Investigations on the Phylogeny of the Angiosperms, No. 4. The Origin and Dispersal of Herbaceous Angiosperms. Ann. Bot., vol. XXVIII, 1914, pp. 547–600.
[pp. 82, 83.]

Smiles, S. (1878) Robert Dick, Baker, of Thurso. Geologist and Botanist. London, 1878.
[p. 55.]

Smith, J. (1612) in Arber, E. (1884) A Map of Virginia with a description of the Countrey... Written by Captaine Smith. Oxford, 1612. Reprinted in Capt. John Smith, Works, 1608–31. Edited by Edward Arber. Birmingham, 1884.
[p. 45.]

Smith, J. G. (1888) *Buchloe dactyloides*. Bot. Gaz., vol. XIII, 1888, pp. 215–16.
[p. 46.]

Smith, S. G. See **Huskins, C. L. and Smith, S. G. (1932)**.

Smith, W. G. (1918) The Distribution of *Nardus stricta* in Relation to Peat. Journ. Ecology, vol. VI, 1918, pp. 1–13.
[p. 336.]

Smythies, A. (1881) The Male Bamboo. Indian Forester, vol. VII, 1881–2, p. 163.
[p. 71.]

Smythies, A. (1901) Flowering of the Bamboo in the C.P. Indian Forester, vol. XXVII, 1901, pp. 126–7.
[pp. 97, 103.]

Souèges, R. (1924) Embryogénie des Graminées. Développement de l'embryon chez le *Poa annua* L. Comptes rendus de l'acad. d. sci., Paris, vol. CLXXVIII, 1924, pp. 860–2.
[p. 228.]

Sowerby, J. See **Syme, J. T. B.** (afterwards **Boswell**) **(1873)**.

Sowerby, J. E. See **Johnson, C. (n.d. [1857–61])**.

Sprague, T. A. (1928) The Herbal of Otto Brunfels. Journ. Linn. Soc. Lond., Bot., vol. XLVIII, 1928, pp. 79–124.
[p. 14 (Legend of Fig. 7).]

Sprague, T. A. and Nelmes, E. (1931) The Herbal of Leonhardt Fuchs. Journ. Linn. Soc. Lond., Bot., vol. XLVIII, 1931, pp. 545–642.
[pp. 5 (Legend of Fig. 2), 19 (Legend of Fig. 9), 20 (Legend of Fig. 10), 29.]

Spruce, R. (1908) Notes of a Botanist on the Amazon and Andes, 1849–1864. Edited by A. R. Wallace. 2 vols., London, 1908.
[pp. 207, 208.]

Stapf, O. (1897) The Botanical History of the Uva, Pampas Grass and their Allies. Gard. Chron., ser. III, vol. XXII, 1897, pp. 358, 378, 396.
[p. 207.]

Stapf, O. (1898) Gramineae, in Flora of Tropical Africa, vol. IX, pt. I, 1898, edited by the Director, Royal Gardens, Kew.
[pp. 184, 240.]

Stapf, O. (1904[1]) Die Gliederung der Gräserflora von Südafrika. Festschrift zu P. Ascherson's siebzigstem Geburtstage. Berlin, 1904, pt. XXXIV, pp. 391–412.
[p. 344.]

Stapf, O. (1904[2]) On the Fruit of *Melocanna bambusoides*, Trin., an Endospermless, Viviparous Genus of *Bambuseae*. Trans. Linn. Soc. Lond., Bot., ser. II, vol. VI (1901–5), pt. IX, 1904, pp. 401–25.
[pp. 65, 124.]

Stapf, O. (1906) The Oil-grasses of India and Ceylon (*Cymbopogon*, *Vetiveria* and *Andropogon* spp.). Kew Bull., 1906, No. 8, pp. 297–363.
[pp. 55, 405.]

Stapf, O. (1909) The History of the Wheats. Rep. 79th Meeting, Brit. Ass. Adv. Sci., Winnipeg, 1910 (for 1909), pp. 799–807.
[pp. 5, 7, 9.]

Stapf, O. (1914) Townsend's Grass or Ricegrass (*Spartina Townsendii*). Proc. Bournemouth Nat. Sci. Soc., vol. V, 1914 (for 1912–13), pp. 76–82.
[pp. 353, 372.]

Stapf, O. (1917) *Enneapogon mollis* in Ascension Island. Kew Bull., 1917, No. 6, pp. 217–19.
[p. 353.]

Stapf, O. (1921) Kikuyu Grass. (*Pennisetum clandestinum*, Chiov.) Kew Bull., 1921, pp. 85–93.
[p. 47.]

Stapf, O. (1926) *Spartina Townsendii*. Bot. Mag., vol. CLII, 1926, Tab. 9125.
[p. 372.]

Stillingfleet, B. (1759) Observations on Grasses, in Miscellaneous Tracts relating to Natural History, Husbandry, and Physick. London, 1759, pp. 202–30.
[pp. 1, 48.]

Stillingfleet, B. (1811) Observations on Grasses. Literary Life and Select Works, vol. II, pt. I, 1811, pp. 249–357.
[pp. 45, 209.]

Strasburger, E. (1891) Histologische Beiträge. III. Ueber den Bau und die Verrichtungen der Leitungsbahnen in der Pflanzen. Jena, 1891.
[p. 309.]

Sturtevant, E. L. (1881) The Superabundance of Pollen in Indian Corn. Amer. Nat., vol. xv, 1881, p. 1000.
[p. 359.]

Sturtevant, E. L. (1885) Indian Corn and the Indian. Amer. Naturalist, vol. xix, 1885, pp. 225–34.
[p. 29.]

Sturtevant, E. L. (1894) Notes on Maize. Bull. Torr. Bot. Club, vol. xxi, 1894, pp. 319–43.
[pp. 31, 223.]

Swallen, J. R. (1931) The Grass Genus *Amphibromus*. Amer. Journ. Bot., vol. xviii, 1931, pp. 411–15.
[p. 344.]

Syme, J. T. B. (afterwards Boswell) (1873) Sowerby's English Botany, Edition iii, vol. xi, Gramina, 1873.
[pp. 163, 340.]

Takenouchi, Y. (1931) Morphologische und entwicklungs-mechanische Untersuchungen bei japanischen Bambus-Arten. Mem. Coll. Sci., Kyoto Imp. University, ser. B, vol. vi, No. 3, 1931, pp. 109–60.
[pp. 69, 70, 73, 75, 79, 80.]

Theophrastus (Hort, A.) (1916) Enquiry into Plants. With an English Translation by Sir Arthur Hort. 2 vols. London and New York, 1916.
[pp. 81, 307.]

Thoenes, H. (1929) Morphologie und Anatomie von *Cynosurus cristatus* und die Erscheinungen der Viviparie bei ihm. Botanisches Archiv, vol. xxv, 1929, pp. 284–346.
[p. 186.]

Thompson, D'Arcy W. (1917) On Growth and Form. Cambridge, 1917.
[pp. 61, 82, 382.]

Thompson, J. McLean (1924) Studies in Advancing Sterility. Part I. The Amherstieae. Publications of the Hartley Bot. Laboratories. No. 1, Liverpool, 1924.
[p. 180.]

Tieghem, P. van (1897) Morphologie de l'embryon et de la plantule chez les Graminées et les Cypéracées. Ann. d. sci. nat., sér. 8, Bot., vol. iii 1897, pp. 259–309.
[pp. 228, 237.]

Tincker, M. A. H. (1925) The Effect of Length of Day upon the Growth and Reproduction of some Economic Plants. Ann. Bot., vol. xxxix, 1925, pp. 721–54.
[p. 214.]

Tingle, A. (1904) The Flowering of the Bamboo. Nature, vol. lxx, Aug. 11, 1904, p. 342.
[p. 99.]

Toole, E. H. (1924) The Transformations and Course of Development of Germinating Maize. Amer. Journ. Bot., vol. XI, 1924, pp. 325–50.
[p. 226.]

Toumey, J. W. (1891) Peculiar forms of proliferation in Timothy. Bot. Gaz., vol. XVI, 1891, pp. 346–7.
[p. 385.]

Tragus, H. (Bock, J.) (1552) De Stirpium. Strasburg, 1552.
[pp. 14 (Fig. 6), 44 (Fig. 17), 58, 335 (Fig. 175).]

Trevor, C. G. (1927) Growth of Bamboos. Indian Forester, vol. LIII, 1927, pp. 693–5.
[p. 63.]

Trimen, H. (1879) *Phyllorachis*, a new genus of Gramineae from Western Tropical Africa. Journ. Bot., vol. XVII (N.S., vol. VIII), 1879, pp. 353–5.
[pp. 136, 289.]

Trimen, H. (1900) A Handbook to the Flora of Ceylon, pt. v. London, 1900.
[p. 349.]

True, R. H. (1893) On the Development of the Caryopsis. Bot. Gaz., vol. XVIII, 1893, pp. 212–26.
[p. 162.]

Tschirch, A. (1882) Beiträge zu der Anatomie und dem Einrollungsmechanismus einiger Grasblätter. Pringsheim's Jahrb. f. wiss. Bot., vol. XIII, 1882, pp. 544–68.
[p. 301.]

Turesson, G. (1926) Studien über *Festuca ovina* L. I. Normalgeschlechtliche, halb- und ganzvivipare Typen nordischer Herkunft. Hereditas, vol. VIII, 1926, pp. 161–206.
[p. 393.]

Turesson, G. (1930) Studien über *Festuca ovina* L. II. Chromosomenzahl und Viviparie. Hereditas, vol. XIII, 1930, pp. 177–84.
[p. 393.]

Turesson, G. (1931) Studien über *Festuca ovina* L. III. Weitere Beiträge zur Kenntnis der Chromosomenzahlen viviparer Formen. Hereditas, vol. XV, 1931, pp. 13–16.
[p. 393.]

Turner, D. M. See **Burr, S. and Turner, D. M. (1933).**

Turner, E. (1828) Chemical Examination of Tabasheer. Edinburgh Journ. of Sci., vol. VIII, 1828, pp. 335–8.
[p. 60.]

Turpin, P. J. F. (1819) Mémoire sur l'inflorescence des Graminées et des Cypérées. Mém. du mus. d'hist. nat., Paris, vol. V, 1819, pp. 426–92.
[pp. 176, 279.]

Turpin, P. J. F. (1820) Essai d'une Iconographie...des Végétaux, Paris, 1820 (forming vol. III of Poiret, J. L. M., Leçons de Flore).
[pp. 381, 382.]

Uittien, H. (1928) Ueber den Zusammenhang zwischen Blattnervatur und Sprossverzweigung. Rec. des travaux botaniques néerlandais, vol. xxv, 1928, pp. 390–483.
[p. 331.]

Vavilov, N. I. (1927) Geographical regularities in the distribution of the genes of cultivated plants. Bull. Applied Bot., Gen. and Plant Breeding, vol. xvii, 1927, No. 3, pp. 411–28 (Russian with English summary).
[p. 41.]

Vavilov, N. I. (1928) Geographische Genzentren unserer Kulturpflanzen. Verhandl. d. V. Internat. Kongresses Vererbungswissenschaft. Berlin, 1927. Supplementband I der Zeitschrift für induktive Abstammungs- und Vererbungslehre, 1928, pp. 342–69.
[pp. 9, 41.]

Vavilov, N. I. (1931[1]) The rôle of Central Asia in the origin of cultivated plants. Bull. Applied Bot., Gen. and Plant Breeding, vol. xxvi, 1931, pp. 3–44 (Russian with English summary).
[p. 41.]

Vavilov, N. I. (1931[2]) Mexico and Central America as the principal centre of origin of cultivated plants of New World. Bull. Applied Bot., Gen. and Plant Breeding, vol. xxvi, 1931, pp. 135–99 (Russian with English summary).
[p. 41.]

Vavilov, N. I. (1931[3]) The Problem of the Origin of the World's Agriculture in the Light of the Latest Investigations, No. 6 (10 pp. and map) in *Science at the Cross Roads* (papers by U.S.S.R. Delegates, Intern. Hist. of Sci. Congress, London, 1931), Kniga (England), Ltd., 1931. (Also published as a separate pamphlet.)
[p. 41.]

Vavilov, N. I. and Bukinich, D. D. (1929) Agricultural Afghanistan. Supp. 33 to Bull. Applied Bot., Gen. and Plant Breeding, Leningrad, 1929.
[pp. 3, 8, 13, 41, 288, 333.]

Vavilov, N. I. and Kouznetsov, E. S. (1922) On the genetic Nature of Winter and Spring Varieties of Plants. Annals Agron. Fac. Univ. Saratov, vol. i, 1922, 25 pp. (Russian with English summary).
[p. 38.]

Vélain, C. (1878) Étude microscopique des verres résultant de la fusion des cendres de graminées. Bull. de la Soc. Minéral. de France, 1878, No. 7, pp. 113–24.
[p. 258.]

Vries, H. de (1900) Sur la Fécondation Hybride de l'Endosperme chez le Maïs. Revue gén. de bot., vol. xii, 1900, pp. 129–37.
[pp. 224, 263.]

Wallich, N. (1832) Plantae Asiaticae Rariores. London, 1830–32, vol. iii, 1832.
[p. 57.]

Ward, H. Marshall (1901) Grasses: A Handbook for use in the Field and Laboratory Cambridge, 1901.
[pp. 258, 294, 336.]

"Wathôn" (1903) The Flowering of *Bambusa polymorpha*. Indian Forester, vol. XXIX, 1903, pp. 244–5.
[p. 98.]

Watkins, A. E. (1930) The Wheat Species: a Critique. Journ. Genetics, vol. XXIII, 1930, pp. 173–263.
[p. 7.]

Watkins, A. E. (1933) The Origin of Cultivated Plants. Antiquity, vol. VII, no. 25, 1933, pp. 73–80.
[p. 9.]

Watt, G. (1889–96) A Dictionary of the Economic Products of India. 6 vols. Calcutta, 1889–96.
[pp. 53, 54, 59, 60, 78.]

Watt, G. (1904) Coix spp. or Job's Tears. Agricult. Ledger, vol. XI, 1905 (for 1904), no. 13, 1904, pp. 189–229.
[pp. 34, 208.]

Watt, W. L. (1925) Kikuyu Grass. Kew Bull., 1925, p. 403.
[p. 47.]

Weatherwax, P. (1916) Morphology of the Flowers of *Zea Mays*. Bull. Torr. Bot. Club, vol. XLIII, 1916, pp. 127–44.
[p. 355.]

Weatherwax, P. (1917) The development of the spikelets of *Zea Mays*. Bull. Torr. Bot. Club, vol. XLIV, 1917, pp. 483–96.
[p. 357.]

Weatherwax, P. (1918) The Evolution of Maize. Bull. Torr. Bot. Club, vol. XLV, 1918, pp. 309–42.
[pp. 355, 364, 366, 367, 368.]

Weatherwax, P. (1919[1]) Gametogenesis and Fecundation in *Zea Mays* as the basis of xenia and heredity in the endosperm. Bull. Torr. Bot. Club, vol. XLVI, 1919, pp. 73–90.
[p. 362.]

Weatherwax, P. (1919[2]) The Ancestry of Maize—a reply to criticism. Bull. Torr. Bot. Club, vol. XLVI, 1919, pp. 275–8.
[p. 372.]

Weatherwax, P. (1920) Position of scutellum and homology of coleoptile in Maize. Bot. Gaz., vol. LXIX, 1920, pp. 179–82.
[p. 235.]

Weatherwax, P. (1922[1]) The Popping of Corn. Proc. Indiana Acad. Sci., 1921, publ. 1922, pp. 149–53.
[p. 33.]

Weatherwax, P. (1922[2]) A Rare Carbohydrate in Waxy Maize. Genetics, vol. VII, 1922, pp. 568–72.
[p. 226.]

Weatherwax, P. (1923) The Story of the Maize Plant. University of Chicago Science Series, 1923, xv + 247 pp.
[p. 359.]

Weatherwax, P. (1925[1]) The Reported Origin of Indian Corn from Teosinte. Proc. Indiana Acad. Sci., vol. xxxiv, 1925 (1924), pp. 225–8.
[p. 372.]

Weatherwax, P. (1925[2]) Anomalies in maize and its relatives. Bull. Torr. Bot. Club, vol. LII, 1925, pp. 87–92, 167–70.
[p. 363.]

Weatherwax, P. (1926) Comparative Morphology of the Oriental Maydeae. Indiana University Studies, No. 73, vol. XIII, 1926, 18 pp.
[pp. 196, 197.]

Weatherwax, P. (1928[1]) Cleistogamy in two species of *Danthonia*. Bot. Gaz., vol. LXXXV, 1928, pp. 104–9.
[pp. 203, 206.]

Weatherwax, P. (1928[2]) The Morphology of the Spikelets of Six Genera of Oryzeae. Amer. Journ. Bot., vol. XVI, 1928, pp. 547–55.
[p. 184.]

Weatherwax, P. (1929) Cleistogamy in *Poa Chapmaniana*. Torreya, vol. XXIX, 1929, pp. 123–4.
[pp. 203, 206.]

Weatherwax, P. (1930) The Endosperm of *Zea* and *Coix*. Amer. Journ. Bot., vol. XVII, 1930, pp. 371–80.
[p. 226.]

Weatherwax, P. (1931) The Ontogeny of the Maize Plant. Bull. Torr. Bot. Club, vol. LVII, 1931, pp. 211–19.
[pp. 162, 180, 226.]

Weaver, J. E. (1919) The Ecological Relations of Roots. Carnegie Institution of Washington, Publication No. 286, 1919.
[p. 248.]

Weaver, J. E. (1920) Root Development in the Grassland Formation. Carnegie Institution of Washington, Publ. No. 292, 1920, 151 pp.
[pp. 46, 248.]

Weddell, H. A. (1875) Les Calamagrostis des Hautes Andes. Bull. de la soc. bot. de France, vol. XXII, 1875, pp. 153–60.
[p. 46.]

Weideman, M. (1927) A Contribution to the Genetics and the Morphology of Barley. (On the genetic nature of the lateral Spikelets or Barley.) Bull. Applied Bot., Gen. and Plant Breeding, vol. XVII, No. 2, 1927, pp. 3–70 (Russian with English summary).
[p. 199.]

Werth, E. (1922) Zur experimentellen Erzeugung eingeschlechtiger Maispflanzen und zur Frage: Wo entwickeln sich gemischte (androgyne) Blutenstände am Mais? Ber. d. deutsch. Bot. Gesellsch., vol. XL, 1922, pp. 69–77.
[pp. 364, 365.]

Wheeler, W. M. (n.d. [1923]) Social Life among the Insects. [U.S.A. 1923.]
[p. 398.]

Willis, J. C. and Burkill, I. H. (1893) Observations on the Flora of the Pollard Willows near Cambridge. Proc. Camb. Phil. Soc., vol. VIII, 1892–5, pt. II, 1893, pp. 82–91.
[p. 351.]

Wilson, A. S. (1873[1]) On *Lolium temulentum*, L. (Darnel). Trans. and Proc. Bot. Soc. Edinburgh, vol. XI, 1873, pp. 457–65.
[p. 50.]

Wilson, A. S. (1873[2]) Further Experiments with Darnel (*Lolium temulentum*). Trans. and Proc. Bot. Soc. Edinburgh, Dec. 1873, in vol. XII, 1876, pp. 38–44.
[p. 50.]

Wilson, A. S. (1874) On the Fertilisation of Cereals. Trans. and Proc. Bot. Soc. Edinburgh, Feb. 1874, in vol. XII, 1876, pp. 84–95.
[pp. 160, 180.]

Wilson, A. S. (1875) Wheat and Rye Hybrids. Trans. and Proc. Bot. Soc. Edinburgh, April 1875, in vol. XII, 1876, pp. 286–8.
[p. 13.]

Winton, A. L. (1903) Ueber amerikanische Weizen-Ausreuter. Zeitschr. f. Untersuchung der Nahrungs- und Genussmittel, vol. VI, pt. X, 1903, pp. 433–47.
[p. 301.]

Woodruffe-Peacock, E. A. (1916) *Sieglingia decumbens* in Lincolnshire. Journ. Bot., vol. LIV, 1916, pp. 359–60.
[p. 352.]

Woodruffe-Peacock, E. A. (1917) The Means of Plant Dispersal. I. Storm-columns. Selborne Magazine, vol. XXVIII, 1917, pp. 40–4.
[p. 349.]

Woodward, A. Smith (1909) Presid. Address to Section C, Geology. Rep. Brit. Ass. Adv. Sci., Winnipeg, 1910 (for 1909), pp. 462–71.
[pp. 84, 91, 93.]

Worsdell, W. C. (1915–16) The Principles of Plant-Teratology. 2 vols. Ray Soc., London, 1915–16.
[pp. 308, 366.]

Worsdell, W. C. (1916) The Morphology of the Monocotyledonous Embryo and of that of the Grass in particular. Ann. Bot., vol. XXX, 1916, pp. 509–24.
[p. 235.]

Yapp, R. H. (1908) Sketches of Vegetation at Home and Abroad. IV. Wicken Fen. New Phyt., vol. VII, 1908, pp. 61–81.
[p. 283.]

Yonge, C. M. (1932) The influence of the processes of feeding and digestion upon evolution. Rep. Brit. Ass. Adv. Sci., York, 1932, pp. 333–4.
[p. 93.]

York, H. H. See **Johnson, D. S. and York, H. H. (1915).**

Yule, H. and Burnell, A. C. (1903) Hobson-Jobson: A Glossary of Anglo-Indian Words. New ed., W. Crooks, London, 1903.
[pp. 58, 61.]

Zalenski, W. von (1902) Ueber die Ausbildung der Nervatur bei verschiedenen Pflanzen. Ber. d. deutsch. Bot. Gesellsch., vol. xx, 1902, pp. 433–40.
[p. 305.]

Zuderell, H. (1909) Ueber das Aufblühen der Gräser. Sitzungsber. d. k. Akad. d. Wissens., Wien, Math.-Naturw. Kl., vol. cxviii, pt. 1, 1909, pp. 1403–26.
[pp. 149, 152, 157, 159.]

INDEX

[Names of authors will not, in general, be found here, since page references are given in connection with the titles in the bibliography, which thus serves as an index of authors; but the names of those who have given information by letter, which is cited in the text or footnotes, are included.]

CAMBRIDGE: PRINTED BY W. LEWIS, M.A., AT THE UNIVERSITY PRESS